Evolutionary Ecology of Social and Sexual Systems

Crustaceans as Model Organisms

Edited by

J. Emmett Duffy and Martin Thiel

With Illustrations by Marco Leon

OXFORD
UNIVERSITY PRESS
2007

Oxford University Press, Inc., publishes works that further
Oxford University's objective of excellence
in research, scholarship, and education.

Oxford New York
Auckland Cape Town Dar es Salaam Hong Kong Karachi
Kuala Lumpur Madrid Melbourne Mexico City Nairobi
New Delhi Shanghai Taipei Toronto

With offices in
Argentina Austria Brazil Chile Czech Republic France Greece
Guatemala Hungary Italy Japan Poland Portugal Singapore
South Korea Switzerland Thailand Turkey Ukraine Vietnam

Published by Oxford University Press, Inc.
198 Madison Avenue, New York, New York 10016

www.oup.com

Library of Congress Cataloging-in-Publication Data
Evolutionary ecology of social and sexual systems : crustaceans as model organisms /
edited by J. Emmett Duffy and Martin Thiel; with illustrations by Marco Leon.
p. cm.
Includes bibliographical references.
ISBN 978-0-19-517992-7
1. Crustacea—Behavior. 2. Crustacea—Sexual behavior. 3. Social behavior in
animals. I. Duffy, J. Emmett, 1960– II. Thiel, Martin, 1962–
QL435.E96 2007
595.3′156—dc22 2006031700

Front cover images:
Upper row of small images are (left to right): mating pair of the littoral amphipod *Parhyalella ruffoi*, photo by I. Hinojosa, Universidad Católica del Norte, Chile; female isopod *Limnoria chilensis* at the entrance of its burrow, photo by I. Hinojosa; female crab *Cancer setosus* incubating embryo mass, photo by I. Hinojosa; king crab *Lithodes santolla* from southern Chile, photo by G. Fösterra and V. Häussermann, Huinay Scientific Field Station, Chile; female bromeliad crab *Metopaulias depressus* in her breeding pool, photo by R. Diesel, ScienceMedia, Germany. Main Pictures are: snapping shrimp *Synalpheus regalis* fighting on surface of host sponge, photo by J.E. Duffy, Virginia Institute of Marine Science, USA; pair of mating rock shrimp *Rhynchocinetes typus*, photo by I. Hinojosa.

9 8 7 6 5 4 3 2 1

Printed in the United States of America
on acid-free paper

To our teachers and students

PREFACE

The idea for this book was spawned about 10 years ago, after each of us had studied social biology of crustaceans for several years. We had independently explored the growing but scattered literature in this area and noticed the curiously low profile of crustaceans in the literature on sexual and social evolution, which draws primarily from insects and vertebrates. Taking inspiration from our own observations, from the many crustacean stories published in the literature, and from the fascinating lore of field biologists revealed around the bar at conferences, we became convinced that these amazing creatures have much to offer the general student of social and sexual evolution.

Perhaps inevitably, these realizations led us to the idea for this book.

As is often true, after the excitement around the idea's birth, a long time passed before it took concrete form. The turning point was in June 2003, when we organized a symposium on behavioral ecology and sociobiology of Crustacea at the annual meeting of the Crustacean Society. The symposium was a great success—at least in our eyes—and uncovered a strong sense of excitement and ferment in the field. The enthusiasm shown by contributors and attendants at the symposium, and by colleagues we subsequently discussed it with, finally convinced us that the time was ripe for this book and overcame our initial apprehension (subsequently confirmed!) at the prospective work load of editing it. Many of the symposium participants eagerly agreed to contribute to the book, and we are delighted to see how their enthusiasm has borne fruit here. To round out the subject matter, we invited contributions from a few other colleagues who could not make it to the symposium.

In editing this book, we were fortunate to begin with a strong foundation. The important book *Sexual Biology of Crustaceans* (edited by R.T. Bauer and J.T. Martin, 1991) was published more than 15 years ago now, and we are pleased that some of its contributors have also joined us in the present volume. Many contributions in *Sexual Biology of Crustaceans* remain required reading for any student of crustacean sexual and social behavior today—we hope that our book will be seen as a worthy successor.

The contributors to the present volume span a broad range of fields—from neurobiology and genetics to field behavior and ecology—and hail from around the world. They include young researchers beginning their careers as well as distinguished senior scientists who pioneered study of crustacean social and sexual biology and who can provide comprehensive reviews of influential career-long research programs. It is thanks largely to the efforts of this dedicated group of researchers that crustaceans increasingly claim their deserved place in the general literature on evolutionary ecology.

Necessarily, the present book includes only a sample of what is known about crustacean sexual and social biology. We strived to include contributions that capture the broad range of problems and topics being studied and to provide a concise overview of the most exciting and important recent research on crustacean evolutionary ecology. But we are well aware that such a book cannot be complete, and we hope that readers will be enticed to hunt for the gems present in the more specialized crustacean literature, for which these chapters should provide an entrée. One valuable function of a collection like this is that it illuminates not only what we know but, equally important, the holes in our current understanding. We hope that making these holes visible will stimulate the next generation to explore them. Finally, a major goal in assembling the book was to foster integration—among the disparate topics covered herein, with other animal taxa, and with the larger field of evolutionary ecology. To this end, we urged all contributors to reach beyond their own specialties to seek parallels and contrasts between their own work and what is known about other animals. We hope that the finished product will illustrate the value of crustaceans as model organisms for addressing a variety of problems in evolutionary ecology, reveal some of the exciting opportunities awaiting a new generation of students, and inspire them to embrace the research that will produce the next book like this one.

ACKNOWLEDGMENTS

We are indebted to a large number of friends, colleagues, and institutions, without whom this book would have been impossible. In particular, we thank all those who served as reviewers of individual contributions—their comments and suggestions provided critical perspective for us and for the authors. We are grateful to Stefan Dennenmoser, who masterfully prepared the book's index. J.E.D. acknowledges a history of generous research funding by the National Science Foundation and the Smithsonian Institution that laid the groundwork for undertaking this project, as well as the support of the Virginia Institute of Marine Science and, most important, the support of Liz and Conor, who have long indulged his eccentric fascination with sea "bugs." M.T. acknowledges continuing financial support from the Chilean Comisión Nacional de Investigación Científica y Tecnológica (CONICYT) and, in particular, the intellectual input from an enthusiastic gang of students who has accompanied him over the years in the adventures on crustacean social behavior.

CONTENTS

Part III. Mating and Courtship

Part IV. Social Systems

Part V. Synthesis

CONTRIBUTORS

Jelle Atema
Boston University Marine Program
Marine Biological Laboratory
Woods Hole, MA 02543
USA

J. Antonio Baeza
Department of Biology
University of Louisiana
Lafayette, LA 70504
USA

Raymond T. Bauer
Department of Biology
University of Louisiana
Lafayette, LA 70504
USA

Michael J. Childress
Department of Biological Sciences
Clemson University
Clemson, SC 29634
USA

John H. Christy
Smithsonian Tropical Research Institute
Apartado Postal 0843–03092
Balboa, Ancón
Republic of Panama

Rickey D. Cothran
Department of Zoology
University of Oklahoma
730 Van Vleet Oval
Norman, OK 73019
USA

Bernard J. Crespi
Department of Biosciences
Simon Fraser University
Burnaby, BC V5A 1S6
Canada

Rudolf Diesel
ScienceMedia
Römerweg 5
D 82346 Andechs-Erling
Germany

J. Emmett Duffy
School of Marine Science & Virginia Institute of Marine Science
The College of William and Mary
Gloucester Point, VA 23062
USA

Jens Herberholz
Department of Psychology & Neuroscience and Cognitive Science Program
University of Maryland
College Park, MD 20742
USA

Veijo Jormalainen
Department of Biology
University of Turku
FIN-20014 Turku
Finland

Karl Eduard Linsenmair
Theodor-Boveri-Institut für
Biowissenschaften
Universität Würzburg
Am Hubland
D-97074 Würzburg
Germany

Brian Mahon
Department of Biology
University of Louisiana
Lafayette, LA 70504
USA

Paul A. Moore
Department of Biological Sciences
Bowling Green State University
Bowling Green, OH 43403
USA

Joseph Neigel
Department of Biology
University of Louisiana
Lafayette, LA 70504
USA

Alastair M.M. Richardson
School of Zoology
University of Tasmania
Private Bag 05
Hobart, Tasmania 7001
Australia

Bernard Sainte-Marie
Fisheries Science and Aquaculture Branch
Maurice Lamontagne Institute
Fisheries and Oceans Canada
850 route de la Mer
C.P. 1000 Mont-Joli (Québec) G5H 3Z4
Canada

Christoph D. Schubart
Biologie 1
Universität Regensburg
93040 Regensburg
Germany

Stephen M. Shuster
Department of Biological Sciences
Northern Arizona University
Flagstaff, AZ 86011–5640
USA

Molly A. Steinbach
Boston University Marine Program
Marine Biological Laboratory
Woods Hole, MA 02543
USA

Martin Thiel
Facultad Ciencias del Mar
Universidad Catolica del Norte
Larrondo 1281
Coquimbo
Chile

Thijs Christiaan van Son
Centre for Marine Resource Management
Norwegian College of Fishery Science
University of Tromsø
9037 Tromsø
Norway

Gary A. Wellborn
Department of Zoology
University of Oklahoma
730 Van Vleet Oval
Norman, OK 73019
USA

PART I

Conceptual Background and Context

The Behavioral Ecology of Crustaceans
A Primer in Taxonomy, Morphology, and Biology

Martin Thiel

J. Emmett Duffy

When I look at a crayfish I envy it, so rich is it in organs with which to do all that it has to do. From the head to the tail, it is crowded with a large assortment of executive appendages. In this day of multiplicity of duties, if we poor human creatures only had the crayfish's capabilities, then might we hope to achieve what lies before us.

—**Anna Botsford Comstock**

1

By any measure—taxonomic, morphological, ecological—the Crustacea must be reckoned one of the greatest evolutionary radiations in the animal kingdom. Crustaceans encompass an astonishingly diverse assemblage of taxa, from minute microscopic plankton, to deep-sea spider crabs with a leg span measured in meters, to saclike parasites scarcely recognizable as animals. Indeed, the morphological diversity of Crustacea is greater than that of any other class of animals on Earth (Martin and Davis 2001). Although all species have arisen from a common body plan, and many still share its basic features, others are highly modified (Fig. 1.1). The latter include the sessile barnacles that dominate many rocky shores, and some parasites so unusual that they were not recognized as Crustacea until molecular data were available (the former phylum Pentastomida; Abele et al. 1989). Crustaceans have colonized all of Earth's major habitats, from hot hydrothermal vents and the surrounding frigid ocean abyss, to tropical coral reefs, to the soil and canopies of forests, rivers and lakes, mountain tops, and even deserts. The diversity of lifestyles and habitats alone suggests that the Crustacea should provide especially fertile ground for discovering new social and sexual systems and for rigorous comparative analysis of their ecology and evolution. What adaptations have enabled crustaceans to colonize this range of habitats, and what are the ecological and

Figure 1.1 Crustaceans from various aquatic and terrestrial environments. From left to right: top row, ghost crab on sandy beach, king crab on mud bottom (photo courtesy of Verena Häussermann), stomatopod with spearing claw (photo courtesy of Roy Caldwell); middle row, mating pair of rock shrimp, benthic barnacles on hard bottom (photo courtesy of V. Häussermann), pelagic barnacles on floating algae, pill bugs (isopod) on forest soil; bottom row, boring isopod in its burrow, beach hoppers (amphipods) aggregated on food, mating pair of amphipods, male caprellid amphipod.

behavioral consequences of those adaptations? The chapters in this book begin to address these questions in the specific context of the behavioral ecology of crustacean mating and social systems, but the sometimes surprising behavioral adaptations of crustaceans can be understood only with some appreciation of the functional morphology, physiology, and the evolutionary history that has produced them. In this chapter we provide, for those less familiar with these fascinating animals, a brief overview of the basic structural and functional biology of crustaceans. We point out some intriguing parallels and contrasts with other taxa, and we attempt to place these features in the context of general theory of social and mating behavior.

Some Essential Taxonomy and Classification

The deep phylogeny of crustaceans has been a subject of long debate, due in part to the group's remarkable morphological diversity and sheer size. As Joel Martin confessed in the Preface to the 2001 classification of Crustacea, "For anyone with interests in a group of organisms as large and diverse as the Crustacea, it is difficult to grasp the enormity of the entire taxon at one time" (Martin and Davis 2001). However, recent molecular and morphological analyses are beginning to elucidate the history of crustacean diversification and relationships. There is now strong evidence that Crustacea is in fact a paraphyletic taxon and that the insects (Hexapoda) are nested within a pancrustacean clade that diverged initially in the Precambrian (Regier et al. 2005). Thus, crustaceans are not water-breathing insects (see chapter 20)—rather, insects are air-breathing crustaceans! Closer to the tips of the crustacean tree, there has been much recent progress in resolving species-level phylogenetic relationships, and these are proving valuable in illuminating the evolution of social and mating behaviors (see chapter 3).

Among the major lineages of Crustacea, the Malacostraca is the most species-rich and morphologically diverse and includes most of the large-bodied, ecologically and economically important species (Table 1.1). Other currently recognized major groups of Crustacea (Table 1.1) include the Branchiopoda (e.g., cladocerans such as *Daphnia*), which are dominant zooplankton in freshwater habitats; the Copepoda, which dominate the marine zooplankon and are among the most abundant metazoans on Earth; the Cirripedia (barnacles), familiar and ecologically important members of rocky shore and fouling communities throughout the world's oceans (Fig. 1.1); and the Ostracoda, small bivalved creatures unfamiliar to the layperson but abundant in marine soft bottoms and of major importance as microfossils. The biology and, in particular, the mating behaviors of some of these species are fascinating in themselves (see, e.g., Barnes et al. 1977, Morin and Cohen 1991), but little is known about their social behaviors beyond the interactions between one male and one female.

As it happens, the chapters in this book deal exclusively with members of the class Malacostraca. Most of our present knowledge about the social behavior of crustaceans comes from species in this class, probably because they include relatively large, conspicuous, and often commercially important species. Major clades within the Malacostraca include the Stomatopoda, Decapoda, and Peracarida. The behavioral ecology of stomatopods ("mantis shrimps"), which carry a powerful claw and live in benthic marine habitats, has been described in numerous studies by Roy Caldwell

Table 1.1. The main crustacean taxa, with their approximate number of described species, environments, habitats, lifestyles, and feeding and reproductive biology.

Taxa		Environment			Habitat			Lifestyle			Feeding biology					Development	
Subphylum / Class / Subclass / Superorder / Order / Suborder / Infraorder	Number of Species	Marine	Freshwater	Terrestrial	Soft Bottoms	Hard Bottoms	Water Column	Mobile	Semisessile	Sessile	Suspension Feeders	Deposit Feeders	Grazers	Predators/Scavengers	Symbiotic	Planktonic	Direct
Crustacea																	
Remipedia	12	++			+			+						+			(+)
Cephalocarida	10	++			+			+				+				+	
Branchiopoda	900	(+)	++		+			+			++	+	+	+		+	+
Maxillopoda																	
Thecostraca																	
Cirripedia	1,285	+++				+				+	+					++	
Copepoda	12,000	++	+	(+)	+	+	++	+			+	+	+	+	+	++	
Ostracoda	13,000	++	+	(+)	+			+			+	+	+		+	++	+
Malacostraca																	
Phyllocarida																	
Leptostraca	36	++			+			+			+	+		+		+	

Eumalacostraca																
Hoplocarida																
Stomatopoda	350	+++			+	+		+	+				++		++	
Syncarida	200		++		+			+		++		+			+	
Eucarida																
Euphausiacea	90	+++					+	+		++			+		++	
Decapoda																
Dendrobranchiata	450	+++			+		+	+		+	+	+	+		++	
Pleocyemata																
Caridea	**2,500**	+++	+		+	+		+	+	+	+	+	+	+	++	+
Brachyura	**10,500**	+++	+	+	+	+		+	+	+	+	+	++	+	++	+
Anomura	2,500	+++	+	+	+	++		+		+	+	+	+	+	++	+
Astacidea	**660**	++	++	+	++	+		+	+		+	+	+		+	+
Palinura	**170**	+++			+	++		+				+	+		++	
Thalassinidea	560	+++			++			+	+		+				++	+
Peracarida																
Mysida	1,000	+++					+	+		+						++
Amphipoda	**8,000**	+	+	+	++	++	+	+	+	+	+	+	+	+		+++
Isopoda	**10,000**	+	+	+	+	++	+	+	+	+	+	+	+	+		+++
Tanaidacea	1,500	+++			+	+			+	+	+	+				+++
Cumacea	1,300	+++			++			+		+	+					+++

+, trait has been reported; ++, trait is common; +++, this trait is predominant for this taxon. Taxa addressed in this volume are shown in boldface. Taxa with representatives in marine, freshwater, and terrestrial environments are shaded gray. Data based mostly on information from Martin and Davis (2001) and Brusca and Brusca (2002).

and collaborators (e.g., Caldwell 1992, Cronin et al. 2001, Mazel et al. 2004). While stomatopods are restricted to marine environments, the species-rich decapods and peracarids are ecologically dominant groups in marine, freshwater, and even some terrestrial environments, which makes these groups most interesting and valuable for comparative analysis of the evolution of social and mating systems. For this reason, we strive in this book to provide a balanced suite of chapters dealing with the social behavior of decapod and peracarid crustaceans from all these major environments.

The Decapoda includes most of the larger crustacean species familiar to the general public and seafood gourmands, namely, the shrimp, lobsters, crayfish, and crabs. Shrimp are characterized by a large abdomen with powerful muscles and swimming appendages. Most shrimp species are agile swimmers that live close to the bottom, but several lineages have adopted endobenthic lifestyles, living in burrows, in other dwellings, or in symbiotic associations with other animals, and adults of these species rarely swim. Many species occur in large conspecific aggregations, groups, or swarms—the basis of shrimp trawl fisheries worldwide—but some have assumed a solitary or monogamous lifestyle. Contributions in this book mainly concern caridean shrimp, which include some of the most colorful crustaceans familiar to divers and aquarists (see chapters 11, 18). Lobsters are essentially large, bottom-dwelling shrimp, with a large abdomen and strong walking legs, and adults usually do not swim. They may live in groups or alone (see chapters 6, 13). Crayfish are morphologically similar to small, clawed lobsters and inhabit freshwater and semiterrestrial environments. Many crayfish species are powerful diggers that live in self-excavated burrows in rubble or soil. These burrows can be inhabited by solitary individuals, heterosexual pairs, or entire families (see chapter 15). Aggressive interactions may arise over ownership of burrows and other important resources such as food or mates (see chapters 4, 5). The final major group of decapods is the crabs, which feature a reduced abdomen folded underneath the carapace (in the case of hermit crabs, the inflated abdomen is hidden in a gastropod shell). Crabs are found in marine, freshwater, and terrestrial habitats and support major fisheries worldwide. Brachyuran crabs are common and ecologically important predators in many benthic marine ecosystems. Habits of brachyuran and anomuran crabs range from highly mobile and free-ranging to a more sedentary lifestyle, inhabiting burrows or other dwellings. Social interactions are as diverse and complex as their lifestyles (see chapters 9, 10, 12, 17).

The Peracarida consist of a diverse group of orders, including the Amphipoda, Isopoda, Tanaidacea, and Cumacea. Although little known to the layperson, species in this group are diverse, ubiquitous, often extraordinarily abundant, and hence ecologically important in many aquatic (and some terrestrial) ecosystems. Most peracarid crustaceans are of the same size order as insects, ranging from a few up to about 20 mm in body length. A feature common to all peracarids, with far-reaching implications for ecology and behavior, is that they brood their offspring to late larval or early juvenile stages, with no planktonic dispersal of larvae (Johnson et al. 2001). Contributions herein deal primarily with the amphipods and isopods. However, interesting social behaviors can also be expected from the tanaids, many of which inhabit self-constructed tubes and occur in dense assemblages (Drumm 2005). Some tanaids feature complex sequences of sex change (Gutu and Sieg 1999), and although little is known about the factors responsible, the density-dependent appearance of males suggests interesting sexual selection dynamics (Highsmith 1983). The amphipods (sometimes called scuds or sand fleas) and isopods (sowbugs or woodlice) are ubiquitous in marine, freshwater, and terrestrial habitats. Most peracarid species are epibenthic, but some are efficient burrowers that excavate small dwellings in diverse substrata, and others are highly modified parasites of larger animals. The social behaviors of peracarids described in this book range from mating interactions (see chapters 2, 7, 8) to complex parent–offspring interactions (see chapters 14, 16).

General Functional Morphology and Anatomy

Body Organization

Like other arthropods, crustaceans feature a segmented body covered by a chitinous cuticle, which is shed and replaced repeatedly during growth. In most species, the body is subdivided into three main regions: the head, thorax, and abdomen. In the Decapoda, the head and the anterior thoracic region are fused to form a cephalothorax, which in some species is heavily sclerotized to form a protective carapace covering these body regions. Beyond these commonalities, there is a tremendous diversity of form and functional morphology among crustacean taxa. The carapace may be heavily armored with spines or serrations and invested with calcium carbonate, improving its strength and protective function against predators. The abdomen is large and muscular (supporting swimming movements) in shrimp, lobsters, and crayfish but comparatively small in most crabs, in which it has little or no importance for movement. In peracarids, the distinction between the thorax and abdomen is less obvious to casual inspection, but the body has seven pairs of thoracic walking legs and three to five pairs of abdominal swimming appendages.

Crustaceans vary in size from species measured in millimeters to large species whose body sizes (including the legs) may exceed 1 m. Many smaller crustacean species, such as calanoid copepods, are planktonic drifters, adapted to a life spent entirely in the water column. Others such as harpacticoid copepods are primarily benthic, though some make frequent excursions into the water column. Intermediate-sized species, such as most peracarid and many shrimp species, lead a benthic life but retain high mobility, enabling short bouts of swimming. Larger, heavier species, not surprisingly, usually are bottom dwellers. Most of these species are poor swimmers and thus must rely on their body armor and claws for defense or live cryptically in burrows or crevices inaccessible to most enemies.

A fundamental characteristic of crustaceans, which has served as a "toolbox" for their evolutionary radiation into diverse lifestyles and habitats, is the set of biramous appendages (with an outer branch, the exopod, and an inner branch, the endopod) that have diverged to serve multiple functions in various taxa (Fig. 1.2). All species possess two pairs of antennae and three pairs of "true" mouthparts (mandibles and two pairs of maxillae). These are followed by the thoracic appendages (pereopods), some of which may be fused with the head region and support the mouthparts during feeding. The posterior thoracic appendages are variously modified for walking, grasping, or swimming. Abdominal appendages (pleopods) may also be highly modified. In shrimp and in many peracarid species, the pleopods serve as swimming paddles. In many decapods and isopods, the pleopods are modified for reproductive tasks.

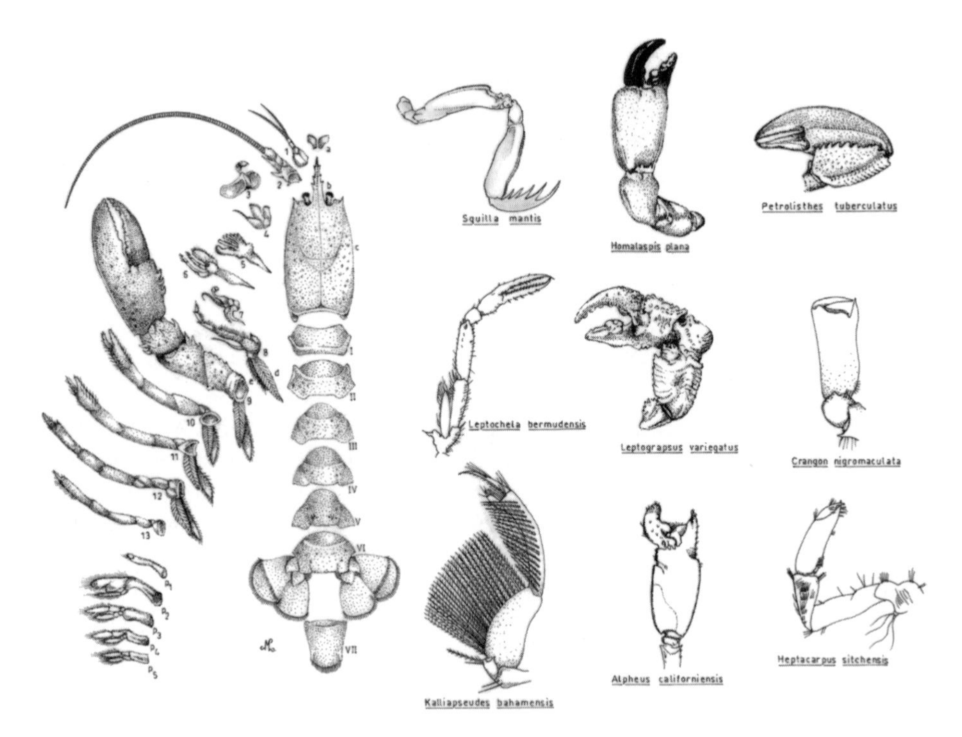

Figure 1.2 Left, a crayfish with all appendages dissected and laid out (a, metastoma; b, eye; c, carapace; d, gill; e, epipodite; x, exopodite; I–VII, abdominal segments; 1, antenna 1; 2, antenna 2; 3, mandible; 4 and 5, maxillae; 6 a 8, maxillipeds; 9, chela (= pereopod 1); 10 a 13, pereopods 2 a 5). Right, first claws of different malacostracan crustaceans.

The two pairs of antennae have primary sensory function, containing numerous chemo- and mechanoreceptors, and are of central importance in interactions among individual crustaceans (see chapters 4–6). The second antennae, in particular, can be very long, sometimes exceeding the total body length of the animal. In many species, antennae are sexually dimorphic, with the much larger antennae of males presumably reflecting the adaptation for efficient mate searching in roving males. In some species, antennae are also used to capture food particles, excavate sediments, facilitate swimming, shove away opponents, or grasp mating partners.

True mouthparts are relatively conservative among the different crustacean taxa, although there is some variance in form related to feeding biology (e.g., Watling 1993). In general, the mandibles are used to bite or rasp the food, and the maxillae to transfer it into the mouth. In many crustaceans, the first two pairs of thoracic appendages have been modified into maxillipeds, which assist feeding movements. These also may be used during intraspecific interactions in some species. The mouthparts of the parasitic copepods, isopods, and Branchiura (fish lice) are highly modified for blood sucking and, in some species, are lacking entirely.

The thoracic legs (pereopods) of most crustaceans are relatively simple extremities that are variously modified for swimming, grasping, crushing, filtering, digging, or walking. The first pereopods are commonly chelate (clawed) or subchelate and are referred to as gnathopods or chelae. Depending on species, these may be used as filtering spoons, pincers, or powerful crushers and weapons (e.g., Mariappan et al. 2000). The extreme diversity of chela (claw) morphology (Fig. 1.2), and their often highly specialized function hints at the importance of chelae for the behavioral ecology of crustaceans, as illustrated by several chapters in this volume. The role of the chelae in mating systems is reflected in the pronounced sexual dimorphism of this structure in many crustaceans (Lee 1995), with fiddler crabs being an extreme example (see chapter 10).

A distinctive feature of crustaceans, with important consequences for their functional morphology, ecology, and behavior, is the diverse array of setae on their appendages, which fulfill a wide range of tasks (Fig. 1.3). Setae are stiff hairlike or bristle-like outgrowths of the cuticle, which, depending on their function, may be simple, serrate, denticulate, plumose or scaled (among others), short and stout, or long and delicate (Garm 2004). Setae assist in activities related to digging, feeding, filtering, cleaning, sensing, mating, and incubation of developing offspring. Groups, rows, or batteries of diverse types of setae, which support feeding activities, are found on the pereopods and mouthparts. In suspension-feeding crustaceans, these setae form fine-meshed filters that are cleaned of food particles by comblike setae on neighboring appendages. In deposit-feeding crustaceans, rows of stout setae on the appendages assist in digging. Setae may also increase the functional surface of the pereopods and pleopods, enhancing swimming efficiency or burrow ventilation. Many crustaceans also have setae-armored appendages that are specialized for grooming of the antennae, body surface, gills (all decapods), or the brooded embryos (e.g., Felgenhauer et al. 1989, Förster and Baeza 2001). Due to their mechano- and chemosensory function, setae are often concentrated on the antennae, near the mouth, and on the pereopods, but they also occur on other body regions (see also "Nervous System and Sensory Biology," below).

Internal Anatomy

Crustaceans have a digestive system composed of a chitin-lined foregut, a midgut, and a chitin-lined hindgut. The foregut is a complex apparatus involved in the initial mechanical maceration and sorting of food: heavily calcified toothlike structures aid in grinding larger food particles, after which the resulting food slurry is sorted by several sets of setal filters, letting only tiny particles pass into the digestive caecae, where enzymatic digestion and absorption take place. Particles retained by these filters pass directly into the intestine, where feces are formed and, in many crustaceans, surrounded by a peritrophic membrane to form a fecal pellet. Feces exit via the chitinous hindgut through the anus, which is located in the posteriormost part of the abdomen. Since the entire foregut region is replaced during each molt, most crustaceans cannot feed immediately before and after molting.

The circulatory system of all crustaceans contains a dorsal heart that pumps the blood from the respiratory organs to the other body organs and tissues. Small crustaceans such as most amphipods and isopods have no arteries, whereas large decapods have a well-defined system of blood vessels channeling blood directly to the internal organs and to the bases of all large appendages, such as antennae and legs. Small crustaceans (e.g., copepods) also lack specialized respiratory organs, but most malacostracans have gills,

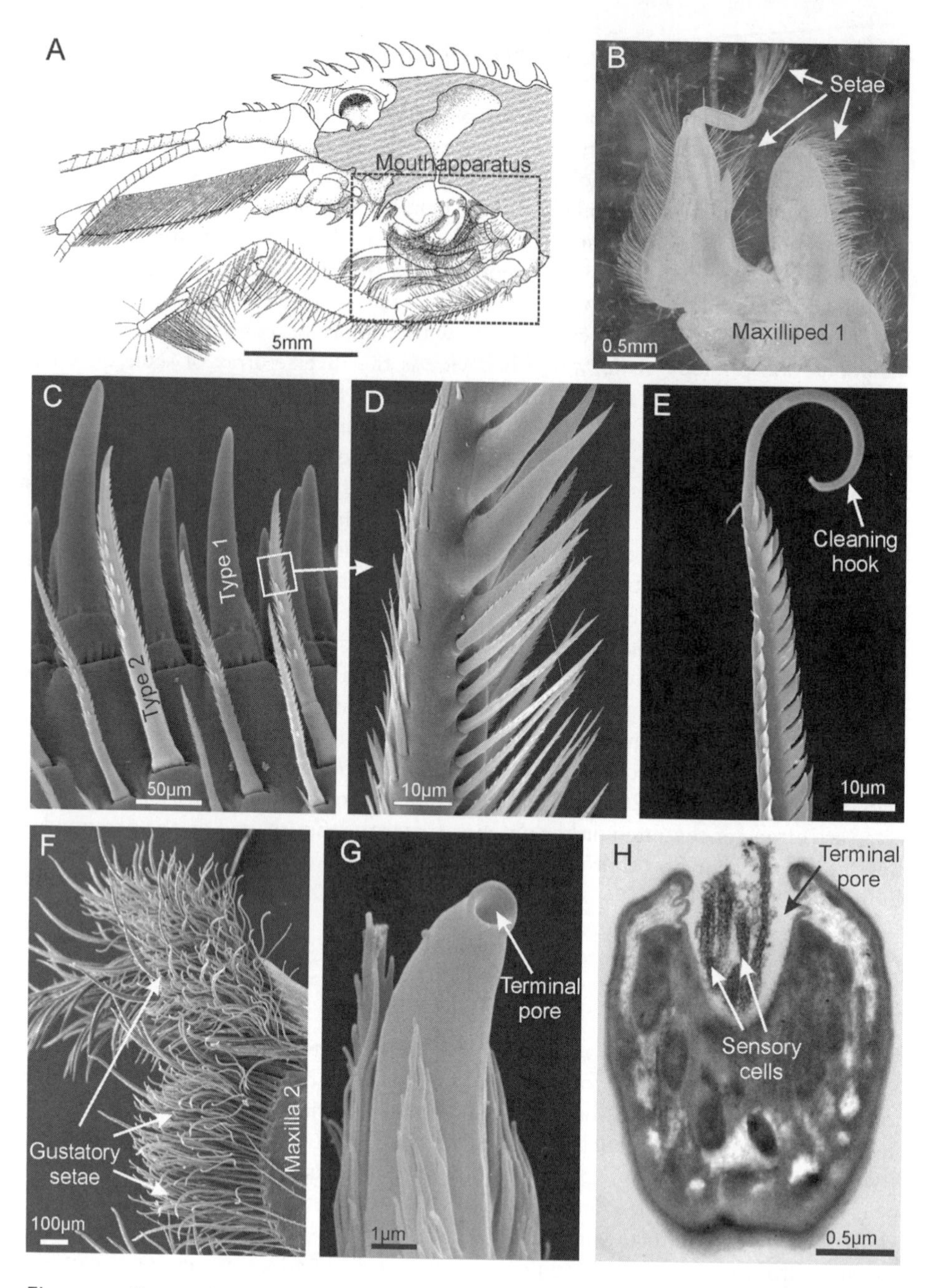

Figure 1.3 Crustacean setae. (A) Side view of the mouth apparatus of the cleaner shrimp, *Stenopus hispidus*. (B) One of the mouthparts (maxilliped 1) from a crayfish. (C) Scanning electron micrograph of two types of setae. (D) Most setae have some outgrowths, and in many cases they resemble small saw-teeth. (E) Many of the setae employed in self-cleaning have a hook at the tip. (F) Many setae have olfactory or gustatory functions, and the center of the taste function in crustacean mouthparts is a dense population of setae on the second maxillae. (G) Chemosensory setae have a terminal pore through which the sensory cells contact the outside world. (H) The sensory cells protrude through the pore. All figures from Anders Garm (copyright A. Garm 2006).

which may be covered by protective structures, as in most decapods and some isopods, or protrude unprotected between the pereopods or pleopods, as in amphipods, stomatopods, and some thalassinideans.

Excretion is via nephridial glands that are connected to nephridiopores in the head region, at the base of either the antennae or maxillae. As in many mammals and insects, some crustaceans utilize controlled and directional urine excretion to convey chemical signals during intraspecific communication (see chapters 4–6). Many freshwater and terrestrial crustaceans have strong osmoregulatory capacities, while estuarine and littoral crustaceans range from osmoconformers to osmoregulators (Mantel and Farmer 1983).

Nervous System and Sensory Biology

The crustacean nervous system is complex, consisting of cephalic ganglia and a ventral nerve cord. New nerve cells are continuously produced during life, and neurogenesis may even be affected by social interactions (Beltz and Sandeman 2003). Cephalization is pronounced in all but highly modified parasitic crustaceans, particularly in large decapods (see chapter 4). Many crustaceans feature giant neurons that are considered important for rapid escape responses (Faulkes 2004).

Crustaceans obtain information about their environment through a diverse array of sensory organs, including the eyes and the antennae with multiple chemo-, mechano-, and proprioceptors. Many species also have chemo- and mechanoreceptors on other body parts, for example, near the mouth or on their pereopods. Antennal receptors are very important for crustaceans in obtaining information about their environment. Rheoreceptors on the antennae permit determination of velocity and direction of water currents. Chemoreceptors (specialized setae, often with hollow tips) detect substances transported to the antennae via water currents. Some larger crustaceans also actively draw water toward their antennae by controlled movements of the ventilatory appendages. Some of the most important chemosensory setae are the aesthetascs, which are found on the antennae of a wide variety of crustaceans. These are stimulated by waterborne chemicals, and the antennae are frequently moved ("flicked") to reduce the thickness of the boundary layer around them, facilitating chemoreception (Goldman and Patek 2002). In many species, important chemical information is based on contact pheromones on the cuticle, and individuals have to touch others with their antennae in order to confirm their sex and reproductive status (e.g., Caskey and Bauer 2005). Interestingly, this is the case with many chemicals involved in mating interactions. Many sensory setae fulfill both chemo- and mechanoreceptor functions (Fig. 1.3). Setae with exclusive mechanoreceptor function appear to be rare among crustaceans, underlining the importance of chemical cues in this group.

Acoustic signals are employed by some terrestrial crustaceans during social (Popper et al. 2001) and mating interactions (see chapter 17). Signals may be transmitted via air or the substratum and can reportedly be perceived over distances of several meters. Certain aquatic crustaceans also produce acoustic signals, notably the snapping or pistol shrimp (Alpheidae), whose characteristic popping provides a constant sonic background on tropical reefs. In social alpheids, coordinated snapping may provide a warning to competitors or other enemies (Tóth and Duffy 2005). In spiny lobsters, acoustic signals appear important in deterring fish predators (e.g., Patek and Oakley 2003). Structures often referred to as stridulatory ridges are reported from a variety of aquatic crustaceans, but their functions are poorly understood.

Like other arthropods, crustaceans have compound eyes, the size of which varies from minute or absent in many deepwater or subterranean species to enormous in some pelagic species such as hyperiid amphipods, in which the entire head consists essentially of a compound eye. The eyes are covered by the cuticle and may be submerged in the head as in amphipods and isopods or raised on movable eyestalks as in stomatopods and decapods. In some species, stalked eyes can be lowered into protective cavities in the carapace and when raised, oriented toward objects. Compound eyes are best suited to detect movement, and it is thus not surprising that visual displays during intraspecific interactions often involve complex movements of the chelae (see chapter 10). Some of the most complex eyes in the animal kingdom are found in stomatopods, in which visual cues are important for both prey detection and intraspecific interactions (Cronin 1986, Mazel et al. 2004). In some semiterrestrial fiddler crabs, color marks are even used to recognize individual neighbors (Detto et al. 2006). Claw color has been suggested as an honest signal (a character reliably indicating health and strength of an individual) during social interactions of semiterrestrial crabs (Detto et al. 2004). Vision seems to be of central importance in some burrow-dwelling malacostracan crustaceans from shallow, clear water and from terrestrial environments. These species use vision to recognize moving objects in their surroundings (conspecifics, predators), and when away from their home burrow they employ path integration to determine their position in relation to their burrow (Zeil and Hemmi 2006). While the spectacular color patterns of many crustaceans may not be directly linked to their own visual perception, and also serve other functions (e.g., crypsis or aposematic coloration), the commonness of bright colors among crustaceans indirectly suggests a role for color vision in these animals.

Reproductive System

The crustacean reproductive system generally consists of paired gonads leading via the gonoducts to gonopores located on the sternites (ventral side of the thoracic segment) or the coxae (basal segments) of the walking legs (pereopods). Males often have a complex vas deferens in which ripe sperm are held before mating. In decapods, the mid and posterior regions of the vas deferens secrete seminal fluids contributing to the formation of spermatophores (Hinsch 1991). In most species, fertilization is external, and spermatophores are transferred via the modified male pleopods to the ventral part of the female body. In brachyuran crabs and some other decapods, however, fertilization is internal, and spermatophores are placed directly by the male into female sperm receptacles (e.g., chapter 9). Male amphipods have a pair of short flexible penes that inject sperm directly into the female's marsupium. In isopods, males possess modified pleopods used in a similar way as those of decapod males to transfer sperm into the female oviducts.

Several hormone-producing glands influence gonad development. The sinus gland, located in the eyestalk or near the basis of the eye, releases the gonad-inhibiting and gonad-stimulating hormones, which control oocyte maturation in the ovaries

(Charniaux-Cotton 1985). Spermatocyte development is induced by androgenic hormones produced in the androgenic gland, which is located near the end of the vas deferens. These hormones are also involved in the production of secondary sexual characters and courtship (Sagi and Khalaila 2001).

Development, Growth, and Molting

The mode of development strongly influences the genetic and kin structure of conspecific groups and thus has important implications for social and sexual systems. Many non-malacostracans and most decapods and stomatopods release planktonic larvae (see frontispiece) into the plankton, where they may drift for weeks or even months before settling to the bottom and metamorphosing into the adult form. In some species, however, planktonic stages are reduced and fully developed juveniles emerge from the egg. Amphipods, for example, brood their offspring to the juvenile stage, whereas in isopods and tanaids, late larval stages (often called "manca") with juvenile characteristics hatch from the egg (Johnson et al. 2001). During the planktonic phase, larvae of decapods pass through several larval stages involving gradual changes in size and form (i.e., anamorphic development) and may be dispersed over considerable distances, depending on currents and duration of larval development. Before settlement to the benthos, the larvae undergo metamorphosis, which is usually accompanied by dramatic changes in size and form. At this time, the larval pool has generally been well mixed, making it unlikely that adjacent individuals are closely related. In species with direct development (all amphipods and isopods, as well as some decapods; see chapters 15, 17, 18), juveniles hatch from the egg in the immediate vicinity of their mother, which may result in the development of closely related kin groups.

During postsettlement ontogeny, crustaceans pass through several molt stages that are commonly accompanied by morphological and physiological changes. Intervals between molts are variable, depending mainly on environmental variables such as food supply and temperature. Most malacostracans have indeterminate growth, but some species have a terminal molt stage, after which they cease growth and molting (Hartnoll 2001). Molting frequency declines with size and age: early juvenile stages may molt several times within a month, whereas adults of some large decapods can have intermolt intervals of several years. Crustaceans also have a high regenerative capacity, and lost or damaged appendages may be repaired or replaced during molts. Regenerated appendages are small and weak after the first molt but during successive molts grow faster than the other limbs. Individuals lacking limbs are probably handicapped during intraspecific interactions (Juanes and Smith 1995) and may abstain temporarily from reproduction. Many decapods are also capable of autotomy, intentionally and rapidly separating limbs when caught by predators or conspecific opponents (Wasson and Lyon 2005).

Molting, which is under strong hormonal control, may be induced by internal or external stimuli (Hartnoll 2001). Hormones involved in molting are produced in the Y-organ and the neurosecretory system (Chang 1995). The Y-organ produces ecdysone and other ecdysteroids (molting hormones), which initiate the molting process (Chang 1993). Neurohormones released from the sinus gland (molt-inhibiting hormones) suppress the molting process (e.g., Subramoniam 2000). As decreasing titers of molt-inhibiting hormones permit expression of molting hormones, the molting process begins. Shortly before and after the molt, crustaceans are restricted in their activities, and many species do not feed. Furthermore, due to the softness of the new exoskeleton, recently molted crustaceans are relatively defenseless and vulnerable to both cannibalism and predation (Chang 1995). The risk of cannibalism after molting may have contributed to the relatively solitary lifestyles among many crustaceans. Adult females often enjoy protection by male mating partners during their reproductive molts. Mate guarding in crustaceans may thus serve two primary functions: (1) to reduce the risk of sperm competition (e.g., G.A. Thompson and McLay 2005) and (2) to protect recently molted (and fertilized) females from predation (Wilber 1989). Although molting represents both physiological and ecological challenges, it also offers benefits. During the intermolt period, many aquatic crustaceans accumulate loads of diverse epibionts, which may also foul the gills and sensory structures, thereby compromising their efficient function. Since all these vital organs are covered by the chitinous cuticle, their surface is also shed and renewed during molting (Shields 1992).

Feeding Biology

The feeding ecology of crustaceans has profound consequences for social and mating interactions. The abundance, quality, and physical nature of the food determine the strength and nature of competition, which may manifest in aggression or resource-centered mating systems, and also influences the intimacy of the crustacean's association with its food, which has pervasive ecological and evolutionary consequences (Price 1980, J.N. Thompson 1994). Aggressive interactions are more commonly observed among predators and scavengers that live at high abundances and feed on scarce and monopolizable food sources. In these species, dominant individuals may exclude subordinates from valuable food resources.

As a group, crustaceans feed on nearly all types of living and dead organic material, from tiny suspended detritus particles and microbes to vertebrate carcasses. Suspension feeders, such as caprellid amphipods and porcellanid crabs, often aggregate on vantage points in a favorable current environment. Contests over the best feeding positions may negatively affect growth and health of subordinate individuals (Donahue 2004). Deposit-feeding crustaceans commonly inhabit burrows in soft sediments. They either ingest bulk sediments in the depth of their burrows (e.g., some thalassinidean shrimp) or emerge partially (many amphipods) or wholly (fiddler and ghost crabs) to sift surface sediments for food particles. Most crustacean grazers are generalists, feeding on a variety of algae, seagrasses, marshgrasses, mangroves, and other plants, but some tropical species specialize on particular, usually chemically defended, algal species, on which they experience reduced predation (Hay et al. 1989, 1990). Detritus feeders are common in accumulations of plants or algae drifting in shallow subtidal waters or deposited on the shore. Some brachyuran crabs, hermit crabs, and semiterrestrial crayfish compete intensively for fresh plant detritus, which they may drag into their burrows to be consumed in isolation from predators and competitors (Fratini et al. 2000). Many larger decapods and stomatopods are predators, including the spectacular stomatopod spearers and the snapping shrimp, which are reported to kill fishes with sudden strikes of their powerful chelae. These chelae are also frequently employed during intraspecific conflicts (Montgomery and Caldwell 1984, Tóth and Duffy 2005), and the risk of receiving serious wounds during these interactions may have selected for the complex communication systems of some species (Hughes 1996). Most crustacean predators also feed on carrion, a fact that is put to use in commercial trap fisheries for various decapods (see also chapter 9). Finally, symbiotic crustaceans live in association with a variety of other organisms, which may provide food, shelter, or both. The commonness of specialized host associations among symbiotic, parasitic, and herbivorous crustaceans has important ramifications for population and family structure and thus for mating and social interactions (see chapters 12, 18).

One particularity among many marine crustaceans is the sedentary lifestyle of adults made possible by their feeding ecology. Many suspension feeders, deposit feeders, some predators, and most symbionts reside for long time periods in their dwellings, because their food supply is constantly replenished via water currents, liberating them from the need to forage for food. In contrast, most foraging crustaceans have no permanent home dwelling and hide between feeding bouts in ephemeral shelters. Exceptions are found in some littoral and terrestrial crustaceans, which return to home burrows after foraging excursions (e.g., chapter 16).

Reproductive Biology

Sexual Systems and Sexual Maturity

Several aspects of crustacean reproductive biology have important consequences for sexual and social systems. These include the nature of sex determination, internal versus external fertilization, lifetime schedule of reproduction, mode of larval development, and patterns of investment in eggs and juveniles. Parthenogenetic reproduction is commonly found among freshwater branchiopods such as brine shrimp (*Artemia* spp.) and *Daphnia* species and in the ostracods but has also recently been reported from a decapod (Vogt et al. 2004). Whereas most crustaceans have separate sexes, there are a few sequential and even some simultaneous hermaphrodites, particularly among shrimp (chapter 11). Self-fertilizing hermaphrodites have been reported from the Branchiopoda (Zucker et al. 2001). Sex determination in most species is under genetic control but is determined by environmental factors in some species (Ginsburger-Vogel and Charniaux-Cotton 1982). There is still much debate about control of sex change in hermaphrodites, which can involve environmental factors (e.g., population density or season; Charnov and Anderson 1989, Bergström 2000) and social interactions (Baeza and Bauer 2004).

A wide range of reproductive syndromes can be found among crustaceans, including both semelparity and iteroparity. Many crustaceans continue to grow after achieving sexual maturity, and available energy must be balanced between investment in growth and reproduction. Costs of reproduction are reflected in lower growth rates in some crustaceans (Karplus 2005, Hartnoll 2006). In some amphipods and isopods, morphological changes coincide with becoming reproductively active: during oocyte formation and incubation, the female's digestive system is reduced, and mouthparts become vestigial such that the animal is unable to feed (Borowsky 1996). After a reproductive event, these individuals molt back into a nonreproductive morphology, and during the following intermolt period they feed intensively, with most surplus energy allocated to growth. Thus, at the next reproductive event, these large individuals usually have a very high reproductive potential. In some species (majid crabs and a few others), the problems of balancing energy investment between growth and reproduction are circumvented by the fact that they reach sexual maturity only with the terminal molt, at which point growth is halted. All available surplus energy is then channeled into reproduction and may be invested in a single or several subsequent events. The prospect or absence of future reproductive opportunities has a great influence on the present reproductive investment (and thus the mating behavior) of crustaceans (e.g., Zimmer 2001).

Mating Behavior

The operational sex ratio (OSR) in natural populations, determined by population density, sex ratio, and the proportion of individuals ready for mating (Emlen and Oring 1977), is a relatively good predictor of crustacean mating behaviors (see chapters 2, 7, 8). In most crustaceans, as indeed in most animals, the reproductive potential of males is higher than that of females, resulting in male-biased OSR. However, female-biased OSRs have also been reported, in particular, when males become very scarce in local populations, mostly due to sex-selective mortalities from predation (McCurdy et al. 2005) or fishing (see chapter 9).

In females of most crustacean species, mating and fertilization are closely associated with molting. Females molt, mate shortly thereafter, and release the eggs, which are fertilized in the outer female reproductive tract or on the ventral side of the female. Generally, females are receptive for only minutes to a few days after molting (Bauer and Abdalla 2001, G.A. Thompson and McLay 2005), and mating must take place during this time period; otherwise, eggs will be aborted or reabsorbed (Sastry 1983). Females of some species, though, maintain the ability to mate (repeatedly and with different males) throughout most of their reproductive cycle (Hartnoll 2000). This may allow females to be more selective with respect to the males that father their offspring and sets the stage for sperm competition.

The effects on the OSR of the brief receptivity of crustacean females may be further influenced by the degree of synchrony among females of a local population. In some crustacean species, most females become receptive synchronously and are thus available for mating simultaneously, reducing the opportunity for males to mate with multiple females (Sastry 1983). In contrast, females of other species become receptive separately, resulting in a male-biased OSR and strong competition for mating partners among males. While the influence of female synchrony (or asynchrony) on the potential for polygamy and male–male competition is well recognized (e.g., Orensanz et al. 1995), relatively little is known about the ultimate factors that have led to the evolution of either synchrony scenario in crustaceans (see, e.g., discussions in Knowlton 1979, Mathews 2002).

Courtship behaviors in crustaceans range from nonexistent to quite sophisticated and can involve chemical, acoustic, tactile, and visual signals. Individuals may employ signals to attract mating partners and, once direct contact is established, to evaluate mate quality. In most crustacean species, the males are the searching sex, and females may emit signals to attract males (e.g., Krång and Baden 2004). However, in some species these roles are inverted. Female searching is particularly common in species where males hold important reproductive resources such as a safe dwelling. In these species, the males occupy the resource (shelter, self-constructed burrow, or a host) to which they attract the female using chemical, acoustic, or visual signals (Popper et al. 2001; see also chapters 2, 6, 10). In many crustacean species, mates are attracted by specific signals, whereas in other species, encounters between males and females appear to occur by chance. Usually, in these latter species the males are highly mobile, roaming actively in search of females (Conlan 1991), and direct contact is necessary to reveal the reproductive status of the female.

When encountering a female approaching the reproductive molt (and thus receptivity), males may engage in precopulatory mate guarding—this is commonly observed in amphipods (chapter 7), some isopods (chapter 8), and decapod crabs. Females may resist advances of small males, potentially gaining male mates of higher quality (chapter 8). During precopulatory mate guarding, males may also confront male competitors for their females. These confrontations among males can be promoted by females. For example, Sneddon et al. (2003) suggested that female crabs, when guarded by small males, release pulses of pheromone in presence of a large male, thereby provoking aggressive interactions between the males and takeover by the large male. Although increasing evidence supports such mate selection by both males and females, little is known about the factors promoting mate selection in crustaceans and the mechanisms involved. The opportunity for additional matings (i.e., the environmental potential for polygamy; see chapter 2) of both sexes seems to be of key importance in molding mating systems and precopulatory associations (chapters 7, 8). In species where male–female associations are of very short duration, receptive individuals appear to mate with any individual of the opposite sex they encounter, while in species where associations are long-lasting, either sex may remain selective during the premating phase and exchange mates if better partners are encountered (Shuster and Caldwell 1989, Jivoff and Hines 1998).

Mating itself is generally a short interaction during which males transfer sperm to females. In species with internal fertilization (e.g., brachyuran crabs, penaeoidean shrimp, isopods) males transfer sperm packages directly into the female reproductive tract, where sperm may either be utilized shortly afterward or retained for months or years in spermathecae until later usage. In most species with external fertilization (e.g., hermit crabs, caridean shrimp, lobsters), the male places spermatophores on the ventral part of the female's body. Usually, shortly after spermatophore placement, the female starts to ovulate. The eggs pass in close proximity to or in contact with the spermatophore, and fertilization occurs before the eggs become attached to the female's pleopods (e.g., Hess and Bauer 2002). In amphipods, the male injects sperm into the female's marsupium just before the female ovulates. Oocytes are fertilized in the marsupium and incubated there during embryonic development. Due to the short time period between sperm transfer and ovulation, it is generally assumed that female amphipods receive sperm from only one male during a reproductive cycle. However, it has been reported for at least one species (*Jassa marmorata*) that a female can mate with several males (Clark and Caudill 2001). Females of some crab and shrimp species are known to mate multiply, leaving ample opportunity for sperm competition

(Bilodeau et al. 2005, Brockerhoff and McLay 2005; see also chapter 9). Males of these species may employ diverse tactics (postcopulatory mate guarding, sperm plugs, sperm dilution) to reduce the risk of additional inseminations and sperm competition.

Investment in Reproductive Products

Reproductive investment is highly variable among crustacean species and, as in most animals, generally covaries with body size. Large decapods can produce >100,000 eggs in a single clutch (e.g., Hartnoll 2006), whereas clutch size in amphipods and isopods typically varies between a few up to about 100 eggs (Sainte-Marie 1991). Also, female investment in individual eggs is highly variable. Generally, in species that produce many eggs, individual eggs are small, while in species with few eggs, these are relatively large. Male decapods may require several days (or weeks) to recuperate their sperm reserves after a successful mating (Kendall et al. 2001; see also chapter 9). If the male proportion in local populations falls below a threshold, sperm limitation may compromise reproductive output (Hines et al. 2003, Sato et al. 2005).

Most crustaceans are iteroparous. Semelparity is known for some isopods (Holdich 1968, Shuster 1991) and may also occur in some amphipods, but it is rare among decapods. Whether the number of reproductive events influences reproductive investment in crustaceans in similar ways as suggested for other arthropod taxa (Tallamy and Brown 1999) is unknown, but in at least one isopod, semelparity appears to have resulted in the evolution of long-lasting biparental care for offspring (chapter 16). Reproductive rates of iteroparous species can vary with environment and body size. In many tropical species, reproduction is continuous, while in polar species, reproductive events are usually limited to the short summer periods. In small species, females may produce subsequent clutches on the order of weeks, while in large species, it may take several months or even years.

Parental Care

Most crustaceans exhibit some form of parental care, which is almost exclusively the task of females. During incubation of broods, parents may incur substantial costs, either directly due to care behavior or indirectly due to lost feeding opportunities (Hartnoll 2006). Recent studies of decapod crabs and peracarids have shown that, besides time investments, females also have important energy costs during incubation of broods (Fernandez and Brante 2003, Lardies et al. 2004).

Brooding female crustaceans engage in a variety of care activities. They incubate developing embryos on their body, for example, under their abdomen in most decapods, in special brood structures (the marsupium of most peracarid crustaceans), or even within their ovaries in some isopods. One of the main activities of females during incubation is oxygen provisioning (Dick et al. 2002, Fernandez and Brante 2003). Females may also clean the brood, eliminating moribund embryos (Förster and Baeza 2001), and females of some terrestrial amphipods also manage the osmotic environment in their brood pouches (possibly by urine release into the pouch—Morritt and Richardson 1998). Transfer of nourishment to developing embryos via marsupial fluids has been suggested for some terrestrial isopods (Hoese 1984). When embryonic development is completed, females may seek out sites and time periods favorable for offspring release (see chapter 10). For example, some grazing amphipods release offspring on particular algae thought to be most suitable as food for small juveniles (Poore and Steinberg 1999). In many species with direct development, females (and males if present) may also further care for their juvenile offspring after these have been released from the brooding structures on or in their body. This is commonly observed in many peracarid crustaceans (chapters 14, 16), some snapping shrimp (chapter 18), many terrestrial crabs (chapter 17), and crayfish (chapter 15), and it may have profound consequences for the social interactions in the resulting family groups.

Sexual Selection, Sexual Dimorphism, and Mating Systems

Sexual dimorphism is common among crustaceans, usually manifested in the larger body size and more powerful claws of males, which are employed in combat with male competitors (Conlan 1991, Lee 1995). In species where males search for females, they may also have larger antennae or eyes than females (Bertin and Cezilly 2003). Contrary to the general rule, in some highly mobile species (e.g., some palaemonid shrimp), males engage in scramble competition for females and are usually smaller than females (Bauer 2004). The most pronounced size dimorphism is found in endosymbiotic pinnotherid crabs and isopods, where females are many times larger than the minute males.

Recent work has revealed that female decisions may be of equal (or greater) importance to those of males in molding mating systems of crustaceans (e.g., Brockerhoff and McLay 2005). Mating decisions of both sexes appear to be based on the availability of mating opportunities (chapter 12), mate quality (chapter 8), and environmental conditions (e.g., predation risk; chapter 7). Estimating the environmental potential for polygamy appears particularly promising in order to gain better understanding of the evolution of crustacean mating systems (chapter 2).

Summary and Synthesis

In comparison to most other organisms, crustaceans have supernumerary appendages that are modified for a variety of functions. While terrestrial vertebrates feature only two pairs of legs, insects have three, arachnids have four, but most crustaceans have five or more pairs. In most crustacean taxa, three or four pairs of legs are sufficient for walking, liberating the remaining ones for other functions. There are many examples where some of the thoracic legs have been highly modified to assume particular, highly specific functions. These appendages have also gained a tremendous importance for intraspecific interactions. In particular, the impressive claws featured by many crustacean species are frequently employed in aggression and combat, often inflicting serious wounds or loss of appendages on opponents (see chapter 5). The importance of weapons in evolution of advanced social organization has been emphasized (Starr 1979, Anderson 1984, Alexander et al. 1991), and the heavily armored soldiers or venomous stings of many social insects give testimony to this. Teeth and specialized mandibles found in many social organisms also allow pull-and-tear contests with the possibility of seriously wounding opponents. What is the role that the powerful claws of many crustaceans play in social evolution? Is there a need to develop specific communicational signals in order to avoid the potentially severe consequences of escalated fights (see chapter 5)?

The aquatic lifestyle of most crustaceans has important implications for communication (see chapter 4). In many aquatic environments, visibility is limited, and it is thus not surprising that chemical cues have gained overriding importance. Multiple and diverse chemosensory structures provide evidence that crustaceans continuously scan their environment for chemical information. Waterborne chemicals are also frequently used for intraspecific communication. However, signal transmission via chemical cues in the aquatic environment is dramatically different from that in the terrestrial environment. While clouds of airborne chemicals primarily suffer from dilution effects, in the aquatic environment chemical signals may also be scavenged by suspended particles and bacteria. Thus, long-distance communication via chemical cues may require strong signals and high initial concentrations produced by the sender. These decay processes have many implications for social interactions in the aquatic environment. For example, dominance disputes in several aquatic crustaceans are settled by emission of chemicals (see chapters 5, 6)—only the strongest individuals may be able to sustain high concentrations of these chemicals. Waterborne chemical signals may thus play an important role as honest signals in aquatic crustaceans, being more reliable than visual signals, which lend themselves to cheating (see, e.g., Backwell et al. 2000).

Many crustaceans carry chemical information on their cuticle, which allows rapid recognition of the status of conspecifics when these are touched with the antennae. These chemical badges are also useful for recognition of family members in both aquatic and terrestrial crustaceans (see chapters 16, 18). Interestingly, chemical badges are carried by the animals themselves, and scent marks placed on solid surfaces (as found in many vertebrate systems) seem to play a minor role in crustacean communication. Why have territorial scent marks, so important in social interactions of many other organisms, not evolved in crustaceans? The aquatic environment in which the major radiation of all present crustacean clades has taken place may not favor the usage of scent marks. Interestingly, semi- or fully terrestrial crustaceans that hold feeding territories around their burrows employ visual cues as landmarks (Linsenmair 1979, Christy et al. 2001) (Fig. 1.4).

Crustaceans are usually regarded as aquatic organisms, and they are certainly best adapted to this environment. One of the most important features of crustaceans is the diversity of their development modes. The insects and vertebrates most familiar to behavioral ecologists tend to be sedentary as larvae or juveniles and more mobile as adults. In many crustaceans, the opposite is true, as larvae are generally dispersed by water currents over much larger distances than an adult is likely to travel. The mode of dispersal has profound consequences for behavioral ecology because it largely dictates the structure of kin groups and populations, and the behavioral constraints and opportunities that arise thereof.

The environment in which a crustacean lives affects not only its communication and reproductive systems but also the context in which social communication develops. Because of the high energy and protein content of crustacean tissues and their general lack of chemical defenses, crustaceans are important prey for nearly all organisms larger than themselves, thus serving as central players in nearly all marine

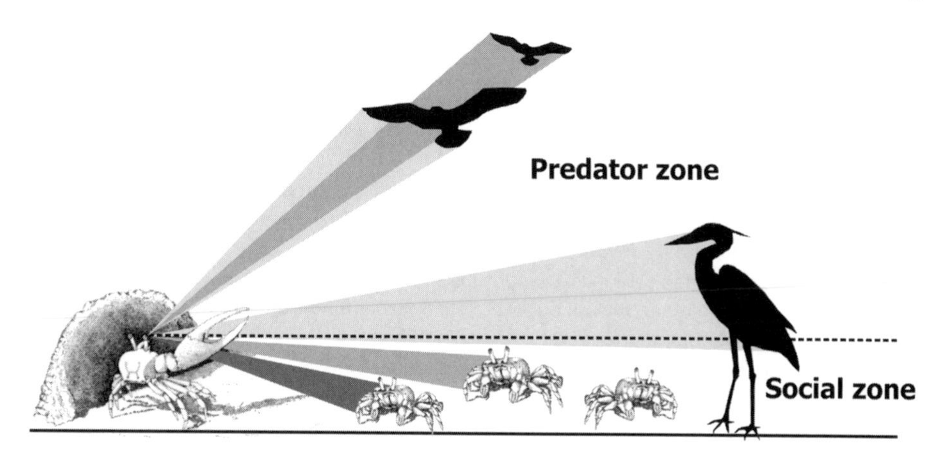

Figure 1.4 The visual environment of fiddler crabs is separated into two main zones, one where cues from predators are important, and the other where social interactions are mediated by visual cues (landmarks) and signals (figure modified after Zeil and Hemmi 2006).

and freshwater food webs. As a result, predation is an ever-present threat for most crustaceans (as is true of most small animals), and this threat has figured importantly in evolution of their social and sexual systems (e.g., chapters 7, 10). For example, predation risk reduces the environmental potential for polygamy (see chapters 2, 12) and—if occurring in conjunction with extended parental care—may favor the evolution of cooperative breeding in crustaceans and other organisms (chapter 20).

The enormous morphological and behavioral diversity among crustaceans opens exciting opportunities but, by the same token, complicates easy generalizations with respect to social evolution. Present knowledge of crustacean social behavior is approaching levels that permit comparisons across taxa, especially within particular environments. Many of the chapters in this book and the multispecies analyses therein offer testimony to this, underlining the value of crustaceans as model systems. We invite readers to explore the pages that follow for the exciting opportunities that crustaceans offer for a better understanding of social and sexual evolution.

Acknowledgments We are especially grateful to Anders Garm for his willingness to share previously unpublished figures with us. We also thank Antonio Baeza, Darryl Felder, Jeffrey Shields, and Les Watling for highly constructive comments.

References

Abele, L.G., W. Kim, and B.E. Felgenhauer. 1989. Molecular evidence for inclusion of the phylum Pentastomida in the Crustacea. Molecular Biology and Evolution 6:685–691.

Alexander, R.D., K.M. Noonan, and B.J. Crespi. 1991. The evolution of eusociality. Pages 3–44 *in*: P.W. Sherman, J.U.M. Jarvis, and R.D. Alexander, editors. The biology of the naked mole-rat. Princeton University Press, Princeton, N.J.

Anderson, M. 1984. The evolution of eusociality. Annual Review of Ecology and Systematics 15:165–189.

Backwell, P.R.Y., J.H. Christy, S.R. Telford, M.D. Jennions, and N.I. Passmore. 2000. Dishonest signalling in a fiddler crab. Proceedings of the Royal Society of London, Series B 267:719–724.

Baeza, J.A., and R.T. Bauer. 2004. Experimental test of socially mediated sex change in a protandric simultaneous hermaphrodite, the marine shrimp *Lysmata wurdemanni* (Caridea: Hippolytidae). Behavioral Ecology and Sociobiology 55:544–550.

Barnes, H., M. Barnes, and W. Klepal. 1977. Studies on the reproduction of cirripedes. I. Introduction: copulation, release of oocytes, and formation of the egg lamellae. Journal of Experimental Marine Biology and Ecology 27:195–218.

Bauer, R.T. 2004. Remarkable shrimps: adaptations and natural history of the carideans. University of Oklahoma Press, Norman, Okla.

Bauer, R.T., and J.A. Abdalla. 2001. Male mating tactics in the shrimp *Palaemonetes pugio* (Decapoda, Caridea): precopulatory mate guarding vs. pure searching. Ethology 107:185–199.

Beltz, B.S., and D.C. Sandeman. 2003. Regulation of life-long neurogenesis in the decapod crustacean brain. Arthropod Structure and Development 32:39–60.

Bergström, B.I. 2000. The biology of *Pandalus*. Advances in Marine Biology 38:55–245.

Bertin, A., and F. Cezilly. 2003. Sexual selection, antennae length and the mating advantage of large males in *Asellus aquaticus*. Journal of Evolutionary Biology 16:698–707.

Bilodeau, A.L., D.L. Felder, and J.E. Neigel. 2005. Multiple paternity in the thalassinidean ghost shrimp, *Callichirus islagrande* (Crustacea: Decapoda: Callianassidae). Marine Biology 146:381–385.

Borowsky, B. 1996. Laboratory observations on the life history of the isopod *Sphaeroma quadridentatum* Say, 1818. Crustaceana 69:94–100.

Brockerhoff, A.M., and C.L. McLay. 2005. Mating behaviour, female receptivity and male-male competition in the intertidal crab *Hemigrapsus sexdentatus* (Brachyura: Grapsidae). Marine Ecology Progress Series 290:179–191.

Brusca, R.C., and G.J. Brusca. 2002. Invertebrates. Sinauer Associates, Sunderland, Mass.

Caldwell, R.L. 1992. Recognition, signalling and reduced aggression between former mates in a stomatopod. Animal Behaviour 44:11–19.

Caskey, J.L., and R.T. Bauer. 2005. Behavioral tests for a possible contact sex pheromone in the caridean shrimp *Palaemonetes pugio*. Journal of Crustacean Biology 25:571–576.

Chang, E.S. 1993. Comparative endocrinology of molting and reproduction: insects and crustaceans. Annual Review of Entomology 38:161–180.

Chang, E.S. 1995. Physiological and biochemical changes during the molt cycle in decapod crustaceans: an overview. Journal of Experimental Marine Biology and Ecology 193:1–14.

Charniaux-Cotton, H. 1985. Vitellogenesis and its control in malacostracan Crustacea. American Zoologist 25:197–206.

Charnov, E.L., and P.J. Anderson. 1989. Sex change and population fluctuations in pandalid shrimp. American Naturalist 134:824–827.

Christy, J.H., P.R.Y. Backwell, and S. Goshima. 2001. The design and production of a sexual signal: hoods and hood building by male fiddler crabs *Uca musica*. Behaviour 138:1065–1083.

Clark, R., and C.C. Caudill. 2001. Females of the marine amphipod *Jassa marmorata* mate multiple times with the same or different males. Marine and Freshwater Behaviour and Physiology 34:131–138.

Conlan, K.E. 1991. Precopulatory mating behavior and sexual dimorphism in the amphipod Crustacea. Hydrobiologia 223:255–282.

Cronin, T.W. 1986. Optical design and evolutionary adaptation in crustacean compound eyes. Journal of Crustacean Biology 6:1–23.

Cronin, T.W., N.J. Marshall, and R.L. Caldwell. 2001. Tunable colour vision in a mantis shrimp. Nature 411:547–548.

Detto, T., J. Zeil, R. Magrath, and S. Hunt. 2004. Sex, size and colour in the semaphore crab *Heloecius cordiformis*. Journal of Experimental Marine Biology and Ecology 302:1–15.

Detto, T., P.R.Y. Backwell, J.M. Hemmi, and J. Zeil. 2006. Visually mediated species and neighbour recognition in fiddler crabs (*Uca mjoebergi* and *Uca capricornis*). Proceedings of the Royal Society of London, Series B 273:1661–1666.

Dick, J.T.A., R.J.E. Bailey, and R.W. Elwood. 2002. Maternal care in the rockpool amphipod *Apherusa jurinei*: developmental and environmental cues. Animal Behaviour 63:707–713.

Donahue, M.J. 2004. Size-dependent competition in a gregarious porcelain crab *Petrolisthes cinctipes* (Anomura: Porcellanidae). Marine Ecology Progress Series 267:196–207.

Drumm, D.T. 2005. Comparison of feeding mechanisms, respiration, and cleaning behavior in two kalliapseudids, *Kalliapseudes macsweenyi* and *Psammokalliapseudes granulosus* (Peracarida: Tanaidacea). Journal of Crustacean Biology 25:203–211.

Emlen, S.T., and L.W. Oring. 1977. Ecology, sexual selection and the evolution of mating systems. Science 197:215–223.

Faulkes, Z. 2004. Loss of escape responses and giant neurons in the tailflipping circuits of slipper lobsters, *Ibacus* spp. (Decapoda, Palinura, Scyllaridae). Arthropod Structure and Development 33:113–123.

Felgenhauer, B., L. Watling, and A. Thistle, editors. 1989. Functional morphology of feeding and grooming in Crustacea. Crustacean Issues 6:1–225.

Fernandez, M., and A. Brante. 2003. Brood care in brachyuran crabs: the effect of oxygen provision on reproductive costs. Revista Chilena de Historia Natural 76:157–168.

Förster, C., and J.A. Baeza. 2001. Active brood care in the anomuran crab *Petrolisthes violaceus* (Decapoda: Anomura: Porcellanidae): grooming of brooded embryos by the fifth pereiopods. Journal of Crustacean Biology 21:606–615.

Fratini, S., S. Cannicci, and M. Vannini. 2000. Competition and interaction between *Neosarmatium smithi* (Crustacea: Grapsidae) and *Terebralia palustris* (Mollusca: Gastropoda) in a Kenyan mangrove. Marine Biology 137:309–316.

Garm, A. 2004. Mechanical functions of setae from the mouth apparatus of seven species of decapod crustaceans. Journal of Morphology 260:85–100.

Ginsburger-Vogel, T., and H. Charniaux-Cotton. 1982. Sex determination. Pages 257–281 *in*: D.E. Bliss, editor. The biology of Crustacea, Volume 2. Academic Press, New York.

Goldman, J.A., and S.N. Patek. 2002. Two sniffing strategies in palinurid lobsters. Journal of Experimental Biology 205:3891–3902.

Gutu, M., and J. Sieg. 1999. Ordre des Tanaidacés (Tanaidacea Hansen, 1895). Mémoires de l'Institut Océanographique, Monaco 19:353–389.

Hartnoll, R.G. 2000. Evolution of brachyuran mating behavior: relation to the female molting pattern. Crustacean Issues 12:519–525.

Hartnoll, R.G. 2001. Growth in Crustacea. Hydrobiologia 449:111–122.

Hartnoll, R.G. 2006. Reproductive investment in Brachyura. Hydrobiologia 557:31–40.

Hay, M.E., J.R. Pawlik, J.E. Duffy, and W. Fenical. 1989. Seaweed herbivore predator interactions: host plant specialization reduces predation on small herbivores. Oecologia 81:418–427.

Hay, M.E., J.E. Duffy, and W. Fenical. 1990. Host plant specialization decreases predation on a marine amphipod: an herbivore in plant's clothing. Ecology 71:733–743.

Hess, G.S., and R.T. Bauer. 2002. Spermatophore transfer in the hermit crab *Clibanarius vittatus* (Crustacea, Anomura, Diogenidae). Journal of Morphology 253:166–175.

Highsmith, R.C. 1983. Sex reversal and fighting behavior: coevolved phenomena in a tanaid crustacean. Ecology 64:719–726.

Hines, A.H., P.R. Jivoff, P.J. Bushmann, J. van Montfrans, S.A. Reed, D.L. Wolcott, and T.G. Wolcott. 2003. Evidence for sperm limitation in the blue crab, *Callinectes sapidus*. Bulletin of Marine Science 72:287–310.

Hinsch, G.W. 1991. Structure and chemical content of the spermatophores and seminal fluid of reptantian decapods. Pages 290–307 *in*: R.T. Bauer and J.W. Martin, editors. Crustacean sexual biology. Columbia University Press, New York.

Hoese, B. 1984. The marsupium in terrestrial isopods. Symposium Zoological Society London 53:65–76.

Holdich, D.M. 1968. Reproduction, growth and bionomics of *Dynamene bidentata* (Crustacea: Isopoda). Journal of Zoology 156:137–153.

Hughes, M. 1996. Size assessment via a visual signal in snapping shrimp. Behavioral Ecology and Sociobiology 38:51– 57.

Jivoff, P., and A.H. Hines. 1998. Female behaviour, sexual competition and mate guarding in the blue crab, *Callinectes sapidus*. Animal Behaviour 55:589–603.

Johnson, W.S., M. Stevens, and L. Watling. 2001. Reproduction and development of marine peracaridans. Advances in Marine Biology 39:105–260.

Juanes, F., and L.D. Smith. 1995. The ecological consequences of limb damage and loss in decapod crustaceans: a review and prospectus. Journal of Experimental Marine Biology and Ecology 193:197–223.

Karplus, I. 2005. Social control of growth in *Macrobrachium rosenbergii* (De Man): a review and prospects for future research. Aquaculture Research 36:238–254.

Kendall, M.S., D.L. Wolcott, T.G. Wolcott, and A.H. Hines. 2001. Reproductive potential of individual male blue crabs, *Callinectes sapidus*, in a fished population: depletion and recovery of sperm number and seminal fluid. Canadian Journal of Fisheries and Aquatic Sciences 58:1168–1177.

Knowlton, N. 1979. Reproductive synchrony, parental investment, and the evolutionary dynamics of sexual selection. Animal Behaviour 27:1022–1033

Krång, A.S., and S.P. Baden. 2004. The ability of the amphipod *Corophium volutator* (Pallas) to follow chemical signals from con-specifics. Journal of Experimental Marine Biology and Ecology 310:195–206.

Lardies, M.A., I.S. Cotoras, and F. Bozinovic. 2004. The energetics of reproduction and parental care in the terrestrial isopod *Porcellio laevis*. Journal of Insect Physiology 50:1127–1135.

Lee, S.Y. 1995. Cheliped size and structure: the evolution of a multifunctional decapod organ. Journal of Experimental Marine Biology and Ecology 193:161–176.

Linsenmair, K.E. 1979. Untersuchungen zur Soziobiologie der Wüstenassel *Hemilepistus reaumuri* und verwandter Isopodenarten (Isopoda, Oniscoidea): Paarbindung und Evolution der Monogamie. Verhandlungen Deutsche Zoologische Gesellschaft 1979:60–72.

Mantel, L.H., and L.L. Farmer. 1983. Osmotic and ionic regulation. Pages 53–161 *in*: D.E. Bliss, editor. The biology of Crustacea, Volume 5. Academic Press, New York.

Mariappan, P., C. Balasundaram, and B. Schmitz. 2000. Decapod crustacean chelipeds: an overview. Journal of Biosciences 25:301–313.

Martin, J.W., and G.W. Davis. 2001. An updated classification of the recent Crustacea. Natural History Museum of Los Angeles County, Science Series 39:1–123.

Mathews, L.M. 2002. Tests of the mate-guarding hypothesis for social monogamy: does population density, sex ratio, or female synchrony affect behavior of male snapping shrimp (*Alpheus angulatus*)? Behavioral Ecology and Sociobiology 51:426–432.

Mazel, C.H., T.W. Cronin, R.L. Caldwell, and N.J. Marshall. 2004. Fluorescent enhancement of signaling in a mantis shrimp. Science 303:51.
McCurdy, D.G., M.R. Forbes, S.P. Logan, D. Lancaster, and S.I. Mautner. 2005. Foraging and impacts by benthic fish on the intertidal amphipod *Corophium volutator*. Journal of Crustacean Biology 25:558–564.
Montgomery, E.L., and R.L. Caldwell. 1984. Aggressive brood defense by females in the stomatopod *Gonodactylus bredini*. Behavioral Ecology and Sociobiology 14:247–251.
Morin, J.G., and A.C. Cohen. 1991. Bioluminescent displays, courtship, and reproduction in ostracodes. Pages 1–16 *in*: R.T. Bauer and J.W. Martin, editors. Crustacean sexual biology. Columbia University Press, New York.
Morritt, D., and A.M.M. Richardson. 1998. Female control of the embryonic environment in a terrestrial amphipod, *Mysticotalitrus cryptus* (Crustacea). Functional Ecology 12:351–358.
Orensanz, J.M., A.M. Parma, D.A. Armstrong, J. Armstrong, and P. Wardrup. 1995. The breeding ecology of *Cancer gracilis* (Crustacea: Decapoda: Cancridae) and the mating system of cancrid crabs. Journal of Zoology 235:411–437.
Patek, S.N., and T.H. Oakley. 2003. Comparative tests of evolutionary trade-offs in a palinurid lobster acoustic system. Evolution 57:2082–2100.
Poore, A.G.B., and P.D. Steinberg. 1999. Preference-performance relationships and effects of host plant choice in an herbivorous marine amphipod. Ecological Monographs 69:443–464.
Popper, A.N., M. Salmon, and K.W. Horch. 2001. Acoustic detection and communication by decapod crustaceans. Journal of Comparative Physiology A 187:83–89.
Price, P.W. 1980. Evolutionary biology of parasites. Princeton University Press, Princeton, N.J.
Regier, J.C., J.W. Shultz, and R.E. Kambic. 2005. Pancrustacean phylogeny: hexapods are terrestrial crustaceans and maxillopods are not monophyletic. Proceedings of the Royal Society of London, Series B 272:395–401.
Sagi, A., and I. Khalaila. 2001. The crustacean androgen: a hormone in an isopod and androgenic activity in decapods. American Zoologist 41:477–484.
Sainte-Marie, B. 1991. A review of the reproductive bionomics of aquatic gammaridean amphipods—variation of life-history traits with latitude, depth, salinity and superfamily. Hydrobiologia 223:189–227.
Sastry, A.N. 1983. Ecological aspects of reproduction. Pages 179–270 *in*: D.E. Bliss, editor. The biology of Crustacea, Volume 8. Academic Press, New York.
Sato, T., M. Ashidate, S. Wada, and S. Goshima. 2005. Effects of male mating frequency and male size on ejaculate size and reproductive success of female spiny king crab *Paralithodes brevipes*. Marine Ecology Progress Series 296:251–262.
Shields, J.D. 1992. Parasites and symbionts of the crab *Portunus pelagicus* from Moreton Bay, eastern Australia. Journal of Crustacean Biology 12:94–100.
Shuster, S.M. 1991. Changes in female anatomy associated with the reproductive moult in *Paracerceis sculpta*, a semelparous isopod crustacean. Journal of Zoology 225:365–379.
Shuster, S.M., and R.L. Caldwell. 1989. Male defense of the breeding cavity and factors affecting the persistence of breeding pairs in the stomatopod, *Gonodactylus bredini* (Manning) (Crustacea, Hoplocarida). Ethology 82:192–207.
Sneddon, L.U., F.A. Huntingford, A.C. Taylor, and A.S. Clare. 2003. Female sex pheromone-mediated effects on behavior and consequences of male competition in the shore crab (*Carcinus maenas*). Journal of Chemical Ecology 29:55–70.
Starr, C.K. 1979. Origin and evolution of insect sociality: a review of modern theory. Pages 35–79 *in*: H.R. Hermann, editor. Social insects, Volume 1. Academic Press, New York.
Subramoniam, T. 2000. Crustacean ecdysteroids in reproduction and embryogenesis. Comparative Biochemistry and Physiology C 125:135–156.

Tallamy, D.W., and W.P. Brown. 1999. Semelparity and the evolution of maternal care in insects. Animal Behaviour 57:727–730.

Thompson, G.A., and C.L. McLay. 2005. Mating behaviour of *Heterozius rotundifrons* (Crustacea: Brachyura: Belliidae): is it a hard or soft shell mater? Marine and Freshwater Research 56:1107–1116.

Thompson, J.N. 1994. The coevolutionary process. University of Chicago Press, Chicago, Ill.

Tóth, E., and J.E. Duffy. 2005. Coordinated group response to nest intruders in social shrimp. Biology Letters 1:49–52.

Vogt, G., L. Tolley, and G. Scholtz. 2004. Life stages and reproductive components of the marmorkrebs (marbled crayfish), the first parthenogenetic decapod crustacean. Journal of Morphology 261:286–311.

Wasson, K., and B.E. Lyon. 2005. Flight or fight: flexible antipredatory strategies in porcelain crabs. Behavioral Ecology 16:1037–1041.

Watling, L. 1993. Functional morphology of the amphipod mandible. Journal of Natural History 27:837–849.

Wilber, D.H. 1989. The influence of sexual selection and predation on the mating and postcopulatory guarding behavior of stone crabs (Xanthidae, *Menippe*). Behavioral Ecology and Sociobiology 24:445–451.

Zeil, J., and J.M. Hemmi. 2006. The visual ecology of fiddler crabs. Journal of Comparative Physiology A 192:1–25.

Zimmer, M. 2001. Why do male terrestrial isopods (Isopoda: Oniscidea) not guard females? Animal Behaviour 62:815–821.

Zucker, N., B. Stafki, and S.C. Weeks. 2001. Maintenance of androdioecy in the freshwater shrimp, *Eulimnadia texana*: relative longevity of males to hermaphrodites. Canadian Journal of Zoology 79:393–401.

The Evolution of Crustacean Mating Systems

Stephen M. Shuster

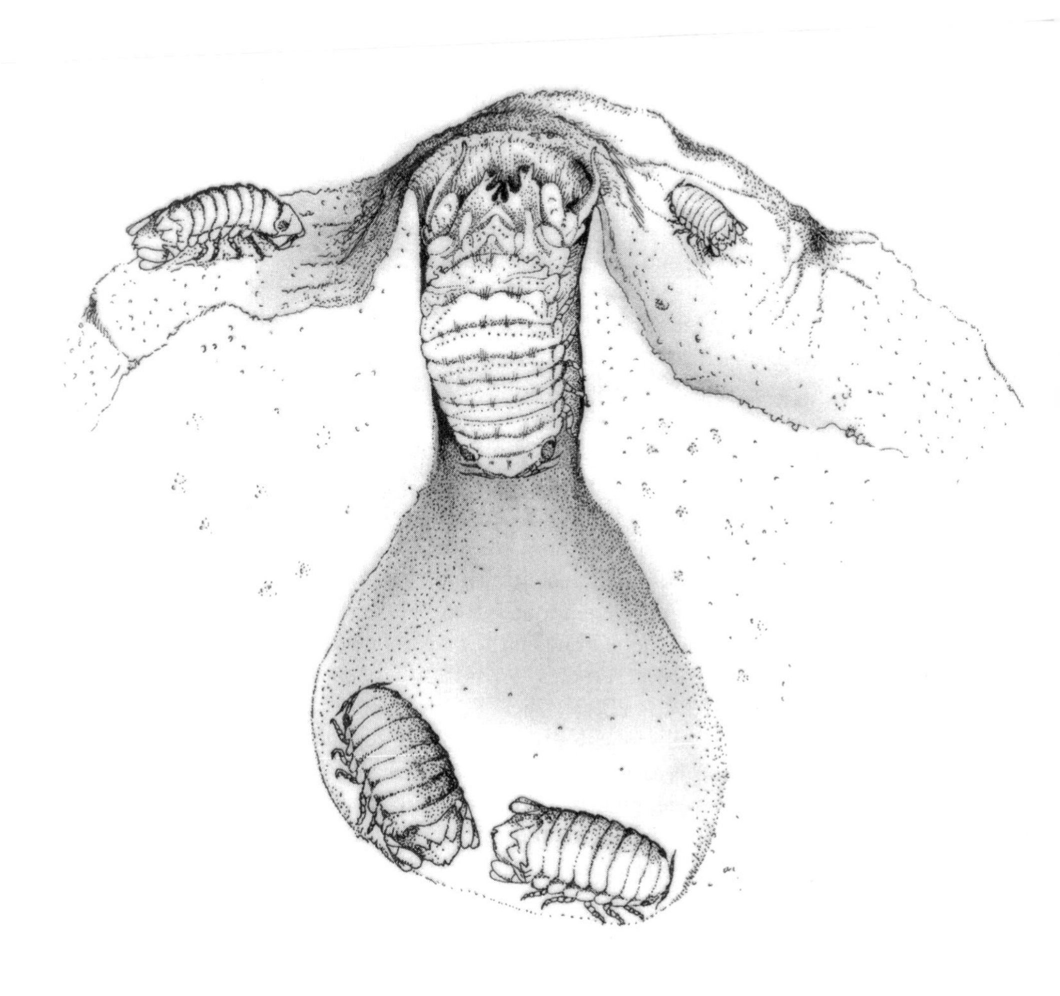

2

There are two central issues in the study of animal mating systems: (1) the source of sexual selection and (2) the intensity of sexual selection. These issues are important because the approach researchers take to explore them determines (a) the processes that are presumed to cause sexual selection, (b) the procedures that are undertaken to observe these processes, and (c) the variables that are measured in hypothesis testing. The analysis of animal mating systems, until recently, has been based on the hypothesis that sex differences in parental investment are the source of sexual selection (reviewed in Shuster and Wade 2003; see also chapters 7, 9, 12). This emphasis was crystallized by Trivers (1972), who, following discussions by Darwin (1874), Bateman (1948), and Williams (1966), proclaimed, "What governs the operation of sexual selection is the relative parental investment of the sexes in their offspring" (p. 141).

Indeed, most studies of mating systems are consistent with parental investment theory (PIT). According to this view, female reproduction is limited by the availability of resources required for parental investment, and because these resources vary in their abundance in space and in time, male reproduction is limited by the spatial distribution of resources and by the temporal distribution of sexually receptive females. To estimate the degree to which these latter distributions influence the intensity of selection, Emlen and Oring (1977) defined two measures: the operational sex ratio (OSR) and the environmental potential for polygamy (EPP).

The OSR was originally defined by Emlen (1976, p. 283) as "the ratio of potentially receptive males to receptive females at any time." There have been multiple interpretations of this description, but in its simplest form, OSR = $R_o = N_♂/N_♀$, where $N_♂$ and $N_♀$ equal the number of males and females, respectively (Shuster and Wade 2003). With OSR > 1, females are rare and competition for mates is presumed to be intense, although this assumption depends on the degree to which male mating success or failure is consistent among males throughout the breeding season. With OSR < 1, females are abundant and competition for mates is presumably relaxed, although again, depending on the cause of a female-biased sex ratio, such conditions may still allow certain males to contribute disproportionately to the next generation (Shuster and Wade 2003). The EPP measures the degree to which social and ecological conditions allow males to monopolize females. However, appropriate methods for quantifying female distributions, and the scale on which EPP should be measured were never defined. As a result, while serving as a conceptual proxy for the intensity of sexual selection, the uncertain relationship between EPP and selection intensity makes comparisons within and among species imprecise (Shuster and Wade 2003).

Researchers emphasizing PIT have encountered further difficulties in putting its assumptions to rigorous empirical tests. Despite Trivers's (1972) prediction, a sex difference in relative parental investment has proven extremely difficult to compare within and among species. Not only are the relative amounts of energy, cost, and risk associated with relative parental investment difficult to quantify (Strohm and Linsenmair 1999; see also chapters 7–9), but also, the correlation between sex differences in parental investment and sexual dimorphism is dismal, particularly in species with sex role reversal (S.M. Shuster and M.J. Wade, unpublished data). Measures of sexual selection intensity based on PIT require laboratory conditions that are rarely encountered in nature (e.g., potential reproductive rate; Clutton-Brock and Vincent 1991) or make assumptions that underestimate the variance in mating success among individuals (e.g., Q, which focuses on individuals "qualified" to mate; Ahnesjö et al. 2001). Moreover, like other research paradigms grounded in optimality theory (reviewed in Cheverud and Moore 1998), PIT has an unfortunate tendency to emphasize adaptive outcomes, wherein researchers declare to what "should" evolve due to sex differences in parental investment or in expected fitness returns, and then proceed to look for evidence of adaptations consistent with their initial predictions.

In this chapter, I explain the utility of a quantitative approach for measuring the source and intensity of sexual selection (see also Shuster and Wade 2003). Using ecological, life history, and behavioral data that are commonly available for sexual species, here, using crustaceans as examples, I show how the magnitude of the sex difference in fitness variance, estimated using measurements of *actual* male and female fitness, can be used to classify mating systems. I also show how the sex difference in fitness variance is influenced, by the spatial and temporal crowding of receptive females, by female life history, by male and female reproductive behavior, and by runaway processes in various forms. My goal is to suggest an empirical framework for the study of sexual species that measures the selective forces responsible for sex differences in adult phenotype.

The Sex Difference in Fitness Variance

Most research on sexual selection and its effects on mating systems, including that of Darwin (1874) himself, has focused either on the *context* in which sexual selection occurs (i.e., via male combat or female choice) or on the evolutionary *outcome* of sexual selection (i.e., on descriptions of sexual dimorphism or mating behavior). These approaches, while interesting in their own right, consider neither the process nor the degree to which sexual selection may achieve its evolutionary effects. Shuster and Wade (2003) asked, "How can sexual selection be strong enough to counter the combined, opposing forces of male and female viability selection?" (p. 10). This Quantitative Paradox is resolved by measuring the fitness variance for males and females within and among species. This method illustrates when and why sexual selection can be strong enough to overwhelm the effects of natural selection and therefore how it produces the phenotypes its researchers find so compelling.

Consider a hypothetical crustacean population consisting of 20 individuals. If each of the 10 females in the population produces a clutch of 10 ova, and if each clutch is fertilized by a single male, then the total number of offspring produced is N_{ototal} = (10 ova) × (10 females) = 100. Because each mating pair produces 10 offspring, the total offspring produced by all females, $N_{o♀}$, equals the total offspring produced by all males, $N_{o♂}$ = 100. Because there are 10 females and 10 males in our population, the average number of offspring per female, $O_♀ = N_{ototal}/N_♀$ is 10, which equals the average number of offspring per male, $O_♂ = N_{ototal}/N_♂$. Also, because each individual produces the same number of offspring (10), no variance in offspring numbers exists for either sex. Thus, $V_{o♀} = V_{o♂} = 0$. Separately calculating the mean and variance in offspring numbers for each sex shows how differences in mate numbers between the sexes may influence these parameters.

Now consider a case in which 1 of the 10 males secures more than one mate (e.g., 3) as might occur in harpacticoid copepods (Stancyk and Moreira 1988) or in cypridinid ostracods (Morin and Cohen 1991). The total offspring produced by our population, $N_{ototal} = 100$, remains unchanged. Similarly, because $N_♂ = N_♀ = 10$, the average offspring number per male, $O_♂ = N_{ototal}/N_♂ = 10$, equals the average offspring number per female, $O_♀ = N_{ototal}/N_♀ = 10$. Because each female still secures one mate with whom she produces a single brood, the variance in offspring numbers for females, $V_{o♀}$, is 0. However, because one male has three mates, two males must be excluded from mating. That is, for every *k* mates obtained by one male, there must be *k*—1 males who are unable to mate. When this happens, the variance in offspring numbers among males, $V_{o♂}$, must increase.

How can we quantify this increase in fitness variance among males? If paternity data are available, we could simply calculate the statistical variance in offspring numbers for males (Shuster and Wade 1991). Unfortunately, such data can be difficult to obtain (see chapter 9). An alternative approach is to partition the variance in offspring numbers within and among the classes of mating and nonmating males. The data necessary to do this, the mean and variance in mate numbers for males, and the mean and variance in offspring numbers for females, are often available in standard life history analyses. Why should we do this? This approach allows us to measure the *fitness variance* within each sex, which is proportional to the *intensity of selection*. Measures of fitness variance provide direct estimates of selection intensity, and the sex difference in selection intensity estimates the degree to which the sexes will diverge in phenotype.

We begin by identifying the classes of mating males and their population frequencies. There are three such classes: males who do not mate, p_o (= 2/10 males = 0.2), males who mate once, p_1 (= 7/10 males = 0.7), and males who mate three times, p_3 (= 1/10 males = 0.1). Here, we represent the proportion of the male population in each mating class as p_j, where *j* represents the number of females in the *j*th mating class. The sum of all mating classes, $\Sigma p_j = (0.2 + 0.7 + 0.1) = 1$. We next use these values to identify the average number of offspring produced by males in each *j*th mating class, $O_{♂j}$, as well as the average number of offspring produced by all males, $O_♂$. The average number of offspring that males in each mating class produce equals the average number of offspring per female, $O_♀$, multiplied by the number of mates, *j*, that males in each *j*th mating class secures, or $O_{♂j} = j(O_♀)$. Thus, the average number of offspring produced by males who do not mate, $O_{♂o}$, is (0)(10) = 0. For males who mate once, $O_{♂1} = (1)(10) = 10$. And, for males who mate three times, $O_{♂3} = (3)(10) = 30$. The average number of offspring produced by *all* males, $O_♂$, is equal to the number of offspring produced by the average female, $O_♀$, multiplied by the number of females mated by males in each mating class, *j*, multiplied by the fraction of the males belonging to that mating class, p_j, and summed over all mating classes, or

$$O_♂ = \Sigma p_j j(O_♀) \tag{1}$$

Using the values in our example above, $O_♂ = 10$. Note that although mates are unevenly distributed among males, the average number of offspring among all males remains unchanged compared to the initial case in which all males have equal mate numbers.

The distribution of females across all classes of mating males is equal to the population sex ratio, R (= $1/R_o$, where R_o = OSR; Shuster and Wade 2003), which is calculated as the number of females mated by males in each mating class, j, multiplied by the fraction of the males in each mating class, p_j, and summed over all classes of males, or, $R = \Sigma j\ p_j = 1$. Because the distribution of all females with all males equals the average mates per male, R also equals $N_♀/N_♂ = 1$. Thus, by substitution, we see that the average number of offspring per male, $O_♂$, equals the average mates per male, R, multiplied by the average number of offspring per female, $O_♀$, or $O_♂ = RO_♀ = 10$. Note, again, that although the distribution of females is now uneven among males, the average mates per male, R, the average offspring number per female, $O_♀$, and the average offspring number per male, $O_♂$, *remain unchanged* relative to our initial mating conditions.

With these expressions defined, we can now express the total variance in offspring numbers for males, $V_{o♂}$. As in analyses of variance (ANOVA), the total variance in male fitness equals the sum of two components: (1) the average variance in offspring numbers for males *within* the classes of males who sire offspring, and (2) the variance in average number of offspring sired by males *among* these same categories (Shuster and Wade 2003). The first component of variance in male offspring numbers is calculated in three steps. First, for each mating class of males, the variance in female offspring numbers, $V_{o♀}$, is multiplied by the number of mates obtained by males in each jth mating class. Next, this product is multiplied by the proportion of males in the population, p_j, that belong to each jth mating class. Last, these products are summed over all mating classes. Thus, the variance in offspring numbers within the classes of mating males equals

$$V_{o♂(\text{within})} = \Sigma p_j(jV_{o♀}) \tag{2}$$

In this example, because there is no variance in offspring numbers among females ($V_{o♀} = 0$, and all females produce 10 offspring), the variance in offspring numbers within the classes of mating males is zero ($V_{o♂(\text{within})} = 0$). We will return to this point shortly.

The second component of variance in male offspring numbers, the variance in the average number of offspring sired by males *among* these same categories, is calculated in four steps. First, for each mating class of males, we calculate the difference between the average number of offspring per male, $O_♂$, and the average number of offspring produced by that mating class, $O_{♂j}$ ($= O_♂ - O_{♂j}$). Next, we square each difference. Third, we multiply each squared difference by the fraction of males belonging to each mating class, p_j, and last, we sum across all classes to obtain

$$V_{o♂(\text{among})} = \Sigma p_j(O_♂ - O_{♂j})^2 \tag{3}$$

Substituting in the values from above, we have $V_{o♂(\text{among})} = 60$. Thus, the total variance in offspring numbers among males is the sum of the within- and among-male components in offspring numbers, or

$$V_{o♂} = \Sigma p_j(jV_{o♀}) + \Sigma p_j(O_♂ - O_{♂j})^2 \tag{4}$$

Because there is no variance in offspring numbers for females, $V_{o♀} = 0$, and the first term in Eq. 4 drops out. Because $V_{o♂(\text{among})} = V_{o♂}$, we can easily see that the variance in fitness among males goes from 0 to 60 when a single male mates with three females instead of one. Note, too, that the increase in fitness variance comes *entirely* from the among-male component of total fitness variance (Wade and Shuster 2004). If one male mates with all 10 of the females in this population, the mean and variance in offspring numbers for females, again, remain unchanged ($O_♀ = 10$, $V_{o♀} = 0$). Also, there is no change in either the sex ratio, $R = 1$, or the average number of offspring per male, $O_♂ = 10$. However, because one male mates 10 times, nine males do not mate at all (here $k = 10$, so $k - 1 = 9$); thus, $p_{♂0} = 9/10 = 0.9$, $p_{♂1}$ to $p_{♂9} = 0$, and $p_{♂10} = 1/10 = 0.1$. When these values are substituted into Eq. 4, we see $V_{o♂}$ now increases 15-fold to 900.

Three Rules

This exercise shows three simple rules. First,

> When the sex ratio equals 1, both sexes have equal average fitness.

This is true whether or not individual males and females have different mate numbers. It means that the average mate numbers, as well as the average offspring numbers, must be equal for each sex (Wade and Shuster 2002, 2005). When the sex ratio does not equal 1, the average fitness of the minority sex will increase (Eq. 1). However, as explained below, biases in sex ratio are only one component of sexual selection. This is an important consideration for studies of crustacean mating systems in which fluctuating or biased sex ratios are common (Shuster et al. 2001; see also chapter 7). As discussed below, this is also why, contrary to most mating system analyses conducted in accord with PIT, it is not sufficient to measure OSR alone to understand the intensity of sexual selection.

The second rule is:

> When some individuals are excluded from mating, the variance in offspring numbers within that sex will increase.

This increase in fitness variance indicates that selection is occurring within that sex. Such selection can lead to the evolution of specialized behaviors or structures associated with mating. For example, in rhizocephalan barnacles, parasitic copepods, and epicaridean isopods, only a small fraction of females locate hosts successfully (Høeg 1991; see also chapter 12). Such extreme variance in female fitness appears to favor high fecundity and grotesquely large body size. Because females in these species tend to be widely dispersed in space, only a small fraction of males successfully locate females and mate (Kabata and Cousens 1973). Extreme variance in male fitness appears to favor rapid maturation and the ability to locate and remain with individual females. The extreme fitness variance in both sexes appears to explain the remarkable sexual dimorphism in many of these crustaceans, even in species in which apparent monogamy occurs (Shuster and Wade 2003; see also chapter 12).

The third rule is:

> If the fraction of individuals excluded from mating is larger in one sex than it is in the other, a sex difference in the variance in offspring numbers will arise.

This sex difference in the variance in offspring numbers is the source of sexual selection. In the above example, $V_{o♂} - V_{o♀} = 900 - 0 = 900$. Because the variance in offspring numbers is proportional to the strength of selection, the magnitude of this sex difference in offspring numbers determines the intensity of sexual selection. Strong, single-sex selection leads to sexual dimorphism because traits associated with high fitness are disproportionately transmitted to the next generation. Strong, single-sex selection also represents a functional bias in sex ratio (Shuster et al. 2000). Such biases are often equalized by the evolution of alternative mating strategies (Shuster and Wade 1991, 2003). The existence of nonmating individuals of one phenotype creates a "mating niche" that can be invaded by individuals expressing alternative mating phenotypes, for example, in isopods (*Paracerceis sculpta*; Shuster 1992), amphipods (*Microdeutopus gryllotalpa*; Borowsky 1980), and decapods (*Rhynchocinetes typus*; Correa et al. 2000). Such invasions act to reduce the functional bias in sex ratio and thereby reduce the sex difference in fitness variance.

The Opportunity for Sexual Selection

Crow (1958) noted that the variance in absolute fitness, V_W, divided by the squared average fitness, W^2, equals the variance in *relative* fitness, or $V_W/W^2 = V_w$. Crow also called this value, I, the "opportunity for selection." This ratio provides a dimensionless, empirical estimate of selection's maximum strength, placing an upper bound not only on the change in mean fitness due to selection but also on the change in the standardized mean of every other trait (Wade 1979, Shuster and Wade 2003). As stated above, it is the sex difference in the variance in fitness that determines whether and to what degree the sexes will diverge in character, because fitness variance is proportional to selection intensity. For this reason, the opportunity for selection approach is useful for understanding the strength of selection within each sex. The value of I for each sex is expressed as the ratio of the variance in offspring numbers, V_o, to the squared average in offspring numbers, O^2, among members of that sex. Thus, $I_{♂} = V_{o♂}/O_{♂}^2$ and $I_{♀} = V_{o♀}/O_{♀}^2$. These expressions are linked through the sex ratio and mean fitness, which must be equal for both sexes (Wade and Shuster 2002). Thus, there is a fundamental algebraic relationship between the opportunity for selection on males, $I_{♂}$, and the opportunity for selection on females, $I_{♀}$.

How can we express this relationship for a natural population? Rewriting Eq. 4, substituting values from Eqs. 2 and 3, and rearranging terms (Shuster and Wade 2003), we have

$$V_{o♂} = RV_{o♀} + O_{♀}^2 V_{\text{mates}} \tag{5}$$

When $R = 1$, Eq. 5 shows that the variance in fitness for males, $V_{o♀}$, equals the variance in fitness for females, $V_{o♀}$, *plus* the quantity $O_{♀}^2 V_{\text{mates}}$. This latter term equals the average female fitness squared, $O_{♀}^2$, multiplied by the variance in mate numbers among males, V_{mates} $[=\Sigma p_j(R-j)^2]$. For the above example, $O_{♀}^2 V_{\text{mates}} = 900$, which shows that the sex difference in fitness variance is due to the fitness effects of a sex difference in the variance in mate numbers (Wade 1979). Recall that $I = V_W/W^2$ (Crow 1958). We can obtain an analogous expression for the variance in fitness for males in terms of offspring numbers, $V_{o♂}$, by dividing Eq. 5 by $(RO_♀)^2$, that is, by the squared average offspring number for males. We now have

$$I_♂ = (1/R)(I_♀) + I_{\text{mates}} \qquad (6)$$

or $I_♂ = (R_o)(I_♀) + I_{\text{mates}}$ because R equals 1/OSR (= $1/R_O$; Shuster and Wade 2003).

These expressions show, contrary to recent discussions of mating systems based on PIT (Reynolds 1996, Correa et al. 2000, Ahnesjö et al. 2001), that the sex ratio is only *part* of the total opportunity for selection. In particular, OSR (= R_o = $1/R$) has its strongest influence on the sex difference in fitness when $I_{\text{mates}} = 0$. When the sex ratio equals 1 ($R = 1/R_o = 1$), subtracting $I_♀$ from both sides of Eq. 6 yields $I_♂ - I_♀ = I_{\text{mates}}$, demonstrating that the sex difference in the opportunity for selection, that is, the opportunity for *sexual* selection, is due to differences in mate numbers between the sexes (Wade 1979, Shuster and Wade 2003).

Inserting the values from the above example into this latter equation shows that when males and females have equal mate numbers, $I_{\text{mates}} = 0$. When males vary in mate numbers, $I_♀$ still equals 0, so *all* of the opportunity for selection on males is due to sexual selection, $I_♂ = I_{\text{mates}}$. If $V_{o♀}$ becomes nonzero, $I_♀$ increases and I_{mates} will be eroded, to a degree determined by the relative magnitudes of $I_♂$ and $I_♀$ (see below). The point is this: I_{mates} can be estimated for *any* population for which the mean and variance in offspring numbers among females, and the mean and variance in mate numbers among males are known (for worked examples using the marine isopod *Paracerceis sculpta*, see Shuster and Wade 2003). However, because data on offspring and mate numbers may be every bit as difficult to obtain as parentage data, yet another type of data may be used to estimate the sex difference in the opportunity for selection.

The Spatial and Temporal Crowding of Mates

Emlen and Oring (1977) observed that female reproductive ecology determines the degree to which males may monopolize females or the resources on which breeding females depend. Wade (1995) used mean crowding, Lloyd's (1967) ecological measure of density-dependent competition, to relate the clustering of receptive females at resources, to the strength of sexual selection. When females are patchily distributed on resources, and when males defend patches to mate with the females on them, the mean and the variance in harem size increase as females become increasingly clumped in space. For this reason, the mean spatial crowding of females, m^*, provides a *direct* estimate of the opportunity for sexual selection, I_{mates}. In short, m^* equals I_{mates}.

Calculating m^* is straightforward. The average density of females per patch, m, equals the sum of all females over all m_i patches, divided by the total number of patches containing one or more females, M, or $m = \Sigma m_i/M$. The variance in the number of females per patch equals the squared difference of average female density and the density of females on the ith patch [= $(m - m_i)^2$], multiplied by the fraction of the total patches, p_i, that each ith patch comprises, summed over all patches, or $V_m = \Sigma p_i(m - m_i)^2$. The mean spatial crowding of females on resource patches, m^*, equals $m + [(V_m/m) - 1]$ (Wade 1995). The value of the variance, V_m, relative to the mean number of females per patch, m, indicates whether females are clumped (m^* large; $V_m > m$), dispersed (m^* small; $V_m < m$), or randomly distributed in space ($m^* = m$; $V_m = m$).

The *temporal* distribution of female sexual receptivity also affects the sex difference in the opportunity for selection. Shuster and Wade (2003) derived t^*, an expression similar to m^*, for describing the mean crowding of female sexual receptivity over intervals of the breeding season. When the breeding season is divided into intervals of duration equal to the average duration of female receptivity, the mean temporal crowding of sexually receptive females during the breeding season, t^*, is $t + [(V_t/t) - 1]$, where t and V_t equal the average and variance in the number of receptive females per interval, respectively. The mean temporal crowding of females, t^*, quantifies the number of other receptive females the average female experiences for the period when she herself is sexually receptive. The value of the variance, V_t, relative to the mean number of females per interval, t, indicates whether females are clumped (t^* large; $V_t > t$), dispersed (t^* small; $V_t < t$), or randomly distributed in time ($t^* = t$; $V_t = t$).

Both m^* and t^* provide direct estimates of the opportunity for sexual selection, I_{mates}. But, the relationship between the spatial patchiness of receptive females and I_{mates} is linear, whereas the relationship between the temporal crowding of receptive females and I_{mates} is reciprocal (Shuster and Wade 2003). Thus, when females become *synchronous* in their sexual receptivity, the ability of certain males to mate with multiple females decreases, as it does in mass-mating cumaceans (*Mancocuma*; Guewuch and Croker 1973) or in pair-bonded snapping shrimp (*Alpheus*; Knowlton 1980). In contrast, when females become *asynchronous* in their sexual receptivity, it is possible for certain males to become serially polygynous, as occurs in brine shrimp (*Branchinecta*; Belk 1991) and in lobsters (*Homarus*; Cowan 1991).

Because of the different relationships between m^*, t^*, and I_{mates}, temporal variations in the OSR fail to describe the actual intensity of sexual selection. When female receptivities are asynchronous, OSR measured at any time can be large and sexual selection may *seem* strong. However, such measurements of OSR do not quantify the *consistency* of male mating success over time, that is, the covariance among intervals in male mating success. It is only when *particular* males mate successfully across the breeding season that high OSR leads to strong sexual selection (see chapter 10). When a *different* male mates with each female that becomes receptive, apparently intense sexual selection (high OSR) is actually diminished.

The I_{mates} Surface

When either m^* or t^* accurately reflects the variance in mate numbers among males, its measurement alone is sufficient to estimate I_{mates}. However, because female spatiotemporal distributions can lead to highly dynamic responses by males (e.g., in the isopod *Paracerceis sculpta*; Shuster 1992), it is often necessary to measure m^* and t^* simultaneously to visualize the actual intensity of sexual selection. Multiple measurements of m^* and t^* throughout a breeding season create a three-dimensional surface describing I_{mates} for a particular species, a surface whose shape and orientation will vary depending on how the spatial and temporal distributions of females interact within and among seasons (Fig. 2.1a). Species-specific differences in the "ellipsoids" of points that appear on the "I_{mates} surface" are likely to be identifiable (Fig. 2.1b), and because I_{mates} is dimensionless (Shuster and Wade 2003), comparisons of its value within and among breeding seasons for particular species, as well as in phylogenetic comparisons, are permitted. Because changes in the spatiotemporal distributions of females change the intensity, as well as the evolutionary effects of sexual selection, related species are expected to show more similar I_{mates} ellipsoids, whereas ellipsoids for species within larger taxa should be predictably divergent in three-dimensional I_{mates} space (Fig. 2.1c).

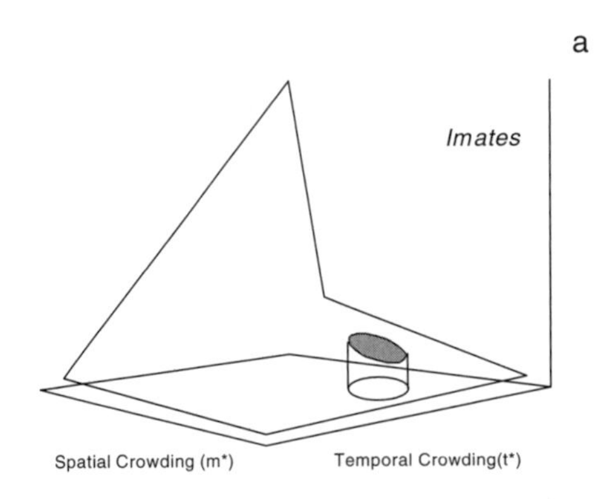

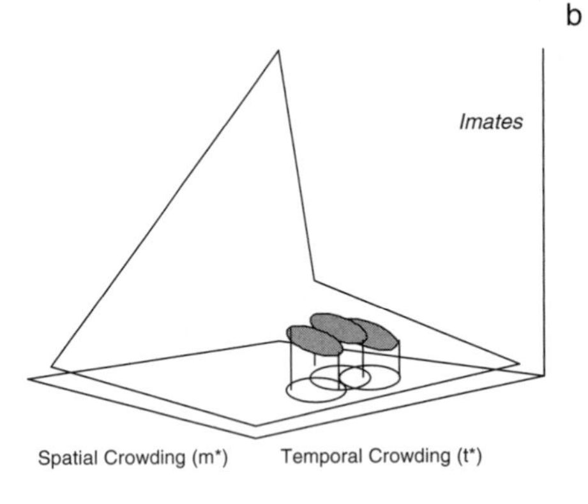

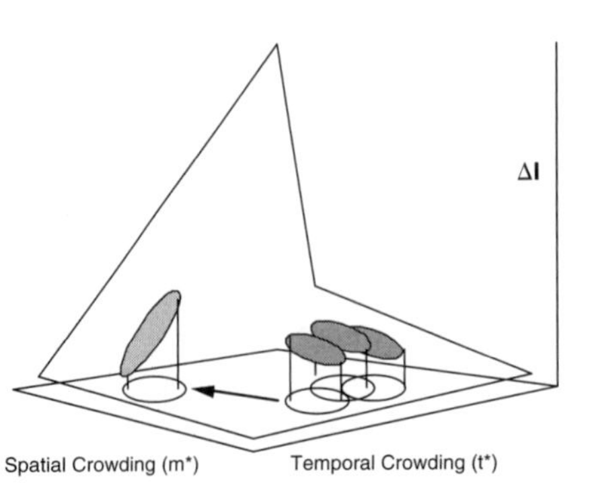

Figure 2.1 (a) Multiple measurements of m^* and t^* throughout a breeding season creates a three-dimensional surface describing I_{mates} for a particular species, a surface whose shape and orientation will vary depending on how the spatial and temporal distributions of females interact within and among seasons. (b) Species-specific differences in the "ellipsoids" of points that appear on the "I_{mates} surface" are likely to be identifiable. (c) Changes in the spatial and temporal distributions of females, the effects of female life history, and behavioral responses of each sex to these distributions may make mating system evolution a dynamic process. Thus, related species are expected to show similar ΔI ellipsoids, whereas ellipsoids for divergent species within larger taxa should be predictably divergent in value and position in three-dimensional space. Redrawn from Shuster and Wade (2003).

The marine isopod *Paracerceis sculpta* provides the single published example in which the values of m^* and t^* are calculated (see chapters 2 and 3 in Shuster and Wade 2003). The spatial distribution of females in this species was easily estimated because females breed within clearly defined territories (e.g., spongocoels of intertidal sponges). The temporal distribution of *P. sculpta* females was also easily estimated, based on the known duration of female receptivity and the rate at which females proceed through the stages of their reproductive cycle. This information allowed the approximate number of females in the population that were receptive during each interval (~24 hours) within the breeding season to be estimated with considerable accuracy. However, the I_{mates} surface defined in terms of m^* and t^* provides information only on how the spatial and temporal distributions of matings influences I_{mates}. It says nothing about how fitness variance among females influences the total opportunity for selection, or how behavioral responses of each sex to these distributions make mating system evolution a dynamic process.

Sources of Fitness Variance for Males and Females

As explained above, when males mate more than once, other males are excluded from mating and variance in offspring numbers among males appears. For females, fitness variance arises from three sources: (1) females mate either once or more than once (monandry vs. polyandry), (2) females reproduce either once or more than once (semelparity vs. iteroparity), and (3) iteroparous females reproduce either repeatedly within a single season or repeatedly within multiple seasons. When a female mates once and produces only one clutch of offspring, she awards her entire complement of ova to a single male. However, when a female mates more than once, barring rigid patterns of sperm precedence, she divides her clutch of eggs into subclutches, equal in number to the number of males who succeed in fertilizing ova. The overall effect on I_{mates} of polyandrous matings by females is, when each mating male sires a fraction of the offspring of each mate he secures, the variance in fitness among males is *reduced* (see chapter 10).

Reduction in the variance in mate numbers that results from multiple matings by females is the likely context for the evolution of male mate guarding (Shuster and Wade 2003; see also chapters 7, 8). Males who defend their mates for the duration of their receptivity ensure their exclusive paternity of that female's brood, whereas male promiscuity, particularly when sperm mixing occurs, is usually favored only when the rate at which males may encounter and mate unguarded, receptive females is extremely high. A surprising prediction of this hypothesis is that male mate guarding in some form will be favored for nearly all spatial and temporal distributions of females. Mate guarding effectively prevents sperm competition; thus, where it does not occur, other forms of male paternity assurance appear to have evolved. In calanoid copepods, males attach individual spermatophores that cover female genitalia and prevent reinsemination (Subramoniam 1993). Female brachyuran crabs may receive multiple mates and/or store sperm, but males place secretions within females' spermathecae that seal off these other ejaculates (Diesel 1991; see also chapter 9). There are few behaviors more characteristic of crustacean sexuality than mate guarding (Jormalainen 1998; see also chapter 8). The near ubiquity of this behavior despite considerable variation in crustacean mating systems (see chapter 12) lends support to these predictions.

However, there is also abundant evidence that multiple mating occurs and that sperm compete for fertilizations (Diesel 1991, Koga et al. 1993; see also chapter 9). Shuster and Wade (2003) showed that the intensity of sperm competition can be quantified as the mean crowding of ejaculates within females, m^*_P, which is directly affected by female promiscuity (see chapter 4 in Shuster and Wade 2003). For sperm competition to lead to sexual selection, males who mate with disproportionate success must *also* have sperm that are disproportionately successful within *each* of the females with whom they mate. Otherwise, multiple mating weakens rather than intensifies sexual selection. Also, while both sexes must have equal average fitness, when males gain more offspring by repeated matings than do females, a sex difference in the covariance between promiscuity and offspring numbers can exist. Sex differences in this covariance are the likely source of perception that "males are promiscuous and females are coy" (see chapter 12). In fact, because the average fitnesses of males and females must be equal (Eq. 1), a sex difference in average promiscuity cannot exist (Wade and Shuster 2002, 2005).

When a female is semelparous, that is, when she produces only one clutch of offspring in her lifetime, no variance exists *within* females in the number of offspring produced. All of the variance exists *among* females. However, when a female produces more than one clutch, the variance in offspring numbers can be partitioned into within- and among-female components. Just as male fitness is influenced by mate numbers and offspring per mating, the corresponding two components of female fitness are clutch numbers and offspring per clutch. Each of these sources of variance acts to decrease the sex difference in fitness variance (Eq. 6). Thus, multiple reproductive episodes by females erode I_{mates} because as clutch number increases, I_{mates} becomes a smaller fraction of the total variance in offspring numbers.

These patterns generate specific predictions about the overall effects of female life history on the opportunity for sexual selection. In particular, the effects of sexual selection (= sexual dimorphism) will be proportional to the magnitude of I_{mates}. In general, I_{mates} will be eroded least in monandrous, semelparous species and eroded most in polyandrous, iteroparous species. Indeed, within the sphaeromatid Isopoda, the most extreme sexual dimorphism appears in genera in which females are semelparous (*Dynamene*, Holdich 1968; *Paracerceis*, Shuster 1992; *Cymodopsis*, Hurley and Jansen 1977), whereas sexual dimorphism is reduced in genera in which females are iteroparous (*Sphaeroma* and *Parasphaeroma*, Hurley and Jansen 1977; *Thermosphaeroma*, Jormalainen et al. 1999). Yet even when $V_{o♀}$ seems large, its effects may be dwarfed by I_{mates}. In *P. sculpta*, sexual selection on males is nearly 20 times stronger than natural selection on females (Wade and Shuster 2004). Such conditions reduce the ability of females to respond evolutionarily to sexual conflict, even when the negative consequences of conflict on female fitness seem severe.

Factors Affecting ΔI

The value of I_{mates}, after the effects of female life history are considered, equals ΔI, the total opportunity for sexual selection (Shuster and Wade 2003). However, additional influences on the sex difference in the opportunity for selection are possible that make mating system evolution a dynamic process. ΔI is enhanced by any female tendency to copy the behavior of other females. Mutual attraction to patchily distributed sources of food or shelter may cause females to become more spatially clumped. Although male mate guarding restricts female opportunities to engage in mate copying, genetic covariance between female tendencies to copy each other and males tendencies to guard their mates can favor, depending on the values of m^* and t^*, either the explosive breeding aggregations observed in land crabs (*Geocarcinoides*; Seeger 1996) or the formation of female aggregations that are defended by large males as in freshwater prawns (*Macrobrachium*; Kuris et al. 1987). A similar runaway process may favor males who display in groups, as well as females who are attracted to these signals, as in bioluminescent ostracods (*Vargula*; Morin and Cohen 1991) or in structure-building fiddler crabs (*Uca*; Christy 1983, Kim et al. 2004; see also chapter 10).

Similar processes may lead males to defend feeding sites, nesting sites, or display sites that are conspicuous to females. When males defend such sites, males are likely to become sedentary and females may become mobile. Nest site defense will depend on the degree to which nest control influences male mate numbers. When male mating effort and male parental effort both enhance offspring number for males, the opportunity for sexual selection on males can become extreme, and males may attempt to attract the attention of transient females by visual, auditory, or chemical displays. However, when males have few options for multiple mating or when particular males become highly successful at mating, male–female pairs are likely to arise and persist (see chapter 12).

Future Directions

The framework of Shuster and Wade (2003) combines these various processes to generate a classification scheme that defines mating systems in terms of the causal processes that produce them, rather than in terms of the presumed outcomes of male competition and female mate choice. The method begins by summarizing the spatial and temporal distributions of sexually receptive females. An estimate of the opportunity for sexual selection on males, I_{mates}, is obtained from (A) the mean spatial crowding of receptive females, m^*, and (B) the mean temporal crowding of receptive females, t^*. As explained above, each pair of m^* and t^* coordinates generates a unique value of I_{mates} arising from the spatial and temporal distribution of matings for each species (Fig. 2.1). The effects of female life history characters on the opportunity for sexual selection are then summarized using two additional parameters: (C) the opportunity for selection due to the effects on female clutch size of matings by individual sires, $I_{cs,sires}$, and (D) the opportunity for selection due the effects on female clutch size of multiple reproductive episodes by females, $I_{cs,clutch}$ (see chapter 5 in Shuster and Wade 2003).

From the resulting value of ΔI, it is possible to predict specific details in behavior and morphology that allow each combination of traits to be classified as a mating system. These predictions include (1) the degree to which sperm competition may occur, based on whether males are likely or unlikely to guard their mates; (2) whether female mate copying is likely, based on the spatial and temporal distribution of females; and (3) the estimated magnitude of the adjusted opportunity for sexual selection, ΔI, arising from combinations of factors 1 and 2 and factors A–D above. Empirical estimates of ΔI go beyond mere verbal predictions based on assumptions of optimality. Because they estimate the strength of selection directly,

Table 2.1. Representative examples of the major categories and subcategories of crustacean mating systems (see Shuster and Wade 2003).

Major Category/Subcategory	Taxon	Reference
Sedentary pairs		
Eumonogamy	*Hemilepistus* (Isopoda)	Baker 2004
Persistent pairs	*Spongicola* (Stenopodidea)	Hayashi and Ogawa 1987
Sequential pairs	*Gonodactylus* (Stomatopoda)	Shuster and Caldwell 1989
Itinerant pairs		
Attendance polygyny	*Eulimnadia* (Anostraca)	Belk 1991
Attendance polygynandry	*Euterpina* (Copepoda)	Stancyk and Moreira 1988
Attendance androdioecy	*Triops newberryi* (Notostraca)	Sassaman et al. 1997
Attendance polyandry	*Salmincola* (Copepoda)	Kabata and Cousens 1973
Coercive polygynandry	*Thermosphaeroma* (Isopoda)	Jormalainen et al. 1999
Mass mating		
Semelparous mass mating	*Mancocuma* (Cumacea)	Guewuch and Croker 1973
Mass mating with male parental care	*Pullosquilla* (Stomatopoda)	Jutte 1998
Iteroparous mass mating	*Geocarcinoides* (Brachyura)	Seeger 1996
Polygamy		
Cursorial polygyny	*Moina* (Branchipoda)	Forró 1993
Polygamy	*Pandalus* (Caridea)	Charnov 1982
Iteroparous classic leks	NA	
Male dominance		
Dominance polygyny	*Dynamene* (Isopoda)	Holdich 1968
Dominance polygynandry	*Macrobrachium* (Caridea)	Barki et al. 1992
Social pairs		
Pair-bond polygyny	NA	
Pair-bond polygamy	NA	
Pair-bond polygynandry	*Alpheus* (Caridea)	Knowlton 1980
Mating swarms		
Eumonogamy	NA	
Persistent pairs	NA	
Polyandrous mating swarms	NA	
Polygynous mating swarms	*Vargula* (Ostracoda)	Morin and Cohen 1991
Polygynandrous mating swarms	*Uca* (Brachyura)	Kim et al. 2004
Leks		
Semelparous exploded leks	NA	
Semelparous classic leks	NA	

Major Category/Subcategory	Taxon	Reference
Iteroparous exploded leks	NA	
Iteroparous classic leks	NA	
Feeding sites		
Semelparous feeding site polygyny	NA	
Semelparous classic leks	NA	
Iteroparous feeding site polygyny	NA	
Iteroparous classic leks	NA	
Nesting sites with female care		
Semelparous harem polygynandry	*Elaphognathia* (Isopoda)	Tanaka and Aoki 1999
Iteroparous harem polygynandry	*Microdeutopus* (Amphipoda)	Borowsky 1980
Nesting sites with male care		
Semelparous nest site polygynandry	NA	
Iteroparous nest site polygynandry	NA	
Polyandrogyny		
Eumonogamy	*Lernaeodiscus* (Rhizocephala)	Høeg 1991
Cursorial polyandrogyny	NA	
Mass mating with male parental care	NA	
Dominance polyandrogyny	NA	
Pair-bond polyandrogyny	*Synalpheus* (Caridea)	Duffy and MacDonald 1999
	Sacculina (Rhizocephala)	Høeg 1991
Harem polyandrogyny	*Leidya* (Isopoda)	Markham 1992

NA, no data available.

they allow precise predictions about (4) the likely form of male–female associations at breeding sites, (5) the degree and form of sexual dimorphism, (6) the tendency for males to provide parental care, (7) whether and how sexual conflict may arise between the sexes, and (8) whether as well as in what form alternative mating strategies are likely to exist. With this information it is possible to assign (9) a detailed descriptive category that summarizes each suite of male and female adaptations to each mating system, before, lastly, (10) classifying the mating system under this scheme (Table 2.1).

Although only a single attempt has been made to use this framework to classify crustacean mating systems (*Paracerceis sculpta*), a preliminary review of the literature suggests that crustacean representatives appear in nearly all 12 major mating system categories (Table 2.1; see also chapters 7–10, 12). I refer the interested reader directly to chapter 9 of Shuster and Wade (2003) for details on why I have classified some of these species as I have. This list is not exhaustive, and I hope that the above discussion stimulates research designed to fill the conspicuous gaps, such as the apparent lack of crustacean leks. My tentative hypothesis in this regard is that mate guarding occupies individual males sufficiently that extreme variance in male mating success is prevented. I invite those sufficiently motivated regarding why I have made my less obvious choices either to direct further discussion toward refining this scheme or, better yet, to use the methods discussed above to measure the components of ΔI and classify these species themselves.

Summary

In this chapter, I describe a quantitative approach for mating system analysis that measures the source and intensity of sexual selection. Using data commonly available from ecological, life history, and behavioral analyses and using crustaceans as specific examples, I show how the magnitude of the sex difference in fitness variance can be used to classify the mating systems of any sexual species. I also show how a sex difference in the opportunity for selection is influenced by the spatial and temporal crowding of matings, variation in female life history, male and female reproductive behavior, and runaway processes in various forms. My goal is to suggest an empirical framework for the study of crustacean and other mating systems that emphasizes the measurement of selective forces responsible for the evolution of male–female differences, an approach that is easier to test and interpret than current frameworks emphasizing optimality or parental investment theory.

References

Ahnesjö, I., C. Kvarnemo, and S. Merilaita. 2001. Using potential reproductive rates to predict mating competition among individuals qualified to mate. Behavioral Ecology 12:397–401.

Baker, M.B. 2004. Sex biased state dependence in natal dispersal in desert isopods, *Hemilepistus reaumuri*. Journal of Insect Behavior 17:579–598.

Barki, A., I. Karplus, and M. Goren. 1992. Effects of size and morphotype on dominance hierarchies and resource competition in the freshwater prawn, *Macrobrachium rosenbergii*. Animal Behaviour 44:547–555.

Bateman, A.J. 1948. Intra-sexual selection in *Drosophila*. Heredity 2:349–368.

Belk, D. 1991. Anostracan mating behavior: a case of scramble-competition polygyny. Pages 111–125 *in*: R.T. Bauer and J.W. Martin, editors. Crustacean sexual biology. Columbia University Press, New York.

Borowsky, B. 1980. The pattern of tube-sharing in *Microdeutopus gryllotalpa* (Crustacea: Amphipoda). Animal Behaviour 28:790–797.

Charnov, E.L. 1982. The theory of sex allocation. Princeton University Press, Princeton, N.J.

Cheverud, J.M., and A.J. Moore. 1998. Quantitative genetics and the role of the environment provided by relatives in behavioral evolution. Pages 67–100 *in*: C.R.B. Boake, editor. Quantitative genetic studies of behavioral evolution. University of Chicago Press, Chicago, Ill.

Christy, J.H. 1983. Female choice in the resource defense mating system of the sand fiddler crab, *Uca pugilator*. Behavioral Ecology and Sociobiology 12:160–180.

Clutton-Brock, T.H., and A.C.J. Vincent. 1991. Sexual selection and the potential reproductive rates of males and females. Nature 351:58–60.

Correa, C., J.A. Baeza, E. Dupre, I.A. Hinojosa, and M. Thiel. 2000. Mating behaviour and fertilization success of three ontogenetic stages of male rock shrimp *Rynchocinetes typus* (Decapoda: Caridea). Journal of Crustacean Biology 20:628–640.

Cowan, D.F. 1991. Courtship and chemical signals in the American lobster. Journal of Shellfish Research 10:284.

Crow, J.F. 1958. Some possibilities for measuring selection intensities in man. Human Biology 30:1–13.

Darwin, C.R. 1874. The descent of man and selection in relation to sex, 2nd edition. Rand, McNally and Co., New York.

Diesel, R. 1991. Sperm competition and the evolution of mating behavior in Brachyura, with special reference to spider crabs (Decapoda, Majidae). Pages 145–163 *in*: R.T. Bauer and J.W. Martin, editors. Crustacean sexual biology. Columbia University Press, New York.

Duffy, J.E., and K.S. MacDonald. 1999. Colony structure of the social snapping shrimp *Synalpheus filidigitus* in Belize. Journal of Crustacean Biology 19:283–292.

Emlen, S. T. 1976. Lek organization and mating strategies in the bullfrog. Behavioral Ecology and Sociobiology 1:283–313.

Emlen, S.T., and L.W. Oring. 1977. Ecology, sexual selection, and the evolution of mating systems. Science 197:215–223.

Forró, L. 1993. Mating behaviour in *Moina brachiata* (Jurine, 1820) (Crustacea, Anomopoda). Hydrobiologia 360:153–159.

Guewuch, W.T., and R.A. Croker. 1973. Microfauna of northern New England marine sand: I. The biology of *Mancocuma stellifera* Zimmer, 1943 (Crustacea: Cumacea). Canadian Journal of Zoology 51:1011–1020.

Hayashi, K.-I., and Y. Ogawa. 1987. *Spongicola levigata* sp. nov., a new shrimp associated with a hexactinellid sponge from the East China Sea (Decapoda, Stenopodidae). Zoological Science 4:367–373.

Høeg, J.T. 1991. Functional and evolutionary aspects of the sexual system in the Rhizocephala (Thecostraca: Cirrepedia). Pages 208–227 *in*: R.T. Bauer and J.W. Martin, editors. Crustacean sexual biology. Columbia University Press, New York.

Holdich, D.M. 1968. Reproduction, growth and bionomics of *Dynamene bidentata* (Crustacea: Isopoda). Journal of Zoology 156:136–153.

Hurley, D.E., and P.K. Jansen. 1977. The marine fauna of New Zealand: family Sphaeromatidae (Crustacea: Isopoda: Flabellifera). New Zealand Oceanographic Institution, Memoirs 63:1–95.

Jormalainen, V. 1998. Precopulatory mate guarding in crustaceans—male competitive strategy and intersexual conflict. Quarterly Review of Biology 73:275–304.

Jormalainen, V., S.M. Shuster, and H. Wildey. 1999. Reproductive anatomy, sexual conflict and paternity in *Thermosphaeroma thermophilum*. Marine and Freshwater Behavior and Physiology 32:39–56.

Jutte, P. 1998. The ecology, behavior, and visual systems of *Pullosquilla litoralis* and *P. thomassini*, two monogamous species of stomatopod crustacean. Unpublished Ph.D. thesis, University of California, Berkeley.

Kabata, Z., and B. Cousens. 1973. Life cycle of *Salmincola californiensis* (Dana 1852) (Copepoda: Lernaeopodidae). Journal of the Fisheries Research Board of Canada 30:881–903.

Kim, T.W., K.W. Kim, R.B. Srygley, and J.C. Choe. 2004. Semilunar courtship rhythm of the fiddler crab *Uca lactea* in a habitat with great tidal variation. Journal of Ethology 22:63–68.

Knowlton, N. 1980. Sexual selection and dimorphism in two demes of a symbiotic, pair-bonding snapping shrimp. Evolution 34:161–173.

Koga, T., Y. Henmi, and M. Murai. 1993. Sperm competition and the assurance of underground copulation in the sand-bubbler crab, *Scopimera globosa* (Brachyura: Ocypodidae). Journal of Crustacean Biology 13:134–137.

Kuris, A.M., Z. Ra'anan, A. Sagi, and D. Cohen. 1987. Morphotypic differentiation of male Malaysian giant prawns, *Macrobrachium rosenbergii*. Journal of Crustacean Biology 7:219–237.

Lloyd, M. 1967. Mean crowding. Journal of Animal Ecology 36:1–30.

Markham, J.C. 1992. The Isopoda Bopyridae of the eastern Pacific—missing or just hiding? Proceedings of the San Diego Society of Natural History 17:1–5.

Morin, J.G., and A.C. Cohen. 1989. Bioluminescent displays, courtship and reproduction in ostracodes. Pages 1–16 *in*: R.T. Bauer and J.W. Martin, editors. Crustacean sexual biology. Columbia University Press, New York.

Reynolds, J.D. 1996. Animal breeding systems. Trends in Ecology and Evolution 11:68–72.

Sassaman, C., M.A. Simovich, and M. Fugate. 1997. Reproductive isolation and genetic differentiation in North American species of *Triops* (Crustacea: Branchiopoda: Notostraca). Hydrobiologia 359:125–147.

Seeger, E. 1996. Christmas crabs. Wildlife Conservation 99:28–35.

Shuster, S.M. 1992. The reproductive behaviour of α, β-, and γ-males in *Paracerceis sculpta*, a marine isopod crustacean. Behaviour 121:231–258.

Shuster, S.M., and R.L. Caldwell. 1989. Male defense of the breeding cavity and factors affecting the persistence of breeding pairs in the stomatopod, *Gonodactylus bredini* (Crustacea: Hoplocarida). Ethology 82:192–207.

Shuster, S.M., and M.J. Wade. 1991. Equal mating success among male reproductive strategies in a marine isopod. Nature 350:606–610.

Shuster, S.M., and M.J. Wade. 2003. Mating systems and strategies. Princeton University Press, Princeton, N.J.

Shuster, S.M., J.O.W. Ballard, G. Zinser, C. Sassaman, and P. Keim. 2001. The influence of genetic and extrachromosomal factors on population sex ratio in *Paracerceis sculpta*. Crustacean Issues 13:313–326.

Stancyk, S.E., and G.S. Moreira. 1988. Inheritance of male dimorphism in Brazilian populations of *Euterpina acutifrons* (Dana) (Copepoda: Harpacticoida). Journal of Experimental Marine Biology and Ecology 120:125–144.

Strohm, E., and K.E. Linsenmair. 1999. Measurement of parental investment and sex allocation in the European beewolf *Philanthus triangulum* F. (Hymenoptera: Sphecidae). Behavioral Ecology and Sociobiology 47:76–88.

Subramoniam, T. 1993. Spermatophores and sperm transfer in marine crustaceans. Advances in Marine Biology 29:129–214.

Tanaka, K., and M. Aoki. 1999. Spatial distribution patterns of the sponge-dwelling gnathiid isopod *Elaphognathia cornigera* (Nunomura) on an intertidal rocky shore of the Isu Peninsula, southern Japan. Crustacean Research 28:160–167.

Trivers, R.L. 1972. Parental investment and sexual selection. Pages 136–179 *in*: B. Campbell, editor. Sexual selection and the descent of man. Aldine Press, Chicago, Ill.

Wade, M.J. 1979. Sexual selection and variance in reproductive success. American Naturalist 114:742–764.
Wade, M.J. 1995. The ecology of sexual selection—mean crowding of females and resource-defense polygyny. Evolutionary Ecology 9:118–124.
Wade, M.J., and S.M. Shuster. 2002. The evolution of parental care in the context of sexual selection: a critical reassessment of parental investment theory. American Naturalist 160:285–292.
Wade, M.J., and S.M. Shuster. 2004. Sexual selection: harem size and the variance in male reproductive success. American Naturalist 164:E83–E89.
Wade, M.J., and S.M. Shuster. 2005. Don't throw Bateman out with the bathwater! Integrative and Comparative Zoology 45:945–951.
Williams, G.C. 1966. Adaptation and natural selection. Princeton University Press, Princeton, N.J.

Molecular Approaches in Crustacean Evolutionary Ecology

Joseph Neigel

Brian Mahon

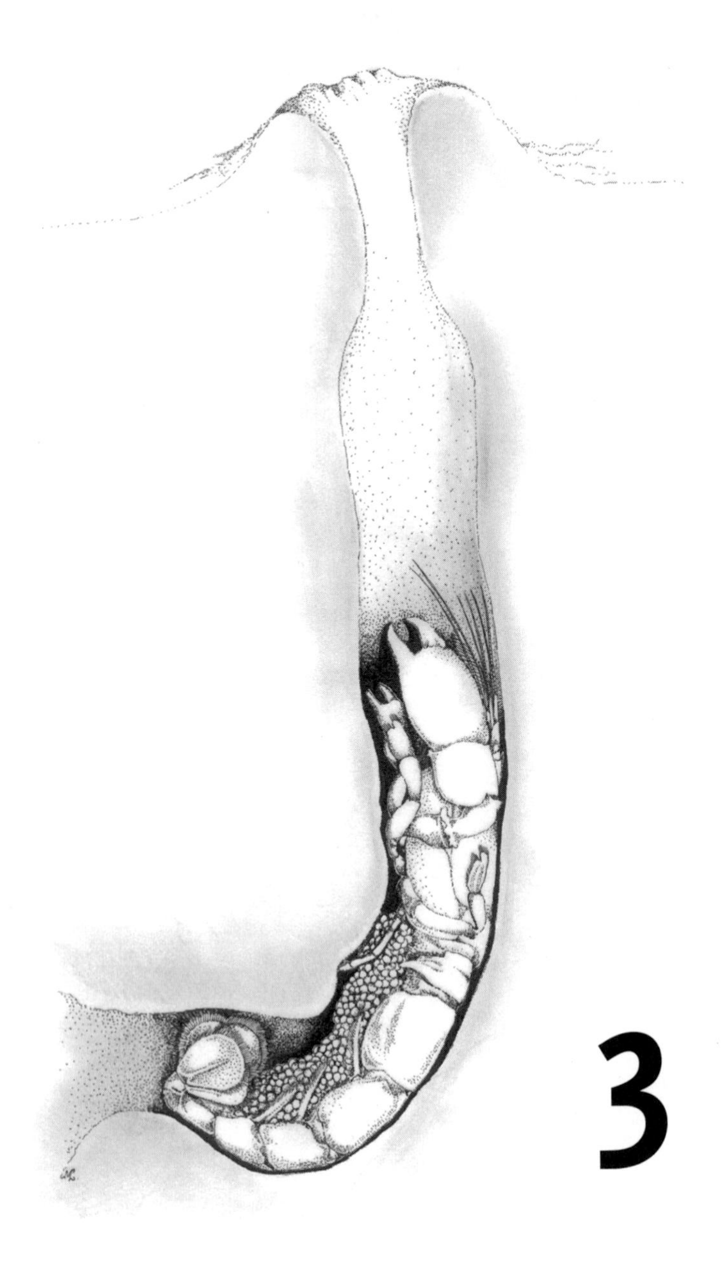

3

This chapter examines the uses of molecular markers to analyze relationships of descent among both individuals and taxa in the Crustacea. Molecular markers can be highly effective for these purposes but vary considerably in cost, ease of use, and suitability for specific applications (Avise 2004). The power of molecular markers to reveal parentage and kinship has revolutionized the study of social behavior in animals, producing such dramatic results as the discovery of widespread polyandry among animals (Zeh and Zeh 2003). Molecular markers have also allowed us to investigate mating systems even when it has not been practical to observe mating behavior. For example, it was possible to demonstrate multiple paternity in the thalassinidean ghost shrimp, *Callichirus islagrande* (Bilodeau et al. 2005), although mating of thalassinideans has never been observed because it occurs deep within their burrows. Molecular markers have also revolutionized phylogenetic analysis by providing a rich set of shared characters that enable powerful approaches to phylogenetic inference. This revolution in phylogenetics has in turn allowed the widespread adoption of phylogenetic comparative methods that use phylogenies to test hypotheses about the evolution of characters. This chapter is intended both as a critical assessment of these uses of molecular markers in crustacean evolutionary ecology and as an introduction for those considering their use.

Relationships Between Individuals

Genetic Markers for Individuals

Since the mid-1990s, microsatellite loci have become the preferred markers for inference of parentage and relatedness (A.G. Jones and Ardren 2003) and have become important for population genetics and genetic mapping. Another class of marker, the amplified fragment length polymorphism (AFLP) (Vos et al. 1995), has been used for many of the same applications, although they provide data that are fundamentally different. A comparison of these two classes of markers serves to illustrate some of the important considerations that arise in the selection of a genetic marker for studies of parentage and relatedness.

Microsatellites

Microsatellite sequences are short, tandemly repeated sequences that are scattered throughout the genomes of higher organisms (Tautz 1989). When a microsatellite sequence is replicated, either "replication slippage" or unequal crossovers can change the number of repeats, and successive changes can produce a large number of alleles that differ in length (Eisen 1999). Detection of microsatellite length variation is straightforward: polymerase chain reaction (PCR) is used to amplify the entire microsatellite sequence, and the length of the amplicon is determined by gel electrophoresis. Heterozygotes yield amplicons of two distinct sizes, while homozygotes yield a single amplicon. The mutation rate for replication slippage can be as high as 10^{-2} per replication (Cronn et al. 2002), and microsatellite loci tend to be much more polymorphic than most other genetic markers (Tautz 1989).

The development of microsatellite markers can be tedious and expensive, and the equipment and expertise needed to perform the initial isolation of microsatellite loci can exceed that found in laboratories primarily concerned with behavioral or organismal questions. After PCR primers for a microsatellite locus are developed, they must be tested to determine whether they amplify the correct sequence, whether the locus is polymorphic, and if possible, whether the locus follows Mendelian inheritance. It is not unusual for the development of microsatellite markers to stretch to a year, although after this process has become routine, a month or two is usually sufficient. An increasingly feasible alternative to in-house development is to contract commercial laboratories to perform many of the steps of microsatellite development (Selkoe and Toonen 2006). Brief descriptions of new microsatellites are regularly published in *Molecular Ecology Notes*. However, microsatellite primers designed for one species are not always useful for related species; they may fail to amplify microsatellite loci, or the loci can be less polymorphic.

Several types of artifacts can compromise the quality of microsatellite data. Null alleles fail to amplify because their flanking sequences do not match the primers (Callen et al. 1993). Primers might also amplify more than one microsatellite locus, especially if the loci are within repetitive sequence families (Harris and Crandall 2000). Allele "dropouts" can occur in PCR reactions when alleles with small numbers of repeats outcompete alleles with more repeats (Jensen and Bentzen 2004). Replication slippage during PCR can produce byproducts a few repeats shorter or longer than the original sequence (Shinde et al. 2003), which can make it difficult to judge the sequence's true size. Artifacts can often be detected as departures from Hardy-Weinberg proportions of genotype frequencies. For example, null alleles produce overall heterozygote deficiencies (e.g., Shaw et al. 1999), while allele dropouts result in heterozygotes deficiencies for specific combinations of alleles (Miller et al. 2002).

AFLPs

The AFLP technique is based on selective amplification of restriction fragments. The amplicons that result from this process, typically 50–100 in number, are resolved by gel electrophoresis as a pattern of bands. Alleles that produce bands are dominant, and so the exact genotype (homozygote or heterozygote) of individuals with a dominant phenotype cannot be determined (Vos et al. 1995). In comparison to microsatellites, AFLPs are relatively easy to develop. However, the use of AFLPs for some applications, such as population genetics, has been a subject of controversy (Sunnucks 2000). The black box nature of AFLP variation makes it difficult to generalize about underlying mutation processes or their rates. Errors in genotyping appear to be higher for AFLPs than for microsatellites (Bonin et al. 2004), and the dominance of AFLP bands makes it impossible to detect artifacts or bands produced by contaminating organisms as departures from Hardy-Weinberg proportions. These criticisms are valid, although some can be addressed by controls that demonstrate the reproducibility and Mendelian inheritance of AFLPs.

A little history might help to explain why the AFLP method has been criticized so strongly. Five years before the introduction of the AFLP method, the random amplified polymorphic DNA (RAPD) method was introduced (J.G.K. Williams et al. 1990). Like AFLPs, RAPD polymorphisms are dominant and scored as PCR products of specific sizes. When the RAPD method first appeared, it generated a great deal of enthusiasm because it was simple and quickly produced abundant data. However, disappointment followed when it became clear that RAPD bands were often not reproducible (Riedy et al. 1992, Ellsworth et al. 1993, Ayliffe et al. 1994, Perez et al. 1998). The AFLP method is now unfavorably and perhaps unfairly compared with the RAPD method. An investigator considering the use of either RAPDs or AFLPs for population genetics or estimation of relatedness should be aware of the biases against these techniques by both funding agencies and journals.

Allozymes (Lewontin 1991, Avise 2004) remain a viable choice for analysis of parentage or relatedness when precision is not required and sufficient amounts of fresh or frozen tissue are available. Allozymes have less power to resolve relationships than do either microsatellites or AFLPs. However, for some purposes it is enough to estimate an average degree of relatedness for many pairs of individuals (Blouin 2003). For example, Duffy (1996; see also chapter 18) used allozymes to demonstrate that the degree of relatedness between individuals within colonies of the alpheid shrimp *Synalpheus* matched that expected for full siblings.

All types of genetic markers are prone to significant occurrences of artifacts and errors, which include both those inherent to the techniques themselves and those due to human errors. There is a growing consensus that studies based on genetic marker data should acknowledge the inevitability of artifacts and errors and incorporate strategies for reduction of errors, automation and blind controls to reduce subjectivity in the collection of data, and estimates of error rates in data analysis (Bonin et al. 2004).

Analysis of Relationships Between Individuals

Parentage Analysis

Parentage analysis can be used to address fundamental questions in behavioral and evolutionary ecology (Avise 2004). The Crustacea offer many interesting problems associated with mate choice, sperm competition, and mechanisms of paternity assurance that can now be approached by analysis of paternity (see Bauer and Martin 1991; see also chapters 7–12). A classic example of how this approach can be integrated with an understanding of a species' reproductive biology is provided by studies of mating in the majid crab *Chionoecetes opilio* (D.M. Taylor et al. 1985, Urbani et al. 1998, Sainte-Marie et al. 1999; see also chapter 9).

Several recent reviews have provided important practical considerations for analysis of parentage with genetic markers (Gerber et al. 2000, Van de Casteele et al. 2001, A.G. Jones and Ardren 2003). The simplest approach is a process of elimination: Mendelian principles are used to exclude all potential parents except the two that are the true parents of the propositus, or focal individual. The probability that a nonparental individual can be excluded depends on the number of marker loci, the effective number of alleles at each locus, and whether alleles are dominant or codominant. Exclusion probabilities greater than 99% can be achieved with about five highly polymorphic microsatellite loci (see equation 4 in Gerber et al. 2000). In contrast, with dominant markers (AFLPs or RAPDs), it is not possible to exclude any individuals as potential parents. If the genotype of one parent is known, this further limits the possible genotypes of the other parent, and there is a greater probability that a nonparent can be excluded. Dominant markers can also be used to exclude individuals in this situation, although each locus has less discriminatory power than a locus with codominant alleles. The most advantageous situation for reconstruction of parentage is when it is known which pairs of individuals mated so that potential parents can be considered as pairs rather than single individuals. In this case, a single microsatellite locus could be sufficient to exclude more than 99% of nonparental pairs.

Identification of parents by unambiguous exclusion of all nonparents is attractive in its simplicity but not always practical. In some cases, the available markers cannot exclude all nonparents. A second possibility is that the true parents will be incorrectly excluded as the result of errors or artifacts. In situations with multiple alternative hypotheses that cannot be rejected by tests of statistical significance, likelihood is a useful alternative (Edwards 1992). Likelihood methods for parentage analysis have been developed for both dominant (Meagher and Thompson 1986) and codominant markers (Gerber et al. 2000). Likelihood can be used simply to assign progeny to their most probable parents; however, for some purposes, it is more useful to assign "fractions of progeny" to each possible parent. For example, if the total reproductive success of an individual were to be estimated, fractional assignments provide less biased estimates than those from all-or-none assignments (Neff et al. 2001). Fractional assignments are generally based on Bayesian estimates of the probability of parentage.

Multiple Mating

Polyandry (mating with multiple males) is widespread among animals, although it is often unclear how it benefits polyandrous females (Reynolds 1996; but see chapter 9). Polyandry is typically defined at the level of a brood (Neff and Pitcher 2002). Detection of polyandry is straightforward with codominant genetic markers. If a female's genotype is known, nonmaternal alleles represented in her brood are assumed to be of paternal origin. If a female's genotype is unknown, it is assumed that there are at most two alleles per parent in a brood. For accurate estimates of the frequency of polyandry, two or more highly polymorphic codominant marker loci are needed, and for accurate estimation of the number of sires per brood, larger numbers of highly polymorphic loci are required. Microsatellites are the obvious choice for most investigations of polyandry (Neff and Pitcher 2002).

Genetic markers can also be used to detect polygyny (mating with multiple females). However, in the typical situation in which broods are found with females, the process is more complicated than detection of polyandry. This is because it must be shown that the same male sired multiple broods, which requires that the parentage of each brood be determined. Since the occurrence of polygyny itself is generally considered unremarkable, it is most often considered in the context of male reproductive success (e.g., Zamudio and Sinervo 2000).

Genetic markers have proved to be useful in the detection of polyandry in crustaceans. Within the order Decapoda, polyandry at the level of broods has been detected in the nephropid lobsters *Homarus americanus* (Nelson and Hedgecock 1977, M.W. Jones et al. 2003, Gosselin et al. 2005) and *Nephrops norvegicus* (Streiff et al. 2004), the cambarid crayfish *Orconectes placidus* (Walker et al. 2002), the porcellanid crab *Petrolisthes cinctipes* (Toonen 2004), the callianassid ghost shrimp *Callichirus islagrande* (Bilodeau et al. 2005), the cancrid crab *Cancer pagurus* (Burfitt 1980), and the palaemonid shrimp *Palaemonetes pugio* (Baragona et al. 2000). In the majid crab *Chionoecetes opilio*, females can carry stored sperm from multiple males, but generally only one male sires each brood (Sevigny and Sainte-Marie 1996, Urbani et al. 1998, Sainte-Marie et al. 1999; see also chapter 9). The only nondecapod crustacean example of polyandry that we are aware of is the porcellionid isopod *Porcellio scaber* (Sassaman 1978).

Genealogical Relationships Other Than Parentage

Measurements of relatedness based on molecular markers can serve as the basis for estimates of heritability, the number of breeders in a population, variance in reproductive success, and tests of kin selection theory. Although there are few examples of measurements of relatedness for crustaceans, there are many for social insects. Microsatellite markers were used to demonstrate that males are produced by queens rather than workers in the apid bee *Schwarziana quadripunctata* (Tóth et al. 2003) and the vespid wasp *Brachygastra mellifica* (Hastings et al. 1998). For the primitively eusocial vespid wasp *Ropalidia revolutionalis*, microsatellite markers revealed that queens almost always mate singly, which is of interest because it creates the potential for conflicts over the production of males (Henshaw and Crozier 2004). In a study of the multiple-queen formicid ant *Leptothorax acervorum* (Bourke et al. 1997), used microsatellite markers to show that queens mated singly and were usually closely related to coexisting queens, and that because of a high turnover of queens, workers were usually not the offspring of the current queens. Other examples can be found in Ross (2001).

Many different statistical methods have been developed for estimation of relatedness from genetic markers (Milligan 2003). Methods exist for both codominant and dominant markers. Relatedness can be estimated as a continuous variable or by assignment to categories of relationship. Accurate estimation of relatedness for pairs of individuals generally requires a large number of loci; however, even a few loci can be sufficient to estimate average relatedness within groups (e.g., Duffy 1996). A good recent overview of these methods can be found in Blouin (2003).

Phylogenetic Relationships and the Comparative Method

Phylogenies as a Framework for Comparative Analysis

The parallel or convergent evolution of similar traits (i.e., character states) in unrelated species that are subject to similar environmental conditions is evidence that the traits are adaptations to those conditions. In contrast, traits shared by related species in similar environments cannot be considered independent instances of adaptation because they could have arisen once in a common ancestor and are now shared due to "phylogenetic inertia". Phylogenetic comparative methods (PCMs) are intended to separate shared history from independent evolution by analyzing the phylogenetic distributions of traits (Brooks and McLennan 1991, Harvey and Pagel 1991).

Phylogenies have been used to investigate whether similar behavioral traits have evolved independently in crustacean lineages. Kitaura et al. (1998) used a phylogeny based on mitochondrial DNA sequences to trace the evolution of mud-using territorial behavior in the semaphore crabs, genus *Ilyoplax*. Three behavioral traits—burrow plugging, barricade building, and fence building—were placed on this phylogeny.

Under the assumption that gains and losses of traits were equally likely, two most parsimonious reconstructions of ancestral traits were found. In one reconstruction, barricade building evolved three times, and in the other, it evolved twice and was lost once (Fig. 3.1a). Schubart et al. (1998) used a phylogeny based on mitochondrial sequences to examine the origins of adaptation to terrestrial life in endemic Jamaican land crabs. The phylogeny included related marine species from the Americas and Southeast Asia as well as those from Jamaica. The Jamaican species formed a single

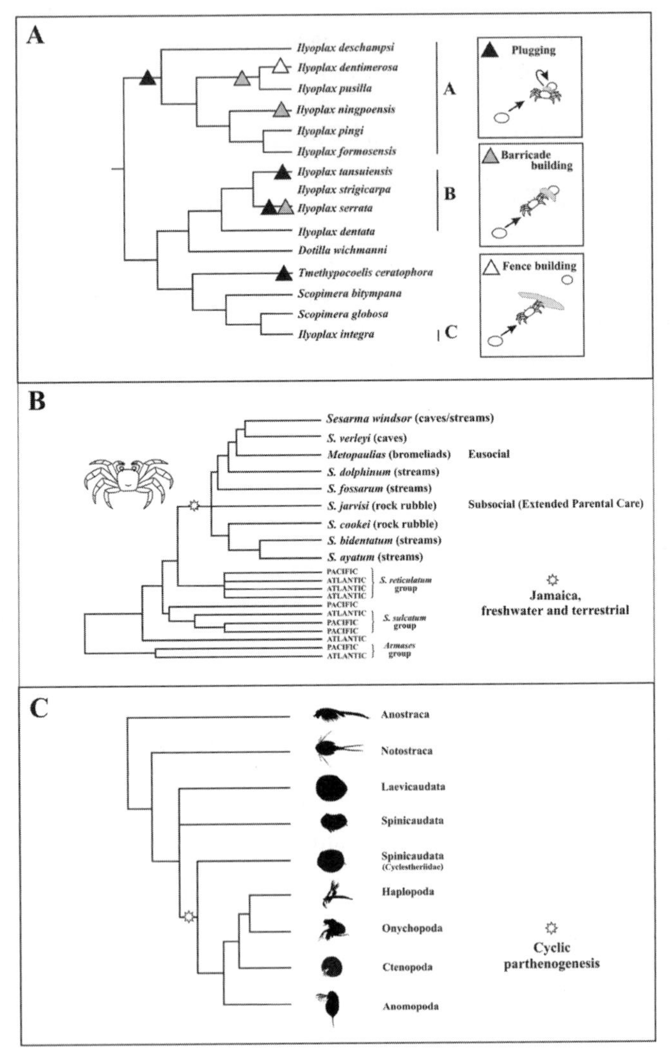

Figure 3.1 Phylograms of three crustacean taxa on which traits were mapped to infer their likely origins. (A) Intertidal crab species from the family Ocypodidae and three different territorial behaviors. (B) Crab species from the family Sesarmidae of different geographic origins, including the freshwater and terrestrial species of *Sesarma* from Jamaica. (C) Branchiopod crustacean taxa, including those taxa that exhibit cyclic parthenogenesis. Figures modified after Kitaura et al. (1998), Schubart et al. (1998), and D.J. Taylor et al. (1999).

monophyletic group, which is consistent with a single adaptive radiation from a marine ancestor (Fig. 3.1b). D.J. Taylor et al. (1999) used a phylogeny based on a combination of nuclear and mitochondrial DNA sequences and morphological characters to address questions about the origins of cyclic parthenogenesis in the shrimplike Branchiopoda. The five orders with cyclical parthenogens formed a monophyletic group, in support of the view that cyclical parthenogenesis arose once within the class (Fig. 3.1c).

The most powerful applications of PCMs are tests of hypotheses about causative relationships among traits or between traits and environmental factors. Duffy et al. (2000; see also chapter 18) used a phylogeny based on both mitochondrial DNA sequences and morphological characters to investigate the origins of eusociality in the sponge-dwelling alpheid shrimps (genus *Synalpheus*) and to test the hypothesis that eusociality has led to ecological dominance. Eusocial taxa were distributed among three distinct clades, each of which also contained noneusocial taxa. By the principle of parsimony, this provides evidence for three separate origins of eusociality within the genus. Phylogenetically independent contrasts (Felsenstein 1985) revealed a significant correlation between eusociality and the tendency for species to predominate within their host sponges. This uniquely marine example of the evolution of eusociality is reviewed by Duffy in chapter 18. Species of spiny lobsters (Palinuridae) are often characterized by gregarious behaviors such as shelter sharing and group migration. In chapter 13, Childress uses a phylogeny for Palinuridae based on mitochondrial DNA sequences to explore potential relationships between specific ecological and life history traits and the evolution of social behavior within this family.

There have been two forms of criticism against PCMs (Freckleton et al. 2002). The first is directed against the argument that traits shared by related species are not independent. The counterargument is that, irrespective of their origins, traits must be maintained by selection, and this occurs independently in every species. The second form of criticism concerns the assumptions of PCMs. All PCMs require assumptions about character evolution, typically that characters evolve randomly and are selectively neutral. Such assumptions are inconsistent with the use of PCMs to detect selective forces that result in directional evolution. If their assumptions are invalid, PCMs can have less power to detect significant patterns or can even generate incorrect results (Bjorklund 1997, Ackerly and Donoghue 1998, Cunningham et al. 1998). Not all PCMs are subject to the same criticisms. For example, Hansen (1997) developed a PCM that considers not only the independent origins of traits but also their maintenance by stabilizing selection. Although some PCMs assume that character evolution can be described by a random walk, others use models in which character evolution is directed or constrained by selection (Martins 2000). In general, most PCMs require that there be at least a correlation between phylogenetic relatedness and phenotypic similarity. Statistical tests can be used to determine if these correlations are significant (Cheverud et al. 1985, Ackerly and Donoghue 1998, Diniz et al. 1998), and the strength of the correlation can serve as a guide to how a PCM should be applied (Freckleton et al. 2002). In one simulation study (Martins et al. 2002), PCMs generally outperformed nonphylogenetic methods, even when the PCM's assumptions were violated.

There are several questions that should be answered before undertaking a phylogenetic comparative analysis. Are the traits best represented as discrete or continuous characters? Are there enough species to provide the statistical power to detect patterns of character evolution? Is a suitable phylogeny already available, or will it be necessary to first construct one? The answers to these questions will determine the feasibility of the analysis and provide some indication of how to proceed. Before a PCM (or computer program to perform a PCM) is chosen, we recommend a fresh review of the most recent literature on this very dynamic subject rather than reliance on earlier studies as models.

Most PCMs assume that phylogenetic relationships are known with certainty, although this is never the case. Phylogenetic reconstructions can be treated as estimates of the true phylogeny, and as such, they are subject to error. There are two approaches to the problem of phylogenetic error in comparative analysis. The first is to attempt to minimize it by the use of reliable data and accurate methods of phylogenetic inference. The second is to allow for the occurrence of error with analyses that consider all plausible phylogenies rather than just a single "best" tree (Schultz and Churchill 1999, Huelsenbeck and Bollback 2001, Ronquist 2004). Both approaches imply that the phylogenies used for comparative analysis should meet high standards if we are to avoid erroneous or ambiguous conclusions. Since the support of future comparative analyses is a common justification for work in systematics, it is important to apply these standards as broadly as possible. Below we offer our suggestions on how this could be achieved for molecular systematics of crustaceans.

Problems and Solutions in Molecular Phylogenetics

Most molecular phylogenies of crustaceans have been based on single gene sequences or at best sequences of a few genes. Although such studies have been useful, trees based on single genes are unlikely to be entirely accurate. One reason is that gene trees are shaped not only by speciation events but also by genealogical processes within species (Neigel and Avise 1986, N.A. Rosenberg 2003). Thus, even if a gene tree is known with complete accuracy, it is unlikely to be an exact representation of the true species tree. Other sources of errors arise from the complexities of sequence evolution. Most commonly used phylogenetic methods assume that sequence evolution is represented by one of a limited set of models with parameters that can be estimated from the data (Felsenstein 1981, Huelsenbeck et al. 2001); use of an inappropriate model can strongly bias results (Lemmon and Moriarty 2004). Another source of error is the inadvertent use of paralogous sequences, such as duplicated genes or pseudogenes. Studies of crustacean mitochondrial gene sequences suggest this problem can be quite serious. A mitochondrial large-subunit ribosomal RNA (16S rRNA) nuclear pseudogene in the menippid crab *Menippe* was detected only because it coamplified with the functional mitochondrial gene; there were no telltale sequence characteristics that would have identified it as a pseudogene (Schneider-Broussard and Neigel 1997). Within the alpheid shrimp genus *Alpheus*, cytochrome oxidase I pseudogenes were common, often impossible to identify by sequence criteria alone, and sometimes amplified preferentially over the true mitochondrial sequence (S.T. Williams et al. 2001).

Theoretical considerations suggest that accurate phylogenetic reconstruction requires data from many independently segregating loci (Wu 1991). This requirement has been supported by empirical studies with varying numbers of loci. For example,

Table 3.1. Selected crustacean studies using DNA sequence data.

Sequence	Product	Reference	Group(s)
Mitochondrial Sequences			
COI	Protein	Folmer et al. 1994	Metazoa
COII	Protein	Perez-Losada et al. 2004	Aeglidae
ND5	Protein	Colbourne et al. 1998	Cladocera
12S	rRNA	Colbourne and Hebert 1996	Cladocera
16S	rRNA	Cunningham et al. 1992	Anomura
Nuclear Sequences			
EF-1α	Protein	Regier and Shultz 1997	Arthropoda
EF-2	Protein	Regier and Shultz 2001	Arthropoda
GPI	Protein	Williams et al. 2001	*Alpheus*
H3	Protein	Colgan et al. 1998	Arthropoda
Pol II	Protein	Shultz and Regier 2000	Arthropoda
18S	rRNA	Spears et al. 1992	Brachyura
28S	rRNA	Taylor et al. 1999	Branchiopoda
ITS1	rRNA	Schwenk et al. 2000	Cladocera
ITS2	rRNA	Schwenk et al. 2000	Cladocera

Abbreviations: COI, mitochondrial cytochrome C oxidase subunit I; COII, mitochondrial cytochrome C oxidase subunit II; ND5, mitochondrial NADH dehydrogenase; 12S, mitochondrial small-subunit ribosomal RNA; 16S, mitochondrial large-subunit RNA; EF-1α, nuclear elongation factor 1α; EF-2, nuclear elongation factor 2; GPI, nuclear glucose 6-phosphate isomerase; H3, nuclear histone H3; Pol II, nuclear RNA polymerase II; 18S, small-subunit nuclear RNA; 28S, large-subunit nuclear RNA; ITS1, first internal transcribed spacer of nuclear RNA; ITS2, second internal transcribed spacer of nuclear RNA.

a recent study with seven species of yeast demonstrated that, on average, 8–20 independently segregating loci were needed to achieve 100% bootstrap support of phylogenetic relationships (Rokas et al. 2003). At present, there are a limited number of loci in use for crustacean systematics. Although there are nominally 14 such loci (Table 3.1), the five mitochondrial genes represent only one segregating unit, as do the four nuclear ribosomal sequences; there are thus only seven independently segregating loci. The usefulness of some of these loci has been demonstrated only for some taxonomic levels or specific groups, for example, GPI and EF-1α in *Alpheus* (S.T. Williams et al. 2001) and EF1-α, EF-2, and POL II for higher taxonomic levels of the Arthropoda (Regier and Shultz 1997, 2001, Shultz and Regier 2000).

In addition to a dependence on the number of loci, the accuracy of molecular phylogenetic reconstruction depends on the number of taxa sampled. The addition of taxa provides more information about the states of internal nodes in a phylogeny (Graybeal 1998) and better estimates of substitution rates at particular sites in DNA sequences (Pollock and Bruno 2000). There has been some debate over the relative merits of sampling more taxa versus more loci (Hillis et al. 2003, M.S. Rosenberg and Kumar 2003), but there are certainly examples in which well-supported but incorrect phylogenies were obtained when the number of loci was high but the number of sampled taxa was low (Soltis et al. 2004).

One of the central problems in molecular systematics is the high degree of homoplasy (parallel or convergent evolution of the same character state) in nucleotide substitutions and small insertions and deletions (indels) (Broughton et al. 2000). However, there are other types of molecular characters that are relatively free from homoplasy. Rokas and Holland (2000) reviewed the use of rare genomic changes that include indels of entire introns, unique indels in protein or RNA sequences (signature sequences), retroposon events (transposable elements), gene order rearrangements in organelle genomes, and variations in genetic codes. Mitochondrial gene rearrangements have proven useful for crustacean systematics. For example, C.L. Morrison et al. (2002) used them to construct a phylogeny in which it appears that carcinization evolved independently in the decapod lineages Brachyura, Porcellanidae, Lomisidae, Lithodidae, and the paguroid crab genus *Birgus*. Lavrov et al. (2004) used mitochondrial gene order to determine the close affiliation of the Pentastomida (tongue worms) with the Cephalocarida (horseshoe shrimps) and the Maxillopoda (ostracods, copepods, and barnacles).

Sequence alignment is a critical step in molecular phylogenetic analysis. Alignments represent assumptions about the homology of characters, and phylogenetic inference can be very sensitive to changes in alignments (D.A. Morrison and Ellis 1997). Alignment algorithms insert gaps only if they improve the overall alignment score by more than the value of the "gap penalty" (Setubal and Meidanis 1997). However, gap penalties are usually set arbitrarily and often do not accurately represent how sequences actually evolve (Gu 1995). Because of the complexity of the problem of multiple alignment (Bonizzoni and Vedova 2001), heuristic algorithms must be used that are not guaranteed to find alignments with the best scores. These limitations suggest that overly complex alignments are a dubious foundation for phylogenetic inference.

Alignments of noncoding sequences that are extremely variable in length (e.g., structural RNAs, introns, and intergenic sequences) are usually the most problematic. Unfortunately, such sequences represent nearly half of the gene sequences used in crustacean systematics (Table 3.1). These sequences tend to have numerous indels, runs of repeated nucleotides, and variable numbers of tandem repeats (e.g., microsatellite sequences). In such ambiguous cases, the conservative approach is to eliminate problematic regions from the data used for phylogenetic analysis, although this can also result in elimination of many otherwise informative sites (D.A. Morrison and Ellis 1997).

We hope that we have made a strong case for adding more loci, especially nuclear loci, to the current set available for crustacean molecular systematics. These might include both protein coding loci and conserved noncoding sequences. Many candidates can be found in public sequence databases, although considerable effort is needed to develop PCR primers that work reliably across a range of taxa. As new loci are identified for crustacean systematics, it will be important to demonstrate (to the extent that it is possible) the orthology of sequences from different taxa. The large set of loci that have been tested in phylogenetic studies of insects can serve as a guide to what is likely to be useful for the Crustacea. As of 2000, around 40 protein coding loci had been used for insect systematics as well as all of the major ribosomal RNA genes and numerous noncoding regions (Caterino et al. 2000).

Phylogenetic Analysis of Sequence Data

Overview

Considerable progress has been made over the past 20 years in the development of methods for phylogenetic analysis of DNA sequences (Swofford et al. 1996), although this has been accompanied by intense debate over these methods. There has been no final answer to the question of which method is most likely to produce the "true tree"; however, they can be objectively compared as statistical estimators that differ with respect to robustness, consistency, and efficiency. Here, we present a brief overview of methods for phylogenetic inference and consider application of these methods to comparative studies.

Distance Methods

Distance-based, or phenetic, methods use algorithms to cluster sequences into trees that reflect pairwise measures of distance between DNA sequences (Swofford et al. 1996). These distances are based on models of sequence evolution that correct for multiple substitutions at individual sites. The neighbor-joining algorithm has become the standard for building trees from distance data because it is robust and efficient in comparison to other distance methods (Saitou and Nei 1987). Distance methods are considerably faster than the others considered here, but simulation studies have shown that they are less likely to produce the correct tree (Huelsenbeck 1995a, 1995b).

Maximum Parsimony

Maximum parsimony (MP) represents an optimality criterion to compare trees rather than an algorithm to build trees (Swofford et al. 1996). Phylogenetic trees are considered as hypotheses; the tree that requires the fewest character state changes is the preferred hypothesis. MP has been justified on the basis of William of Occam's famous dictim, "Entities should not be multiplied unnecessarily," as well as the fact that it does fairly well at finding the true tree in tests with simulated sequence data (Huelsenbeck 1995a, 1995b). MP also works with any type of sequence characters, including indels, and allows molecular and morphological data to be combined. Although MP is still a respected and widely used approach, it suffers from several disadvantages that limit its range of application. For more than about 10 taxa, a prohibitively large number of trees must be examined to guarantee that the most parsimonious tree is found. This necessitates the use of "heuristic" methods that are not guaranteed to find the best tree. Under some circumstances, MP is statistically inconsistent; with more data, it will converge on an incorrect estimate of phylogeny in which rapidly evolving taxa are grouped together even if they are unrelated (Felsenstein 1978). A more fundamental limitation for many PCMs is that MP does not provide estimates of branch lengths.

Maximum Likelihood

Maximum likelihood (ML) is another criterion to compare trees, but unlike parsimony, it is based on statistical theory. The tree with the ML can be considered to be an estimate of the true tree, rather than simply a preferred hypothesis (Felsenstein 1981). Likelihood provides a measure of how much the data support an estimate or hypothesis and can be used to construct confidence limits (Edwards 1992). As in the case of distance methods, the calculation of likelihoods must be based on models of sequence evolution. With the correct model, ML tends to outperform most other methods at finding the true tree, but with the wrong model, it can perform poorly (Huelsenbeck 1995a, 1995b). Generally, the simplest model that can explain the data according to one or more criteria is preferred (Posada and Crandall 1998). As with MP, a prohibitively large number of trees would need to be examined to guarantee that the ML tree is found, but in addition, the parameters of the model of sequence evolution must also be estimated. For this reason, ML analyses are computationally the most demanding.

Bayesian Inference

Bayesian methods of phylogenetic inference use the same models of sequence evolution as distance methods and ML but are unique in estimating the probabilities of phylogenetic trees (Huelsenbeck et al. 2001). Although it is generally impossible to evaluate the probabilities of all possible trees, the distribution can be approximated with the Markov chain Monte Carlo (MCMC) method. From this distribution, a consensus tree with the probability of each branch evaluated can be constructed (Larget and Simon 1999). This use of the MCMC method makes the Bayesian approach less computationally demanding than ML (Huelsenbeck et al. 2001).

Bayesian methods offer an attractive solution to the problem of phylogenetic uncertainty in comparative analysis. Trees can be sampled according to their posterior probabilities so that the results of the analysis are not based on a single tree but are weighted to reflect the most probable trees (Huelsenbeck and Bollback 2001). Bayesian approaches have also been developed for the reconstruction of ancestral character states (Schultz and Churchill 1999) and can accommodate both phylogenetic uncertainty and uncertainty in the reconstruction of ancestral character states (Huelsenbeck and Bollback 2001).

Prospects for the Future

Within the Crustacea, diverse and often puzzling reproductive adaptations have evolved that are apparent in morphology and behavior (Bauer and Martin 1991; see also chapters 7–12). Molecular markers are tools that we can use to examine the effects of these adaptations on individual fitness and the factors that have influenced their evolution. However, the power and promise of molecular markers also represent a new set of challenges. The effort, expertise, and expense required to collect specimens that are suitable for DNA analysis, perform the necessary laboratory work, and perform sophisticated data analyses are often beyond the means of an individual investigator. Collaboration has become essential. Some forms of collaboration are already well established, such as collaborations between classically trained morphologists and molecular biologists. However, broader forms of collaboration will be needed to establish community resources that will allow us to reap the full potential of our efforts. Repositories of information and knowledge are needed, as are physical repositories of voucher specimens, DNA samples, and PCR primers. Small aliquots of genomic DNA from important specimens should be made available as a matter of course, as should novel PCR primers. Efforts should be made to coordinate the development of new loci for systematics to avoid the "Tower of Babel" situation that has developed in insect systematics (Caterino et al. 2000). Along with these shared resources, there is a need for recognized standards of quality that reflect both immediate needs and the long-term utility of our work.

Summary

Molecular markers provide powerful means to analyze relationships of descent both among individuals and taxa. Microsatellite loci have become the standard for studies of paternity and kinship because they are highly polymorphic and codominant, properties that provide statistical power and facilitate the detection of artifacts. They have proven to be useful for the analysis of crustacean mating systems, although their full potential has yet to be realized.

Phylogenetic comparative methods are intended to separate instances of convergent or parallel evolution from shared evolutionary history by analyzing the phylogenetic distributions of traits. They have been criticized for the assumptions they make about how traits evolve, although not all of these methods make the same assumptions. Most assume an accurate phylogeny is known, which implies that a high standard should be required of phylogenies that will be used for comparative analysis. Crustacean phylogenies have mostly been based on small number of sequences that do not have the most desirable properties for phylogenetic inference. This situation is likely to be remedied by the development of PCR primers that amplify additional independently segregating nuclear loci.

Acknowledgments We thank J.E. Duffy, M. Thiel, and two anonymous reviewers for useful comments on the manuscript. We are also thank the National Science Foundation (OCE-0326383) for their support.

References

Ackerly, D.D., and M.J. Donoghue. 1998. Leaf size, sapling allometry, and Corner's rules: phylogeny and correlated evolution in maples (*Acer*). American Naturalist 152:767–791.

Avise, J.C. 2004. Molecular markers, natural history and evolution. Sinauer Associates, New York.

Ayliffe, M.A., G.J. Lawrence, J.G. Ellis, and A.J. Pryor. 1994. Heteroduplex molecules formed between allelic sequences cause nonparental RAPD bands. Nucleic Acids Research 22:1632–1636.

Baragona, M.A., L.A. Haig-Ladewig, and S.Y. Wang. 2000. Multiple paternity in the grass shrimp *Palaemonetes pugio*. American Zoologist 40:935–935.

Bauer, R.T., and J.W. Martin, editors. 1991. Crustacean sexual biology. Columbia University Press, New York.

Bilodeau, A.L., D.L. Felder, and J.E. Neigel. 2005. Multiple paternity in the thalassinidean ghost shrimp, *Callichirus islagrande*. Marine Biology 146:381–385.

Bjorklund, M. 1997. Are "comparative methods" always necessary? Oikos 80:607–612.

Blouin, M.S. 2003. DNA-based methods for pedigree reconstruction and kinship analysis in natural populations. Trends in Ecology and Evolution 18:503–511.

Bonin, A., E. Bellemain, P. Bronken Eidesen, F. Pompanon, C. Brochmann, and P. Taberlet. 2004. How to track and assess genotyping errors in population genetics studies. Molecular Ecology 13:3261–3273.

Bonizzoni, P., and G.D. Vedova. 2001. The complexity of multiple sequence alignment with SP-score that is a metric. Theoretical Computer Science 259:63–79.

Bourke, A.F.G., H.A.A. Green, and M.W. Bruford. 1997. Parentage, reproductive skew and queen turnover in a multiple-queen ant analysed with microsatellites. Proceedings of the Royal Society of London, Series B 264:277–283.

Brooks, D.R., and D.A. McLennan. 1991. Phylogeny, ecology and behavior: a research program in comparative biology. University of Chicago Press, Chicago, Ill.

Broughton, R., S. Stanley, and R. Durrett. 2000. Quantification of homoplasy for nucleotide transitions and transversions and a reexamination of assumptions in weighted phylogenetic analysis. Systematic Biology 49:617–627.

Burfitt, A.H. 1980. Glucose phosphate isomerase inheritance in *Cancer pagurus* L broods as evidence of multiple paternity (Decapoda: Brachyura). Crustaceana 39:306–310.

Callen, D.F., A.D. Thompson, Y. Shen, H.A. Phillips, R.I. Richards, J.C. Mulley, and G.R. Sutherland. 1993. Incidence and origin of null alleles in the (AC)N microsatellite markers. American Journal of Human Genetics 52:922–927.

Caterino, M.S., W. Cho, and F.A.H. Sperling. 2000. The current state of insect molecular systematics: a thriving tower of Babel. Annual Review of Entomology 45:1–54.

Cheverud, J.M., M.M. Dow, and W. Leutenegger. 1985. The quantitative assessment of phylogenetic constraints in comparative analyses: sexual dimorphism in body-weight among primates. Evolution 39:1335–1351.

Colbourne, J.K., and P.D.N. Hebert. 1996. The systematics of North American *Daphnia* (Crustacea: Anomopoda): a molecular phylogenetic approach. Philosophical Transactions of the Royal Society of London, Series B 351:349–360.

Colbourne, J.K., T.J. Crease, L.J. Weider, P.D.N. Hebert, F. Dufresne, and A. Hobaek. 1998. Phylogenetics and evolution of a circumarctic species complex (Cladocera: *Daphnia pulex*). Biological Journal of the Linnean Society 65:347–365.

Colgan, D., A. McLachlan, G. Wilson, S. Livingston, G. Edgecombe, J. Macaranas, G. Cassis, and M. Gray. 1998. Histone H3 and U2 snRNA DNA sequences and arthropod molecular evolution. Australian Journal of Zoology 46:419–437.

Cronn, R., M. Cedroni, T. Haselkorn, C. Grover, and J.F. Wendel. 2002. PCR-mediated recombination in amplification products derived from polyploid cotton. Theoretical and Applied Genetics 104:482–489.

Cunningham, C.W., N.W. Blackstone, and L.W. Buss. 1992. Evolution of king crabs from hermit crab ancestors. Nature 355:539–542.

Cunningham, C.W., K.E. Omland, and T.H. Oakley. 1998. Reconstructing ancestral character states: a critical reappraisal. Trends in Ecology and Evolution 13:361–366.

Diniz, J.A.F., C.E.R. De Santana, and L.M. Bini. 1998. An eigenvector method for estimating phylogenetic inertia. Evolution 52:1247–1262.

Duffy, J.E. 1996. Eusociality in a coral-reef shrimp. Nature 381:512–514.

Duffy, J.E., C.L. Morrison, and R. Rios. 2000. Multiple origins of eusociality among sponge-dwelling shrimps (*Synalpheus*). Evolution 54:503–516.

Edwards, A.W.F. 1992. Likelihood. Johns Hopkins University Press, Baltimore, Md.

Eisen, J. 1999. Mechanistic basis for microsatellite instability. Pages 34–48 *in*: D.B. Goldstein and C. Schlotterer, editors. Microsatellites: evolution and applications. Oxford University Press, Oxford.

Ellsworth, D.L., K.D. Rittenhouse, and R.L. Honeycutt. 1993. Artifactual variation in randomly amplified polymorphic DNA banding patterns. Biotechniques 14:214–217.

Felsenstein, J. 1978. Cases in which parsimony and compatibility methods will be positively misleading. Systematic Zoology 27:401–410.

Felsenstein, J. 1981. Evolutionary trees from DNA sequences: a maximum likelihood approach. Journal of Molecular Evolution 17:368–376.

Felsenstein, J. 1985. Phylogenies and the comparative method. American Naturalist 125:1–15.

Folmer, O., M. Black, W. Hoeh, R. Lutz, and R. Vrijenhoek. 1994. DNA primers for amplification of mitochondrial cytochrome c oxidase subunit I from diverse metazoan invertebrates. Molecular Marine Biology and Biotechnology 3:294–297.

Freckleton, R.P., P.H. Harvey, and M. Pagel. 2002. Phylogenetic analysis and comparative data: a test and review of evidence. American Naturalist 160:712–726.

Gerber, S., S. Mariette, R. Streiff, C. Bodenes, and A. Kremer. 2000. Comparison of microsatellites and amplified fragment length polymorphism markers for parentage analysis. Molecular Ecology 9:1037–1048.

Gosselin, T., B. Sainte-Marie, and L. Bernatchez. 2005. Geographic variation of multiple paternity in wild American lobster, *Homarus americanus*. Molecular Ecology 14:1517–1525.

Graybeal, A. 1998. Is it better to add taxa or characters to a difficult phylogenetic problem? Systematic Biology 47:9–17.

Gu, X. 1995. The size distribution of insertions and deletions in human and rodent pseudogenes suggests the logarithmic gap penalty for sequence alignment. Journal of Molecular Evolution 40:464–473.

Hansen, T.F. 1997. Stabilizing selection and the comparative analysis of adaptation. Evolution 51:1341–1351.

Harris, D.J., and K.A. Crandall. 2000. Intragenomic variation within ITS1 and ITS2 of freshwater crayfishes (Decapoda: Cambaridae): implications for phylogenetic and microsatellite studies. Molecular Biology and Evolution 17:284–291.

Harvey, P.H., and M.D. Pagel. 1991. The comparative method in evolutionary biology. Oxford University Press, Oxford.

Hastings, M.D., D.C. Queller, F. Eischen, and J.E. Strassmann. 1998. Kin selection, relatedness, and worker control of reproduction in a large-colony epiponine wasp, *Brachygastra mellifica*. Behavioral Ecology 9:573–581.

Henshaw, M.T., and R.H. Crozier. 2004. Mating system and population structure of the primitively eusocial wasp *Ropalidia revolutionalis*: a model system for the evolution of complex societies. Molecular Ecology 13:1943–1950.

Hillis, D.M., D.D. Pollock, J.A. McGuire, and D.J. Zwickl. 2003. Is sparse taxon sampling a problem for phylogenetic inference? Systematic Biology 52:124–126.

Huelsenbeck, J.P. 1995a. Performance of phylogenetic methods in simulation. Systematic Biology 44:17–48.

Huelsenbeck, J.P. 1995b. The robustness of two phylogenetic methods: four taxon simulations reveal a slight superiority of maximum likelihood over neighbor joining. Molecular Biology and Evolution 12:843–849.

Huelsenbeck, J.P., and J.P. Bollback. 2001. Empirical and hierarchical Bayesian estimation of ancestral states. Systematic Biology 50:351–366.

Huelsenbeck, J.P., F. Ronquist, R. Nielsen, and J.P. Bollback. 2001. Bayesian inference of phylogeny and its impact on evolutionary biology. Science 294:2310–2314.

Jensen, P.C., and P. Bentzen. 2004. Isolation and inheritance of microsatellite loci in the Dungeness crab (Brachyura: Cancridae: *Cancer magister*). Genome 47:325–331.

Jones, A.G., and W.R. Ardren. 2003. Methods of parentage analysis in natural populations. Molecular Ecology 12:2511–2523.

Jones, M.W., P.T. O'Reilly, A.A. McPherson, T.L. McParland, D.E. Armstrong, A.J. Cox, K.R. Spence, E.L. Kenchington, C.T. Taggart, and P. Bentzen. 2003. Development, characterization, inheritance, and cross-species utility of American lobster (*Homarus americanus*) microsatellite and mtDNA PCR-RFLP markers. Genome 46:59–69.

Kitaura, J., K. Wada, and M. Nishida. 1998. Molecular phylogeny and evolution of unique mud-using territorial behavior in ocypodid crabs (Crustacea: Brachyura: Ocypodidae). Molecular Biology and Evolution 15:626–637.

Larget, B., and D.L. Simon. 1999. Markov chain Monte Carlo algorithms for the Bayesian analysis of phylogenetic trees. Molecular Biology and Evolution 16:750–759.

Lavrov, D.V., W.M. Brown, and J.L. Boore. 2004. Phylogenetic position of the Pentastomida and (pan)crustacean relationships. Proceedings of the Royal Society of London, Series B 271:537–544.

Lemmon, A.R., and E.C. Moriarty. 2004. The importance of proper model assumption in Bayesian phylogenetics. Systematic Biology 53:265–277.

Lewontin, R.C. 1991. Twenty-five years ago in genetics: electrophoresis in the development of evolutionary genetics: milestone or millstone? Genetics 128:657–662.

Martins, E.P. 2000. Adaptation and the comparative method. Trends in Ecology and Evolution 15:296–299.

Martins, E.P., J.A.F. Diniz, and E.A. Housworth. 2002. Adaptive constraints and the phylogenetic comparative method: a computer simulation test. Evolution 56:1–13.

Meagher, T.R., and E. Thompson. 1986. The relationship between single parent and parent pair genetic likelihoods in genealogy reconstruction. Theoretical Population Biology 29:87–106.

Miller, C.R., P. Joyce, and L.P. Waits. 2002. Assessing allelic dropout and genotype reliability using maximum likelihood. Genetics 160:357–366.

Milligan, B.G. 2003. Maximum-likelihood estimation of relatedness. Genetics 163:1153–1167.

Morrison, C.L., A.W. Harvey, S. Lavery, K. Tieu, Y. Huang, and C.W. Cunningham. 2002. Mitochondrial gene rearrangements confirm the parallel evolution of the crab-like form. Proceedings of the Royal Society of London, Series B 269:345–350.

Morrison, D.A., and J.T. Ellis. 1997. Effects of nucleotide sequence alignment on phylogeny estimation: a case study of 18S rDNAs of Apicomplexa. Molecular Biology and Evolution 14:428–441.

Neff, B.D., and T.E. Pitcher. 2002. Assessing the statistical power of genetic analyses to detect multiple mating in fishes. Journal of Fish Biology 61:739–750.

Neff, B.D., J. Repka, and M.R. Gross. 2001. A Bayesian framework for parentage analysis: the value of genetic and other biological data. Theoretical Population Biology 59:315–331.

Neigel, J.E., and J.C. Avise. 1986. Phylogenetic relationships of mitochondrial DNA under various demographic models of speciation. Pages 515–534 *in*: S. Karlin and E. Nevo, editors. Evolutionary processes and theory. Academic Press, New York.

Nelson, K., and D. Hedgecock. 1977. Electrophoretic evidence of multiple paternity in lobster *Homarus americanus* (Milne-Edwards). American Naturalist 111:361–365.

Perez, T., J. Albornoz, and A. Dominguez. 1998. An evaluation of RAPD fragment reproducibility and nature. Molecular Ecology 7:1347–1357.

Perez-Losada, M., G. Bond-Buckup, C.G. Jara, and K.A. Crandall. 2004. Molecular systematics and biogeography of the southern South American freshwater "crabs" *Aegla* (Decapoda: Anomura: Aeglidae) using multiple heuristic tree search approaches. Systematic Biology 53:767–780.

Pollock, D.D., and W.J. Bruno. 2000. Assessing an unknown evolutionary process: effect of increasing site-specific knowledge through taxon addition. Molecular Biology and Evolution 17:1854–1858.

Posada, D., and K.A. Crandall. 1998. MODELTEST: testing the model of DNA substitution. Bioinformatics 14:817–818.

Regier, J.C., and J.W. Shultz. 1997. Molecular phylogeny of the major arthropod groups indicates polyphyly of crustaceans and a new hypothesis for the origin of hexapods. Molecular Biology and Evolution 14:902–913.

Regier, J.C., and J.W. Shultz. 2001. Elongation factor-2: a useful gene for arthropod phylogenetics. Molecular Phylogenetics and Evolution 20:136–148.

Reynolds, J.D. 1996. Animal breeding systems. Trends in Ecology and Evolution 11:68–72.

Riedy, M. F., W.J. Hamilton, and C.F. Aquadro. 1992. Excess of non-parental bands in offspring from known primate pedigrees assayed using RAPD PCR. Nucleic Acids Research 20:918–918.

Rokas, A., and P.W.H. Holland. 2000. Rare genomic changes as a tool for phylogenetics. Trends in Ecology and Evolution 15:454–459.

Rokas, A., B.L. Williams, N. King, and S.B. Carroll. 2003. Genome-scale approaches to resolving incongruence in molecular phylogenies. Nature 425:798–804.

Ronquist, F. 2004. Bayesian inference of character evolution. Trends in Ecology and Evolution 19:475–481.

Rosenberg, M.S., and S. Kumar. 2003. Taxon sampling, bioinformatics, and phylogenomics. Systematic Biology 52:119–124.

Rosenberg, N.A. 2003. The shapes of neutral gene genealogies in two species: probabilities of monophyly, paraphyly, and polyphyly in a coalescent model. Evolution 57:1465–1477.

Ross, K.G. 2001. Molecular ecology of social behaviour: analyses of breeding systems and genetic structure. Molecular Ecology 10:265–284.

Sainte-Marie, B., N. Urbani, J.M. Sevigny, F. Hazel, and U. Kuhnlein. 1999. Multiple choice criteria and the dynamics of assortative mating during the first breeding season of female snow crab *Chionoecetes opilio* (Brachyura, Majidae). Marine Ecology Progress Series 181:141–153.

Saitou, N., and M. Nei. 1987. The neighbor-joining method: a new method for reconstructing phylogenetic trees. Molecular Biology and Evolution 4:406–425.

Sassaman, C. 1978. Mating systems in porcellionid isopods: multiple paternity and sperm mixing in *Porcellio scaber* Latr. Heredity 41:385–397.

Schneider-Broussard, R., and J.E. Neigel. 1997. A large subunit mitochondrial ribosomal DNA sequence translocated to the nuclear genomes of two stone crabs. Molecular Biology and Evolution 14:156–165.

Schubart, C.D., R. Diesel, and S.B. Hedges. 1998. Rapid evolution to terrestrial life in Jamaican crabs. Nature 393:363–365.

Schultz, T.R., and G.A. Churchill. 1999. The role of subjectivity in reconstructing ancestral character states: a Bayesian approach to unknown rates, states, and transformation asymmetries. Systematic Biology 48:651–664.

Schwenk, K., D. Posada, and P.D.N. Hebert. 2000. Molecular systematics of European *Hyalodaphnia*: the role of contemporary hybridization in ancient species. Proceedings of the Royal Society of London, Series B 267:1833–1842.

Selkoe, K. A., and R. J. Toonen. 2006. Microsatellites for ecologists: a practical guide to using and evaluating microsatellite markers. Ecology Letters 9:615–629.

Setubal, J., and J. Meidanis. 1997. Introduction to computational molecular biology. PWS Publishing, Boston, Mass.

Sevigny, J.M., and B. Sainte-Marie. 1996. Electrophoretic data support the last male sperm precedence hypothesis in the snow crab *Chionoecetes opilio* (Brachyura: Majidae). Journal of Shellfish Research 15:437–440.

Shaw, P.W., G.J. Pierce, and P.R. Boyle. 1999. Subtle population structuring within a highly vagile marine invertebrate, the veined squid *Loligo forbesi*, demonstrated with microsatellite DNA markers. Molecular Ecology 8:407–417.

Shinde, D., Y.L. Lai, F.Z. Sun, and N. Arnheim. 2003. Taq DNA polymerase slippage mutation rates measured by PCR and quasi-likelihood analysis: (CA/GT)(n) and (A/T)(n) microsatellites. Nucleic Acids Research 31:974–980.

Shultz, J.W., and J.C. Regier. 2000. Phylogenetic analysis of arthropods using two nuclear protein-encoding genes supports a crustacean plus hexapod clade. Proceedings of the Royal Society of London, Series B 267:1011–1019.

Soltis, D.E., V.A. Albert, V. Savolainen, K. Hilu, Y.-L. Qiu, M.W. Chase, J.S. Farris, S. Stefanovic, D.W. Rice, J.D. Palmer, and P.S. Soltis. 2004. Genome-scale data, angiosperm relationships, and "ending incongruence": a cautionary tale in phylogenetics. Trends in Plant Science 9:477–483.

Spears, T., L.G. Abele, and W. Kim. 1992. The monophyly of brachyuran crabs: a phylogenetic study based on 18S ribosomal RNA. Systematic Biology 41:446–461.

Streiff, R., S. Mira, M. Castro, and M.L. Cancela. 2004. Multiple paternity in Norway lobster (*Nephrops norvegicus* L.) assessed with microsatellite markers. Marine Biotechnology 6:60–66.

Sunnucks, P. 2000. Efficient genetic markers for population biology. Trends in Ecology and Evolution 15:199–203.

Swofford, D.L., G.J. Olsen, P.J. Waddell, and D.M. Hillis. 1996. Phylogenetic inference. Pages 407–514 *in*: D.M. Hillis, C. Moritz, and B.K. Mable, editors. Molecular systematics. Sinauer Associates, Sunderland, Mass.

Tautz, D. 1989. Hypervariability of simple sequences as a general source for polymorphic markers. Nucleic Acids Research 17:6463–6571.

Taylor, D.J., T.J. Crease, and W.M. Brown. 1999. Phylogenetic evidence for a single long-lived clade of crustacean cyclic parthenogens and its implications for the evolution of sex. Proceedings of the Royal Society of London, Series B 266:791–797.

Taylor, D.M., R.G. Hooper, and G.P. Ennis. 1985. Biological aspects of the spring breeding of snow crabs *Chionoecetes opilio*, in Bonne Bay, Newfoundland (Canada). U.S. National Marine Fisheries Service Bulletin 83:707–711.

Toonen, R.J. 2004. Genetic evidence of multiple paternity of broods in the intertidal crab *Petrolisthes cinctipes*. Marine Ecology Progress Series 270:259–263.

Tóth, E., J.E. Strassmann, V.L. Imperatriz-Fonseca, and D.C. Queller. 2003. Queens, not workers, produce the males in the stingless bee *Schwarziana quadripunctata*. Animal Behaviour 66:359–368.

Urbani, N., B. Sainte-Marie, J.M. Sevigny, D. Zadworny, and U. Kuhnlein. 1998. Sperm competition and paternity assurance during the first breeding period of female snow crab (*Chionoecetes opilio*) (Brachyura: Majidae). Canadian Journal of Fisheries and Aquatic Sciences 55:1104–1113.

Van de Casteele, T., P. Galbusera, and E. Matthysen. 2001. A comparison of microsatellite-based pairwise relatedness estimators. Molecular Ecology 10:1539–1549.

Vos, P., R. Hogers, M. Bleeker, M. Reijans, T. van de Lee, M. Hornes, A. Frijters, J. Pot, J. Peleman, M. Kuiper, and M. Zabeau. 1995. AFLP: a new technique for DNA fingerprinting. Nucleic Acids Research 23:4407–4414.

Walker, D., B.A. Porter, and J.C. Avise. 2002. Genetic parentage assessment in the crayfish *Orconectes placidus*, a high-fecundity invertebrate with extended maternal brood care. Molecular Ecology 11:2115–2122.

Williams, J.G.K., A.R. Kubelik, K.J. Livak, J.A. Rafalski, and S.V. Tingey. 1990. DNA polymorphisms amplified by arbitrary primers are useful as genetic markers. Nucleic Acids Research 18:6531–6535.

Williams, S.T., N. Knowlton, L.A. Weigt, and J.A. Jara. 2001. Evidence for three major clades within the snapping shrimp genus *Alpheus* inferred from nuclear and mitochondrial gene sequence data. Molecular Phylogenetics and Evolution 20:375–389.

Wu, C.-I. 1991. Inferences of species phylogeny in relation to segregation of ancient polymorphisms. Genetics 127:429–435.

Zamudio, K.R., and B. Sinervo. 2000. Polygyny, mate-guarding, and posthumous fertilization as alternative male mating strategies. Proceedings of the National Academy of Sciences, USA 97:14427–14432.

Zeh, J.A., and D.W. Zeh. 2003. Toward a new sexual selection paradigm: polyandry, conflict and incompatibility. Ethology 109:929–950.

PART II

Communication

The Neural Basis of Communication in Crustaceans

Jens Herberholz

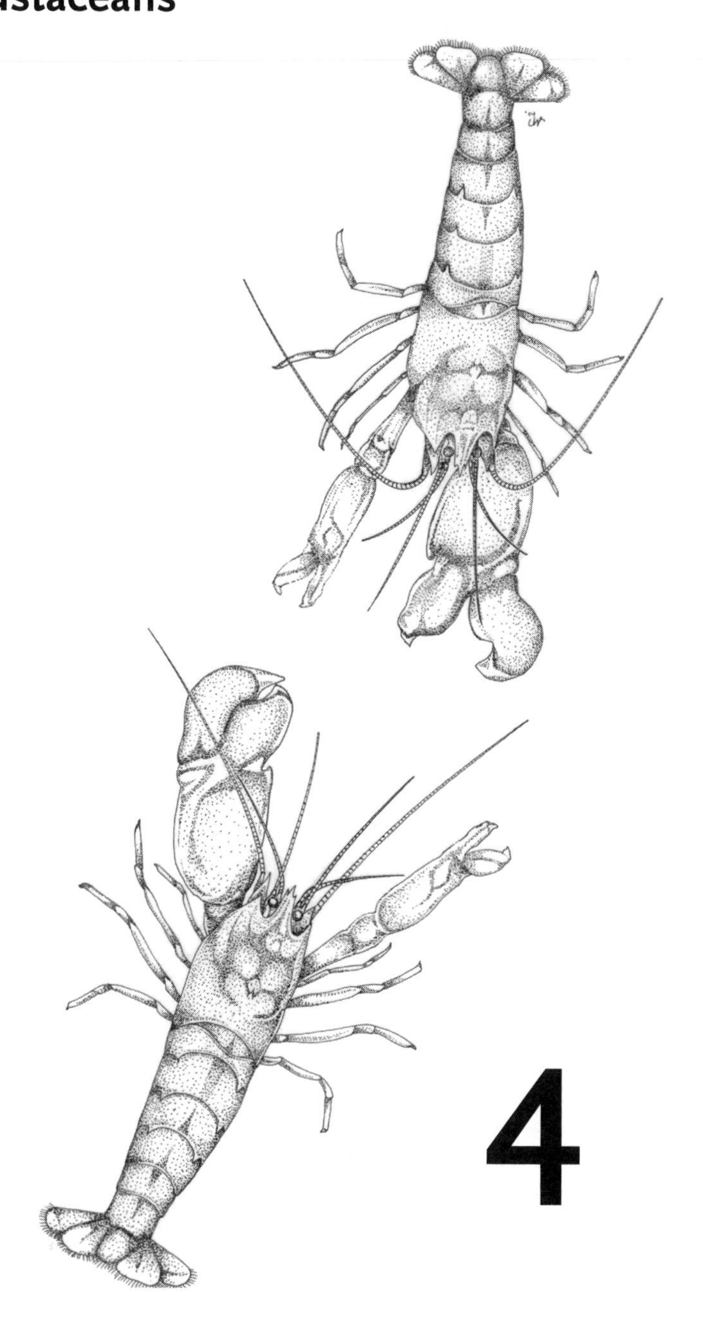

4

The great diversity of communication tools utilized by crustaceans during intra- and interspecific interactions has attracted the attention of biologists for many years. Numerous studies have generated substantial material with detailed descriptions of the behaviors displayed during aggressive competition for resources (see chapter 5), in response to predatory attacks (e.g., Herberholz et al. 2004), and during courtship and mating (see chapter 13). The identities of acoustic, visual, tactile, and chemical signals released by the sender are partially known, and some of the receiving structures that perceive these signals have been identified. Specialized receptors that are tuned to allow communication among members of the same species or across different species have been analyzed with the help of modern biological techniques.

However, communication is not only the generation and transmission of signals by the sender and their reception by the receiver, but also includes neural processes that lead to beneficial behavioral responses (Greenfield 2002). Currently, we still have a poor understanding of how the central nervous system (CNS) of crustaceans integrates these signals, and many of the underlying neural mechanisms that allow and control communication are unknown. While we are reasonably well informed about the neural processes that occur in the periphery of the nervous system, that is, on the level of the receptors, and have some understanding of the projection pathways that are involved in signal transmission, we lack knowledge about higher order processes that take place in the nervous system. This may seem surprising, since neurobiologists typically consider crustaceans as model systems for neurophysiological studies because they possess a nervous system of tractable complexity that is readily accessible for electrophysiological investigations. Despite these potentialities, the experimental transition from the periphery where communicative signals are detected into areas of the brain where these signals are processed to orchestrate adaptive behaviors has only begun.

Organization of the CNS

To understand communication, we must know about the neural structures that underlie the production of communication signals, receive and integrate signals, and form behavioral responses. Thus, in this section I introduce our current knowledge of the anatomy of the crustacean CNS. Since nervous systems are as variable in morphology and function as their bearers, I limit my remarks mostly to the decapod crustaceans.

Ground Plan

The primitive form of the crustacean CNS is often organized like a ladder, with multiple ganglia linked longitudinally by connectives and across the midline via commissures (Sandeman 1982). The ventral nerve cord consists of a variable number of ganglia in the thorax and abdomen that are fused across the midline but separated longitudinally in many long-bodied decapod crustaceans (Fig. 4.1a). In short-bodied decapods, the thoracic ganglia are fused and the abdominal ganglia are reduced (Sandeman 1982). In the decentralized CNS of crustaceans, ganglia of the ventral nerve cord play an important role in the production of behaviors, some of which are executed without any participation of higher centers of the CNS.

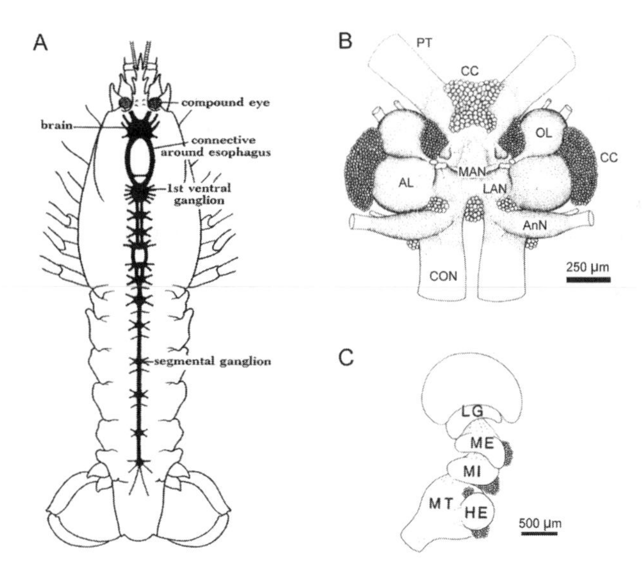

Figure 4.1 The CNS of decapod crustaceans. (A) Nervous system of a lobster. The brain lies dorsally and is joined with the subesophageal ganglion by two connectives that pass around the esophagus. The ventral nerve cord consists of five thoracic and six abdominal ganglia that are linked to each other by connectives. The left and right ganglia in each segment are fused. All ganglia give rise to paired sensory and motor nerves. From Buchsbaum et al. (1987). (B) The outer surface of the crayfish brain showing nerve roots, cell clusters, and neuropils. AL, accessory lobe; AnN, antenna II neuropil; CC, cell cluster; CON, connectives; LAN, lateral antennular neuropil; MAN, median antennular neuropil; OL, olfactory lobe; PT, protocerebral tract. (C) Optic ganglia in the eyestalks of crayfish. HE, hemiellipsoid body; LG, lamina ganglionaris; ME, medulla externa; MI, medulla interna; MT, medulla terminalis. B and C modified after Sandeman et al. (1988).

The brain lies anterior and dorsal to the ventral nerve cord and is made from several fused ganglia (Fig. 4.1b). It is joined with the ventral nerve cord by two connectives that stretch around the esophagus. Areas where communication signals are integrated and more complex behaviors are formed lie within the brain (also referred to as the cerebral or supraesophageal ganglion).

Organization of the Brain in Decapod Crustaceans

The brain of adult decapod crustaceans is only a few millimeters in size and contains some 10,000 neurons that are organized into distinct neuropils, cell clusters, and neural tracts (Fig. 4.1b; Sandeman et al. 1992). The brain is subdivided into three regions reflecting the three ganglia that have fused to form it: (1) the protocerebrum, (2) the deutocerebrum, and (3) the tritocerebrum. The protocerebrum is further subdivided into three parts: the optic ganglia and the lateral and median protocerebrum. The optic ganglia and the more proximal located lateral protocerebrum are both situated in each of the two eyestalks (Fig. 4.1c) in most but not all decapod crustaceans that have stalked eyes. Three optic ganglia, the lamina ganglionaris, the medulla externa, and the medulla interna, are all involved in visual processing. The medulla terminalis and the hemiellipsoid body are part of the lateral protocerebrum and integrate olfactory and other sensory inputs (Fig. 4.1c). The median protocerebrum contains a pair of anterior and posterior lobes, the protocerebral bridge, and the central body that lies between the lobes. The deutocerebrum includes the bilateral olfactory lobes that receive primary chemosensory input from the antennules, and the lateral and median antennular neuropils (Fig. 4.1b) that receive information from the statocysts and from antennular chemo- and mechanoreceptors. The paired accessory lobes (Fig. 4.1b) are not innervated by primary afferents and vary substantially in size among the eureptantian decapod crustaceans. They are also part of this brain division along with some associated commissures and tracts. The tritocerebrum contains the tegumentary neuropil, which receives afferent input through the tegumentary nerve from the dorsal carapace and the antenna II neuropils (also referred to as "antennal lobes"; Fig. 4.1b), which receive mechanosensory information from the antennae.

The brains of decapod crustaceans exhibit certain species-specific differences in morphology, and a nomenclature for homologous structures among groups of decapod crustaceans has been introduced by Sandeman et al. (1992). The peripheral and central sensory pathways as well as the connections between different neuropils in the brain are described to some extent, mostly in the olfactory and visual system. However, the function of many neuroanatomical structures within the brain is still unknown, and their role in processing and integrating signals, especially those that are used during communication, is not fully understood (see below).

Communication Modes

Crustaceans use several different channels (acoustic, tactile and hydrodynamic, visual, chemical, and multimodal) for intra- and interspecific communication. This section covers both the variety and uniqueness of communication signals received and produced by crustaceans, the behaviors evoked by these signals, and the neural processes that take place both in the periphery of the nervous system and in the higher centers of the CNS.

Acoustic Communication

Members in more than 20 families of the Crustacea, mostly among the aquatic and terrestrial Malacostraca, possess sound-producing devices (Schmitz 2002). Crustaceans can rattle, snap, chirp, rasp, stridulate, and generate sound in other ways. However, little is known about their capability of sound detection. In fact, there is an ongoing debate among experts in the field regarding whether crustaceans can actually *hear* sound.

Behavioral and physiological experiments have shown that some crustaceans are receptive to vibration and others respond to the near-field displacement of fluid particles when the signal is emitted from close range. Semiterrestrial crabs use acoustic signals for communication (Popper et al. 2001). Males produce sounds during the night to attract females to their burrows and to deter competitors. The signals are detected by conspecifics with specialized receptor organs (Barth's myochordotonal organ) in the walking legs that are sensitive to mechanical vibrations. Sensory input is conducted via mechanosensory interneurons to a paired neuropil located in the dorso-medial area of the tritocerebrum where the information is processed (Hall 1985a, 1985b). Sensory activation is typically elicited by substrate-borne vibrations, although members of the family Ocypodidae (ghost crabs) are able to detect airborne sounds by use of the same receptor organ.

While detection of substratum-borne and airborne signals is documented for semiterrestrial crustaceans, no conclusive evidence has yet been presented that *aquatic* crustaceans have the ability to detect and utilize the pressure component of sound. The lack of resonating structures supports this skepticism (Breithaupt 2002). The identified receptors (i.e., the statocysts, sensory hairs on the body surface, and proprioceptive organs) that can be stimulated by acoustic signals respond to the hydrodynamic component and not to the pressure component generated by sound under water. This has been shown with physiological and behavioral experiments in which crayfish and lobster responded only to acoustic stimuli that were generated at very short distances (Goodall et al. 1990, Breithaupt and Tautz 1990). Neural mechanisms that underlie the detection of waterborne sounds are sparsely studied, and the pathways of neural transmission and the areas of higher order processing in the CNS have been explored only in preliminary fashion (Hall 1985b, Yoshino et al. 1983).

Why, then, did all these sound-producing devices evolve in aquatic crustaceans if they can't hear each other?

First, the main purpose of sound production may not lie predominantly in intraspecific communication (unless sender and receiver are within very close range) but may be more important during interspecific communication, for example, when used as a defense mechanism against predators that can detect the sound and are warned off by it. This has been shown in several crustacean species, including spiny lobsters (Palinuridae), which rub an extension of the antenna over the antennular plate to produce loud rasps. Interestingly, spiny lobsters generate the sound by means of frictional interactions between soft tissues, unlike other arthropods (e.g., crickets) in which similar sounds are produced by rubbing hard surfaces against each other (Patek 2001). This allows sound production throughout the molt cycle when the exoskeleton softens and spiny lobsters are under increased risk of predation.

Second, the emitted sound could be a mere "byproduct" without obvious biological relevance generated in connection with other signals that are important for intraspecific communication. This seems to be the case for members of the family Alpheidae (snapping shrimp) that produce loud acoustic signals when rapidly closing a modified snapper claw. The sound-generating mechanisms differ among species and can result from the claw closure itself or from the formation and collapse of a cavitation bubble during claw closure (Versluis et al. 2000). The rapid closure of the claw has two important consequences: the production of a short and loud snapping sound, and the production of a fast and powerful water jet. Snapping is common during intra- and intersexual encounters with conspecifics and is also frequently used as a predatory mechanism. The water jets can be powerful enough to injure and kill small prey but are used for nonviolent communication among members of the same species during territorial and courtship behaviors (Herberholz and Schmitz 1998). The acoustic signals, on the other hand, have no apparent effect on the behavior of the shrimps and evoke no response when presented as playbacks (Hazlett 1972).

The role of the snapping sounds as defense mechanisms against predators has not been examined, but during intraspecific communication they might simply be a side effect of the water jet production (Schmitz 2002). On the other hand, Tóth and Duffy (2005) have documented coordinated snapping among social shrimp (*Synalpheus*) and suggested that it might provide a signal warning intruders that the host sponge is occupied by a social colony; this hypothesis could involve auditory or mechanical signals but remains to be tested.

Clearly, we are only at the beginning of our exploration of acoustic communication in crustaceans. Additional field studies that avoid problems faced in laboratory experiments (e.g., signal distortion due to reflecting boundaries) are needed to measure differences in behavioral responses to acoustic signals that are transmitted through air, water, or substrate. Once the receivers' responses are characterized, we will be able to take advantage of the accessibility to the crustacean CNS for further neurophysiological studies under controlled conditions.

Tactile and Hydrodynamic Communication

Many crustaceans possess numerous mechanoreceptors on almost every part of their body that are sensitive to touch (tactile stimulus) or water movement (hydrodynamic stimulus). In addition, changes in the position of the body or the position of the appendages that are elicited by movement or muscular activity are constantly monitored with internal proprioceptors.

Hermit crabs (Paguridae) occupy shells of gastropods and fight aggressively over ownership. During these agonistic interactions, the attacker raps its own shell against that of the defender in a series of bouts that are separated by pauses during which the initiator tries to pull the opponent out of the shell (Briffa et al. 2003). The chances to displace the defender increase with more powerful bouts of rapping, and the probability of a successful eviction depends on the physical fitness of the attacker. Both temporal rate and magnitude of rapping are subject to fatigue, and only hermit crabs in good physical condition are successful in defeating and evicting the opponent (Briffa et al. 2003). While this is a unique and fascinating example of intraspecific agonistic communication, it has been explored only on the behavioral level. Thus, future work should be directed toward the identification of the receptors that are involved in the perception of the signals expressed during shell rapping.

When foraging for phytoplankton, Antarctic krill (Euphausidae) form large and densely packed schools. Swimming in well-arranged schools presumably reduces energy costs during long periods of foraging and can also confuse or deter possible predators (Hamner and Hamner 2000). Within schools, all individuals swim at considerable speed and in close proximity to each other (Wiese and Marschall 1990). Locomotion by each member of the school is characterized by continuous beating of the pleopods, which produce a propulsion jet flow and provide conspecifics that swim behind and below in close range with a hydrodynamic communication signal. Electrophysiological recordings from the antennular nerve have revealed a class of sensory cells that respond to artificial hydrodynamic stimuli similar to those produced by the swimming krill itself. The sensory cells are part of proprioceptors that are located in the hinge of the antennular flagellum and measure the flexion and extension of it (Wiese 1996).

When snapping shrimp (Alpheidae) come in contact with a predator, a potential prey object, or a conspecific, they generate fast and well-focused water jets by rapid closure of a specialized claw. The shrimp rarely injure conspecifics with the water jets, but small prey can be stunned, injured, and killed. During intraspecific encounters, opponents of both sexes exchange snaps from a larger distance that is kept constant throughout the contest, and they usually aim at the armored claw of the opponent (Schmitz and Herberholz 1998). The receiver analyzes the hydrodynamic signal that accompanies the water jets with help of mechanosensory hairs on the claws. Depletion of these hairs disrupts the ritualized exchange of snaps between conspecifics (Herberholz and Schmitz 1998). The signal probably allows the sender to communicate its strength and reduces the probability that the interaction escalates to more violent and damaging stages. The snapping behavior and signal itself also contain information about the sex of the producer because females generate less powerful jets than do males of equal size and snapper claw length (Herberholz and Schmitz 1999).

Agonistic interactions in clawed lobster and crayfish escalate to stages where individuals of equal size exchange antennal taps and grasp each other with the claws to push, drag, or turn the opponent over. Higher intensity levels of unrestrained fighting are potentially damaging to the combatants and are typically reached only when prior signaling during the early and more ritualized stages failed to halt the interaction and none of the opponents was willing to retreat (see chapter 5). We must assume that during these stages of physical contact many extero- and proprioceptors are activated and that the information is integrated into a behavioral choice that determines the progression of the interaction, if not its outcome. During the later stages of fighting in pairs of socially naive crayfish, both opponents produce offensive tailflips that differ clearly in behavior and neural activity from those produced for escape (Herberholz et al. 2001). Offensive tailflips are generated by rapid tail flexions while the animals are interlocked and thrust the tailflipping crayfish up into the water column above the opponent, which is dragged along. Offensive tailflips do not cause injury to an opponent (unlike aggressive tailflips performed during lobster fights) but rather signal the strength of the producer. Crayfish that eventually win the encounter always generate more offensive tailflips than their rivals and are the last ones to do so before the loser terminates the fight by escaping (Herberholz et al. 2001). How and where crayfish perceive and analyze the strong tactile and hydrodynamic signals that accompany offensive tailflips remain to be explored. Equally important, we need to determine why an offensive tailflip is perceived (and responded to) differently than a tailflip performed by a conspecific that escapes a predator. Only such a comparative approach will allow us to isolate the signals' identities that are characteristic for each type of tailflip. This will then enable us to test (combinations of) different stimuli independently and measure the corresponding mechanisms of neural processing.

Visual Communication

Long-distance visual communication is complicated under water because light has a lower intensity and contains a narrower spectrum of wavelengths, and contrast is reduced. Moreover, many aquatic crustaceans inhabit environments that are murky or turbid. Thus, with some conspicuous exceptions, communication via visual signals is more commonly represented among terrestrial and semiterrestrial crustacean species than in those that live in aquatic environments.

Not surprisingly, visual signaling has been carefully investigated in fiddler crabs (Ocypodidae) that inhabit intertidal mud flats and sand flats. The males use one enlarged claw to attract females to their burrow with a waving display but also to fight potential male intruders. Unfortunately, while much is known about the behavioral mechanisms during territorial competition, predator avoidance, and courtship (see chapter 10), we have little knowledge of the neural processes that underlie the responses to visual signals emitted by intruders or mates. Nor can we explain the astonishing neural capacity these crabs demonstrate during intraspecific behaviors: foraging crabs can estimate the distance between a competitor and their own burrow even without seeing the burrow entrance (Hemmi and Zeil 2003). When a potential intruder moves within a certain range, the crabs run back and defend their home. It was proposed that the animals achieve this impressive performance by combining information about their own position with respect to their burrow (path integration) and retinal information that determines the position of the intruder (Hemmi and Zeil 2003). How the CNS manages to combine positional cues and visual input that leads to the well-directed motor response has yet to be determined.

An elevated body posture together with raised and spread chelipeds are visual signals used by lobsters, crayfish, and crabs to demonstrate aggressiveness, while a lowered, extended body posture and lowered claws represent visual signals correlated with submissiveness (Bruski and Dunham 1987). Meral spread displays have been described in several aquatic crustaceans as powerful visual signals used during intra- and interspecific encounters. This threat display, which is produced by exposing armored and often colorful appendages to predators or conspecifics, is known to deter opponents from engaging in more violent agonistic interactions. Mantis shrimp (Stomatopoda) are equipped with a set of powerful raptorial appendages that are spread during intraspecific contests for underwater burrows or cavities. The display is common during the earlier stages of agonistic competition and is also used by freshly molted, soft animals that are unable to strike but instead "bluff" their opponents with the visual display (Steger and Caldwell 1983). In display mode, mantis shrimp reveal many colorful and often species-specific markings. The eyes of some stomatopod crustaceans contain polychromatic retinas with at least 14 photoreceptor classes (most other crustaceans only possess one or two), some of which would allow the perception of colored markings displayed by an intraspecific opponent (Cronin and Hariyama 2002). Mantis shrimp can also be trained to learn the orientation of polarized light, and the capacity for polarized vision may be important for intraspecific recognition and communication since certain areas of the body show species-specific patterns of polarization (Marshall et al. 1999). Most recently, Mazel et al. (2004) reported that some species of mantis shrimp also display strongly fluorescent markings and that several classes of their photoreceptors are tuned to recognize fluorescence. This is especially useful for underwater signaling at greater depth (where some members of these species live), because the intensity and visibility of colored signals are enhanced by the fluorescence. However, spectral sensitivity of photoreceptors has mostly been measured by means of spectrophotometry in isolated retinas, and conclusive evidence that color is an important signal during communication can be obtained only with additional electrophysiological and behavioral experiments.

The display of visual signals for communication purposes in aquatic crustaceans other than stomatopods has also been investigated. It was hypothesized that visual display of large colorful claws plays a role during hierarchy formation in the freshwater prawn *Macrobrachium* (Barki et al. 1992). The adult male prawns develop through three morphotypes that differ in size and color. Blue-clawed males have longer and heavier claws than the orange-clawed males and the small "colorless" males. The blue-clawed morphotype dominates both other types through aggressive interactions. Claw size rather than body size determines dominance status, and claw coloration might serve as a visual signal that allows prawns to assess fighting ability and form stable dominance hierarchies (Barki et al. 1992). Occasionally, visual signals occur in combination with other nonvisual signals, and co-occurrence is required to evoke behavioral responses from the receiver (see "Communication via Multiple Concurrent Signals," below).

Primarily confined to decapod crustaceans, the integration of visual signals in the higher centers of the nervous system has been mainly investigated by means of electrophysiological experiments. These experiments were largely restricted to the periphery of visual processing, that is, the optic lobes and especially the optic nerve (protocerebral tract), which are most accessible for these studies. Different classes of interneurons were discovered in each species, distinguished by their receptive fields and named after the respective "trigger" stimulus for their activation, for example, sustaining, dimming, and motion (Wiersma et al. 1982). These visual interneurons project to the more central brain areas, where they make connections with inputs from other sensory systems (see below) or with neurons that control motor output such as the oculomotor reflex or the visually evoked defense reflex. A set of "jittery movement detector" neurons found in the optic nerve of crayfish excites command interneurons in the connectives between the brain and the next posterior ganglion. Stimulation of these command interneurons elicits the defense reflex, a behavior that comprises raising the thorax and extending the open claws (Glantz 1977, Wiersma et al. 1982). The same neural mechanisms may underlie the aforementioned meral spread displays but have not yet been studied in that context. As mentioned above, crustaceans commonly use the same or similar behavioral patterns in several (and often notably different) social situations. Since these patterns might be controlled by shared neural circuitry, they present a promising means for studying and comparing them in different contexts, including communication.

Higher order processing of visual inputs has been studied only in a preliminary fashion. While current knowledge is insufficient to make assumptions about the integration of visual communication signals, recent experiments provide important first steps for identification of brain structures and brain mechanisms that are involved in visual processing. For example, implantations of chronic electrodes in the eyestalk and in the brain of crayfish were used to simultaneously record neural activity in both areas and to measure latencies of spike trains that travel from the periphery to the centers of the brain (Serrato et al. 1996). Stimulation of the eye photoreceptors with light pulses revealed that visual information is integrated in both regions, first within the optic ganglia that lie in the eyestalks and again in the more central areas of the brain. Recently, Tomsic et al. (2003) identified a class of neurons in the optic ganglia of the crab *Chasmagnathus granulatus* that show short- and long-term visual memory of a moving object. The cell bodies of these motion detector neurons are located in a cell cluster beneath the medulla interna and arborize extensively in the medulla interna and the lateral protocerebrum. Intracellular recordings made in intact animals have shown that these neurons learn and memorize a visual stimulus that evokes escape behavior in freely behaving crabs. Crabs that are exposed to a visual danger stimulus (illustrating a possible predator) exhibit short- or long-term memory of the threat display, depending on the presentation regime. Rapid repetitions of the stimuli evoke a decrease in the escape responses without retention, while timely spaced repetitions evoke decreases during training that are memorized for at least 24 hours. Neural responses in the identified motion detector neurons in the optic ganglia of the crab resemble the observed behavioral changes. They show a decrease in activity during mass training and recover to initial response levels when stimulated again on the next day but retain low response activity for at least one day after spaced training. Thus, visual learning and memory are accomplished in a class of neurons that are located in the eyestalks (Tomsic et al. 2003). Whether the capacity of these neurons is due to intrinsic features alone or is the result of an integrated response from higher order neurons that make contact with these cells remains unknown at this point.

Chemical Communication

An aquatic environment favors the exchange of chemical signals over most others, and that is why chemical communication is more developed among crustaceans than any other form of communication (see chapter 6). I restrict my contribution here to a few examples that describe current advances in understanding the mechanisms of neural processing in the higher centers of the nervous system. The integration of olfactory information in the crustacean brain is clearly among the better understood.

We have some comprehension of the organization of neural pathways that transmit chemical information from the receptors to the central areas of the brain. Distinct neuroanatomical areas in the crustacean brain that receive and process olfactory inputs have been identified, and the projections between these neuropils and other regions of higher orders in the brain have recently been described in some detail.

Crustaceans receive chemosensory input from sensory cells that are located on almost all body surfaces (Laverack 1988, Derby 1989, Cate and Derby 2002). The receptors are most abundant on the appendages, particularly on the antennules, which are considered to serve as the primary olfactory organs (Hallberg et al. 1992). The antennules are composed of two flagella, each containing chemo- and mechanosensory hairs. Chemical signals are perceived by chemoreceptive structures on these flagella and transmitted to the brain via two distinct neural pathways, one originating from the aesthetasc receptor neurons located only on the lateral flagellum and projecting to the olfactory lobes, and the other coming from non-aesthetasc receptors projecting to the lateral and median antennular neuropils (Schmidt et al. 1992, Schmidt and Ache 1992). These two pathways remain separated on the next level of neural integration. Ascending projections from the olfactory lobe and the lateral antennular neuropil terminate in different areas of the medulla terminalis, which is part of the lateral protocerebrum and located in the eyestalks (Sullivan and Beltz 2001a). The medulla terminalis is presumably involved in the recognition and integration of visual and chemical stimuli related to food items. Despite our detailed anatomical knowledge about these two neural pathways, we lack concrete experimental evidence for their involvement in particular olfaction-guided behaviors. Behavioral studies revealed an overlap in the function of both the aesthetasc and non-aesthetasc pathways for odor-activated search behavior, odor learning, and discrimination (Steullet et al. 2002). Recently, Horner et al. (2004) tested the ability of spiny lobsters to locate a distant odor source in a seawater flume. After ablating specific receptor hairs that transmit olfactory information through one or the other neural pathway, they showed that lobsters can locate the food source equally well with only the aesthetasc or the non-aesthetasc pathway left intact. Thus, the importance of having two neuroanatomically distinct sensory pathways in the lobster brains for processing olfactory information may not be linked to feeding behaviors but possibly to other behaviors such as communication (see chapter 6). It seems conceivable that decapod crustaceans process chemical information that is exchanged during intraspecific interactions (e.g., fighting, courtship, aggregation) by means of only one (specialized) pathway that is different from the one used for food-related behaviors. Further analysis of the functional role played by the two olfactory pathways in mediating communication signals is needed.

Our knowledge of neural processes involved in chemical communication among crustaceans has greatly improved in recent years. However, it still compares poorly to other arthropods, especially insects, in which the molecular properties of hundreds of chemical components (pheromones) used for communication have been characterized. Consequently, higher order processing of these components and how they generate response behaviors are well understood in insects (Hansson 1995). While the identification of gaseous signals is straightforward by use of common techniques, waterborne signals are more difficult to isolate (Zimmer and Butman 2000). To close the gap, we will need to characterize crustacean pheromones so that their effects on the crustacean CNS and their role in chemical communication can be tested experimentally (see chapter 6).

Communication via Multiple Concurrent Signals

Communication with combined signals of different sensory qualities is common among crustaceans. In some cases, the communication signal is simply supported, enhanced, or transported with the help of another potentially important signal. This has been reported in several crustacean species (e.g., lobster, crayfish, and shrimp) where the distribution of chemical signals is facilitated and directed with co-occurring water currents that are generated by movements of appendages. Agonistic intraspecific interactions most often involve chemical communication, and the chemical signal itself is presumably contained in the urine that both contestants exchange during fights (Breithaupt and Eger 2002). The urine-borne signal is released through a pair of nephropores that are situated on the ventral side of the head and is then actively forwarded and directed toward the opponent by water currents that are generated with the frontal appendages or the scaphognathites. The chemical signal can also be directed posteriorly by the beating pleopods (see chapter 6). The most important function of these water currents is probably the transport of chemical signals from sender to receiver, but they may also act as hydrodynamic signals over close distances. During intraspecific fights in snapping shrimp, water jets are often combined with strong anterior-directed gill currents. Both the number of released water jets and gill currents are positively correlated with agonistic success; that is, shrimp that will eventually win the contest always produce more of both compared with the animals that end up defeated. The function of the gill currents is probably the transport of chemicals signals, and it may allow the receiver to associate the physical strength of the opponent (by measuring water jet intensity) with its chemical identity, thus preventing subsequent encounters between familiar opponents from escalating (Herberholz and Schmitz 2001, Obermeier and Schmitz 2003a). The hydrodynamic component of the gill currents has been shown to be of minor relevance for the receiver, and it is predominantly used to transport urine-borne signals from one shrimp to the other (Obermeier and Schmitz 2003b).

Sometimes, the co-occurrence of two different signals is actually required to elicit a behavioral response in the receiver or it changes the response significantly. This is the case during mate search in the crayfish *Austropotamobius pallipes* (Acquistapace et al. 2002). Males do not respond to female odors alone, but when the odor is paired with a visual signal (the sight of a female or another male crayfish), they spend less time hiding under a shelter and instead increase their locomotion behavior. They also adopt a more elevated posture during concurrent visual and chemical stimulation compared with a control situation (Acquistapace et al. 2002). Snapping shrimp do not respond to olfactory cues from the opposite sex alone but change their behavior once the odor signals are combined with visual stimuli (Hughes 1996). Male snapping shrimp respond significantly less often with a threat display when an (isolated) open snapper claw is presented in combination with female odors than when presented in combination with male odors. Conversely, females respond more aggressively when the visual stimulus occurs in combination with female odors than with male odors, although the differences in the responses from females are less distinct. We have yet to understand how these multiple sensory inputs are combined, organized, and processed in the brains of crustaceans.

An intriguing example of sex-specific communication via signals of different modality has recently been described in rock shrimp (Diaz and Thiel 2004). Here, each sex of the same species utilizes only one communication signal and disregards the other one during mate search. Males are incapable of locating females by use of olfactory cues but are guided visually toward receptive females. The visual signals that provide males with the sensory cues for mate localization are delivered by aggregations of other (subordinate) males around the females. Females, on the other hand, locate males via olfactory cues as shown by measuring the responses to male odor in a Y-maze apparatus. Although dominant reproductive males are characterized by a distinct morphology, females neglect these visual signals and rely on olfactory senses for recognition. Males instead choose visual cues to find a mating partner, although it is unclear whether female rock shrimp actually release sex-specific pheromones that could be detected by the males (Diaz and Thiel 2004). Obviously, both sexes pursue different search strategies in the same environment (where none of the communication signals are favored), and it will be important to understand why these sex-specific strategies for mate localization have evolved.

As mentioned above, higher order processing of multimodal sensory inputs takes place outside the central brain areas in the lateral protocerebrum of some decapod crustaceans. Besides receiving visual inputs and other inputs such as those from the lateral antennular neuropil, the medulla terminalis is also targeted by projection neurons that originate in the deutocerebrum and innervate the olfactory lobes, which receive chemosensory inputs from primary afferents. The adjacent hemiellipsoid bodies, however, receive visual information and inputs from projection neurons that innervate the accessory lobes. Both neural pathways project to discrete regions of the lateral protocerebrum through separate tracts, thus maintaining the original specificity of the two lobes (Sullivan and Beltz 2001b). Within the hemiellipsoid bodies, multimodal sensory interneurons (parsol cells) have been identified. These neurons receive secondary olfactory, tactile, and visual sensory inputs as well as primary visual inputs from the ipsilateral eye. The neurons are characterized by an underlying spontaneous oscillatory activity that is differentially modulated by visual, chemical, or mechanosensory stimulations (Mellon 2000). The functional significance of both the medulla terminalis and the hemiellipsoid body is still unsettled, but both structures seem to play a role in the initiation of feeding behaviors. The importance of the parasol cells may lie in the contextual identification of food items. Whether these structures are also involved in the processing of communication signals needs to be assessed.

The accessory lobes are centers of higher order processing common to eureptantian decapods. They probably evolved with the adaptation to benthic life and are large in crustaceans such as lobsters and crayfish that show strong territorial behavior. Due to secondary reduction, the less territorial anomuran and brachyuran crabs have much smaller accessory lobes (Sandeman et al. 1993). Since the accessory lobes receive secondary multimodal inputs from visual, mechanosensory, and olfactory interneurons, they may play an important role during territorial communication.

Comparison to Other Organisms

For more than a century, crustaceans have been used as model systems in neuroethological research. The reduced complexity of their CNS compared with most vertebrate animals has made crustaceans attractive research objects for studies of sensory–motor integration, neuromuscular control, neural circuitry of fixed action patterns, neurotransmission, neuromodulation, rhythmic pattern generation, and many others, some of which were pioneered in crustacean preparations (Marder 2002, Wiese 2002, Edwards and Herberholz 2005). Important advantages over vertebrate systems include the accessibility of crustaceans in the wild; their small size, sturdy constitution, and sustainability of invasive experimental procedures; parts of the CNS can easily be isolated, intracellular recordings in reduced preparations can last hours to days; and electrodes can be implanted into freely behaving animals. The effects of neuromodulators, some of which are shared with other organisms, including humans, can easily be tested in crustacean preparations (due in part to the lack of a blood–brain barrier), and the physiological response in single neurons or well-described neural circuits can be measured (Yeh et al. 1996, Nusbaum et al. 2001). Most important, complete patterns of behavior can be released by natural or electrical stimulation of identified and recognizable command neurons (Bowerman and Larimer 1976, Edwards et al. 1999).

However, to understand the neural basis of communication (and other social behaviors), we need to know how the neural circuits are orchestrated in the higher centers of the CNS to produce the observed patterns of behavior. Here, our knowledge compares poorly with that of other systems, including other arthropods such as insects. This may be in part because the signals used during crustacean communication are not as well isolated as in other animals. For example, while communication via chemical signals is by far the most prominent communication channel used by crustaceans (see chapter 6), the molecular structure of most chemical signals that crustaceans exchange is still unknown. This is in strong contrast to the insects where communication pheromones have been identified and subsequently assayed in behavioral and physiological experiments (Cardé and Minks 1997). Thus, the nature of crustacean communication signals need to be better isolated and identified; only then can they be tested and corresponding peripheral as well as central neural processing can be investigated.

At the same time, there is evidence that behavioral patterns displayed by crustaceans in more than one social situation are controlled by the same or shared neural circuitry (Herberholz et al. 2003). Thus, a limited number of neurons need to be identified that might explain a variety of social behaviors, including the ones that are expressed during communication.

Conclusions and Future Directions

At this time, we have only an incomplete understanding of how crustaceans process and integrate communication signals that are exchanged during intra- and interspecific interactions and orchestrate the arising patterns of neural activity into adaptive behaviors. Moreover, many of the neural processes are under hormonal control and modified by neuromodulators, adding another level of complexity (see chapter 5).

The basis of communication has been explored using different experimental techniques and asking various questions directed toward the signal identity, the structure and function of the receptive organs, and the mechanisms that generate the resulting behavior. However, the *neural* basis of communication, especially processing of sensory information by higher order neuropils, must be largely considered unknown territory. The complexity that goes along with even the simplest form of communication permits only small experimental advances at a time, and it would be naive to imagine that the observed variety in behaviors displayed during communication, many of which are also species specific, can be explained by the activity of some easily identifiable command neurons. At the same time, crustaceans provide a unique opportunity to investigate neural processes because of the limited number of neurons involved and the accessibility of these neurons to a wide range of experimental techniques.

The interest in the higher centers of neural processing has steadily grown over the last few years. Recent attempts have been successful in supplying us with much needed neuroanatomical and physiological information on neural pathways, projection neurons, and single identifiable neurons involved in sensory signal processing. In many cases, we still need to understand how these higher order neural processes are linked to communication and how they are translated into motor responses that accomplish the formation of adaptive behaviors expressed during social encounters. The application of newly developed techniques in combination with traditional experimental methods ideally will accelerate the search and provide us with insights into the neural basis of communication in crustaceans that will then generalize to many other organisms.

Summary

Crustaceans are used as model systems for studying behavioral and physiological processes common to many animals. Crustaceans are especially attractive to neuroethologists since most of their behavioral repertoire is controlled by a nervous system of relatively low complexity that is readily accessible for a variety of experimental techniques. Many basic neural mechanisms were first discovered in crustacean preparations and have then been generalized to many other organisms.

In several taxa of social crustaceans, communication signals of different modalities are exchanged between conspecifics. Individual fitness is displayed through ritualized behavioral patterns during agonistic encounters that determine social dominance, and once the social relationship is established, dominance and submission are communicated through status-specific signals. Various signals are exchanged during courtship, and the order and quality of these signals often determine mating success. Crustaceans also interact with individuals outside their own species, and here communication can be a matter of life and death.

Crustaceans are well suited to investigate the neural basis of communication. They utilize most sensory modalities and incoming signals are received, relayed, and sometimes integrated by the peripheral nervous system. The underlying mechanisms have been intensively studied and are reasonably well understood. Importantly, the experimental transition from the periphery into central brain areas of higher order processing has now begun. This will significantly improve our understanding of how signals are integrated into adaptive behavioral responses, thus illustrating how nervous systems shape communication.

Acknowledgments I thank Donald H. Edwards and Jeremy M. Sullivan for their critical comments on an earlier version of the manuscript.

References

Acquistapace, P., L. Aquiloni, B.A. Hazlett, and F. Gherardi. 2002. Multimodal communication in crayfish: sex recognition during mate search by male *Austropotamobius pallipes*. Canadian Journal of Zoology 80:2041–2045.

Barki, A., I. Karplus, and M. Goren. 1992. Effects of size and morphotype on dominance hierarchies and resource competition in the freshwater prawn *Macrobrachium rosenbergii*. Animal Behavior 44:547–555.

Bowerman, R.F., and J.L. Larimer. 1976. Command neurons in crustaceans. Comparative Biochemistry and Physiology A 54:1–5.

Breithaupt, T. 2002. Sound perception in aquatic crustaceans. Pages 548–558 *in*: K. Wiese, editor. The crustacean nervous system. Springer-Verlag, Berlin.

Breithaupt, T., and P. Eger. 2002. Urine makes the difference: chemical communication in fighting crayfish made visible. Journal of Experimental Biology 205:1221–1231.

Breithaupt, T., and J. Tautz. 1990. The sensitivity of crayfish mechanoreceptors to hydrodynamic and acoustic stimuli. Pages 114–120 *in*: K. Wiese, W.D. Krenz, J. Tautz, H. Reichert, and B. Mulloney, editors. Frontiers in crustacean neurobiology. Birkhäuser, Basel, Switzerland.

Briffa, M., R.W. Elwood, and J.M. Russ. 2003. Analysis of multiple aspects of a repeated signal: power and rate of rapping during shell fights in hermit crabs. Behavioral Ecology 14:74–79.

Bruski, C.A., and D.W. Dunham. 1987. The importance of vision in agonistic communication of the crayfish *Orconectes rusticus*. I: An analysis of bout dynamics. Behaviour 103:83–107.

Buchsbaum, R., M. Buchsbaum, J. Pearse, and V. Pearse. 1987. Animals without backbones, 3rd edition. University of Chicago Press, Chicago, Ill.

Cardé, R.T., and A.K. Minks. 1997. Insect pheromone research: new directions. Chapman and Hall, New York.

Cate, H.S., and C.D. Derby. 2002. Hooded sensilla homologues: structural variations of a widely distributed bimodal chemomechanosensillum. Journal of Comparative Neurology 444:345–357.

Cronin, T.W., and T. Hariyama. 2002. Spectral sensitivity in crustacean eyes. Pages 499–511 *in*: K. Wiese, editor. The crustacean nervous system. Springer-Verlag, Berlin.

Derby, C.D. 1989. Physiology of sensory neurons in morphologically identified cuticular sensilla of crustaceans. Pages 27–47 *in*: B.E. Felgenhauer, L. Watling, and A.B. Thistle, editors. Crustacean Issues 6: Functional morphology of feeding and grooming in Crustacea. A.A. Balkema, Rotterdam, The Netherlands.

Diaz, E.R., and M. Thiel. 2004. Chemical and visual communication during mate searching in rock shrimp. Biological Bulletin 206:134–143.

Edwards, D.H., and J. Herberholz. 2005. Crustacean models of aggression. Pages 38–61 *in*: R.J. Nelson, editor. The biology of aggression. Oxford University Press, New York.

Edwards, D.H., W.J. Heitler, and F.B. Krasne. 1999. Fifty years of a command neuron: the neurobiology of escape in crayfish. Trends in Neurosciences 22:153–161.

Glantz, R.M. 1977. Visual input and motor output of command interneurons of the defense reflex pathway in the crayfish. Pages 259–274 *in*: G. Hoyle, editor. Identified neurons and behavior of arthropods. Plenum Press, New York.

Goodall, C., C. Chapman, and D. Neil. 1990. The acoustic response threshold of the Norway lobster, *Nephrops norvegicus* (L.) in a free sound field. Pages 106–113 *in*: K. Wiese, W.D. Krenz, J. Tautz, H. Reichert, and B. Mulloney, editors. Frontiers in crustacean neurobiology. Birkhäuser, Basel, Switzerland.

Greenfield, M.D. 2002. Signalers and receivers: mechanisms and evolution of arthropod communication. Oxford University Press, New York.

Hall, J. 1985a. Neuroanatomical and neurophysiological aspects of vibrational processing in the central nervous system of semi-terrestrial crabs. I. Vibration-sensitive interneurons in the fiddler crab, *Uca minax*. Journal of Comparative Physiology A 157:91–104.

Hall, J. 1985b. Neuroanatomical and neurophysiological aspects of vibrational processing in the central nervous system of semi-terrestrial crabs. II. Comparative anatomical and physiological aspects of stimulus processing. Journal of Comparative Physiology A 157:105–113.

Hallberg, E., K.U.I. Johansson, and R. Elofsson. 1992. The aesthetasc concept: structural variation of putative olfactory receptor cell complexes in Crustacea. Microscopy Research and Technique 22:325–335.

Hamner, W.M., and P.P. Hamner. 2000. Behavior of Antarctic krill (*Euphausia superba*): schooling, foraging, and antipredatory behavior. Canadian Journal of Fisheries and Aquatic Sciences 57:192–202.

Hansson, B.S. 1995. Olfaction in Lepidoptera. Experientia 51:1003–1027.

Hazlett, B.A. 1972. Ritualization in marine crustacea. Pages 95–125 *in*: H.E. Winn and B.L. Olla, editors. Behavior of marine animals. Plenum Press, New York.

Hemmi, J.M., and J. Zeil. 2003. Robust judgement of inter-object distance by an arthropod. Nature 421:160–163.

Herberholz, J., and B. Schmitz. 1998. Role of mechanosensory stimuli in intraspecific agonistic encounters of the snapping shrimp (*Alpheus heterochaelis*). Biological Bulletin 195:156–167.

Herberholz, J., and B. Schmitz. 1999. Flow visualisation and high speed video analysis of water jets in the snapping shrimp (*Alpheus heterochaelis*). Journal of Comparative Physiology A 185:41–49.

Herberholz, J., and B. Schmitz. 2001. Signaling via water currents in behavioral interactions of snapping shrimp (*Alpheus heterochaelis*). Biological Bulletin 201:6–16.

Herberholz, J., F.A. Issa, and D.H. Edwards. 2001. Patterns of neural circuit activation and behavior during dominance hierarchy formation in freely behaving crayfish. Journal of Neuroscience 21:2759–2767.

Herberholz, J., M.M. Sen, and D.H. Edwards. 2003. Parallel changes in agonistic and non-agonistic behaviors during dominance hierarchy formation in crayfish. Journal of Comparative Physiology A 189:321–325.

Herberholz, J., M.M. Sen, and D.H. Edwards. 2004. Escape behavior and escape circuit activation in juvenile crayfish during prey-predator interactions. Journal of Experimental Biology 207:1855–1863.

Horner, A.J., M.J. Weissburg, and C.D. Derby. 2004. Dual antennular chemosensory pathways can mediate orientation by Caribbean spiny lobsters in naturalistic flow conditions. Journal of Experimental Biology 207:3785–3796.

Hughes, M. 1996. The function of concurrent signals: visual and chemical communication in snapping shrimp. Animal Behavior 52:247–257.

Laverack, M.S. 1988. The diversity of chemoreceptors. Pages 287–312 *in*: J. Atema, R.R. Fay, A.N. Popper, and W.N. Tavolga, editors. Sensory biology of aquatic animals. Springer-Verlag, New York.

Marder, E. 2002. Non-mammalian models for studying neural development and function. Nature 417:318–321.

Marshall, J., T.W. Cronin, N. Shashar, and M. Land. 1999. Behavioural evidence for polarisation vision in stomatopods reveals a potential channel for communication. Current Biology 9:755–758.

Mazel, C.H., T.W. Cronin, R.L. Caldwell, and N.J. Marshall. 2004. Fluorescent enhancement of signaling in a mantis shrimp. Science 303:51.

Mellon, D., Jr. 2000. Convergence of multimodal sensory input onto higher-level neurons of the crayfish olfactory pathway. Journal of Neurophysiology 84:3043–3055.

Nusbaum, M.P., D.M. Blitz, A.M. Swensen, D. Wood, and E. Marder. 2001. The roles of co-transmission in neural network modulation. Trends in Neurosciences 24:146–154.

Obermeier, M., and B. Schmitz. 2003a. Recognition of dominance in the big-clawed snapping shrimp (*Alpheus heterochaelis* Say 1818) part I: individual or group recognition? Marine and Freshwater Behaviour and Physiology 36:1–16.

Obermeier, M., and B. Schmitz. 2003b. Recognition of dominance in the big-clawed snapping shrimp (*Alpheus heterochaelis* Say 1818) part II: analysis of signal modality. Marine and Freshwater Behaviour and Physiology 36:17–29.

Patek, S.N. 2001. Spiny lobsters stick and slip to make sound. Nature 411:153–154.

Popper, A.N., M. Salmon, and K.W. Horch. 2001. Acoustic detection and communication by decapod crustaceans. Journal of Comparative Physiology A 187:83–89.

Sandeman, D.C. 1982. Organization of the central nervous system. Pages 1–61 *in*: H.L Atwood and D.C. Sandeman, editors. The biology of crustacea, Volume 3. Neurobiology: structure and function. Academic Press, New York.

Sandeman, D.C., R.E. Sandeman, and A.R. Aitken. 1988. Atlas of serotonin-containing neurons in the optic lobes and brain of the crayfish, *Cherax destructor*. Journal of Comparative Neurology 269:465–478.

Sandeman, D.C., R.E. Sandeman, C. Derby, and M. Schmidt. 1992. Morphology of the brain of crayfish, crabs, and spiny lobsters—a common nomenclature for homologous structures. Biological Bulletin 183:304–326.

Sandeman, D.C., G. Scholtz, and R.E. Sandeman. 1993. Brain evolution in decapod crustacean. Journal of Experimental Zoology 265:112–133.

Schmidt, M., and B.W. Ache. 1992. Antennular projections to the midbrain of the spiny lobster. II. Sensory innervation of the olfactory lobe. Journal of Comparative Neurology 318:291–303.

Schmidt, M., L. Vanekeris, and B.W. Ache. 1992. Antennular projections to the midbrain of the spiny lobster. I. Sensory innervation of the lateral and medial antennular neuropils. Journal of Comparative Neurology 318:277–290.

Schmitz, B. 2002. Sound production in Crustacea with special reference to the Alpheidae. Pages 536–547 *in*: K. Wiese, editor. The crustacean nervous system. Springer-Verlag, Berlin.

Schmitz, B., and J. Herberholz. 1998. Snapping behaviour in intraspecific agonistic encounters in the snapping shrimp (*Alpheus heterochaelis*). Journal of Biosciences 23:623–632.

Serrato, J., O.H. Hernandez, and F. Ramon. 1996. Integration of visual signals in the crayfish brain: multiunit recordings in eyestalk and brain. Comparative Biochemistry and Physiology A 114:211–217.

Steger, R., and R.L. Caldwell. 1983. Intraspecific deception by bluffing: a defense strategy of newly molted stomatopods (Arthropoda: Crustacea). Science 221:558–560.

Steullet, P., D.R. Krutzfeldt, G. Hamidani, T. Flavus, V. Ngo, and C.D. Derby. 2002. Dual antennular chemosensory pathways mediate odor-associative learning and odor discrimination in the Caribbean spiny lobster *Panulirus argus*. Journal of Experimental Biology 205:851–867.

Sullivan, J.M., and B.S. Beltz. 2001a. Development and connectivity of olfactory pathways in the brain of the lobster *Homarus americanus*. Journal of Comparative Neurology 441:23–43.

Sullivan, J.M., and B.S. Beltz. 2001b. Neural pathways connecting the deutocerebrum and lateral protocerebrum in the brains of decapod crustaceans. Journal of Comparative Neurology 441:9–22.

Tomsic, D., M. Berón de Astrada, and J. Sztarker. 2003. Identification of individual neurons reflecting short- and long-term visual memory in an arthropod. Journal of Neuroscience 23:8539–8546.

Tóth, E., and J.E. Duffy. 2005. Coordinated group response to nest intruders in social shrimp. Biology Letters 1:49–52.

Versluis, M., B. Schmitz, A. von der Heydt, and D. Lohse. 2000. How snapping shrimp snap: through cavitating bubbles. Science 289:2114–2117.

Wiersma, C.A.G., J.L. Roach, and R.M. Glantz. 1982. Neural integration in the optic system. Pages 2–31 *in*: D.C. Sandeman and H.L. Atwood, editors. The biology of crustacea, Volume 4. Neural integration and behavior. Academic Press, New York.

Wiese, K. 1996. Sensory capacities of euphausiids in the context of schooling. Marine and Freshwater Behaviour and Physiology 28:183–194.

Wiese, K. 2002. Crustacean experimental systems in neurobiology. Springer-Verlag. Berlin.

Wiese, K., and H.P. Marschall. 1990. Sensitivity to vibration and turbulence of water in context with schooling in Antarctic krill (*Euphausia superba*). Pages 121–130 *in*: K. Wiese, W.-D. Krenz, J. Tautz, H. Reichert, and B. Mulloney, editors. Frontiers in crustacean neurobiology. Birkhäuser, Basel, Switzerland.

Yeh, S.-R., R.A. Fricke, and D.H. Edwards. 1996. The effect of social experience on serotonergic modulation of the escape circuit of crayfish. Science 271:366–369.

Yoshino, M., Y. Kondoh, and M. Hisada. 1983. Projection of statocyst sensory neurons associated with crescent hairs in the crayfish *Procambarus clarkii* Girard. Cell and Tissue Research 230:37–48.

Zimmer, R.K., and C.A. Butman. 2000. Chemical signaling processes in the marine environment. Biological Bulletin 198:168–187.

Agonistic Behavior in Freshwater Crayfish
The Influence of Intrinsic and Extrinsic Factors on Aggressive Encounters and Dominance

Paul A. Moore

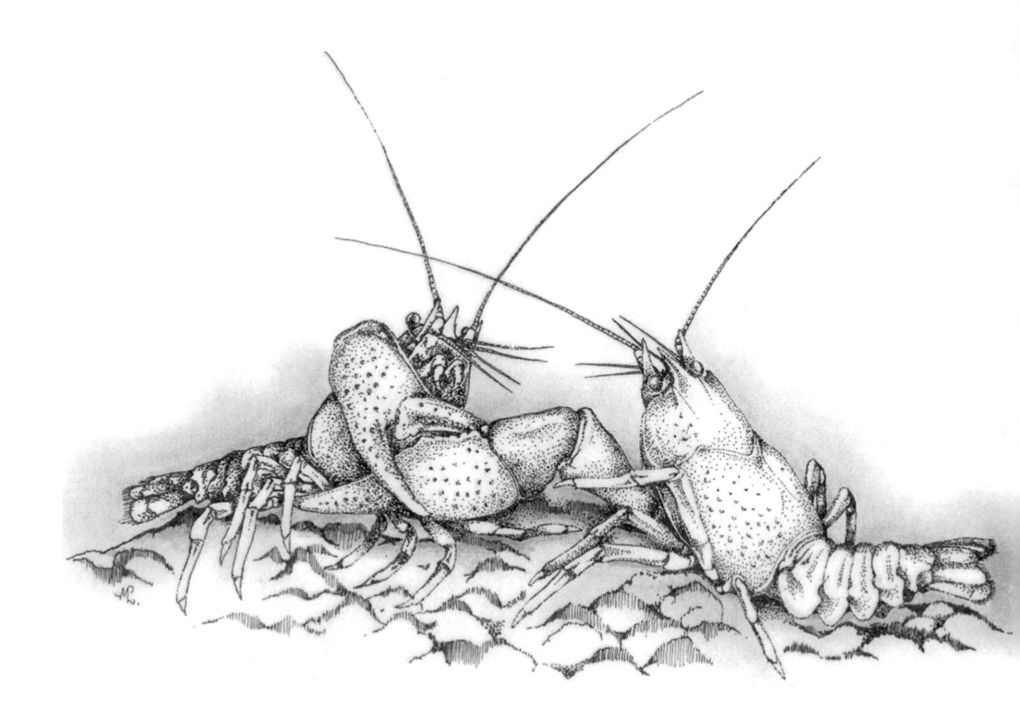

5

Aggressive encounters between animals of the same species have been termed "agonistic behavior" in order to differentiate these social interactions from aggression in predator–prey and other nonsocial interactions. Agonistic interactions occur when individuals display and/or fight over resources such as habitats, shelters, mates, and food. These interactions often follow strictly ritualized displays that usually result in nonlethal consequences. The displays may contain auditory, mechanical, visual, and even chemical signals designed to transfer pertinent information such as size or reproductive state of the combatant. The main result of agonistic interactions is the establishment of a dominance relationship that can alter access to resources. Throughout this chapter, I equate the term "winner" in an agonistic encounter with that animal being dominant. In dominance theory, dominance has distinct evolutionary advantages (i.e., increased reproductive fitness; Ellis 1995). In general terms, dominant individuals can acquire more and/or maintain longer control over critical ecological resources, which can result in more matings and higher reproductive success and thereby increased evolutionary fitness. The ability of an individual to possess and control resources within a population is referred to as resource holding potential (RHP; Wilson 1975).

More specifically, selective pressure should be expected on those factors (e.g., RHP) that influence agonistic behaviors during social interactions. To generate a larger theory incorporating both proximate and ultimate mechanisms underlying aggression and its impact on the evolution of behavior, it is necessary to better understand the extrinsic and intrinsic factors, how these factors interact to influence aggression, and how this leads to dominance. In general, we are asking the question, "What makes an animal dominant?" To begin to answer this broad question, a crustacean system, based on the crayfish, has been adopted as a model for social behavior and aggression.

It is important at this juncture to recognize various levels of sociality. Sociality can range from highly organized groups of individuals, such as those that occur in eusocial insects and primate communities, to organisms that exhibit aggressive interactions during brief periods of the year, such as mating periods. Crayfish exhibit agonistic behavior and social dominance in laboratory settings, but given the lack of direct field observations, it is difficult to label crayfish as a highly social organism.

General Crayfish Ecology

Habitat and Burrows

There are currently more than 500 recognized species of crayfish that inhabit many different freshwater lakes, ponds, streams, rivers, and other aquatic habitats across all continents except Antarctica and Africa (Taylor 2002). Crayfish inhabit areas with all types of substrata, ranging from sandy and muddy substrates to cobble and gravel. Shelters can range from natural assemblages of rocks to constructed burrows in mud or sand. Often these burrows will have a large main entrance and one or two side entrances. For *Orconectes virilis*, one of the species discussed below, home ranges around burrows have been quantified (Hazlett et al. 1974).

Crayfish have been observed fighting over shelters, and the occupant (presumably the owner) usually wins most of these fights (Bergman and Moore 2003). In addition, crayfish have been observed to repeatedly return to burrows even if they are occupied by other crayfish (P.A. Moore and K.C. Fero, personal observations). Some social interactions in burrows do not result in a fight but consist simply of one crayfish temporarily evicting the other from the burrow. We can conclude from these two observations that it is highly likely that crayfish repeatedly encounter the same individuals and even engage these individuals in agonistic encounters for resources in nature. For the most part, the peak time of this social activity occurs during the night, although limited diurnal activity can be observed. In addition to agonistic behavior over shelters, the majority of nocturnal activity is confined to foraging trips, but social interactions over food resources and mates are not infrequent. During the day, crayfish are often found in shelters with their major chelae pointing out of the shelter, possibly in position to defend a resource over which they claim ownership (P.A. Moore, personal observation).

Ecological Role in Food Webs

In most environments, crayfish can play an important role in the structure of the food web. Crayfish are omnivorous, consuming everything from detritus and macrophytes to carrion (Hill and Lodge 1994). Crayfish will also consume benthic macroinvertebrates, and crayfish, in turn, are consumed by fish populations. In some aquatic habitats, crayfish are keystone species based on their central role in food web dynamics (Hill and Lodge 1995). In streams and lakes, crayfish can function as "shredding" organisms that, through foraging, consume and break up terrestrial leaf material, which allows this material to be consumed by other aquatic macroinvertebrates. In addition, adult crayfish can have significant impacts on plant and algal communities in aquatic habitats. The nature of crayfish diets and the exact role that they play in aquatic food webs depend upon the size and age of the crayfish, the distribution of resources available, and the presence and distribution of predators (Lodge and Hill 1994, Momot 1995).

Sensory Capabilities

Crayfish have typical sensory capabilities in that they respond to visual, mechanical, and chemical stimuli in their environments. Crayfish have two well-developed compound eyes, with each eye placed on an eyestalk that can be moved independently from the other. Developed as superposition eyes, they are sensitive to low light levels and also exhibit sensitivity to polarized light (Vogt 2002). Crayfish have numerous appendages sensitive to chemical signals, although the majority of behavioral work has focused on the antennae and antennules that originate near the eyestalks. There are additional chemoreceptors on the major chelae, pereopod chelae, telson (i.e., tail), and maxillipeds (Holdich and Reeve 1988). Receptors sensitive to hydrodynamic stimuli are located on the same appendages as the chemoreceptors as well as on individual receptors across the entire body (Thomas 1970).

Basic Dynamics of Crayfish Fights

Fighting behavior of clawed decapod crustaceans, and crayfish in particular, has been a valuable addition to the behavioral literature on aggression, social behavior, and dominance due to the presence of conspicuous visual displays (Bovbjerg 1953) and a basic understanding of the neural circuitry underlying social behavior (Edwards and Kravitz 1997). This large body of work has shown that when two crayfish interact, their meeting leads to agonistic interactions that progressively escalate until one of the opponents withdraws (Bovbjerg 1956). Through these interactions, a dominance relationship is formed. Here I define a dominance relationship as one in which subsequent social interactions between these pairs is highly predictable (Daws et al. 2002).

As a consequence of this dominance relationship, subordinate animals will rarely seek to engage a dominant opponent (Daws et al. 2002). Dominant animals actively reinforce their status upon subordinates through short, low-intensity interactions (all of the descriptions below arise from laboratory studies). These subsequent interactions are often short in duration with lower intensities, where the dominant animal will either display or approach the subordinate, who adopts a submissive posture or repeatedly retreats (Ameyaw-Akumfi 1979). In hermit crabs, the ritualized nature of this dominance relationship may arise from decreased motivation to engage in further aggressive acts by the subordinate, individual recognition of a proven superior opponent, or the detection of an opponent's relative dominance status (Winston and Jacobson 1978). Whether these three mechanisms (decreased motivation, individual recognition, or status recognition) work in concert or whether one of them is the primary mechanism is unknown. However, all of these mechanisms can affect the outcome of an encounter.

Crayfish agonistic interactions are highly ritualized and can be described by a series of common behaviors of increasing or decreasing levels of aggression (Rubenstein and Hazlett 1974, Zulandt Schneider et al. 2001; Table 5.1). Although the general description that follows is based on a compilation of the fighting behavior

Table 5.1. Crayfish ethogram codes (used to score fight intensity levels).

Intensity Level	Description
−2	Tailflip away from opponent or fast retreat.
−1	Retreat by slowly backing away from opponent.
0	Visually ignore opponent with no response or threat display.
1	Approach without a threat display, walking slowly toward the opponent.
2	Approach with meral spread threat display with the major chelae; antennal (2nd antennae) whips are present, often with maxillipeds creating currents. Antennules (1st antennae) often are seen flicking.
3	Initial major chela use by boxing, pushing, and/or touching with closed chelae. Chelae are not used to grasp but can be opened and pushed. Antennal whips are more vigorous. Antennule (1st antennae) flicking is not seen.
4	Active major chela use by grabbing and/or holding opponent. Crayfish will try to turn opponents over or physically manipulate them, generating force through active major chela use.
5	Unrestrained fighting by pulling at opponent's claws or body parts. Opponents try to pull or tear legs, antennae, or major chelae off of individuals.

of three species of crayfish (*Orconectes rusticus*, *Orconectes virilis*, and *Procambarus clarkii*), they resemble general accounts of fighting in other crayfish species. As with other decapods, when two *O. rusticus* individuals are introduced into a confined space, a typical fight containing common behavioral patterns results. In general, a temporal sequence of these patterns is evident, with intensities increasing as fights increase in duration (Bruski and Dunham 1987, Daws et al. 2002).

The typical encounter begins with a simple approach of crayfish toward each other (level 1). In this first step of aggression, there are no changes in body posture or limb use (see Table 5.1). Level 2 involves an approach that contains threat displays,

Figure 5.1 Two form I male *Orconectes rustiscus* crayfish involved in an agonistic encounter. Both males are engaged in open chela combat, which is considered level 3 (see Table 5.1). Notice that the large antennae of the closer animal are in the position to whip the opponent. The lateral and medial antennules of both individuals are carried lower than during nonfighting postures and are placed to maximize detection of urinary signals being released during the encounter. Photo courtesy of Dan Bergman.

which consist of heightened body posture and spreading of the chelae (or meral spread). This usually occurs while the crayfish are at some distance apart, typically a couple of body lengths. The next level of aggression occurs when the crayfish use their major chelae in a passive manner with the claws either open or closed (level 3). Crayfish will actively push their opponents back and forth using their chelae but will not actively close the chelae or use them in a grasping fashion (Fig. 5.1). Previous researchers have usually called this boxing or restrained claw use. The greatest proportion of time that crayfish interact is spent at this level of aggression (Zulandt Schneider et al. 2001, Bergman et al. 2003). Accompanying this boxing are frequent antennal whips, which are rapid downward strikes against the opponent's carapace with the second antennae.

When combatants increase their level of aggression, they begin grasping their opponent with their chelae (level 4). At this level, chelae are used not only to push opponents but also to flip opponents over or to twist the opponent's chelae. The most intense fights occur when chelae are used in an unrestrained manner (level 5). This occurs when crayfish use their chelae to actively grasp their opponent's body parts (typically major chelae) and attempt to damage or remove these appendages. While infrequent, some interactions may result in injuries or lost limbs.

Agonistic interactions are terminated when one of the individuals displays a simple retreat (level –1) or tailflip (level –2) away from the opponent. As a side note, crayfish have multiple types of tailflips with different neural mechanisms, and some tailflips can be used in an offensive manner during interactions (Herberholz et al. 2001). It is important to note, when documenting fight behavior, the difference between offensive and retreat tailflips. Once dominance is established, subsequent encounters remain highly predictable for several hours, days, or weeks, depending upon the exact context (see below).

This foregoing description of the agonistic behaviors indicates the most visually obvious aspects of crayfish interactions. More recently, it has been demonstrated that crayfish spend periods of quiescence during fights, in which olfactory information is exchanged (Bergman et al. 2005; see below).

Factors Influencing Agonistic Behavior and Dominance

What physical, neural, behavioral, and environmental factors make a crayfish dominant? The level of aggression in crayfish can be influenced by many different factors, which include physical size, ownership of resources, previous social history, and even their neurochemistry (Bovbjerg 1956). As an initial approach, it is possible to categorize these factors as either intrinsic or extrinsic. Intrinsic factors are either physiological or physical features inherent to the crayfish. The most obvious of these is physical size of the crayfish or its major chelae (Gherardi et al. 1999). Extrinsic factors include chemical signals from opponents or the value and type of the resource being contested. I recognize that this distinction of extrinsic and intrinsic is somewhat artificial, but it permits grouping of those factors that arise from an individual crayfish (intrinsic) versus those factors that arise from a crayfish's interactions with their biotic and abiotic environment (extrinsic).

Intrinsic Factors

Body and Major Chela Size

Early studies on agonistic behavior in crayfish focused solely on body size as the most important indicator of dominance in crayfish (Dingle 1983, Edsman and Jonsson 1996). Crayfish do not have a terminal molt; therefore, it is difficult to distinguish the effects of age on aggression from the effects of size. If there are no large-scale differences in chela size between animals (~30% or less difference in size), the crayfish with the larger carapace usually becomes dominant when they fight in pairs. Body size is an excellent predictor of agonistic success (larger animals become dominant in >95% of interactions), if there is a greater than 30% difference in carapace length of the combatants, with all other factors being equal. If the body and chela size difference is less than 10%, then the outcome of the fight becomes random (Pavey and Fielder 1996, Daws et al. 2002).

If carapace size is similar, then weaponry size (major chelae) becomes an important determining factor for dominance (Rutherford et al. 1995). Crayfish with intact chelae that are larger than those of their opponents are more likely to win an agonistic encounter (Schroeder and Huber 2001). In addition to chela size, chela strength is also important. It appears as if dominance in crabs with similar-sized chelae is determined by the strength of the pinching (Sneddon et al. 2000a). During fights, chelae are used as signals of aggression during meral spreads. A meral spread occurs when one crayfish spreads its major chelae, displaying the size of its chelae and body. If viewed from above, the crayfish resembles a Y-shape. The size of the meral spread is thought to be an aggressive signal indicating the size of the crayfish (Zulandt Schneider et al. 1999).

The use of this aggressive signal depends on the physical health of the crayfish. As with most crustaceans (Bliss 1960), appendages are often lost or damaged and can be regrown over a period of time. Male crayfish deprived of one chela have fewer aggressive displays, initiate fewer agonistic encounters, and ultimately rank lower in hierarchies than do crayfish with intact chelae (Gherardi et al. 1999). Once chelae are lost, crayfish increase the frequency of molts. It is during their molt cycle that crustaceans alter their levels of aggression based on changes in hormone levels (Tamm and Cobb 1978). To summarize, in pairwise interactions, crayfish with the larger body or chela size win the majority of encounters and become dominant (Bovbjerg 1956).

Sex

The sex of the individual crayfish is an important factor in determining the outcome of agonistic interactions. Crayfish males are typically dominant over females (Capelli 1975, Peeke et al. 1995, Figler et al. 1999). In many species, males have more robust chelae relative to carapace length than do females. Male–male agonistic interactions are more intense than are male–female or female–female interactions (P.A. Moore, personal observations). Since males participate in agonistic encounters with more intensity than do females, it is hypothesized that the size differences between male and female chelae are due to selection for fighting ability (Stein 1976). In aggregations with both sexes present, males occupy the top spots within the hierarchy even if older females are considerably larger. Inter- and intrasex encounters follow similar patterns of aggression, indicating that agonistic interactions are ritualized regardless of the sex of the combatants. In many species, females have a smaller ratio of chelae to body size than do males of either reproductive form (Reynolds 2002; see below). Indeed, it may be this difference in weaponry size that accounts for differences in hierarchical position in regard to the sexes. Crayfish can recognize the sex of their opponents through multiple sensory channels (Acquistapace et al. 2002). Although it is clear that males are dominant over females, what remains unclear is whether this social distinction is due to variation in size across the sexes, variation in chela-to-carapace ratio, or some underlying neurochemical/neurohormonal difference between the sexes.

Reproductive State

Similar to the importance of sex in determining aggressive outcomes, the reproductive status of both males and females can alter aggression levels. Adult male crayfish in the family Cambaridae have alternating reproductive morphologies (Berril and Arsenault 1984). These two morphologies are a reproductive form (form I) and nonreproductive form (form II; Fig. 5.2). In northern-hemisphere temperate regions, males will molt from form I to form II in early summer (mid-June) and back to form I in late August to early fall (Crocker and Barr 1968, Berril and Arsenault 1984). Male crayfish have allometric growth during the change from form II to form I. Along with the many physiological changes accompanying this molt, form I males have larger chelae relative to body length than do form II males. The cyclical molting between forms is matched by cyclical changes in chela-to-carapace ratios.

Form I males are dominant over form II males, even when males have similar carapace lengths (Guiasu and Dunham 1998). In a series of carefully planned experiments, these authors concluded that the difference in relative chela size was the factor that determined dominance in interform contests. In addition, form I individuals exhibited a higher level of aggression toward form II individuals. Despite changes in physiology and hormones, there was no difference in overall levels of aggression between the two forms during intraform contests. In some North American habitats, form I and form II individuals can co-occur, and differences in chela size may have profound differences in RHP or resource acquisition in mixed-form populations.

The reproductive status of females can reverse the dominance order between sexes that is outlined above. Female maternal crayfish were more aggressive and won more bouts against males and nonmaternal females than did nonmaternal female crayfish. These results clearly show that the presence of unhatched eggs or first instar larvae increased aggression levels in females (Figler et al. 1995). Changes in aggression or social behavior as a consequence of reproductive status have been observed in other species. This is particularly true for mate guarding (see chapters 7, 10, 16). In summary, reproductive status has the potential to increase aggression in male and female crayfish and alter the outcome of agonistic interactions. Reproductively active males are more aggressive than their nonreproductive counterparts, and maternal females are more aggressive than are males and nonmaternal females.

Previous Social Experience

Another important intrinsic factor that regulates aggression and dominance is the previous social experience of a crayfish. Crayfish without any social interactions for a

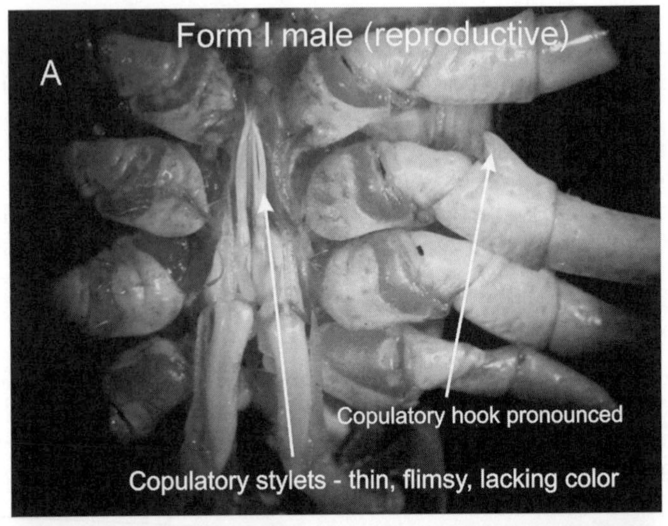

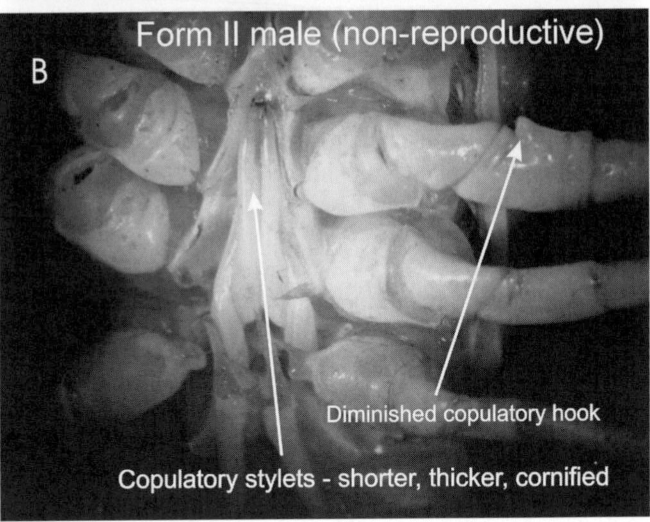

Figure 5.2 Photos showing the ventral surface of form I (A) and form II (B) male *Orconectes rusticus* crayfish. Notice the difference in size of the copulatory stylets and copulatory hooks. Photo courtesy of Jodie Simon.

period of seven days appear to interact as if they are socially naive (Zulandt Schneider et al. 1999). Otherwise, repeated previous social interactions shape the level of aggression and ultimately can influence the outcome of future interactions (Daws et al. 2002). Previous social experience can have two opposing results: the "winner effect" and the "loser effect." When a crayfish has a repeated previous winning experience within pairwise encounters, that individual has an increased likelihood of winning the next encounter. Conversely, a crayfish with a repeated previous losing experience has an increased likelihood of losing the next encounter (Daws et al. 2002). This change in the outcome of subsequent interactions is present whether the next encounter is against familiar or unfamiliar opponents. Winning and losing effects can result from encounters that vary in duration, intensity, and repetition. Interestingly, short-term winner effects can be produced even from a single short encounter that lasts no longer than 30 seconds (Bergman et al. 2003). Stronger and more pronounced winner effects on subsequent social behavior are evident from repeated encounters over extended periods of time (Bergman et al. 2003). The strength and duration of the winner or loser effect probably depends on the subsequent reinforcement through repeated encounters, but the types of encounters necessary to produce specific durations of winner or loser effects have not been systematically studied. The ability of winner and loser effects to alter subsequent social interactions is strong enough to overcome the role of size disparities in opponents that would otherwise serve as an accurate predictor of the outcome (Daws et al. 2002). All of the previous winner effects described above occur only through repeated social encounters.

The impact of winner effects on crayfish social behavior is a time-dependent phenomenon. It appears as if there are two separate processes that underlie winner effects. Winner effects produced from single short interactions (up to 30 seconds) give crayfish an advantage against naive individuals for only 40–60 minutes (Bergman et al. 2003). In fact, the winner effect produced in this manner decays in a linear fashion, with the largest influence appearing within 20 minutes after the first encounter. At 60 minutes, the winner effect is missing, and the crayfish behaves such that the outcome of the subsequent interaction is random or determined by size differences (Bergman et al. 2003).

The mechanisms by which the winner and loser effects arise are not well studied, and there are several theories on how a previous winner's behavior has changed as a result of a positive outcome in an interaction. Winner and loser effects may be a result of intrinsic changes in the motivation to engage in interactions on the part of the winning or losing crayfish (Copp 1986). Changes in motivation may be tied to short-term or long-term changes in the neurochemistry of winners and losers (see below). In addition, changes in social behavior may alter the way an individual crayfish perceives the fighting ability of an opponent. This may be accomplished by associating specific physical characteristics (e.g., carapace or chela size) with an opponent's fighting ability. Conversely, an animal may assess its own RHP compared to the population distribution of fighting abilities (Parker 1974) observed when habitat values vary within a natural setting. Changes associated with becoming a winner or loser may be communicated through extrinsic signals (i.e., visual displays, social pheromones; Copp 1986, Zulandt Schneider et al. 2001). The underlying mechanism of winner effects could also include changes in serotonin levels that have been shown to produce both heightened aggressive states and an increase in agonistic behaviors (Huber et al. 2001; see below). If intrinsic causation is responsible for the results of agonistic interactions, an animal that has had a prior winning experience may function as a "successful" fighter and thereby fight more readily in future agonistic interactions. Although the exact mechanism by which winner and loser effects occur is unknown, there is increasing evidence that crustaceans have a social memory and that this memory, either of individuals or encounters, alters subsequent interactions (Hazlett 1969, Johnson 1977, Karavanich and Atema 1998a, Gherardi and Atema 2005).

Winner and loser effects may also result from extrinsic changes, such as changes in the behavior of the winner's opponent that could be recognized by a conspecific.

Recognition of aggressiveness (dominance) could be accomplished visually (Thorp and Ammerman 1978), through a change in the physiological state of the opponent detected by chemoreception (Zulandt Schneider et al. 1999) or other mechanisms (Rufino and Jones 2001, Obermeier and Schmitz 2003a, 2003b). Changes in aggression and dominance that are due to alterations in previous social experience can occur on much shorter time scales than the other three factors discussed above (size, sex, and reproductive state). This may be due to the underlying neural mechanism associated with previous social experience, which is hypothesized to be due to changes in the neurochemistry of crayfish.

Neurochemistry

From a mechanistic perspective, all behaviors ultimately arise as a result of nervous system activity, and consequently, any changes in behavior are likely due to changes in nervous system function. For crayfish, it has been speculated that the behavioral plasticity associated with differences in aggression and dominance influences nervous system neurochemistry. In other words, becoming a dominant or subordinate crayfish alters serotonin neurochemistry. In addition, these changes in neurochemistry affect the subsequent social behavior of the crayfish by altering future levels of aggression and dominance (Yeh et al. 1997). This neurochemistry/behavior loop has the potential to be a positive feedback loop in that dominant crayfish become more aggressive, which in turn can increase their status level. In particular, biogenic amines are thought to underlie changes in aggression and dominance in decapod crustaceans. Biogenic amines are a family of small neuroactive substances that include several behaviorally active compounds. Serotonin, octopamine, norepinephrine, and dopamine have all been strongly implicated in the control of various forms of aggression. Serotonin has received the most attention in regard to its role in aggression and dominance (Edwards and Kravitz 1997, Huber et al. 1997a, 1997b). Despite all of the attention on serotonin, there is evidence that octopamine and dopamine play a role in dominance and aggression in crustaceans (Sneddon et al. 2000b). Shore crabs with higher levels of dopamine and octopamine were more likely to be become dominant when paired with size-matched individuals. In addition, the levels of circulating dopamine and octopamine dropped and serotonin increased after dominance was established (Sneddon et al. 2000b). Clearly, more work on other neuromodulators needs to be done to have a fuller understanding of the chemical mechanisms involved in dominance.

The hypothesized role for serotonin in aggression is that changes in social status as a result of previous social interactions alter the function of serotonin within the nervous system of crayfish. In lobsters, crayfish, and other decapod crustaceans, an increase in serotonin levels is closely associated with heightened aggression or dominant behavior (Edwards and Kravitz 1997). Changes in excitability (Krasne et al. 1997) and serotonin receptor subtype populations (Yeh et al. 1996, 1997) have been reported as a consequence of dominance. For example, the role that serotonin plays in neurochemistry is different in dominant individuals than in subordinates (Yeh et al. 1996). Social status has been shown to determine concentrations of neuromodulators in blood, the efficacy of modulators at identified synapses (Yeh et al. 1996, 1997, Krasne et al. 1997), and that of monoamines in different nervous system regions.

In addition, dominance and aggression are thought to be a direct result of serotonin levels or activity within the nervous system. Direct injection of serotonin elicits stereotypical agonistic behaviors (Antonsen and Paul 1997) and produces a posture resembling meral spread (Livingstone et al. 1980). Moreover, increased serotonergic function, through injections, decreases the likelihood of retreat (Huber et al. 1997b, Huber and Delago 1998). Neurons within local circuits controlling tailflip, a common retreat behavior (Glanzman and Krasne 1983, 1986), exhibit reduced responsiveness in the presence of this amine (Edwards and Kravitz 1997). This theory can be summarized by stating that serotonergic function alters the "dominance decision point" of agonistic interactions by altering the probability of individuals to tailflip in fights (Huber and Delago 1998). Those crayfish with increased serotonergic function are less likely to tailflip and thus are more likely to become dominant in social interactions. Some recent work has begun to question whether serotonin injections actually change levels of aggression and dominance within crayfish or if it simply alters some basic functions such as locomotion and behavioral postures that may be interpreted as changes in aggression (Tierney 2001). The exact nature of the connection between serotonin and dominance remains uncertain, but it is clear that serotonin and other neurotransmitters are influenced by aggression and social history, and these chemicals have the potential to influence future social interactions. What is missing in these two hypotheses on the role of serotonin in dominance theory may be the possible role that serotonin plays in influencing chemical communication during social interactions.

Motivational State

Different physiological states of a crayfish, such as hunger and molt stage, can alter levels of aggression and even the outcome of aggressive interactions (Hazlett et al. 1975). Starvation can decrease the potential for survival and thus lead to an increase in motivation to engage in agonistic encounters over valuable resources (Capelli and Hamilton 1984). Crayfish with reduced energy reserves engage more often in aggressive interactions than do those that have been fully fed (Hazlett et al. 1975). In addition, starved crayfish increased their rate of escalation in fights, possibly indicating a willingness to take more risks within a contest. Presumably, these decisions occur through a comparison of the intrinsic state of energy reserves and the assessed value of an extrinsic resource. The motivational state of crayfish is also altered during and shortly after molting (Adams and Moore 2003; J.A. Adams and P.A. Moore, unpublished data). Shortly after molting, crayfish have a soft exoskeleton that renders their major chelae ineffective during contests. In contests involving recently molted and hard-shelled crayfish, molted crayfish display meral spreads more frequently. Although molted individuals had a greater number of displays, these same individuals tailflipped more frequently rather than increasing the level of aggression to the point of chela use (Steger and Caldwell 1983; J.A. Adams and P.A. Moore, unpublished data). Although not directly demonstrated for crayfish, these results can be explained within a framework of bluffing and then retreating when the bluff fails to deter an opponent. Molting and aggression are under the influence of hormonal changes that could alter either the neurochemistry or neurophysiology associated with social behavior (Chang et al. 1993). These two intrinsic factors, combined as motivational state, will alter social behavior and act through changes in the underlying hormonal systems that influence neural functioning, or vice versa.

Summary of Intrinsic Factors Regulating Dominance

From the descriptions above, it is apparent that most intrinsic factors studied to date are connected to the physical size of the crayfish. Larger crayfish or those with larger chelae often have advantages in agonistic interactions either through larger aggressive displays (meral spreads) or through greater physical force generated through direct contact. Age, sex, and reproductive status can influence the physical size of the carapace and major chelae. In addition, lost or injured chelae can also influence the dominance status of crayfish in populations. All of these factors may influence the intrinsic neurochemistry of individuals, and in particular, there is evidence that serotonin is involved in the neural pathways that play a role in social behavior. In addition, given the ubiquitous nature of serotonin in the crustacean nervous system, it may be possible that serotonin is also connected to motivation and that changes in aggression due to changes in motivation may be accomplished through alteration of serotonin function. While much evidence shows that serotonin function is altered as a consequence of a crayfish becoming dominant, the connection between serotonin function and future social interactions is not yet clear. It is also possible that serotonin function is altered by extrinsic factors such as chemical or visual communication.

Extrinsic Factors

Chemical Communication

Decapods rely heavily on olfactory signals during social interactions. Olfaction is important in both the recognition and determination of dominance in crayfish (Zulandt Schneider et al. 1999, 2001, Bergman et al. 2003). Antennules, arguably the most important chemosensory organ of crustaceans, are involved in sending mechanical signals and the reception of chemical signals during fighting (Rutherford et al. 1996), the recognition of sex (Ameyaw-Akumfi and Hazlett 1975, Dunham and Oh 1992, Acquistapace et al. 2002), molt state (Adams and Moore 2003), individuals (Crook et al. 2004; see also chapter 6), and dominance status (Zulandt Schneider et al. 1999, 2001, Gherardi and Daniels 2003). In addition, the presence of urine increases olfactory sampling through antennule flicking during fights (Ameyaw-Akumfi and Hazlett 1975, Rutherford et al. 1996). For lobsters, urine is necessary for recognition during agonistic encounters (Karavanich and Atema 1998b). Crayfish create and control water currents during social interactions to actively send urine to or sample urine from opponents (Bergman et al. 2005). This is also seen in banded shrimp (Hughes 1996) and lobsters (see chapter 6). Urinary signals are excreted through nephropores located at the base of the antennae, and urine is almost exclusively released during social interactions (Fig. 5.3). This urine is a likely source of social pheromones in crayfish (Zulandt Schneider et al. 2001, Bergman et al. 2005) and other crustaceans (Bushmann and Atema 1993, 1996, Breithaupt et al. 1999, Breithaupt and Atema 2000). The role of chemical signals in the social behavior of crayfish can be divided into three areas: the recognition of dominance status, the long-term alteration of social status, and the control and manipulation of chemical information.

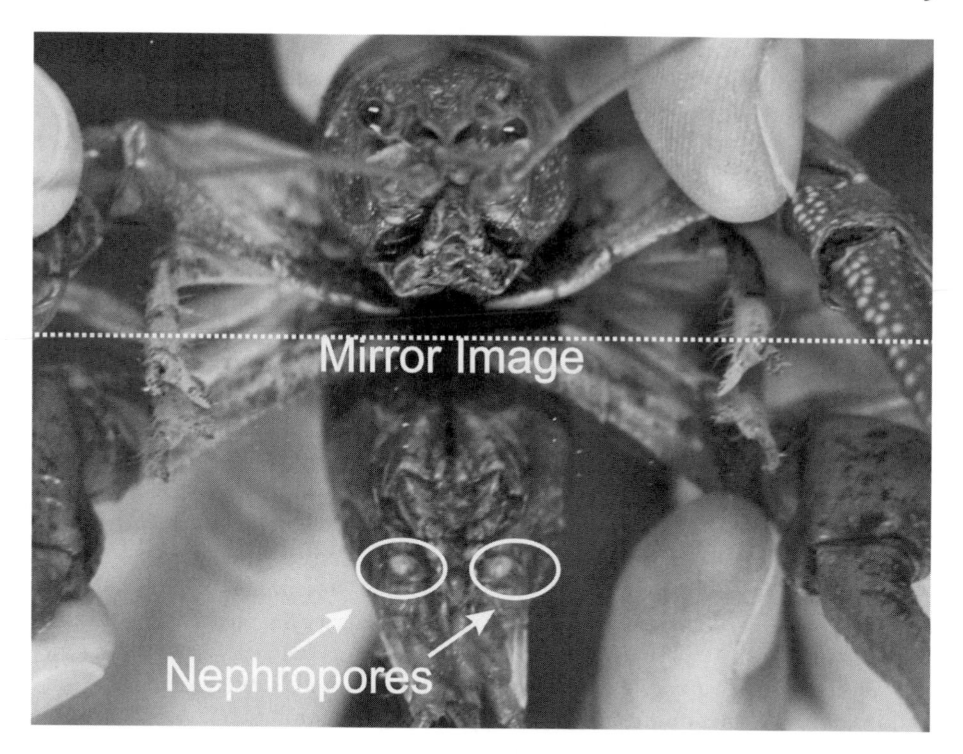

Figure 5.3 Anterior view (with mirror reflection) of a male form I *Orconectes rusticus* crayfish with major chelae held out. Circles show the location of the nephropores that release urine during social interactions. Urine is released as the crayfish controls flow currents that direct the urine outward.

Crayfish can recognize an opponent's social status with their antennae and antennules via chemical signals (Tierney et al. 1984). Chemical information perceived through the antennules alters the fight dynamics of interactions. Removing chemical information from agonistic bouts with crayfish can be accomplished by selectively lesioning the chemoreceptors through osmotic shock or by blocking the release of urine in one of the opponents. When either of these manipulations is performed, agonistic interactions are longer in duration and take longer to escalate to higher levels of aggression (Breithaupt and Eger 2002, Zulandt Schneider et al. 2001). Surprisingly, when crayfish with a winning experience fought against anosmic (unable to smell) naive individuals, the winner effect was eliminated, indicating that chemical signals are involved in the detection of previous social interactions, possibly through recognition of individuals or status (Bergman et al. 2003). The chemical cues involved in this recognition are released in the urine of the crayfish (Zulandt Schneider et al. 1999, 2001). Agonistic battles are longer and reach higher intensity levels when status cues are obstructed (Zulandt Schneider et al. 2001). The presentation of chemosensory cues alone is sufficient to bring about investigative behavior and threat display (Zulandt Schneider et al. 1999). Thus, chemical signals appear to play a role in determining the outcome of social interactions as well as the dynamics of fights themselves.

As well as playing a short-term (hours) role in fight dynamics and outcomes, long-term (days) exposure to social signals can influence subsequent social behavior in crayfish. Crayfish exposed to dominant or subordinate odors adopt a social status that is contrary to the odors to which they were exposed; that is, crayfish exposed to dominant odor become subordinate and vice versa. Crayfish exposed to tank water or odors from naive crayfish do not alter behavior. Thus, previous odor exposure, mediated solely through urinary signals, appears to alter subsequent interactions (Bergman and Moore 2005). The underlying mechanism of changes in social behavior, due to the long-term exposure to social chemicals, could be tied to alteration in the neurochemistry of serotonin (see above).

Finally, since crayfish release urine primarily during social interactions, this strongly suggests that it is a social signal (Breithaupt and Eger 2002, Bergman et al. 2005). Urine release during social interactions shows temporal patterns that are different for dominant and subordinate crayfish (Bergman et al. 2005). In addition, crayfish can create a variety of currents (called information currents) that can either project urine forward or draw an opponent's urine toward their own antennules (Breithaupt 2001, Bergman et al. 2005). During these periods, crayfish are often not engaged in physical contact, and their antennules are rapidly flicking potentially sampling social odors. Crayfish reengage in physical interactions following these short periods of urine exchange. A detailed analysis of the temporal sequence of urine release, current generation, and subsequent behavioral actions indicates that dominant crayfish have different combinations of these three behaviors compared to subordinate crayfish (Bergman et al. 2005). While it is possible to conclude that urine release and current generation are involved in the social behavior of crayfish and in determining dominance, we do not know whether the chemical composition of the urine, the mechanical signal of information currents, or a combined signal of flow and chemical information determines dominance.

In summary, these studies provide a larger picture of the determinants of agonistic interactions in crayfish. These include extrinsic chemical signals containing a dominant pheromone (Zulandt Schneider et al. 2001, Bergman et al. 2005), control of when and where those signals occur during an agonistic interaction, and long-term alteration of social behavior due to exposure to social pheromones. Given the neurophysiological systems of crayfish, the presence of an open circulatory system, and the presence of a bladder system that stores urine, it may be that these urinary signals are connected to the changes in neurochemistry that either precede or accompany changes in dominance status. Serotonin, its precursors, or metabolites that would be excreted via the urine probably form a biochemically "expensive" metabolite, raising the possibility that it might be used for more interesting purposes than solely as an excretory waste product (Weiger 1997). If indeed the social signals are related to serotonin neurochemistry and since social behavior depends on chemical communication, it is possible that a chemical feedback loop is present. In an interaction, one crayfish would become dominant altering both its behavior and chemical signals. These behavioral and chemical signals would alter the social behavior of the receiving crayfish, making it subordinate. This, in turn, would further reinforce the dominant status of the first crayfish (Bergman and Moore 2005).

Visual Communication

Visual signals also play a role in shaping crayfish aggression, particularly during the initial stages of fighting (Bruski and Dunham 1987). This is most evident during approaches with the meral spread display. Crayfish fighting under different light conditions exhibited changes in their fight dynamics consistent with the notion that visual signals are important during agonistic interactions. Crayfish performed visually mediated behaviors under conditions where light was not limited. In particular, behaviors such as tailflipping or retreat were performed by subordinate animals when dominant animals approached or displayed (Bruski and Dunham 1987). In darkened conditions, these behaviors were less evident, suggesting that visually localizing the presence of a dominant crayfish is important for subordinates. Other behaviors mediated through tactile or chemical information were not significantly altered by changes in light conditions. Interestingly, it appears as if visual signals and chemical signals may play different roles in agonistic interactions. Changes in visual information appear to alter fight dynamics, such as the number and type of behaviors or level of aggression, whereas changes in chemical information do not seem to effect these fight dynamics. Conversely, chemical signals appear to play a larger role in determining or communicating the outcome of fights, whereas it is unclear if visual signals are necessary for determining the outcome as opposed to the dynamics of fights.

Mechanical Communication

The final sensory system that plays a role in determining dominance in crayfish is mechanoreception. Given the large role that the physical size of major chelae and antennal whips play during agonistic bouts, it is clear that mechanical information is used during these interactions (Smith and Dunham 1996). In addition, information currents are used during agonistic interactions (Breithaupt 2001), but differently by dominant and subordinate crayfish (Bergman et al. 2005). Compared to the wealth of literature on visual and chemical signals, relatively little is known about the use of mechanical information in agonistic bouts. It is unclear what kind of information is exchanged during antennal whips, chela grasps, and information currents. For example, is the size or strength of an opponent transmitted through force and pressure generated by chelae or by the strength of information currents? Or is this just a mechanism used for physically manipulating the opponent or controlling chemical information? Clearly, more work needs to be done with mechanical signals to understand their role in the social behavior of crayfish.

Resources

Environmental resources are a central part of the RHP theories of dominance and social behavior (Parker 1974). These theories state that the ultimate consequence of dominance is differential access to critical resources like mating territory, mating opportunities, food, or shelters. If true, the perceived value of the resource should influence the level of aggression and the outcome of social interactions. Other crustaceans are more gregarious and exhibit some level of resource sharing, most notably, shelters

(Eggleston and Lipcius 1992, Bushmann and Atema 1997, Duffy et al. 2002). Often dominant crustaceans will control access to shelters or other environments. Dominant lobsters have been observed periodically "evicting" subordinates from their shelters, presumably to reinforce their social status (O'Neill and Cobb 1979, Karnofsky et al. 1989a, 1989b). There is evidence that social crustaceans are attracted to common shelters through chemical signals and that these signals may be related to social status (Nevitt et al. 2000). Unlike these examples, most crayfish do not share shelters and are solitary burrowers (chapter 15).

The presence of food and shelters increases the level of aggression among different crayfish species (Capelli and Hamilton 1984, Usio et al. 2001). Crayfish that perceive "ownership" of the resource are often more likely to increase aggression and to defend their ownership of the resource more vigorously than the intruder crayfish (Peeke et al. 1995). These owner crayfish have a higher probability of winning these encounters and defending their territories or resources (Ranta and Lindstrom 1992, Vorburger and Ribi 1999). In natural settings, agonistic encounters are more intense and last longer on those resources that are considered more valuable. In fights observed in the field (Bergman and Moore 2003), the presence of shelters results in longer and more intense fights than those involving available food resources, such as macrophytes or detritus patches. Furthermore, fights on detritus patches exhibited higher overall intensities and ended with more tailflips away from an opponent than when on macrophyte beds. Because of the differing availability of shelters, detritus patches, and macrophyte beds, Bergman and Moore (2003) concluded that fight intensity and duration were correlated with resource availability. In summary, fighting intensity and levels of aggression are elevated when fights occur over valuable resources. Therefore, the perceived value of the resource to the combatant determines the overall intensity of the fight and influences the outcome of agonistic interactions.

A Comparison of Aggression in Other Crustaceans and Arthropods

A striking similarity in aggressive behavior exists between the American lobster, *Homarus americanus*, and crayfish (for review, see Atema and Voigt 1995; see also chapter 6). Lobster encounters often begin with a meral spread and increase in stepwise fashion to higher intensities of more active chela use. In fact, use of chelae appears to be a common feature in agonistic encounters among several decapod crustaceans. Prawns (Barki et al. 1997), fiddler crabs (Rosenberg 1997), stomatopods (Caldwell 1992), and snapping shrimp (Hughes 1996) actively use and display chelae during agonistic encounters. In all of these cases, larger chelae are often advantageous in the encounters and help confer dominance. This is the case whether the chelae are used actively, such as in grasping or ripping, or in displays. Interestingly, it has been argued that recently molted stomatopods will display their chelae in order to bluff their dominance status or level of aggression (Steger and Caldwell 1983). Although this action has not been documented in crayfish, the concept of bluffing or dishonest signals may be applicable. Finally, snapping shrimp also use their chelae in determination of dominance. Snapping shrimp generate a mechanical signal upon closing their major chelae, and the magnitude of this current is correlated with the size of the chelae (Schmitz and Herberholz 1998; see also chapter 4). Information contained within these signals is used in the determination of dominance.

In addition to these visual and mechanical displays, other crustaceans use chemical signals to transfer information during agonistic encounters. Lobsters will generate information currents that are used to propel water toward or away from the animal during agonistic encounters (Atema and Voigt 1995; see also chapter 6). It is thought that urine is released into these currents and the signals contained within the urine are used in either status recognition or individual recognition. Indeed, stomatopods use chemical signals for individual recognition, and this may modify subsequent social interactions (Caldwell 1979, 1985). Snapping shrimp also produce a series of different currents used to deliver chemical signals during agonistic encounters (Herberholz and Schmitz 2001). In contrast to the many instances where aquatic crustaceans use chemical signals in aggression, terrestrial arthropods appear to be remiss in this category of chemical signals. The terrestrial arthropods have numerous social chemicals used for aggregation, nest recognition, mating, and other social situations (Cardé and Millar 2004), yet despite this diversity of social chemical signals, very few are used for aggression and dominance. Probably the best understood example is the cockroach dominance pheromone, which actually confers dominance on the individual (Moore et al. 1997).

Summary and Future Directions for Research

Fighting success and dominance in crayfish depend on a variety of extrinsic and intrinsic factors (Fig. 5.4). Several intrinsic factors improve the chances of a crayfish becoming dominant, such as physical size (Ranta and Lindström 1992, Daws et al. 2002), larger weapons (Garvey and Stein 1993, Rutherford et al. 1995), the reproductive form (Guiasu and Dunham 1998), previous social experience (Daws et al. 2002), and changes in serotonin function (Huber and Delago 1998). Extrinsic factors that lead to increased dominance include status pheromones in the environment (Bergman et al. 2005), appropriate visual and mechanical signals (Bruski and Dunham 1987, Smith and Dunham 1996), and ownership of valuable resources (Peeke et al. 1995). Despite all this work, relatively little is known about the evolutionary consequences (in terms of overall fitness) of dominance in crayfish (see chapter 7). Presumably, they are similar to those in rock shrimp, where females prefer dominant males over subordinates (Diaz and Thiel 2003). It is unknown whether dominance confers other advantages. While it is fairly easy to list the types of factors that influence aggression and dominance relationships between crayfish, many questions regarding both the proximate and ultimate level of causation and consequences of aggressive behavior remain unanswered.

With a plethora of factors that influence aggression and dominance in crayfish, it is unknown how these factors interact together to produce dominance. Although we are unlikely to identify a single most important factor that determines dominance, it may be possible to investigate how the factors outlined in this chapter interact with each other to make crayfish more or less aggressive and, as a consequence, dominant or subordinate. For example, is it possible that dominance pheromones can make a smaller individual dominant over larger individuals? How much of a winning experience is necessary to establish an individual as dominant over an extended period of time? How effectively can extrinsic factors, such as resource value and ownership, override the influence of intrinsic factors, such as size or serotonin levels? From a

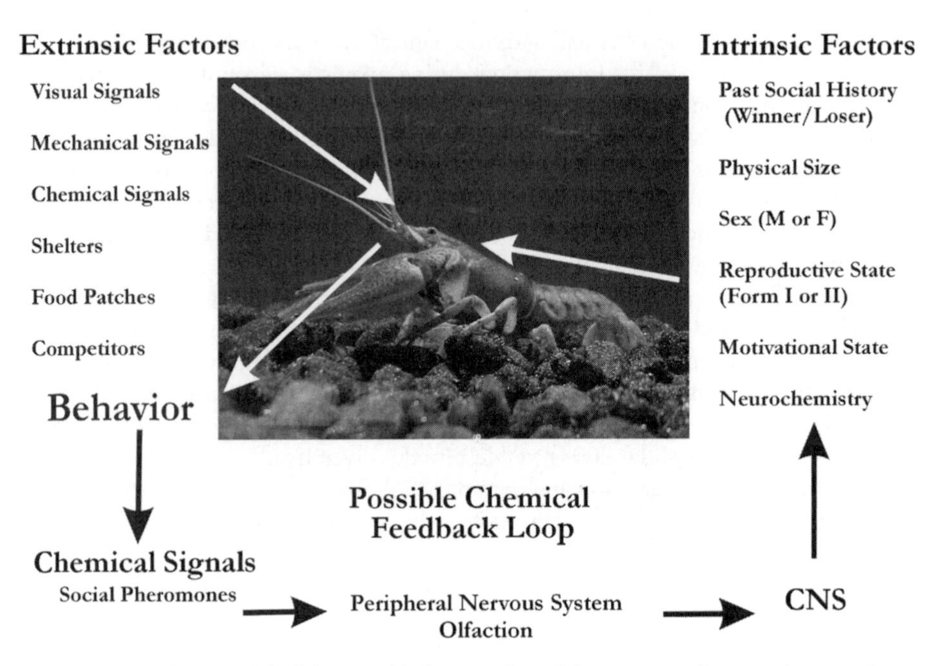

Figure 5.4 Working model of the social behavior of crayfish: summary diagram showing the different influences on the social behavior of the crayfish. The extrinsic and intrinsic factors influence both the behavioral output of the crayfish and the chemical signals used during social interactions. These social signals have the potential to develop a chemical feedback loop by altering neurochemistry, which could in turn alter the composition of chemical signals.

proximate point of view, it is important to establish the relative strengths of intrinsic and extrinsic factors in determining social dominance and levels of aggression.

Most of the studies reviewed in this chapter have been conducted in laboratory settings under controlled conditions. While this allows the establishment of what factors play a role in influencing dominance and aggression, their importance in natural settings is unknown. Researchers have established that crayfish have the capability of individual or status recognition in the lab, but it is uncertain whether this occurs or can occur in natural settings. Crayfish are very mobile in nature, and the role of repeated encounters with the same individual is not well documented. The role of visual, chemical, and mechanical signals in natural habitats needs to be addressed. Crayfish habitats include fast-running streams that can be muddy or swampy, which may limit the effective distances of visual and chemical signals. Many species are nocturnal, which may further limit the usefulness of visual signals. More field studies are needed to establish that the factors that influence aggression and dominance in the lab also influence aggression and dominance in natural settings.

Finally, from an ultimate perspective, the evolutionary consequence of aggression and dominance is assumed to be tied to reproductive success or increased RHP. This assumption needs to be tested in both laboratory and natural settings. Do dominant crayfish get more matings? Do dominant crayfish with more matings have a higher reproductive output? What is the role of female choice and dominance among males? Do females select mates based on hierarchy position, physical size, or a mating pheromone that may be linked to a dominance pheromone? What is the role of genetics in dominance? Are there genetic differences among crayfish in serotonin levels or the types of dominance pheromones produced? In general, all of these questions are designed to determine the evolutionary advantage of being dominant. Only through a multidisciplinary and multilevel approach is it possible to address the question that started this chapter: "What makes a crayfish dominant?"

Acknowledgments I thank all of the students in the Laboratory for Sensory Ecology who contributed their insight, helpful comments, and stimulating discussion in preparing the manuscript. In particular, comments and ideas for this chapter were shaped during a seminar on animal signals and sociobiology that included Kandi Fero, Art Martin, Jodie Simon, and Mary Wolf. Most of the social behavior work performed in my lab and summarized here has been part of the doctoral dissertations of Julie Adams, Daniel Bergman, Rebecca Zulandt Schneider; the postdoctoral work of Ali Daws; the dissertation work of Art Martin, Kandi Fero, and Jodie Simon; and the undergraduate work of Karen Konzen, Natalie Peters, Beth Pradelski, Tom Zulandt, Brittany Raburn, and Jodie Conin.

References

Acquistapace, P., L. Aquiloni, B. Hazlett, and F. Gherardi 2002. Multimodal communication in crayfish: sex recognition during mate search by male *Austropotamobius pallipes*. Canadian Journal of Zoology 80:2041–2045.

Adams, J.A., and P.A. Moore. 2003. Discrimination of conspecific male molt odour signals by male crayfish, *Orconectes rusticus*. Journal of Crustacean Biology 23:7–14.

Ameyaw-Akumfi, C. 1979. Appeasement displays in cambarid crayfish (Decapoda, Astacoidea). Crustaceana Supplement 5:135–141.

Ameyaw-Akumfi, C., and B.A. Hazlett. 1975. Sex recognition in the crayfish *Procambarus clarkii*. Science 190:1225–1226.

Antonsen, B.L., and D.H. Paul. 1997. Serotonin and octopamine elicit stereotypical agonistic behaviors in the squat lobster *Munida quadrispina* (Anomura, Galatheidae). Journal of Comparative Physiology A 181:501–510.

Atema, J., and R. Voigt. 1995. Behavior and sensory biology. Pages 313–348 *in*: J.R. Factor, editor. Biology of the lobster *Homarus americanus*. Academic Press, San Diego, Calif.

Barki, A., S. Harpaz, and I. Karplus. 1997. Contradictory asymmetries in body and weapon size, and assessment in fighting male prawns, *Macrobrachium rosenbergii*. Aggressive Behavior 23:81–91.

Bergman, D.A., and P.A. Moore 2003. Field observations of intraspecific agonistic behavior of two crayfish species, *Orconectes rusticus* and *Orconectes virilis*, in different habitats. Biological Bulletin 205:26–35.

Bergman, D.A., and P.A. Moore. 2005. The prolonged exposure to social odors alters subsequent social interactions in crayfish (*Orconectes rusticus*). Animal Behaviour 70:311–318.

Bergman, D.A., C.P. Kozlowski, J.C. McIntyre, R. Huber, A.G. Daws, and P.A. Moore. 2003. Temporal dynamics and communication of winner-effects in the crayfish, *Orconectes rusticus*. Behaviour 140:805–825.

Bergman, D.A., A.L. Martin III, and P.A. Moore. 2005. The control of information flow by the manipulation of mechanical and chemical signals during agonistic encounters by crayfish, *Orconectes rusticus*. Animal Behaviour 70:485–496.

Berril, M., and M. Arsenault. 1984. The breeding behaviour of the northern temperate orconectid crayfish, *Orconectes rusticus*. Animal Behaviour 32:333–339.

Bliss, D.E. 1960. Autotomy and regeneration. Pages 561–589 *in*: T.H. Watterman, editor. The physiology of Crustacea, Volume 1. Metabolism and growth. Academic Press, New York.

Bovbjerg, R.V. 1953. Dominance order in the crayfish *Orconectes virilis* (Hagan). Physiological Zoology 26:173–178.

Bovbjerg, R.V. 1956. Some factors affecting aggressive behavior in crayfish. Physiological Zoology 29:127–136.

Breithaupt, T. 2001. Fan organs of crayfish enhance chemical information flow. Biological Bulletin 200:150–154.

Breithaupt, T., and J. Atema. 2000. The timing of chemical signaling with urine in dominance fights of male lobsters (*Homarus americanus*). Behavioral Ecology and Sociobiology 49:67–78.

Breithaupt, T., and P. Eger. 2002. Urine makes the difference: chemical communication in fighting crayfish made visible. Journal of Experimental Biology 205:1221–1231.

Breithaupt, T., D.P. Lindstrom, and J. Atema. 1999. Urine release in freely moving catheterized lobsters (*Homarus americanus*) with reference to feeding and social activities. Journal of Experimental Biology 202:837–844.

Bruski, C.A., and D.W. Dunham. 1987. The importance of vision in agonistic communication of the crayfish *Orconectes rusticus*. I. An analysis of bout dynamics. Behaviour 63:83–107.

Bushmann, P.J., and J. Atema. 1993. A novel tegumental gland in the nephropore of the lobster, *Homarus americanus*: a site for the production of chemical signals? Biological Bulletin 185:319–320.

Bushmann, P.J., and J. Atema. 1996. Nephropore rosette glands of the lobster *Homarus americanus*: possible sources of urine pheromones. Journal of Crustacean Biology 16:221–231.

Bushmann, P.J., and J. Atema. 1997. Shelter sharing and chemical courtship signals in the lobster *Homarus americanus*. Canadian Journal of Fisheries and Aquatic Sciences 54:647–654.

Caldwell, R.L. 1979. Cavity occupation and defensive behaviour in the stomatopod *Gonodactylus festae*: evidence for chemically mediated individual recognition. Animal Behaviour 27:294–301.

Caldwell, R.L. 1985. A test of individual recognition in the stomatopod *Gonodactylus festae*. Animal Behaviour 44:11–19.

Caldwell, R.L. 1992. Recognition, signaling and reduced aggression between former mates in a stomatopod. Animal Behaviour 44:11–19.

Capelli, G.M. 1975. Distribution, life history, and ecology of crayfish in northern Wisconsin, with emphasis on *Orconectes propinquus* (Girard). Unpublished Ph.D. thesis, University of Wisconsin, Madison.

Capelli, G.M., and P.A. Hamilton. 1984. Effects of food and shelter on aggressive activity in the crayfish *Orconectes rusticus* (Girard). Journal of Crustacean Biology 4:252–260.

Cardé, R.T., and J.G. Millar. 2004. Advances in insect chemical ecology. Cambridge University Press, Cambridge.

Chang, E.S., M.J. Bruce, and S.L. Tamone. 1993. Regulation of crustacean molting: a multi-hormonal system. American Zoologist 33:324–329.

Copp, N.H. 1986. Dominance hierarchies in the crayfish *Procambarus clarkii* (Girard, 1852) and the question of learned individual recognition (Decapoda, Astacidea). Crustaceana 51:9–24.

Crocker, D.W., and J.E. Barr. 1968. Handbook of the crayfishes of Ontario. Royal Ontario Museum Life Sciences, University of Toronto Press, Toronto.

Crook, R., B. Patullo, and D. MacMillan. 2004. Multimodal individual recognition in the crayfish *Cherax destructor*. Marine and Freshwater Behaviour and Physiology 37:271–285.

Daws, A.G., J. Grills, K. Konzen, and P.A. Moore 2002. Previous experiences alter the outcome of aggressive interactions between males in the crayfish, *Procambarus clarkii*. Marine and Freshwater Behavior and Physiology 35:139–148.

Diaz, E.R., and M. Thiel. 2003. Female rock shrimp prefer dominant males. Journal of the Marine Biological Association of the United Kingdom 83:941–942.

Dingle, H. 1983. Strategies of agonistic behavior in Crustacea. Pages 85–111 *in*: S. Rebach and D.W. Dunham, editors. Studies in adaptation: the behavior of higher Crustacea. John Wiley and Sons, New York.

Duffy, J.E., C.L. Morrison, and K.S. Macdonald. 2002. Colony defense and behavioral differentiation in the eusocial shrimp *Synalpheus regalis*. Behavioral Ecology and Sociobiology 51:488–495.

Dunham, D.W., and J.W. Oh. 1992. Chemical sex discrimination in crayfish *Procambarus clarkii*: role of antennules. Journal of Chemical Ecology 18:2363–2372.

Edsman, L., and A. Jonsson. 1996. The effect of size, antennal injury, ownership, and ownership duration on fighting success in male signal crayfish, *Pacifastacus leniusculus* (Dana). Nordic Journal of Freshwater Research 72:80–87.

Edwards, D.H., and E.A. Kravitz. 1997. Serotonin, social status and aggression. Current Opinion in Neurobiology 7:812–819.

Eggleston, D., and R. Lipcius. 1992. Shelter selection by spiny lobster under variable predation risk, social conditions, and shelter size. Ecology 73:992–1011.

Ellis, L. 1995. Dominance and reproductive success among non-human animals: a cross-species comparison. Ethology and Sociobiology 16:257–333.

Figler, M.H., M. Twum, J.E. Finkelstein, and H.V.S. Peeke. 1995. Maternal aggression in red swamp crayfish (*Procambarus clarkii*, Girard): the relation between reproductive status and outcome of aggressive encounters with male and female conspecifics. Behaviour 132:107–125.

Figler, M.H., H.M. Cheverton, and G.S. Blank. 1999. Shelter competition in juvenile red swamp crayfish (*Procambarus clarkii*): the influences of sex differences, relative size, and prior residence. Aquaculture 178:63–75.

Garvey, J.E., and R.A. Stein. 1992. Evaluating how chela size influences the invasion potential of an introduced crayfish (*Orconectes rusticus*). American Midland Naturalist 129:172–184.

Gherardi, F., and J. Atema. 2005. Memory of social partners in hermit crab dominance. Ethology 111:1–15.

Gherardi, F., and W. Daniels. 2003. Dominance hierarchies and status recognition in the crayfish *Procambarus acutus acutus*. Canadian Journal of Zoology 81:1269–1281.

Gherardi, F., S. Barbaresi, and A. Raddi. 1999. The agonistic behaviour in the red swamp crayfish, *Procambarus clarkii*: functions of the chelae. Freshwater Crayfish 12:233–243.

Glanzman, D.L., and F.B. Krasne. 1983. Serotonin and octopamine have opposite modulatory effects on the crayfish's lateral giant escape reaction. Journal of Neuroscience 3:2263–2269.

Glanzman, D.L., and F.B. Krasne. 1986. 5,7-Dihydroxytryptamine lesions of crayfish serotonin-containing neurons: effect on the lateral giant escape reaction. Journal of Neuroscience 6:1560–1569.

Guiasu, R.C., and D.W. Dunham. 1998. Inter-form agonistic contests in male crayfishes, *Cambarus robustus* (Decapoda, Cambaridae). Invertebrate Biology 117:144–154.

Hazlett, B.A. 1969. Individual recognition and agonistic behavior in *Pagurus bernhardus*. Nature 222:268–269.

Hazlett, B.A., D. Rittschof, and D. Rubenstein. 1974. Behavioral biology of the crayfish *Orconectes virilis*. I. Home range. American Midland Naturalist 92:301–319.

Hazlett, B.A., D. Rubenstein, and D. Rittschof. 1975. Starvation, energy reserves, and aggression in the crayfish *Orconectes virilis* (Hagen, 1870) (Decapoda, Cambaridae). Crustaceana 28:11–16.

Herberholz, J., and B. Schmitz. 2001. Signaling via water currents in behavioral interactions of snapping shrimp (*Alpheus heterochaelis*). Biological Bulletin 201:6–16.

Herberholz, J., F.A. Issa, and D.H. Edwards. 2001. Patterns of neural circuit activation and behavior during dominance hierarchy formation in freely behaving crayfish. Journal of Neuroscience 21:2759–2767.

Hill, A.M., and D.M. Lodge. 1994. Diel changes in resource demand: competition and predation in species replacement among crayfishes. Ecology 75:2118–2126.

Hill, A.M., and D.M. Lodge. 1995. Multi-trophic level impact of sublethal interactions between bass and omnivorous crayfish. Journal of the North American Benthological Society 14:306–314.

Holdich, D.M., and I.D. Reeve. 1988. Functional morphology and anatomy. Pages 11–51 *in*: D.M. Holdich and R.S. Lowery, editors. Freshwater crayfish—biology, management and exploitation. Croom Helm, London, and Timber Press, Portland, Ore.

Huber, R., and A. Delago. 1998. Serotonin alters decisions to withdraw in fighting crayfish, *Astacus astacus*: the motivational concept revisited. Journal of Comparative Physiology A 182:573–583.

Huber, R., M. Orzeszyna, N. Pokorny, and E.A. Kravitz. 1997a. Biogenic amines and aggression: experimental approaches in crustaceans. Brain, Behavior and Evolution 50:60–68.

Huber, R., K. Smith, A. Delago, K. Isaksson, and E.A. Kravitz. 1997b. Serotonin and aggressive motivation in crustaceans: altering the decision to retreat. Proceedings of the National Academy of Sciences, USA, 94:5939–5942.

Huber, R., J. Panksepp, Z. Yue, A. Delago, and P.A. Moore. 2001. Dynamic interactions of behavior and amine neurochemistry in acquisition and maintenance of social rank in crayfish. Brain, Behavior and Evolution 57:271–282.

Hughes, M. 1996. The function of concurrent signals: visual and chemical communication in snapping shrimp. Animal Behaviour 52:247–257.

Johnson, V.R., Jr. 1977. Individual recognition in the banded shrimp *Stenopus hispidus* (Olivier). Animal Behaviour 25:418–428.

Karavanich, C., and J. Atema 1998a. Individual recognition and memory in lobster dominance. Animal Behaviour 56:1553–1560.

Karavanich, C., and J. Atema 1998b. Olfactory recognition of urine signals in dominance fights between male lobster, *Homarus americanus*. Behaviour 135:719–730.

Karnofsky, E.B., J. Atema, and R.H. Elgin. 1989a. Field observations of social behavior, shelter use, and foraging in the lobster, *Homarus americanus*. Biological Bulletin 176:239–246.

Karnofsky, E.B., J. Atema, and R.H. Elgin. 1989b. Natural dynamics of population structure and habitat use of the lobster, *Homarus americanus*, in a shallow cove. Biological Bulletin 176:247–256.

Krasne, F.B., A. Shamsian, and R. Kulkarni. 1997. Altered excitability of the crayfish lateral giant escape reflex during agonistic encounters. Journal of Neuroscience 17:709–716.

Livingstone, M.S., R.M. Harris-Warrick, and E.A. Kravitz 1980. Serotonin and octopamine produce opposite postures in lobsters. Science 208:76–79.

Lodge, D.M., and A.M. Hill. 1994. Factors governing species composition, population size, and productivity of cool-water crayfishes. Nordic Journal of Freshwater Research 69:111–136.

Momot, W.T. 1995. Redefining the role of crayfish in aquatic ecosystems. Review of Fisheries Science 3:33–63.

Moore, P.J., N.L. Reagan-Wallin, K.F. Haynes, and A.J. Moore. 1997. Odour conveys status on cockroaches. Nature 389:25.

Nevitt, G., N. Pentcheff, K. Lohmann, and R. Zimmer. 2000. Den selection by the spiny lobster *Panulirus argus*: testing attraction to conspecific odors in the field. Marine Ecology Progress Series 203:225–231.

Obermeier, M., and B. Schmitz. 2003a. Recognition of dominance in the big-clawed snapping shrimp (*Alpheus heterochaelis* Say 1818) part I: individual or group recognition? Marine and Freshwater Behaviour and Physiology 36:1–16.

Obermeier, M., and B. Schmitz. 2003b. Recognition of dominance in the big-clawed snapping shrimp (*Alpheus heterochaelis* Say 1818) part II: analysis of signal modality. Marine and Freshwater Behaviour and Physiology 36:17–29.

O'Neill, D.J., and J.S. Cobb 1979. Some factors influencing the outcome of shelter competition in lobsters (*Homarus americanus*). Marine Behavioral Physiology 6:33–45.

Parker, G.A. 1974. Assessment strategy and the evolution of fighting behaviour. Journal of Theoretical Biology 47:223–243.

Pavey, C.R., and D.R. Fielder. 1996. The influence of size differential on agonistic behaviour in the freshwater crayfish, *Cherax cuspidatus* (Decapoda: Parastacidae). Journal of Zoology 238:445–457.

Peeke, H.V.S., J. Sippel, and M.H. Figler. 1995. Prior residence effects in shelter defense in adult signal crayfish (*Pacifastacus leniusculus*), results in same- and mixed-sex dyads. Crustaceana 68:873–881.

Ranta, E., and K. Lindström 1992. Power to hold sheltering burrows by juveniles of the signal crayfish, *Pacifastacus leniusculus*. Ethology 92:217–226.

Reynolds, J.D. 2002. Growth and reproduction. Pages 152–192 *in*: D.M. Holdich, editor. Biology of freshwater crayfish. Blackwell Science, Oxford.

Rosenberg, M. 1997. Evolution of shape differences between the major and minor chelipeds of *Uca pugnax* (Decapoda: Ocypodidae). Journal of Crustacean Biology 17:52–59.

Rubenstein, D.I., and B.A. Hazlett. 1974. Examination of the agonistic behaviour of the crayfish *Orconectes virilis* by character analysis. Behaviour 50:193–216.

Rufino, M.M., and D.A. Jones. 2001. Binary individual recognition in *Lysmata debelius* (Decapoda: Hippolytidae) under laboratory conditions. Journal of Crustacean Biology 21:388–392.

Rutherford, P.L., D.W. Dunham, and V. Allison. 1995. Winning agonistic encounters by male crayfish *Orconectes rusticus* (Girard) (Decapoda, Cambaridae)—chela size matters but chela symmetry does not. Crustaceana 68:526–529.

Rutherford, P.L., D.W. Dunham, and V. Allison. 1996. Antennule use and agonistic success in the crayfish *Orconectes rusticus* (Girard, 1852) (Decapoda, Cambaridae). Crustaceana 69:117–122.

Schmitz, B., and J. Herberholz. 1998. Snapping behaviour in intraspecific agonistic encounters in the snapping shrimp (*Alpheus heterochaelis*). Journal of Bioscience 23:623–632.

Schroeder, L., and R. Huber. 2001. Fight strategies differ with size and allometric growth of claws in crayfish, *Orconectes rusticus*. Behaviour 138:1437–1449.

Smith, M.R., and D.W. Dunham. 1996. Antennae mediate agonistic physical contact in the crayfish *Orconectes rusticus* (Girard, 1852) (Decapoda, Cambaridae). Crustaceana 69:668–675.

Sneddon, L.U., F.A. Huntingford, A.C. Taylor, and J.F. Orr. 2000a. Weapon strength and competitive success in the fights of shore crabs (*Carcinus maenas*). Journal of Zoology 250:397–403.

Sneddon, L.U., A.C Taylor, F.A. Huntingford, and D.G. Watson. 2000b. Agonistic behavior and biogenic amines in shore crabs *Carcinus maenas*. Journal of Experimental Biology 203:537–545.

Steger, R., and R.L. Caldwell. 1983. Intraspecific deception by bluffing: a defense strategy of newly molted stomatopods (Arthropoda: Crustacea). Science 221:558–560.

Stein, R.A. 1976. Sexual dimorphism in crayfish chelae: functional significance linked to reproductive activities. Canadian Journal of Zoology 54:220–227.

Tamm, G., and J.S. Cobb. 1978. Behavior and the crustacean molt cycle: changes in aggression of *Homarus americanus*. Science 200:79–81.

Taylor, C.A. 2002. Taxonomy and conservation of native crayfish sticks. Pages 236–257 *in*: D. Holdich, editor. Biology of freshwater crayfish. Blackwell Science, Oxford.

Thomas, W.J. 1970. The setae of *Austropotamobius pallipes* (Crustacea: Astacidae). Journal of Zoology 160:91–142.

Thorp, J.H., and K.S. Ammerman. 1978. Chemical communication and agonism in the crayfish, *Procambarus acutus acutus*. American Midland Naturalist 100:471–474.

Tierney, A.J. 2001. Structure and function of invertebrate 5-HT receptors: a review. Comparative Biochemistry and Physiology 128:791–804.

Tierney, A.J., C.S. Thompson, and D.W. Dunham. 1984. Site of pheromone reception in the crayfish *Orconectes propinquus* (Decapoda, Cambaridae). Journal of Crustacean Biology 4:554–559.

Usio, N., M. Konishi, and S. Nakano. 2001. Species displacement between an introduced and a "vulnerable" crayfish: the role of aggressive interactions and shelter competition. Biological Invasions 3:179–185.

Vogt, G. 2002. Functional anatomy. Pages 53–151 *in*: D.M. Holdich, editor. Biology of freshwater crayfish. Blackwell Science, Oxford.

Vorburger, C., and G. Ribi. 1999. Aggression and competition for shelter between a native and an introduced crayfish in Europe. Freshwater Biology 42:111–119.

Weiger, W.A. 1997. Serotonergic modulation of behaviour: a phylogenetic overview. Biological Reviews 72:61–95.

Wilson, E.O. 1975. Sociobiology. Belknap Press, Harvard University, Cambridge, Mass.

Winston, M.L., and S. Jacobson. 1978. Dominance and effects of strange conspecifics on aggressive interactions in the hermit crab *Pagurus longicarpus* (Say). Animal Behaviour 26:184–191.

Yeh, S.R., R.A. Fricke, and D.H. Edwards. 1996. The effect of social experience on serotonergic modulation of the escape circuit of crayfish. Science 271:366–369.

Yeh, S.R., B.E. Musolf, and D.H. Edwards. 1997. Neuronal adaptations to changes in the social dominance status of crayfish. Journal of Neuroscience 17:697–708.

Zulandt Schneider, R.A., R.W.S. Schneider, and P.A. Moore. 1999. Recognition of dominance status by chemoreception in the red swamp crayfish, *Procambarus clarkii*. Journal of Chemical Ecology 25:781–794.

Zulandt Schneider, R.A., R. Huber, and P.A. Moore. 2001. Individual and status recognition in the crayfish, *Orconectes rusticus*: the effects of urine release on fight dynamics. Behaviour 138:137–153.

Chemical Communication and Social Behavior of the Lobster *Homarus americanus* and Other Decapod Crustacea

Jelle Atema

Molly A. Steinbach

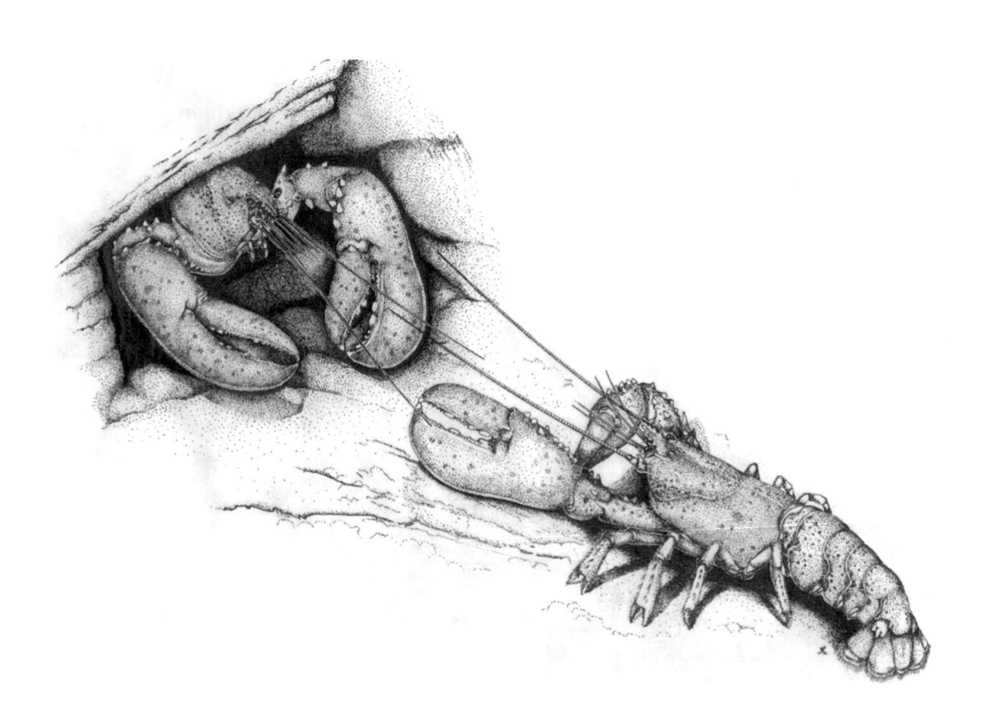

6

Chemical signals are important to the behavior of nearly all organisms in which they have been investigated. Some chemical signals are known as pheromones: chemical communication signals among conspecifics. Pheromones evolve their communication function under natural and sexual selection. We use the broader term "chemical signals" to indicate cases in which there is no evidence that the signals are elaborated by the sender for communication purposes: such signals are released as metabolic waste products that are conveniently detected by anyone with the right receptors and interest, including predators, parasites, and conspecifics. Examples of intraspecific, nonpheromonal chemical signals include certain alarm substances (e.g., Atema and Stenzler 1977) and diet-related excretion (Bryant and Atema 1987).

Pheromones have been classified into four categories based on their biological consequences: *releasers* of behavior and *primers* of (neuro)endocrine processes (Wilson and Bossert 1963) and, recently proposed for humans, *signalers* of information and *modulators* altering outcome probability (Wysocki and Preti 2002, 2004). Marine pheromones are far less studied than are terrestrial pheromones, and most information we have is limited to behavioral evidence for their existence. The pheromones of decapod crustaceans are known mostly from behavioral evidence, which suggests that they can be classified as releasers of stereotyped behaviors such as agonistic responses, mate or gamete attraction, and larval release or settling. We also see signalers of sex, molt state, receptivity, and individual identity. Primers are less known, but dominance regulation and female molt regulation provide examples worth investigating. Modulators have not been studied as such. Urine appears to be the most common vehicle for crustacean pheromones, but contact pheromones have also been suggested.

This review considers the role of chemical signals that regulate the social behavior of decapod crustaceans, primarily dominance and courtship. Reviews of this subject have been published (Dunham 1978, 1988). In the first section, we use the American lobster, *Homarus americanus*, as a model, because the social life of this often nocturnal animal is largely regulated by chemical signals, and there is a large behavioral, ecological and neurobiological database for context (Factor 1995). This includes lab and field studies, signal dispersal, and chemoreception (Atema and Voigt 1995). The lobster example suggests that in other species many more chemical signals will be discovered. In the second section, we review other species that provide both interesting examples of similar signaling and their own unique use of different pheromones. We recognize that the lobster story is by no means complete. Overall, the most glaring gap in knowledge is the lack of information on the chemical identity of the pheromones. The structure determination of bombykol, the major sex attractant of the silk worm moth, *Bombyx mori* (Butenandt et al. 1959) sparked a revolution in the field of chemical ecology. The current knowledge of crustacean pheromones shows that we can expect complexity rivaling that in insects and vertebrates (e.g., Duffy et al. 2002). But unless the chemistry catches up with the biology, we will remain ignorant about the true extent of chemical signaling in crustaceans.

Lobsters as an Example of Complex Chemical Communication in Decapod Crustacea

Chemical Signaling Behavior: Information Currents

To understand the use of chemical signals in the social behavior of any animal, it is important to appreciate the transport mechanisms that carry the signal from sender to receiver. Odor dispersal determines signal detectability. Often, this means understanding the invisible air or water currents, including animal-generated currents that disperse the odors. *Homarus americanus* is a good example of an animal that generates several multipurpose currents. These currents aid not only in locomotion, feeding, and metabolic functions, such as breathing and waste removal, but also in chemical information exchange. They may be thought of as a fluid extension of the animal itself. Juvenile and adult *H. americanus* generate three currents that can operate in isolation or in concert: gill current, fan current, and pleopod current (Fig. 6.1). All three are implicated in chemical communication. In larval and early juvenile lobsters, the currents do not seem to be involved primarily in information transfer, but rather in locomotion and feeding (reviewed in Ennis 1995, Lavalli and Factor 1995).

The gill current is generated by the scaphognathites (leaflike outer branches of the second maxillae) beating inside the gill chambers. Their peristaltic pump action generates a powerful, rapidly pulsating current that jets forward from bilateral "nozzles." This jet reaches distances of seven body lengths (BL) in adults (Atema 1985) (Fig. 6.2) and mean velocities of 3 cm/sec near the nozzle (T. Breithaupt, unpublished data). It is usually a bilateral current, carrying metabolites from the gills. Adult lobsters at warmer temperatures rarely cease producing this ventilating current; however, in winter the current stops for periods of several seconds (J. Atema, personal

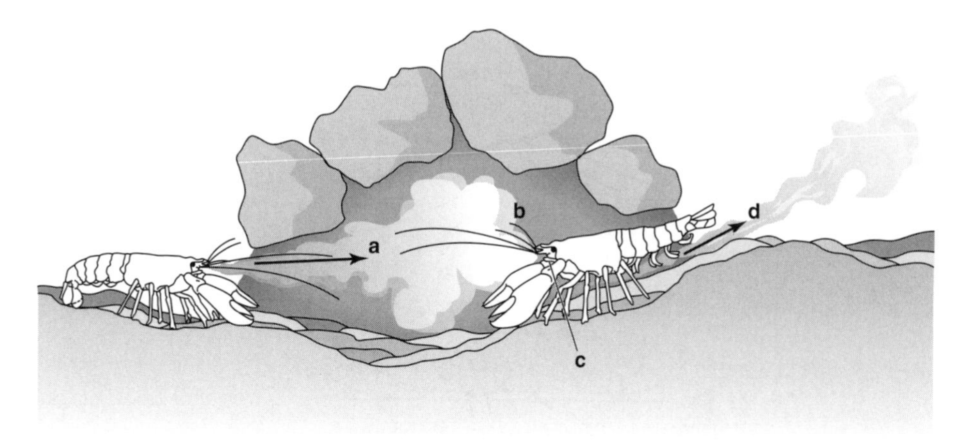

Figure 6.1 Information currents used during lobster courtship as premolt receptive female (left) visits shelter of dominant male. Female jets gill current (a) into shelter. Male (right) retreats to opposite entrance, flicks antennules (b), fans exopodites (c) redirecting his gill current, and beats pleopods under raised abdomen (d) to create posterior "advertising" current outside shelter. From Atema (1986).

observation), presumably reflecting lower metabolic rates. Urine can be released into the gill current from bilateral bladders, through small, ventrally directed nozzles (nephropores) at the base of the large antennae. Lobsters appear to release products of a pair of small nephropore glands into the urine (see "Nephropore Glands," below).

A second current, the fan current, exerts further control over signaling. The stiff exopodite of the first maxilliped can be positioned directly in front of the nozzle, thus deflecting and redirecting the water flow sideways (T. Breithaupt, personal communication). The large, feathery exopodites of the second and third maxillipeds then fan

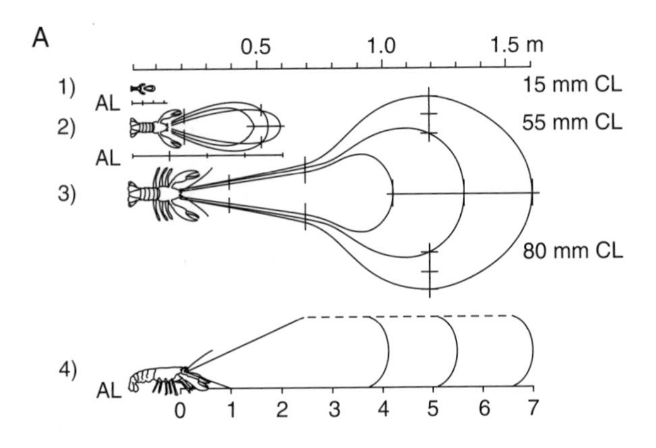

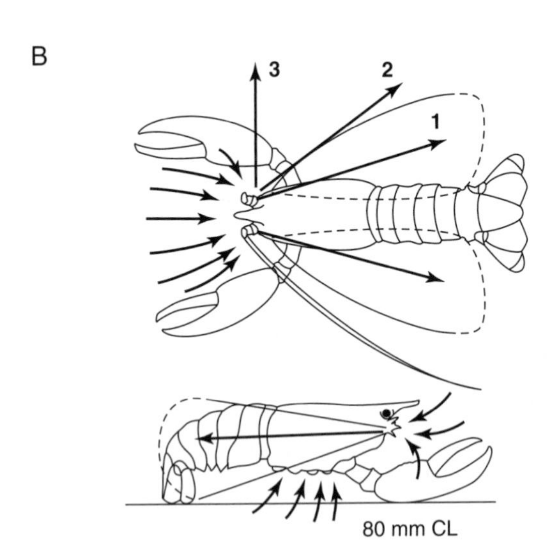

Figure 6.2 Lobster-generated currents. (A) Gill currents with mean and standard deviations: top view of three different-sized animals (1–3; 15, 55, and 80 mm carapace length [CL]) and side view of adult (4; 80 mm CL) animal (broken line: vertical expansion of plume limited by stratification of water). AL: animal body length. (B) Exopodite "fan" current. Top, Small anterior arrows indicate region and direction of water flow drawn toward the lobster. "Wings" indicate areas of turbulent directional current: 1 is most common bilateral; 2 and 3 indicate increasingly unilateral flow fields. Bottom, Arrows in side view indicate ventral water uptake into gill chamber and anterior current resulting from redirected gill current. From Atema (1985).

the deflected water backward. The water displaced by these outgoing currents draws incoming currents from around the head within a radius of about the length of the antennules (Atema 1985), which monitor chemical signals from the environment. The exopodite fan current can be bilateral or unilateral (Fig. 6.2).

The third and most powerful current is the pleopod current, which draws water from below the animal and moves it posteriorly by beating the pleopods—also known as swimmerets—underneath the raised abdomen (Atema 1985). This behavior is seen in the "advertising" of cohabiting males (Cowan and Atema 1990). A good example of the use of these currents in chemical communication is seen in courtship (Fig. 6.1). Much of the information exchange between potential mates occurs when the female approaches the shelter of a male, where both increase urine release (Bushmann and Atema 2000). Females visiting a shelter stand still at its entrance for many seconds (Atema et al. 1979, Karnofsky et al. 1989a, Cowan and Atema 1990). They alternate between fanning and not fanning. Although the current itself is, of course, invisible, the human observer can conclude that *without* fanning the continuous gill current blows into the male shelter. At the time of female visits, the male stands inside and often moves away from the entrance, flicking his antennules (i.e., "sniffing"), fanning his exopodites (thus drawing water toward his antennules and redirecting his gill current backward), and occasionally fanning his pleopods, drawing a large current into his shelter and blowing it out into the environment. Together, male and female currents result in mutual odor signaling. It is not known why, during copulation, both partners beat their pleopods so vigorously and thus disseminate their joint chemical signals into the surrounding environment.

Chemoreception

Lobsters have at least five different major chemoreceptor organs that could be involved in detecting pheromones: the lateral and medial flagella of the antennules; the antennae; the dactyl and propodus segments of the walking legs, particularly the chelated first two pairs; and the endopodite of the third maxillipeds. The antennules and antennae can be grouped as the cephalic receptor organs with neural connections into anterior brain centers, and the others as thoracic receptor organs projecting to centers in the posterior brain and thoracic ganglia (Sandeman et al. 1992). Cephalic chemoreceptor organs function primarily to sample odors in the free stream around the animal's head; the thoracic chemoreceptor organs appear to be primarily involved in feeding behavior. This corresponds roughly to the smell and taste division in vertebrates (Atema 1977). The lateral filament of each antennule flicks (Shepheard 1974, Schmitt and Ache 1979, Berg et al. 1992) and thus samples odor directly in front of the animal. It is the functional equivalent of sniffing and demonstrates again the importance of fluid flow for chemoreception. This organ is involved in the detection of at least some pheromones (Johnson and Atema 2005) and in the tracking of food odor plumes. One of the reasons that more is not known about pheromone receptors is that the compounds are not known. Details on lobster chemoreception have been reviewed elsewhere (Atema and Voigt 1995).

Social Behavior: Dominance, Courtship, and Mating

The courtship and mating of the American lobster are, as in so many animals, based on male dominance and female choice. The necessary information for establishing and maintaining dominance and pair bonds is based largely on chemical signals. Males broadcast their sex, status, identity, and location both to rival males and to potential mates. Females convey their sex, identity, and state of receptivity and may exert chemical control over each other's molt cycles (reproductive suppression), allowing for female molt staggering.

Dominance and Individual Recognition

For lobsters, shelters and mating opportunities are limiting resources. Dominance serves to secure priority access to these resources. Individual recognition and the winning and losing of encounters determine an animal's status within a social group (Dugatkin and Earley 2004). Dominance is established by behavioral displays, generally followed by physical fights. The more closely matched the opponents are, the further the fight escalates. Once established, dominance is maintained mostly by displays and rarely by continued fights. These displays have an important chemical component, perhaps coupled with tactile, hydrodynamic, and visual cues. There appear to be two different kinds of chemical signals that help maintain dominance: status signals and individual recognition signals. We present experimental evidence from behavioral tests for these chemical signals.

We should point out that, although most experiments on dominance involve males, females also have dominance relationships and individual recognition. But, where the function of male dominance seems clear (securing the best shelter and attracting females), the function of female dominance is not as clear. We suggested priority in securing shelters perhaps close to the dominant male, but the expected priority for mating with the dominant male was *not* found (Cowan and Atema 1990).

Much of the evidence for chemical signals in dominance encounters is based on a simple assay: the forced encounter or "boxing match." The first detailed analysis of lobster agonistic behavior comes from Scrivener (1971). Similar results have been obtained under more natural conditions (Stein et al. 1975, Atema et al. 1979, Karnofsky and Price 1989, Cowan and Atema 1990) and in the field (Karnofsky et al. 1989a). Agonistic encounters follow a predictable sequence of escalation from displays such as "meral spread" (the extension and raising of the claws), to physical contact such as pushing with the claws and whipping with the antennae, to damaging behavior such as claw locking, jabbing, and ripping of antennae, antennules, and legs that can lead to limb loss, bleeding, and death. The latter is rare, but occasionally lobsters can be very aggressive, particularly naive animals raised in isolation (Huber and Kravitz 1995). Physical aggression observed in the field and among long-term aquarium residents tends to be infrequent and much less intense than in forced encounters (Stein et al. 1975, Karnofsky et al. 1989a). Social experience reduces aggression. For a detailed description of lobster agonistic behavior, see Atema and Voigt (1995).

The decrease in aggression among familiar animals appears to involve two different chemical signals: a generic dominance status odor combined with the memory of the individual odor of a recent opponent. In boxing matches between two size- and sex-matched adult lobsters, an initially small (if any) difference in aggressiveness develops into a clear winner–loser distinction generally within 15 minutes (although fights lasting hours have been recorded), resulting in a lasting dominant–subordinate relationship. An increasing number of winning or losing experiences affects subsequent fights against other opponents, which suggests a gain or loss in "confidence," known as

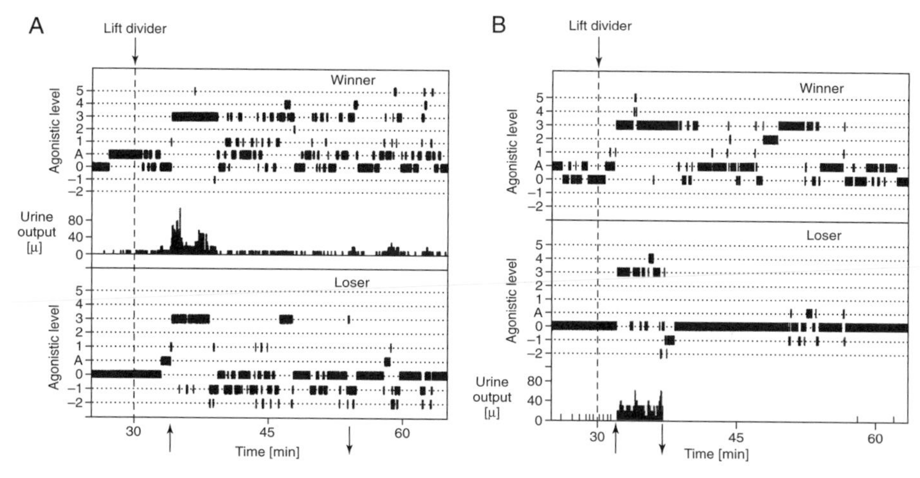

Figure 6.3 Urine release during lobster fights. Two mature male lobsters unfamiliar with each other are brought together in a "boxing tank." Interaction begins when divider is removed (vertical line). Arrows indicate start and end of fight. Agonistic levels scored every 5 seconds for both opponents indicate fight intensity: positive scores, from mere contact to increasingly severe attacks; negative scores, from avoiding to fleeing. (A) The eventual winner is catheterized, and his urine release is measured every 5 seconds. Prior to start of fight (at level 3), the winner releases some urine. Release frequency increases with agonistic level and continues throughout and after the fight. (B) In a different pair, the eventual loser is catheterized. Almost no urine is released prior to the fight; urine release stops abruptly when the loser starts fleeing, signaling the end of the fight. From Breithaupt and Atema (2000).

winner and loser effects. Accounting for effects of body size and claw size, experienced winners win and losers lose significantly more encounters than predicted by chance (Scrivener 1971). Dominance status (or "confidence") appears to affect the rate and amount of urine output at the onset of an encounter: future winners release more in the early fight stages (Breithaupt and Atema 2000) (Fig. 6.3). Status may be signaled through unknown "dominance" compounds in the urine (see "Neuromodulators as Dominance Signals?," below). The impact of a loss or a win on confidence increases until, after five consecutive losses or wins, the probability of a loss or a win, respectively, approaches 95%; winning and losing effects fade with time and future fight results (Morschauser 2002).

In addition to status recognition, there is individual recognition. If two adult lobsters meet for the first time and fight, are then separated for 24 hours (either in isolation or in communal tanks), and subsequently meet again, the previous loser will not challenge his former dominant opponent. However, the loser will challenge and can defeat an unfamiliar opponent, even if the new opponent is the recent winner of another fight (Karavanich and Atema 1998a). These results indicate that lobsters are capable of individual recognition: they distinguish between familiar and unfamiliar dominants. Memory of a specific dominant individual can last for about one week without renewing the acquaintance, whether the week was spent in isolation or among many other individuals (Karavanich and Atema 1998a). This experiment also demonstrates that losing one fight does not turn the loser into a generic subordinate: he or she can win the next fight. Individual recognition was similarly demonstrated in female–female (Atema et al. 1999) and male–female interactions (Berkey and Atema 1999). Male–female aggressive interactions can also change into mating attempts and intermolt mating, perhaps as a result of changing chemical signals.

Individual recognition is critical to the maintenance of lobster social structure. Once the dominant male has established a "mating shelter" large enough for two lobsters, he regularly patrols other nearby shelters, temporarily evicting other animals without a physical fight (O'Neill and Cobb 1979, Karnofsky et al. 1989a). We conclude that this eviction routine must be based on blowing the gill current with urine into the subordinates' shelters. Since memory lasts a week, it may be a mechanism to remind them regularly of his individual odor and status.

Blocking or catheterizing nephropores to prevent urine release into the gill current and lesioning antennules to block olfaction demonstrate that individual odor memory is based on chemical signals in urine mediated by antennular olfaction (Karavanich and Atema 1998b). Olfactory lesions prevent the establishment of a stable hierarchy, and fighting can continue for weeks (Cowan 1991). Individual recognition signals are received specifically by the chemosensory aesthetasc sensilla on the lateral flagella of the antennules (Johnson and Atema 2005). Thus, individual recognition is processed in the olfactory lobe of the brain.

Sex Identification

Male and female sex pheromones are behaviorally evident (Atema and Engstrom 1971, McLeese et al. 1977, Atema and Cowan 1986, Cowan 1991; reviewed in Atema and Cobb 1980, Atema 1986) but, despite some early attempts (McLeese et al. 1977), have not been identified chemically. In laboratory experiments, urine and tank water from the opposite sex cause closing of the seizer claw, a nonaggressive gesture, whereas same-sex stimuli cause claw opening, an aggressive gesture (Atema and Cowan 1986). In choice flume assays, females are more attracted to any male-occupied shelter, dominant or subordinate, than to an empty shelter (Bushmann and Atema 2000). This suggests that males release their sex signal more or less continuously, causing females to be attracted from a distance. Males also attract other males, leading to fights between unfamiliar opponents. In contrast, males do not discriminate between female-occupied and empty shelters (Bushmann and Atema 1997). This suggests either that males are not responding to a continuously emitted female signal or that females do not release their sex signal continuously. If females control the release of their sex signal, they use it in male proximity and to enter male shelters (Bushmann and Atema 2000). At close range, males do discriminate between sexes and even between the receptivity states of females (see "Female Signals and Receptivity," below). Restricted release of sex signals might allow females to conserve signaling compounds and to avoid interfemale aggression.

Generally, lobsters do not accept conspecifics into their shelters. However, when a female of any molt state approaches a male's shelter, she is met with little or no aggression and may be admitted into the shelter. In contrast, another male will be met aggressively, leading to expulsion of either one (Atema et al. 1979, Bushmann and Atema 1997). If the approaching female is catheterized (preventing release of urine with its chemical signals), she is met with high aggression similar to that directed at another male (Bushmann and Atema 2000). Even with normal urine release, not all females are treated equally, depending on their receptive state (see "Female Signals and Receptivity," below).

Male Signals and Female Choice

Naturalistic studies in large aquaria (Atema et al. 1979, Karnofsky and Price 1989, Cowan and Atema 1990) supported by limited field observations (Karnofsky et al. 1989a) show that the dominant male occupies a shelter sufficiently large for two animals and that the shelter of a dominant male becomes a focus of social interactions, including frequent visits by adult, premolt females. These females stop at the male shelter entrance and appear to inspect him, probably chemically. The female's decision to enter seems to be based on information from the male and the female's own internal state (e.g., molt stage, sperm status). During shelter visits both males and females greatly increase urine release (Bushmann and Atema 2000), suggesting an exchange of chemical signals. In odor choice tests, females prefer to associate with dominant males over subordinates (Bushmann and Atema 2000). When the dominant male was catheterized (preventing urine release), females were less attracted to his shelter (Bushmann and Atema 2000). Female lobsters may locate a dominant male by following a plume of dominant male odor to its source, and/or they may learn the location of his shelter from their frequent explorations of the local environment, its shelters, and lobster inhabitants. Lobsters are surprisingly quick in learning the details of their local geography and sociology (J. Atema, personal observation).

Once the female is allowed into the male's shelter, the two animals cohabit until the female molts, after which they mate. During cohabitation, whenever one of a mated pair enters the shelter, they fight mildly (typically "boxing") for a few seconds (Cowan and Atema 1990) suggestive of the time it takes for a chemical signal to disperse and be recognized. This signal may include not only sex information but also an individual signature. Individual recognition in the context of mating has not been investigated. Over the premolt cohabitation period, the frequency of male pleopod fanning increases (see "Chemical Signaling Behavior: Information Currents," above), reaches a maximum during the day of female molting, and wanes in the postmolt cohabitation days. This current carries with it all male and—inevitably—female metabolites (including chemical signals) released by the cohabiting pair (Fig. 6.1). This activity can be seen as male "advertising," since it is positively correlated with visits of other lobsters to the shelter, including premolt females (Cowan and Atema 1990). It is unlikely that the cohabiting female benefits from this advertisement.

Snyder et al. (1993) observed successful courtship and mating when either males and/or postmolt females were catheterized (preventing urine release); they questioned the presence of sex-identifying cues in the urine of *H. americanus* but mentioned that this may be the result of conducting their tests in small tanks. Indeed, in a large flume tank, catheterized males, particularly when their previously collected urine was played back, still attracted females, but with fewer approaches and entry attempts. Females were not attracted to an empty shelter emitting only the urine collected from a dominant male (Bushmann and Atema 2000). This suggests that females use both urine-based cues and other, non-urine-based cues to locate and mate with a preferred male (Bushmann and Atema 2000).

Female Signals and Receptivity

Mating in lobsters can occur at any time throughout the female's molt cycle (Waddy and Aiken 1990). It appears that females can monitor their sperm load and molt state and solicit matings when they need sperm during a long intermolt period. However, long-term, naturalistic aquarium observations showed that nearly all matings occurred shortly (around 30 minutes) after a female molts in the middle of a prolonged cohabitation period (Atema et al. 1979, Cowan and Atema 1990). Females are more likely to be receptive after molting because during the molting process they lose the sperm they may have been storing from a previous mating (Cowan and Atema 1990, Bushmann and Atema 1997). In naturalistic tanks, dominant males allowed only premolt females into their shelter for cohabitation, although females of all molt stages were met with less male aggression than were males (Cowan and Atema 1990). Clearly, males can detect differences in female receptivity.

Intermolt females with depleted sperm stores solicit males, who recognize them and mate with them immediately (Waddy and Aiken 1990, Bushmann and Atema 2000). Even when a group of lobsters is introduced into a new tank, the males take time off from the initial dominance fights to mate with such a female (J. Atema, personal observation). Presence of a sperm plug in the female indicates nonreceptivity. In laboratory experiments, all receptive females that entered male shelters were mounted, and half were mated, whereas nonreceptive females were not mated and few were mounted (Bushmann and Atema 1997). Tank water from freshly molted, isolated (and thus not mated) females is attractive to isolated males, who start searching for the source; it reduces male aggression, including reflexive closing of the seizer claw (Atema and Engstrom 1971, Atema and Cowan 1986). Lobstermen have observed that if a recently molted female finds her way into a trap, the trap fills with males.

Female chemical signals may play a role not only in courtship and cohabitation but also in mating behavior itself. Just before molting, the female faces the male and places her claws on the male's "shoulders" (anterior carapace and claws) in a "knighting" gesture. This curious behavior may be accompanied by urine release as it puts the nephropores almost directly onto the male's antennules (Atema 1986). The signals transmitted through such positioning are unknown but could indicate synchronization for imminent molting and copulation ("Don't leave me now"). Chemical signals may also be transmitted to the male during copulation when he turns over the recently molted female to insert his gonopodia into her seminal receptacle. The appendages the male uses to turn over the female (first two pairs of periopods, and third maxillipeds) are supplied richly with chemo- and mechanoreceptive sensilla (Derby 1982). Contact pheromones have not been studied in *Homarus americanus*.

Female Molt Regulation

The only long-term observations of lobsters in the field (Karnofsky et al. 1989a, 1989b) show that a small group of residents interact with a large number of transient animals. The social interactions between residents and transients are not known.

Almost all of our knowledge concerning molting, mating, and cohabitation comes from laboratory observations in which sex ratios are enforced and immigration and emigration are restricted. Adult female molt staggering is seen in such closed systems—one female at a time cohabits and—directly after molting—mates with the dominant male (Cowan and Atema 1990, Cowan et al. 1991). The mechanism causing molt staggering among females is unknown, but chemical signals (primer pheromones) are known to regulate female reproductive physiology in other animals. In mammals, primer pheromones in male urine can cause estrus synchronization (Ma et al. 1999) and the onset of puberty in females (Novotny et al. 1999). Estrus synchrony (McClintock 1983) and reproductive suppression (e.g., Barrett et al. 1993) can also occur with exposure to female urine. Most dramatically, in social insects such as the honeybee, *Apis mellifera*, chemical signals produced by the queen prevent ovary development in workers (Hoover et al. 2003), producing an effectively sterile worker class. Female reproductive suppression by dominant females has also been observed in zebrafish (Gerlach 2006) and may be more widespread in other fishes. Thus, since female lobsters molt shortly (days) after entering into cohabitation with a dominant male, it seems that his primer pheromones could be accelerating her molting. This would benefit both male and female mating partners. But what prevents the other females from molting and thereby removes their chance to mate with the dominant male? We hypothesize that female lobsters, while not cohabiting with a male, mutually suppress through primer pheromones the final stage of each others' endocrine molt preparation, specifically the well-known premolt ecdysterone peak (reviewed in Hopkins 1983, Quackenbush 1986). The female who then enters the male's shelter, would be released from this inhibition by the absence of the putative female molt-inhibiting pheromone and the exclusive exposure to the putative male molt-accelerating pheromone, allowing her to continue rapidly toward molting and mating. Such a hypothesized mechanism (Atema 1986) would benefit both males and females and result in the observed molt staggering. It could operate equally well in other crustaceans where female molt regulation is important.

Chemical Signals

To date, no lobster pheromone has been identified. We do know, however, that many chemical signals produced by the lobster are released in the urine. A possible source of some of these signals may be the nephropore gland. Logical candidates for female courtship pheromones are compounds related to molt hormones; metabolites of neuromodulators linked to aggression may be connected with dominance. As yet, there is no evidence for these suggested chemical connections.

Urine and Urine Release

Urine is stored in bilateral bladders. A typical 400 gram lobster, when taken out of a tank, can release up to 10 milliliters of urine, 6 milliliters on one side, in one fine stream under considerable pressure (J. Atema, personal observation). Urine is released through paired nephropores on the ventral sides of the basal segments of the antennae. Sphincter muscles suggest control over release (Bushmann and Atema 1996), and behavioral experiments demonstrate this. Urine release increases somewhat in the hours following feeding. It is particularly prominent in social behavior and not due to physical activity; for example, urine is not released when lobsters are chased around the tank (Breithaupt et al. 1999). Male lobsters produce significantly more urine when a second male is introduced into their tank (Breithaupt and Atema 1993, Breithaupt et al. 1999) and also when visiting male and female shelters (Bushmann and Atema 2000). At rest, urine is typically released in occasional approximately 5-second-long squirts of about 1 microliter per second; during fights the release can go up to 10 microliters per second for minutes at a time (Breithaupt and Atema 2000) (Fig. 6.3). Release probability increases with increasing fight intensity; (future) winners but not losers tend to release some urine at the start of a fight. This early release seems to anticipate winning success and may indicate confidence gained from previous winning experience (see "Dominance and Individual Recognition," above). At the end of the fight, losers but not winners stop releasing urine (Breithaupt and Atema 2000) (Fig. 6.3). This urine release behavior strongly suggests its importance in communication.

Of course, urine is a common solvent for chemical signals in many animals. Its metabolic waste disposal function can serve many signaling functions, both for social communication and for discovery by predators. As a waste disposal system, urine could be considered a rather honest signal reflecting the internal state of the animal. As such, metabolites of neuromodulators and molt hormones have been suspected as chemical signals (see below).

Nephropore Glands

The nephropore glands, first described in crabs (Fontaine et al. 1989), are composed of masses of rosette glands located lateral and medial to the ureter, approximately 100 micrometers inside the nephropore (Bushmann and Atema 1993). The glands with their ducts and surrounding muscle tissue appear designed to control release of gland product into the urine. The valve of the nephropore, in turn, controls release of urine, with or without glandular secretion added (Fig. 6.4). Although nephropore glands structurally resemble the rosettes of ubiquitous tegumental glands (Yonge 1932), their large size, highly organized structure, pathway of product release, continuous activity cycle, and histological staining pattern make nephropore glands different from adjacent tegumental rosettes. This suggests a novel function: the nephropore glands are present and active in both males and females, at all molt stages and at all times of the year. Histochemical results indicate that they contain proteinaceous material and mucopolysaccharides (Bushmann and Atema 1996).

Lobster urine contains unusually large amounts of protein, commonly between 100 and 300 micrograms per milliliter (McLaughlin et al. 1999). Releasing such large amounts of protein may indicate a signal function. In addition, urine collected during a fight contained more protein than before or after a fight, and preliminary data show individually distinct patterns of three to six protein bands on gel electrophoresis (J. Atema, M. Mallidis, M. Edattukaran, unpublished observations). The source and function of lobster urine proteins are not known; however, proteins are thought to be excellent chemical signals in aquatic environments, due to their high solubility, potentially short half-lives, and resultant high signal-to-noise ratio (reviewed in Rittschof and Cohen 2004). Proteins mediate individual recognition in vertebrates,

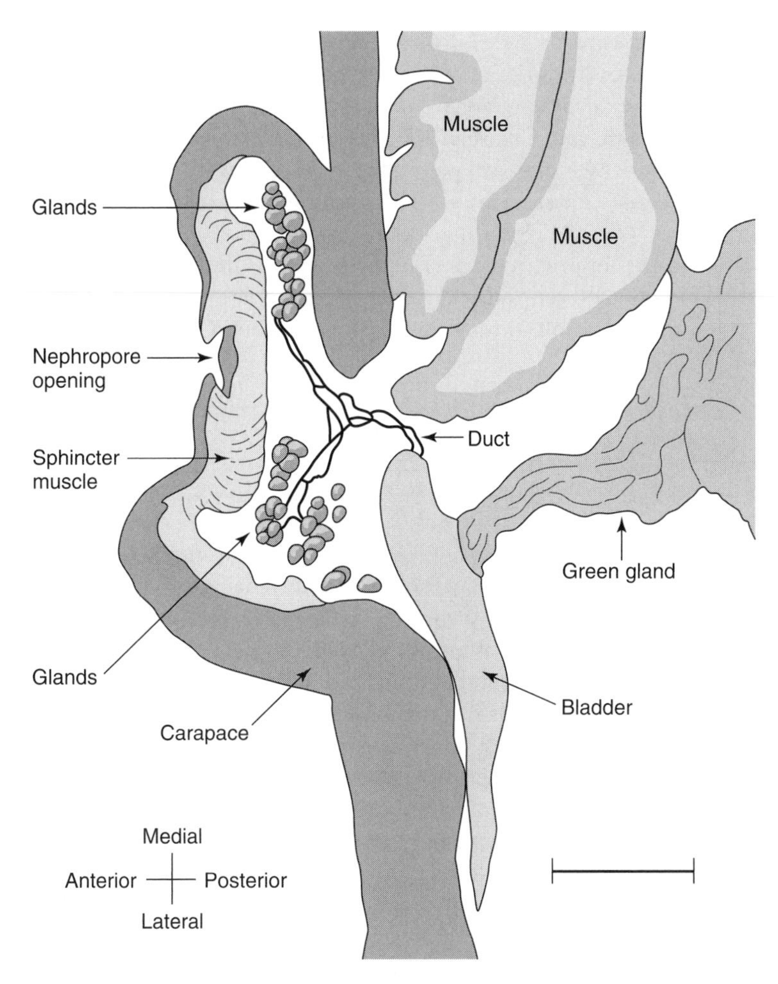

Figure 6.4 Composite drawing of nephropore area of *Homarus americanus*, from horizontal histological sections. Single, large ducts travel from each gland complex and join to become a common duct terminating in the bladder. From Bushmann and Atema (1996).

for example, through the major histocompatibility complex (Apanius et al. 1997) and major urinary proteins (Benyon and Hurst 2004).

Molt Regulation Hormones as Pheromones?

The physiological role of crustecdysone in crustacean molt regulation is well understood; blood titers peak in the days before molting and drop precipitously on the day of molting (reviewed in Hopkins 1983, Quackenbush 1986). Since urine is the primary means of ecdysteroid removal, and since mating usually occurs in concert with molting, it was hypothesized that molting hormones could serve additionally as a crustacean sex pheromone (Kittredge et al. 1971). Several authors questioned this conclusion on theoretical grounds, primarily because crustecdysone would lack sex and species specificity and provide only molt-timing information. This could be useful once the behavioral context of pairing has been accomplished by more specific signals. However, neither crustecdysone nor some of its metabolites elicited sexual behavior in male lobsters (Atema and Gagosian 1973, Gagosian and Atema 1973). Subsequent partial purification of two crab sex pheromones, *Callinectes sapidus* (Gleeson 1980) and *Carcinus maenas* (Hardege et al. 2002), showed that the cue was chemically distinct from crustecdysone. This supports the hypothesis that molting hormones and sex pheromones are not synonymous (Buchholz 1982, Gleeson et al. 1984, Hardege et al. 2002). It does not rule out that crustecdysone and its metabolites could be involved as additional molt timing signals or in other Crustacea (see also "Sex Pheromones," below).

Neuromodulators as Dominance Signals?

As with crustecdysone, it is reasonable to speculate that metabolites of neuromodulators, such as the amines serotonin and octopamine might be excreted in the urine to serve as behavioral signals. Because serotonin has been implied as a regulator of aggressive behavior in lobsters, its metabolites might signal social status or confidence, and thus, serve as dominance pheromones. This link, however suggestive, has not been established despite several attempts. Kravitz (2000) reviewed the work on the connection between serotonin and dominance, using the lobster as a crustacean model. Following original studies of postural effects of serotonin (Livingstone et al. 1980, Kravitz 1988) subordinate lobsters (and crayfish) injected with serotonin showed increased tendency to instigate agonistic encounters with dominants, increased fight duration, and a decreased probability of retreating (Huber et al. 1997a). While such studies indicate a physiological link between the neuromodulator and behavior, it does not indicate a pheromonal effect.

In high-performance liquid chromatography analysis of lobster urine, all radioactive metabolites of serotonin appeared as a single peak that coelutes with the serotonin-O-sulfate standard (Huber et al. 1997b); neither the metabolites nor the standard has been tested for pheromonal activity. Unlike in most vertebrate systems, the primary serotonin metabolite does not appear to be 5-hydroxyindole (Peeke et al. 2000).

It is worth mentioning that, in addition to amines such as serotonin and octopamine, other hormones and peptides are indicated in regulating lobster aggression (reviewed in Kravitz 2000). These include the molting hormone, ecdysone; its active form 20-hydroxyecdysone (= crustecdysone); and a lobster stress hormone, crustacean hyperglycemic hormone (Chang et al. 1999). Crustecdysone levels peak just prior to ecdysis (Hopkins 1983, Quackenbush 1986, Snyder and Chang 1991), coincident with highest levels of aggression (Tamm and Cobb 1978, Atema et al. 1979, Cromarty et al. 1991) and affecting neuromuscular junction potentials in claw and abdominal muscles (Cromarty and Kass-Simon 1998). Crustacean hyperglycemic hormone peptides are produced by neurosecretory neurons, which are in turn influenced by both serotonin and octopamine (Chang et al. 1999).

In sum, despite sound reasoning and tantalizing clues from neurobiology, the chemical or functional links between neuromodulators, hormones, and pheromones are not clear at this point. We therefore do not add to further speculation.

Chemical Communication in Other Decapod Crustacea

Information Currents

The first description of what we now call information currents was given in great detail for *Pagurus arrosor* hermit crabs by Friedrich Brock (1926). He observed that the fan organs, feathery flagella of the exopodites of the maxillipeds, generate water currents that can be directed forward (by beating the exopodites of the maxillipeds bilaterally) and to either side (by unilateral fanning). In unilateral fanning the outgoing current is fed by an incoming current from the opposite side of the head. He then noted that both antennules move sideways to flick in parallel into the incoming current. Forward fan currents are fed by an anterior counter current, and the antennules flick again into this incoming water. In this way, an animal can send and obtain directional chemical information.

Similar to hermit crabs, in the crayfish *Procambarus clarkii* fan organs generate a variety of well-described fluid flows, allowing the animal to send and receive chemical signals from the environment (Breithaupt 2001). These flows can be bilateral or unilateral, drawing fluid horizontally from nearly all directions toward the body, including the antennular chemoreceptors. For *Orconectes rusticus*, Bergman et al. (2005) described posterior currents (i.e., those generated by pleopods) and anterior currents (i.e., those generated by maxilliped fan organs) used in concert with urine release during agonistic behavior, particularly highly aggressive claw grasping, all remarkably similar to currents and behaviors in *H. americanus* (Breithaupt and Atema 2000).

It is interesting in a phylogenetic context that the currents of hermit crabs and crayfish are functionally similar to the ones of *H. americanus* (Atema 1985), but they are generated differently. In the lobster, the anterior currents are driven primarily by the powerful gill current and can be redirected by the exopodite fan organs. The gill current of hermit crabs and crayfish, including *Astacus astacus* and *A. leptodactylus* (J. Atema, unpublished observations), is not as powerful—perhaps as a result of smaller body size and related metabolic demands (see Fig. 6.2a)—and the anterior currents are mostly generated by the fan organs themselves. It would be interesting to see if large crayfish such as *Cherax destructor* or the Tasmanian giant crayfish, *Astacopsis gouldi*, use the fan organs to modulate the gill current as shown in lobsters.

Nephropore Gland

The earliest reference to a nephropore gland is in the green crab, *Carcinus maenas* (Fontaine et al. 1989), and we know of no other published accounts. From its presence in crabs and lobsters, we suspect it will be present in many decapod crustaceans and perhaps in crustaceans in general. The location and structure of the gland appear similar in *H. americanus* and *C. maenas*, and so are the speculations regarding its possible pheromonal function.

Sex Pheromones

The presence of a female sex pheromone in decapod crustaceans was first demonstrated in the crab *Portunus sanguinolentus* (Ryan 1966). Several other species of crabs have since been shown to utilize sex pheromones, but only one has been fully

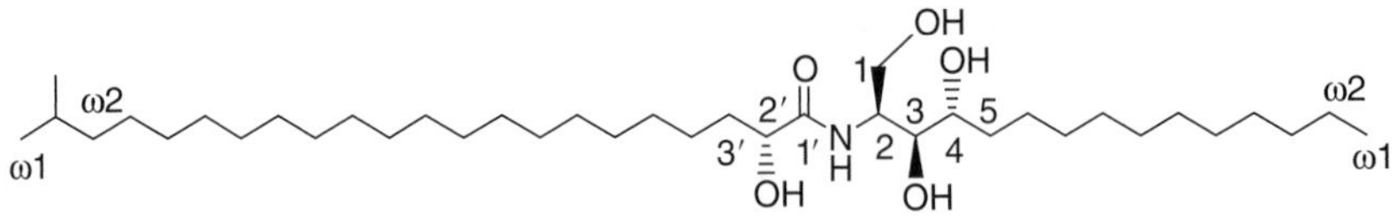

Figure 6.5 Ceramide structure of sex pheromones in the female hair crab, *Erimacrus isenbeckii*. From Asai et al. (2000).

characterized. Pre- and postmolt female hair crabs, *Erimacrus isenbeckii*, use a mixture of ceramides (lipid secondary messengers used in intracellular signaling systems) as a sex pheromone to elicit mating behavior from males; the compounds have been isolated, characterized (Asai et al. 2000), and synthesized (Asai et al. 2001, Masuda et al. 2002). The resultant synthetic compounds (Fig. 6.5) were identical to the natural pheromone. It is not clear if specific mixtures are required or whether any one of the compounds alone can elicit male mating behavior.

Studies on the blue crab, *Callinectes sapidus*, demonstrate that chemical signals are present in the urine of pre- and postpubertal molt females, inciting courtship and mate-carrying behavior in males (Gleeson 1980; reviewed in Gleeson 1991). Visual cues appear to have no effect on courtship or mating behaviors (Gleeson 1980, Bushmann 1999). Fractionation showed that although 20-hydroxyecdysone was present in this urine, it was not in the bioactive fraction (Gleeson et al. 1984), indicating that it is not the pheromone involved in signaling in this context as had been suggested (Kittredge et al. 1971). The active fraction has not been fully characterized.

The presence of a sex pheromone in the green crab, *Carcinus maenas*, was first demonstrated by Eales (1974). Male conspecifics responded to premolt female urine by displaying searching behavior. The pheromone causes male crabs to fight longer and more intensely than in the absence of the signal (Sneddon et al. 2003). Males attempt to mate even with inanimate objects such as stones conditioned with the chemical signal (Hardege et al. 2002). The pheromone may have dose-independent effects (searching for a female) and dose-dependent effects (cradle carrying and stroking the female) (Ekerholm and Hallberg 2005). If antennular chemoreception is blocked, control of sexual behavior is disrupted (Bamber and Naylor 1996). The chemical identity of the female green crab sex pheromone is not known. Hardege et al. (2002) determined it to be a small molecule in premolt and postmolt female urine chemically distinct from 20-hydroxyecdysone. It has a molecular weight of less than 10 kDa, permitting it to cross the renal membrane.

The helmet crab, *Telmessus cheiragonus*, shows two female sex pheromones: a female precopulatory pheromone found in the urine of both pre- and postmolt females that induces male mate guarding (Kamio et al. 2000), and a postmolt female signal (not carried in the urine) that induces copulation (Kamio et al. 2002). Ultrafiltration showed the copulation pheromone to have a molecular size of less than 1 kDa (Kamio et al. 2002).

Studies on different species of crayfish have also indicated the use of sex pheromones. When exposed to the tank water of a conspecific female, male crayfish (*Procambarus clarkii*) behave submissively, displaying feeding behaviors, with lowered claws and curled abdomen. When exposed to male tank water, however, males behave agonistically, raising their claws. These disparate behaviors were seen only in the presence of chemical signals—visual stimulation had no effect (Ameyaw-Akumfi and Hazlett 1975). This sex-specific pheromone was determined to be a carbohydrate, but no further elucidation was done. These results were first questioned by Itagaki and Thorp (1981) but later reaffirmed by Dunham and Oh (1992). Tierney and Dunham (1982) demonstrated the use of pheromones in species recognition as well as sex recognition in the crayfish species *Orconectes propinquus* and *O. virilis*. The two species could perceive the signals of both conspecifics and heterospecifics but were attracted only to signals produced by opposite-sex conspecifics.

Preliminary evidence shows that mature female signal crayfish, *Pacifastacus leniusculus*, release a chemical signal that induces sexual behavior in males (Stebbing et al. 2003). In the crayfish *Austropotamobius pallipes*, males respond only to chemical and visual cues in concert, not to chemical cues alone (Acquistapace et al. 2002).

Females of the New Zealand spiny lobster, *Jasus edwardsii*, prefer to mate with larger males and base their choice partially but not completely on male urine signals. In choice tests, females no longer expressed a clear choice when the urine output of a large male and a small male were exchanged so that the catheterized large male "emitted" the urine of the small male and vice versa. Mate choice was thus thought to be based on visual and tactile senses as well as olfaction (Raethke et al. 2004). However, the same result could also indicate the use of additional chemical signals not carried in the urine, as indicated in *H. americanus* (see "Male Signals and Female Choice," above). Male *J. edwardsii* appear to depend more than do females on olfactory cues in order to mate successfully. Matings between anosmic (i.e., antennule-ablated) males and intact females increased mating delay after female molting and decreased female clutch size. Antennule ablation in females, which are known to use olfaction in mate choice, had no significant effect on fertilization success: it seems they can monitor their own receptivity and do not rely on chemical cues for timing information (Raethke et al. 2004). The female South African rock lobster, *Jasus lalandii*, appears to produce a chemical substance during the time of ecdysis that elicits courtship behavior from males (Rudd and Warren 1976).

In shrimp, early evidence of a dispersing sex pheromone came from the freshwater prawn *Palaemon paucidens*. Tank water from a female undergoing the parturial (i.e., adult) molt elicited searching behavior by males (Kamiguchi 1972). In choice tests, female rock shrimp, *Rhynchocinetes typus*, approached the dominant male significantly more than the subordinate and preferred to mate with dominant ("robustus") males (Diaz and Thiel 2003, Thiel and Hinojosa 2003). To locate dominant males, females followed waterborne chemical cues. Males, however, did not follow a chemical trail to find a female. Rather, they used visual signals—"tumults" created by groups of subordinate males aggregating around the female—to find their mate; they needed physical contact to recognize the female's state of receptivity (Diaz and Thiel 2004), an interesting example of a contact pheromone.

A female contact sex pheromone is also indicated in the grass shrimp *Palaemonetes vulgaris*. It was suggested to be a nondiffusible compound on the female integument detected by the male antennae (Burkenroad 1947). In the highly gregarious *P. pugio*, males recognize receptive females by contact but only during a brief period preceding the female's parturial molt (Berg and Sandifer 1984, Bauer and Abdalla 2001). This close timing seems appropriate when animals already live in dense aggregations.

Similarly, in the hippolytid shrimp *Heptacarpus paludicola*, only a weak male response to receptive females upstream could be observed under experimental conditions; males needed to contact females with their antennae to recognize their receptive state (Bauer 1979). Harlequin shrimp, *Hymenocera picta*, form pair bonds through chemical identification. This information is conveyed by pheromones produced only by females and detected only by males (Seibt 1973). Clearly, in order to form heterosexual pair bonds, sex identification is critical, and chemical signals are commonly used for sex discrimination. We would expect this to be the case in other, yet unstudied pair bonding species.

Dominance and Individual Recognition

Individual recognition is believed to have evolved to allow for stable dominance hierarchies, as well as to unmask individuals who may bluff about fighting ability when faced with an agonistic interaction (Barnard and Burk 1979). It occurs in species with a relatively complex social structure, including pair bonding, and highly aggressive interactions (Halpin 1980). The ability to recognize known individuals within a dominance hierarchy decreases the need to escalate aggressive encounters with known opponents. This strategy decreases costs to both individuals in time and energy expenditure, as well as risk of injury. In decapod Crustacea, individual recognition has arisen in multiple orders and families.

"True" individual recognition is a system in which an individual can identify another by "a unique set of cues defining that individual" (Beecher 1989) and can associate that set of cues with experiences related to that specific individual. This is in contrast to "binary discrimination" in which an individual can recognize a conspecific but is able only to place that individual into a subgroup (dominant/subordinate or familiar/unfamiliar) rather than recognizing it individually (Barrows et al. 1975).

Two species of hermit crabs, *Pagurus bernhardus* and *P. longicarpus*, use individual recognition to establish dominance hierarchies. Hazlett (1969) observed small groups of *P. bernhardus*, noting steadily decreasing aggression between group members over several days of interaction. When an unknown crab was introduced, however, the number of fights observed between group members and the unknown crab increased significantly. Similarly, in *P. longicarpus*, interactions between unfamiliar individuals were more intense than between familiar animals; subordinates were more likely to initiate an interaction with a stranger than with a familiar dominant (Gherardi and Tiedemann 2004a). Carefully designed experiments showed that this individual recognition is chemically mediated—crabs reacted differently, taking longer to approach an empty gastropod shell in the presence of the odor of familiar versus unfamiliar conspecifics (Gherardi and Tiedemann 2004b). In this species, animals recognized individuals before a stable dominance hierarchy was formed, indicating that individual recognition is relatively unconnected with winning and losing experiences (Gherardi and Atema 2005). Initially, individual recognition in *P. longicarpus* was shown to be at least "binary," indicating animals discriminate familiar from unfamiliar but do not distinguish between individuals based on a unique chemical "badge" ("true" individual recognition) (Gherardi and Tiedemann 2004a, Gherardi and Atema 2005). However, more detailed experiments reveal that this species can (1) chemically discriminate between larger crabs in high-quality shells from smaller crabs in lower quality shells as long as the odor donors are familiar to the receiver, (2) associate the odor of an individual crab with the quality of its shell, and (3) change this association when their shell quality changes (Gherardi et al. 2005). This implies that odor information of nearby individuals is fluidly associated with past and new experiences regarding their shell size and shell fit.

Individual recognition is essential in the formation and preservation of pair bonds. Harlequin shrimp, *Hymenocera picta*, almost always form pair bonds. The bonds are maintained by the male's ability to distinguish (and prefer) the odor of his female mate over all other conspecifics (Seibt 1973). If the female of a pair is moved out of the male's territory and restrained elsewhere in a tank, the male will seek out his mate preferentially over remaining in his territory. If other females are available in the tank, he will bypass them, find his mate, and stay with her. The ability persists without visual cues (in darkness or in a Y-maze), confirming that chemical cues are responsible for recognition (Seibt 1973). Males do not distinguish among males, suggesting that individual recognition is not used in dominance interactions in this species, but only in pair bonds. The pheromone is generated only by the female and received only by the male (Seibt 1973). Female behavior was not studied as thoroughly but may contribute to the pair bond: in choice tests (Y-maze), they prefer the odor of their mate to all other conspecifics.

Banded coral shrimp, *Stenopus hispidus*, also display mate recognition in the context of a pair bond. If a pair is separated and placed back together after up to six (but not 33) days, they interact "neutrally" (i.e., at a low level). When separated and subsequently placed with a stranger, they show more intense interactions, indicating that they distinguish between familiar and unfamiliar conspecifics (Johnson 1977). No specific tests were done to show the involvement of chemical cues; however, this seems likely.

The cleaner shrimp *Lysmata debelius* forms pair bonds and can distinguish a partner from a stranger (Rufino and Jones 2001). Individual recognition allows animals to direct aggression toward strangers rather than toward mates. The modality of recognition in this species is not known, nor is the length of time recognition persists. Given the evidence in other pair bonding shrimp, it seems likely that chemical signals are involved.

Some evidence suggests that big-clawed snapping shrimp, *Alpheus heterochelis*, may be capable of individual recognition in pair bonding. Rahman et al. (2001) showed that animals could distinguish former mates from unfamiliar conspecifics. In Y-maze choice tests, individuals also preferred the odor of familiar same-sex conspecifics over the odor of an unfamiliar one. Animals can become conditioned to the odor of a specific individual if exposed to the tank water of that individual (Ward et al. 2004). Snapping shrimp do not, however, appear to use individual recognition in agonistic encounters. Subordinates who meet any dominant animal—familiar or unfamiliar—show *immediate* escape responses, but subordinates who meet and lose to socially inexperienced animals require several days of daily fights to show gradually more subordinate behavior (Obermeier and Schmitz 2003a). The signal appears to be chemical: anosmic (i.e., aesthetasc-ablated) animals did not discriminate between dominant and inexperienced opponents (Obermeier and Schmitz 2003b). Thus, snapping shrimp can recognize by odor the dominance status of an opponent, but not the opponent's identity.

Male but not female snapping shrimp (*Alpheus heterochelis*) use chemical signals to reduce their aggressive response to an open chela display: male odor enhanced and female odor reduced the otherwise aggressive response to this visual signal; females always responded aggressively regardless of any conspecific odor (Hughes 1996).

The colonial, eusocial shrimp *Synalpheus regalis* may well be the crustacean answer to eusocial insects, including dependence on chemical signals. Although not studied yet in specific detail, observations suggest that antennal contact is critical in recognition of nest mates versus intruders; both contact chemicals and dissolved substances may be involved. Queen signals are also indicated (Duffy et al. 2002). In the future, this interesting social environment may well reveal a treasure of pheromone interactions.

Several species of crayfish display status recognition but not individual recognition. A male *Astacus leptodactylus* who has just lost a fight is just as likely to refuse escalating the next fight with a familiar or an unfamiliar dominant; the status signals are present in the urine (Breithaupt and Eger 2002). Similarly, urine carries status information in the crayfish *Orconectes rusticus*: individuals who lost their first fights engaged in much shorter second fights whether their opponent was familiar or unfamiliar (Zulandt Schneider et al. 2001). The same species has been shown to adjust its dominance status as a result of prolonged exposure to tank water from unfamiliar dominant or subordinate conspecifics: a five-day exposure to dominant male odor resulted in subordinate behavior against a new male; dominant behavior resulted from similar exposure to subordinate odor (Bergman and Moore 2005). This result has interesting implications for the study of physiological mechanisms of primer pheromones, but a natural context for such long odor exposure is not known. *Procambarus clarkii* also determine dominance status through olfactory cues (Zulandt Schneider et al. 1999). Status recognition was shown in *Procambarus acutus acutus* (Gherardi and Daniels 2003).

Lowe (1956) showed evidence for individual recognition in *Cambarellus shufeldtii*. When one claw was removed from an established dominant, he would retain his dominant status in the hierarchy. However, when placed with unfamiliar individuals, he would often become subordinate. The large Australian crayfish *Cherax destructor* is both more likely to win fights against familiar than unfamiliar opponents and more attracted to familiar than to unfamiliar conspecifics. This "dear enemy" behavior was demonstrated even when using only visual or only chemical cues (Crook et al. 2004). See chapter 5 for additional information on crayfish social behavior.

Although belonging to the order Stomatopoda, rather than Decapoda, the work done on individual recognition in *Gonodactylus festae* and *G. bredini* requires mention here. Stomatopods are territorial animals that defend their rock crevices against conspecifics. Subordinate *G. festae* avoided crevices that smelled like familiar dominant conspecifics but readily entered and explored crevices without odor or with the odor of an unknown or subordinate animal (Caldwell 1979, 1985). *G. bredini* recognize former mates: males are less likely to evict a former female mate from her crevice, and females are less likely to defend their cavity against a former male mate (Caldwell 1992). This reduction of aggression was mediated by chemical signals and lasted approximately two weeks.

Stress, Aggregation, Larval Release, and Maternal Signals

This section reviews chemical signals that are known in Decapoda, but not in *H. americanus*. Conspecific stress signals were observed and caused male aggressive responses in the crayfish *Procambarus acutus acutus* (Thorp and Ammerman 1978). Aggregation signals are used to form clusters of gregarious animals such as spiny lobsters (see chapter 13). One would not expect this to be the case for the solitary *H. americanus*. Larval release signals are known in some crabs and should be expected in all brooding crustaceans, including *H. americanus*. Maternal signals should occur in species that extend brood care for hatched larvae, such as various crayfish species (see chapters 14, 15).

Starting in late juvenile stages and lasting through adulthood, Caribbean spiny lobsters, *Panulirus argus*, forage solitarily at night but exhibit gregarious shelter use during the day (Eggleston and Lipcius 1992). Shelter choice is often influenced by conspecific density and predator risk (Eggleston and Lipcius 1992). Aggregations are also formed through chemical signals. All but the earliest benthic stages of lobsters produce aggregation odors and are attracted to shelters by conspecific odors (Ratchford and Eggleston 1998). Lobsters produce these aggregation signals near dawn, when conspecifics return to shelters from solitary foraging expeditions. If foraging lobsters are experimentally kept on a different light/dark schedule than the animals producing the signal, they will respond to the signal even if temporally inappropriate; that is, they return to the shelter in the middle of the night rather than at dawn (Ratchford and Eggleston 2000). Field experiments on juvenile spiny lobsters support the use of chemical aggregation cues (Nevitt et al. 2000). Butler et al. (1999) demonstrated that late benthic juveniles of the New Zealand spiny lobster, *Jasus edwardsii*, increasingly strongly begin to aggregate and respond to the odor of larger conspecifics, facilitating gregarious behavior.

In some species, such as the mud crabs *Rhithropanopeus harrisii* (Rittschof et al. 1985) and *Neopanope sayi* (DeVries et al. 1991) and the blue crab, *Callinectes sapidus* (Tankersley et al. 2002), larval release pheromones are emitted by the hatching eggs themselves. These pheromones induce larval release behaviors in females such as compression of the egg mass to help break egg membranes, and abdominal pumping, which release the larvae into the water column (for an overview, see Forward 1987). In several species, for example, *R. harrisii* (Rittschof et al. 1985), the signals consist of a mixture of small peptides (molecular weight < 500 Da)

Early larval stages of several crayfish species, *Orconectes sanborni*, *Cambarus virilis*, and *Procambarus clarkii*, are attracted to chemical cues produced by a brooding female, normally their mother (Little 1975, 1976). The female cue appears at egg deposition, peaks at hatching, and starts disappearing as the larvae become independent. The larvae are no longer attracted, and the female loses her inhibition to feed on them. This maternal pheromone may have originated as a mechanism to prevent cannibalism in brooding animals.

Summary

We have presented the American lobster, *Homarus americanus*, as a model organism of chemical communication in decapod crustaceans. Lobsters use chemical signals to form and maintain dominance hierarchies, to choose and locate mates, to determine reproductive receptivity, to recognize individual conspecifics, and perhaps to regulate their molt cycles. The chemical signals are carried primarily but not completely in the urine and are dispersed into the environment by animal-generated currents. These information currents are also vital for receiving chemical information from the environment, including conspecifics. In addition, the lobsters *H. americanus* and *Panulirus argus* are model organisms for chemosensory physiology, signal transduction, and receptor genetics as well as a number of nonsocial behavior studies such as odor plume tracking and biomimetic robotics. Lobsters and crayfish are important models for crustacean brain anatomy and physiology (see chapter 4). This multidisciplinary database makes them rich study subjects for our understanding of crustacean communication and the evolution of sociality.

However, this account of decapod crustacean behavior influenced by chemical signals also points out the gaping holes in our knowledge: the chemical identity of the signals and their use under field conditions. Many decapod crustaceans use chemical signals for purposes similar to those of *H. americanus*. We must now give high priority to chemical analysis and field studies. The amount of knowledge we possess of the social behavior, physiology, and utilization of chemical signals in *H. americanus* and some other decapod crustaceans is an excellent knowledge base from which to begin identifying the chemical composition of the pheromones. The chemicals used by decapods to communicate sex, receptivity, dominance, and so forth, are likely to bear similarities between species. We anticipate that advances made in pheromone identification will be of great consequence in understanding crustacean chemical communication, just as has been the case in insects. It will help us understand evolutionary and ecological relationships; the neurobiology of signal processing; the costs of signal production, storage, and release; and the physical and chemical constraints on signal design imposed by different aquatic environments.

Acknowledgments We thank Thomas Breithaupt, Gabriele Gerlach, Francesca Gherardi, Paul Moore, and the editors of this volume for thoughtful comments on the manuscript. This work was supported generously by grants from the U.S National Science Foundation, most recently IBN-91358.

References

Acquistapace, P., L. Aquiloni, B.A. Hazlett, and F. Gherardi. 2002. Multimodal communication in crayfish: sex recognition during mate search by male *Austropotamobius pallipes*. Canadian Journal of Zoology 80:2041–2045.

Ameyaw-Akumfi, C., and B.A. Hazlett. 1975. Sex recognition in the crayfish *Procambarus clarkii*. Science 190:1225–1226.

Apanius, V., D. Penn, P. Slev, L.R. Ruff, and W.K. Potts. 1997. The nature of selection on the major histocompatibility complex. Critical Reviews in Immunology 17:179–224.

Asai, N., N. Fusetani, and S. Matsunaga. 2001. Sex pheromones of the hair crab *Erimacrus isenbeckii*. II. Synthesis of ceramides. Journal of Natural Products 64:1210–1215.

Asai, N., N. Fusetani, S. Matsunaga, and J. Sasaki. 2000. Sex pheromones of the hair crab *Erimacrus isenbeckii*. Part 1: Isolation and structures of novel ceramides. Tetrahedron 56:9895–9899.

Atema, J. 1977. Functional separation of smell and taste in fish and Crustacea. Pages 165–174 *in*: J. Le Magnen and P. MacLeod, editors. Olfaction and taste VI. Information Retrieval Ltd., London.

Atema, J. 1985. Chemoreception in the sea: adaptations of chemoreceptors and behavior to aquatic stimulus conditions. Pages 387–423 *in*: M.S. Laverack, editor. Physiological adaptation of marine animals. Society of Experimental Biology Symposium 39, Cambridge.

Atema, J. 1986. Review of sexual selection and chemical communication in the lobster, *Homarus americanus*. Canadian Journal of Fisheries and Aquatic Sciences 43:2283–2290.

Atema, J., and J.S. Cobb. 1980. Social behavior. Pages 409–450 *in*: J.S. Cobb and B. Phillips, editors. The biology and management of lobsters. Academic Press, New York.

Atema, J., and D. Cowan. 1986. Sex-identifying urine and molt signals in the lobster (*Homarus americanus*). Journal of Chemical Ecology 12:2065–2080.

Atema, J., and D. Engstrom. 1971. Sex pheromone in the lobster, *Homarus americanus*. Nature 232:261–263.

Atema, J., and R.B. Gagosian. 1973. Behavioral responses of male lobsters to ecdysones. Marine Behaviour and Physiology 2:15–20.

Atema, J., and D. Stenzler. 1977. Alarm substance of the marine mud snail, *Nassarius obsoletus*: biological characterization and possible evolution. Journal of Chemical Ecology 3:173–187.

Atema, J., and R. Voigt. 1995. Behavior and sensory biology. Pages 313–348 *in*: J.R. Factor, editor. Biology of the lobster *Homarus americanus*. Academic Press, San Diego, Calif.

Atema, J., S. Jacobson, E.B. Karnofsky, S. Oleszko-Szuts, and L. Stein. 1979. Pair formation in the lobster, *Homarus americanus*: behavioral development, pheromones, and mating. Marine Behaviour and Physiology 6:277–296.

Atema, J., T. Breithaupt, A. LeVay, J. Morrison, and M. Edattukaran. 1999. Urine pheromones in the lobster, *Homarus americanus*: both males and females recognize individuals and only use the lateral antennule for this task. Chemical Senses 24:615–616.

Bamber, S.D., and E. Naylor. 1996. Mating behaviour of male *Carcinus maenas* in relation to a putative sex pheromone: behavioural changes in response to antennule restriction. Marine Biology 125:483–488.

Barnard, C.J., and T. Burk. 1979. Dominance hierarchies and the evolution of "individual recognition." Journal of Theoretical Biology 81:65–73.

Barrett, J., D.H. Abbott, and L.M. George. 1993. Sensory cues and the suppression of reproduction in subordinate female marmoset monkeys *Callithrix jacchus*. Journal of Reproduction and Fertility 97:301–310.

Barrows, E.M., W.J. Bell, and C.D. Michener. 1975. Individual odor differences and their social functions in insects. Proceedings of the National Academy of Sciences, USA 72:2824–2828.

Bauer, R.T. 1979. Sex attraction and recognition in the caridean shrimp *Heptacarpus paludicola* Holmes (Decapoda: Hippolytidae). Marine Behaviour and Physiology 6:157–174.

Bauer, R.T., and J. Abdalla. 2001. Male mating tactics in the shrimp *Palaemonetes pugio* (Decapoda, Caridea): precopulatory mate guarding vs. pure searching. Ethology 107:185–199.

Beecher, M.D. 1989. Signaling systems for individual recognition: an information theory approach. Animal Behaviour 38:248–261.

Benyon, R.J., and J.L. Hurst. 2004. Urinary proteins and the modulation of chemical scents in mice and rats. Peptides 25:1553–1563.

Berg, A.B.W., and P.A. Sandifer. 1984. Mating behavior of the grass shrimp *Palaemonetes pugio* Holthuis (Decapoda, Caridea). Journal of Crustacean Biology 4:417–424.

Berg, K., R. Voigt, and J. Atema. 1992. Flicking in the lobster *Homarus americanus*: recordings from electrodes implanted in antennular segments. Biological Bulletin 183:377–378.

Bergman, D.A., and P.A. Moore. 2005. Prolonged exposure to social odours alters subsequent social interactions in crayfish (*Orconectes rusticus*). Animal Behaviour 70:311–318.

Bergman, D.A., A.L. Martin, and P.A. Moore. 2005. Control of information flow through the influence of mechanical and chemical signals during agonistic encounters by the crayfish, *Orconectes rusticus*. Animal Behaviour 70:485–496.

Berkey, C., and J. Atema. 1999. Individual recognition and memory in *Homarus americanus* male-female interactions. Biological Bulletin 197:253–254.

Breithaupt, T. 2001. Fan organs of crayfish enhance chemical information flow. Biological Bulletin 200:150–154.

Breithaupt, T., and J. Atema. 1993. Evidence for the use of urine signals in agonistic interactions of the American lobster. Biological Bulletin 185:318.

Breithaupt, T., and J. Atema. 2000. The timing of chemical signaling with urine in dominance fights of male lobsters (*Homarus americanus*). Behavioral Ecology and Sociobiology 49:67–78.

Breithaupt, T., and P. Eger. 2002. Urine makes the difference: chemical communication in fighting crayfish made visible. Journal of Experimental Biology 205:221–1231.

Breithaupt, T., D.P. Lindstrom, and J. Atema. 1999. Urine release in freely moving catheterized lobsters (*Homarus americanus*) with reference to feeding and social activities. Journal of Experimental Biology 202:837–844.

Brock, F. 1926. Das Verhalten des Einsiedlerkrebses *Pagurus arrosor* Herbst während der Suche und Aufnahme der Nahrung [The behavior of the hermit crab *Pagurus arrosor* Herbst during food search and feeding]. Zeitschrift für Morphologie und Ökologie der Tiere 6:415–552.

Bryant, B.P., and J. Atema. 1987. Diet manipulation affects social behavior of catfish: importance of body odor. Journal of Chemical Ecology 13:1645–1661.

Buchholz, F. 1982. The metabolism of ecdysone and its putative role as the female sex-pheromone in the green shore crab *Carcinus maenas*. Pages 35–46 *in*: Indices biochimiques et milieu marins [Biochemical Indices and the Marine Environment]. Symposium of the French Group for the Advancement of Biochemistry, 18–20 November 1981. Brest, France. CNEXO, Actes des colloques, Paris.

Burkenroad, M. 1947. Reproductive activities of decapod Crustacea. American Naturalist 81:392–398.

Bushmann, P.J. 1999. Concurrent signals and behavioral plasticity in blue crab (*Callinectes sapidus* Rathbun) courtship. Biological Bulletin 197:63–71.

Bushmann, P.J., and J. Atema. 1993. A novel tegumental gland in the nephropore of the lobster, *Homarus americanus*: a site for the production of chemical signals? Biological Bulletin 185:319–320.

Bushmann, P.J., and J. Atema. 1996. Nephropore rosette glands of the lobster *Homarus americanus*: possible sources of urine pheromones. Journal of Crustacean Biology 16:221–231.

Bushmann, P.J., and J. Atema. 1997. Shelter sharing and chemical courtship signals in the lobster *Homarus americanus*. Canadian Journal of Fisheries and Aquatic Sciences 54:647–654.

Bushmann, P.J., and J. Atema. 2000. Chemically mediated mate location and evaluation in the lobster, *Homarus americanus*. Journal of Chemical Ecology 26:883–899.

Butenandt, V., R. Beckmann, D. Stamm, and E. Hecker. 1959. Über den Sexual-Lockstoff des Seidenspinners *Bombyx mori* [The sex attractant of the silk worm moth *Bombyx mori*]. Zeitschrift für Naturforschung Teil B 14:283–284.

Butler, M., A. MacDiarmid, and J. Booth. 1999. The cause and consequence of ontogenetic changes in social aggregation in New Zealand spiny lobsters. Marine Ecology Progress Series 188:179–191.

Caldwell, R.L. 1979. Cavity occupation and defensive behaviour in the stomatopod *Gonodactylus festae*: evidence for chemically mediated individual recognition. Animal Behaviour 27:194–201.

Caldwell, R.L. 1985. A test of individual recognition in the stomatopod *Gonodactylus festae*. Animal Behaviour 33:101–106.

Caldwell, R.L. 1992. Recognition, signaling and reduced aggression between former mates in a stomatopod. Animal Behaviour 44:11–19.

Chang, E.S., S.A. Chang, R. Keller, P.S. Reddy, M.J. Snyder, and J.L. Spees. 1999. Quantification of stress in lobsters: crustacean hyperglycemic hormone, stress proteins, and gene expression. American Zoologist 39:487–495.

Cowan, D.F. 1991. The role of olfaction in courtship behavior of the American lobster, *Homarus americanus*. Biological Bulletin 181:402–407.

Cowan, D.F., and J. Atema. 1990. Moult staggering and serial monogamy in American lobsters, *Homarus americanus*. Animal Behaviour 39:1199–1206.

Cowan, D.F., J. Atema, and A. Solow. 1991. Moult staggering in the American lobster: a statistical analysis. Animal Behaviour 42:863–864.

Cromarty, S.I., and G. Kass-Simon. 1998. Differential effects of a molting hormone, 20-hydroxyecdysone, on the neuromuscular junctions of the claw opener and abdominal flexor muscles of the American lobster. Comparative Biochemistry and Physiology A 120:289–300.

Cromarty, S.I., J.S. Cobb, and G. Kass-Simon. 1991. Behavioral analysis of escape response in the juvenile lobster *Homarus americanus* over the molt cycle. Journal of Experimental Biology 158:565–581.

Crook, R., B.W. Patullo, and D.L. MacMillan. 2004. Multimodal individual recognition in the crayfish *Cherax destructor*. Marine and Freshwater Behaviour and Physiology 37:271–285.

Derby, C.D. 1982. Structure and function of cuticular sensilla of the lobster *Homarus americanus*. Journal of Crustacean Biology 2:1–21.

DeVries, M.C., D. Rittschof, and R.B. Forward. 1991. Chemical mediation of larval release behaviors in the crab *Neopanope sayi*. Biological Bulletin 180:1–11.

Diaz, E., and M. Thiel. 2003. Female rock shrimp prefer dominant males. Journal of the Marine Biological Association of the United Kingdom 83:941–942.

Diaz, E.R., and M. Thiel. 2004. Chemical and visual communication during mate searching in rock shrimp. Biological Bulletin 206:134–143.

Duffy, J.E., C.L. Morrison, and K.S. Macdonald. 2002. Colony defense and behavioral differentiation in the eusocial shrimp *Synalpheus regalis*. Behavioral Ecology and Sociobiology 51:488–495.

Dugatkin, L.A., and R.L. Earley. 2004. Individual recognition, dominance hierarchies and winner and loser effects. Proceedings of the Royal Society of London, Series B 271:1537–1540.

Dunham, P.J. 1978. Sex pheromones in Crustacea. Biological Reviews 53:555–583.

Dunham, P.J. 1988. Pheromones and behaviour in Crustacea. Pages 375–392 *in*: H. Laufer and G. Downder, editors. Endocrinology of selected invertebrate types. Alan R. Liss, New York.

Dunham, D.W., and J.W. Oh. 1992. Chemical sex discrimination in the crayfish *Procambarus clarkii*: role of antennules. Journal of Chemical Ecology 18:2363–2372.

Eales, A. 1974. Sex pheromone in the shore crab *Carcinus maenas* and the site of its release from females. Marine Behaviour and Physiology 2:345–355.

Eggleston, D., and R. Lipcius. 1992. Shelter selection by spiny lobster under variable predation risk, social conditions, and shelter size. Ecology 73:992–1011.

Ekerholm, M., and E. Hallberg. 2005. Primer and short-range releaser pheromone properties of premolt female urine from the shore crab *Carcinus maenas*. Journal of Chemical Ecology 31:1845–1864.

Ennis, G. 1995. Larval and postlarval ecology. Pages 23–46 *in*: J.R. Factor, editor. Biology of the lobster *Homarus americanus*. Academic Press, San Diego, Calif.

Factor, J.R., editor. 1995. Biology of the lobster, *Homarus americanus*. Academic Press, San Diego, Calif.

Fontaine, M.-T., A. Bauchau, and E. Passelecq-Gerin. 1989. Recherche du lieu de synthese de la pheromone sexuelle de *Carcinus maenas* (L.) (Decapoda, Reptantia) [Research on the site of synthesis of the sex pheromone of *Carcinus maenas* (L.)]. Crustaceana 57:208–216.

Forward, R.J. 1987. Larval release rhythms of decapod crustaceans: an overview. Bulletin of Marine Science 41:165–176.

Gagosian, R.B., and J. Atema. 1973. Behavioural responses of male lobsters to ecdysone metabolites. Marine Behaviour and Physiology 2:115–120.

Gerlach, G. 2006. Pheromonal regulation of reproductive success in female zebrafish: female suppression and male enhancement. Animal Behaviour.72:1119–1124.

Gherardi, F., and J. Atema. 2005. Memory of social partners in hermit crab dominance. Ethology 111:271–285.

Gherardi, F., and W. Daniels. 2003. Dominance hierarchies and status recognition in the crayfish *Procambarus acutus acutus*. Canadian Journal of Zoology 81:1269–1281.

Gherardi, F., and J. Tiedemann. 2004a. Binary individual recognition in hermit crabs. Behavioral Ecology and Sociobiology 55:524–530.

Gherardi, F., and J. Tiedemann. 2004b. Chemical cues and binary individual recognition in the hermit crab, *Pagurus longicarpus*. Journal of Zoology 263:23–29.

Gherardi, F., E. Tricario, and J. Atema. 2005. Unraveling the nature of individual recognition by odor in hermit crabs. Journal of Chemical Ecology 31:2877–2896.

Gleeson, R.A. 1980. Pheromone communication in the reproductive behavior of the blue crab, *Callinectes sapidus*. Marine Behaviour and Physiology 7:119–134.

Gleeson, R.A. 1991. Intrinsic factors mediating pheromone communication in the blue crab, *Callinectes sapidus*. Pages 17–32 *in*: R.T. Bauer and J.W. Martin, editors. Crustacean sexual biology. Columbia University Press, New York.

Gleeson, R.A., M.A. Adams, and A.B.I. Smith. 1984. Characterization of a sex pheromone in the blue crab, *Callinectes sapidus*: crustecdysone studies. Journal of Chemical Ecology 10:913–921.

Halpin, Z. 1980. Individual odors and individual recognition: review and commentary. Biology of Behavior 5:233–248.

Hardege, J.D., A. Jennings, D. Hayden, C.T. Müller, D. Pascoe, M.G. Bentley, and A.S. Clare. 2002. A novel behavioural assay and partial purification of a female-derived sex pheromone in *Carcinus maenas*. Marine Ecology Progress Series 244:179–189.

Hazlett, B.A. 1969. "Individual" recognition and agonistic behavior in *Pagurus bernhardus*. Nature 222:268–269.

Hoover, S., C. Keeling, M. Winston, and K. Slessnor. 2003. The effect of queen pheromones on worker honey-bee ovary development. Naturwissenschaften 90:477–480.

Hopkins, P.M. 1983. Patterns of serum ecdysteroids during induced and uninduced proecdysis in the fiddler crab, *Uca pugilator*. General and Comparative Endocrinology 52:350–356.

Huber, R., and E. Kravitz. 1995. A quantitative analysis of agonistic behavior in juvenile American lobsters (*Homarus americanus* L). Brain, Behavior and Evolution 46:72–83.

Huber, R., K. Smith, A. Delago, K. Isaksson, and E.A. Kravitz. 1997a. Serotonin and aggressive motivation in crustaceans: altering the decision to retreat. Proceedings of the National Academy of Sciences, USA 94:5939–5942.

Huber, R., M. Orzeszyna, N. Pokorny, and E.A. Kravitz. 1997b. Biogenic amines and aggression: experimental approaches in crustaceans. Brain, Behavior and Evolution 50:60–68.

Hughes, M. 1996. The function of concurrent signals: visual and chemical communication in snapping shrimp. Animal Behaviour 52:247–257.

Itagaki, H., and J.H. Thorp. 1981. Laboratory experiments to determine if crayfish can communicate chemically in a flow-through system. Journal of Chemical Ecology 7:115–125.

Johnson, M.E., and J. Atema. 2005. The olfactory pathway for individual recognition in the American lobster, *Homarus americanus*. Journal of Experimental Biology 208:2865–2872.

Johnson, V.R., Jr. 1977. Individual recognition in the banded shrimp *Stenopus hispidus* (Olivier). Animal Behaviour 25:418–428.

Kamiguchi, Y. 1972. Mating behavior in the freshwater prawn, *Palaemon paucidens*. A study of the sex pheromone and its effect on males. Journal of the Faculty of Science, Hokkaido University, Series VI, Zoology 18:347–355.

Kamio, M., S. Matsunaga, and N. Fusetani. 2000. Studies on sex pheromones of the helmet crab, *Telmessus cheiragonus*. 1. An assay based on precopulatory mate-guarding. Zoological Science 17:731–733.

Kamio, M., S. Matsunaga, and N. Fusetani. 2002. Copulation pheromone in the crab *Telmessus cheiragonus* (Brachyura: Decapoda). Marine Ecology Progress Series 234:183–190.

Karavanich, C., and J. Atema. 1998a. Individual recognition and memory in lobster dominance. Animal Behaviour 56:1553–1560.

Karavanich, C., and J. Atema. 1998b. Olfactory recognition of urine signals in dominance fights between male lobster, *Homarus americanus*. Behaviour 135:719–730.

Karnofsky, E., and H. Price. 1989. Dominance, territoriality and mating in the lobster, *Homarus americanus*: a mesocosm study. Marine Behaviour and Physiology 15:101–121.

Karnofsky, E.B., J. Atema, and R.H. Elgin. 1989a. Field observations of social behavior, shelter use, and foraging in the lobster, *Homarus americanus*. Biological Bulletin 176:239–246.

Karnofsky, E.B., J. Atema, and R.H. Elgin. 1989b. Natural dynamics of population structure and habitat use of the lobster, *Homarus americanus*, in a shallow cove. Biological Bulletin 176:247–256.

Kittredge, J.S., M. Terry, and F.T. Takahashi. 1971. Sex pheromone activity of the molting hormone, crustecdysone, on male crabs. Fishery Bulletin 69:337–343.

Kravitz, E.A. 1988. Hormonal control of behavior: amines and the biasing of behavioral output in lobsters. Science 241:1775–1781.

Kravitz, E.A. 2000. Serotonin and aggression: insights gained from a lobster model system and speculations on the role of amine neurons in a complex behavior. Journal of Comparative Physiology A 186:221–238.

Lavalli, K.L., and J.R. Factor. 1995. The feeding appendages. Pages 349–394 *in*: J.R. Factor, editor. Biology of the lobster *Homarus americanus*. Academic Press, San Diego, Calif.

Little, E.E. 1975. Chemical communication in maternal behaviour of crayfish. Nature 255:400–401.

Little, E.E. 1976. Ontogeny of maternal behavior and brood pheromone in crayfish. Journal of Comparative Physiology A 112:133–142.

Livingstone, M.S., R. Harris-Warrick, and E.A. Kravitz. 1980. Serotonin and octopamine produce opposite postures in lobsters. Science 208:76–79.

Lowe, M.E. 1956. Dominance-subordinance relationships in the crawfish *Cambarellus shufeldtii*. Tulane Studies in Zoology 4:139–170.

Ma, W., Z. Miao, and M. Novotny. 1999. Induction of estrus in grouped female mice (*Mus domesticus*) by synthetic analogues of preputial gland constituents. Chemical Senses 24:289–293.

Masuda, Y., M. Yoshida, and K. Mori. 2002. Synthesis of (2S,2'R,3S,4R)-2-(2'-hydroxy-21'-methyldocosanoylamino)-1,3,4-pentadecanetriol, the ceramide sex pheromone of the female hair crab *Erimacrus isenbeckii*. Bioscience Biotechnology and Biochemistry 66:1531–1537.

McClintock, M.K. 1983. Modulation of the estrous-cycle by pheromones from pregnant and lactating rats. Biology of Reproduction 28:823–829.

McLaughlin, L.C., J. Walters, J. Atema, and N. Wainwright. 1999. Urinary protein concentration in connection with agonistic interactions in *Homarus americanus*. Biological Bulletin 197:165–166.

McLeese, D., R. Spraggins, A. Bose, and B. Pramanik. 1977. Chemical and behavioral studies of the sex attractant of the lobster (*Homarus americanus*). Marine Behaviour and Physiology 4:219–232.

Morschauser, K.E. 2002. Establishment and maintenance of dominance hierarchies of the American lobster. Unpublished M.Sc. thesis, Boston University Marine Program, Boston, Mass.

Nevitt, G., N.D. Pentcheff, K.J. Lohmann, and R.K. Zimmer. 2000. Den selection by the spiny lobster *Panulirus argus*: testing attraction to conspecific odors in the field. Marine Ecology Progress Series 203:225–231.

Novotny, M.V., W. Ma, D. Wiesler, and L. Zidek. 1999. Positive identification of the puberty-accelerating pheromone of the house mouse: the volatile ligands associating with the major urinary protein. Proceedings of the Royal Society of London, Series B 266:2017–2022.

Obermeier, M., and B. Schmitz. 2003a. Recognition of dominance in the big-clawed snapping shrimp (*Alpheus heterochaelis* Say 1818) part I: individual or group recognition? Marine and Freshwater Behaviour and Physiology 36:1–16.

Obermeier, M., and B. Schmitz. 2003b. Recognition of dominance in the big-clawed snapping shrimp (*Alpheus heterochaelis* Say 1818) part II: analysis of signal modality. Marine and Freshwater Behaviour and Physiology 36:17–29.

O'Neill, D.J., and J.S. Cobb. 1979. Some factors influencing the outcome of shelter competition in lobsters (*Homarus americanus*). Marine Behaviour and Physiology 6:33–45.

Peeke, H.V.S., G.S. Blank, M.H. Figler, and E.S. Chang. 2000. Effects of exogenous serotonin on a motor behavior and shelter competition in juvenile lobsters (*Homarus americanus*). Journal of Comparative Physiology A 186:575–582.

Quackenbush, L.S. 1986. Crustacean endocrinology, a review. Canadian Journal of Fisheries and Aquatic Sciences 43:2271–2282.

Raethke, N., A.B. MacDiarmid, and J.C. Montgomery. 2004. The role of olfaction during mating in the southern temperate spiny lobster *Jasus edwardsii*. Hormones and Behavior 46:311–318.

Rahman, N., D.W. Dunham, and C.K. Govind. 2001. Mate recognition and pairing in the big-clawed snapping shrimp, *Alpheus heterochelis*. Marine and Freshwater Behaviour and Physiology 34:213–226.

Ratchford, S.G., and D.B. Eggleston. 1998. Size- and scale-dependent chemical attraction contribute to an ontogenetic shift in sociality. Animal Behaviour 56:1027–1034.

Ratchford, S.G., and D.B. Eggleston. 2000. Temporal shift in the presence of a chemical cue contributes to a diel shift in sociality. Animal Behaviour 59:793–799.

Rittschof, D., and J.H. Cohen. 2004. Crustacean peptide and peptide-like pheromones and kairomones. Peptides 25:1503–1516.

Rittschof, D., R.B. Forward, and D.D. Mott. 1985. Larval release in the crab *Rhithropanopeaus harrisii* (Gould): chemical cues from hatching eggs. Chemical Senses 10:567–577.

Rudd, S., and F.L. Warren. 1976. Evidence for a pheromone in the South African rock lobster, *Jasus lalandii* (H. Milne Edwards). Transactions of the Royal Society of South Africa 42:103–105.

Rufino, M.M., and D.A. Jones. 2001. Binary individual recognition in *Lysmata debelius* (Decapoda: Hippolytidae) under laboratory conditions. Journal of Crustacean Biology 21:388–392.

Ryan, E. 1966. Pheromone: evidence in decapod Crustacea. Science 151:340–341.

Sandeman, D., R. Sandeman, C. Derby, and M.J. Schmidt. 1992. Morphology of the brain of crayfish, crabs and spiny lobsters—a common nomenclature for homologous structures. Biological Bulletin 183:304–326.

Schmitt, B.C., and B.W. Ache. 1979. Olfaction: responses of a decapod crustacean are enhanced by flicking. Science 205:204–206.

Scrivener, J.C.E. 1971. Agonistic behavior of the American lobster, *Homarus americanus*. Technical Report No. 235. Fisheries Research Board of Canada, Biological Station, Nanaimo, B.C.

Seibt, U. 1973. Sense of smell and pair bond in *Hymenocera picta* Dana. Micronesica 9:231–236.

Shepheard, P. 1974. Chemoreception in the antennule of the lobster, *Homarus americanus*. Marine Behaviour and Physiology 2:261–273.

Sneddon, L.U., F.A. Hungtingford, A.C. Taylor, and A.S. Clare. 2003. Female sex pheromone-mediated effects on behavior and consequences of male competition in the shore crab (*Carcinus maenas*). Journal of Chemical Ecology 29:55–70.

Snyder, M.J., and E.S. Chang. 1991. Ecdysteroids in relation to the molt cycle of the American lobster, *Homarus americanus*. 2. Excretion of metabolites. General and Comparative Endocrinology 83:118–131.

Snyder, M.J., C. Ameyaw-Akumfi, and E.S. Chang. 1993. Sex recognition and the role of urinary cues in the lobster, *Homarus americanus*. Marine Behaviour and Physiology 24:101–116.

Stebbing, P.D., M.G. Bentley, and G.J. Watson. 2003. Mating behaviour and evidence for a female released courtship pheromone in the signal crayfish, *Pacifastacus leniusculus*. Journal of Chemical Ecology 29:465–475.

Stein, L., S. Jacobson, and J. Atema. 1975. Behavior of lobsters (*Homarus americanus*) in a semi-natural environment at ambient temperatures and under thermal stress. Woods Hole Oceanographic Institution Technical Report 75–48:1–49.

Tamm, G.R., and J.S. Cobb. 1978. Behavior and the crustacean molt cycle: changes in aggression of *Homarus americanus*. Science 200:79–81.

Tankersley, R.A., T.M. Bullock, R.B. Forward, and D. Rittschof. 2002. Larval release behaviors in the blue crab *Callinectes sapidus*: role of chemical cues. Journal of Experimental Marine Biology and Ecology 273:1–14.

Thiel, M., and I. Hinojosa. 2003. Mating behavior of female rock shrimp *Rhynchocinetes typus* (Decapoda: Caridea)—indication for convenience polyandry and cryptic female choice. Behavioral Ecology and Sociobiology 55:113–121.

Thorp, J.H., and K.S. Ammerman. 1978. Chemical communication and agonism in the crayfish, *Procambus acutus acutus*. American Midland Naturalist 100: 471–474.

Tierney, A.J., and D.W. Dunham. 1982. Chemical communication in the reproductive isolation of the crayfishes *Orconectes propinquus* and *Orconectes virilis* (Decapoda, Cambaridae). Journal of Crustacean Biology 2:544–548.

Waddy, S.L., and D.E. Aiken. 1990. Intermolt insemination, an alternative mating strategy for the American lobster (*Homarus americanus*). Canadian Journal of Fisheries and Aquatic Sciences 47:2402–2406.

Ward, J., N. Saleh, D.W. Dunham, and N. Rahman. 2004. Individual discrimination in the big-clawed snapping shrimp, *Alpheus heterochelis*. Marine and Freshwater Behaviour and Physiology 37:35–42.

Wilson, E.O., and W.H. Bossert. 1963. Chemical communication among animals. Recent Progress in Hormone Research 19:673–716.

Wysocki, C.J., and G. Preti. 2002. Human pheromones: oxymoron, marketing, Maya, or meaningful messages? Chemosense 5:1–11.

Wysocki, C.J., and G. Preti. 2004. Facts, fallacies, fears, and frustrations with human pheromones. Anatomical Record Part A 281A:1201–1211.

Yonge, C. 1932. On the nature and permeability of chitin. I. The chitin lining in the foregut of decapod Crustacea and the function of the tegumental glands. Proceedings of the Royal Society of London, Series B 111:298–329.

Zulandt Schneider, R.A., R.W.S. Schneider, and P.A. Moore. 1999. Recognition of dominance status by chemoreception in the red swamp crayfish, *Procambarus clarkii*. Journal of Chemical Ecology 25:781–794.

Zulandt Schneider, R.A., R. Huber, and P.A. Moore. 2001. Individual and status recognition in the crayfish, *Orconectes rusticus*: the effects of urine release on fight dynamics. Behaviour 138:137–153.

PART

Mating and Courtship

Ecology and Evolution of Mating Behavior in Freshwater Amphipods

Gary A. Wellborn

Rickey D. Cothran

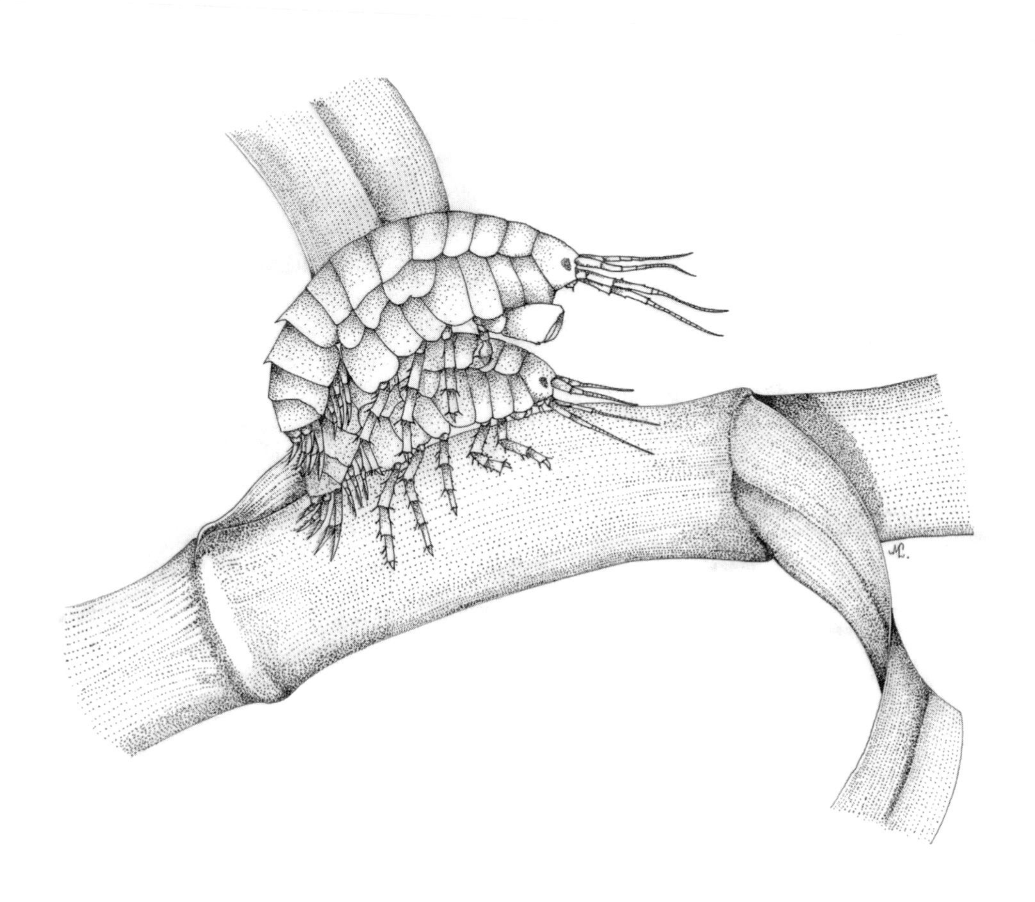

7

Freshwater amphipods have been models for the study of male time-investment strategies and sexual conflict over mate-guarding duration (reviewed in Jormalainen 1998). Other features of their mating biology, most notably, sexual selection and the mechanisms mediating it, have received comparatively little attention despite the important roles these are likely to play in evolution of mating behavior, reproductive traits, and sexual dimorphism (Andersson 1994). We focus on these areas by assessing the state of current knowledge of sexual selection and related issues, discussing the costs and benefits that are likely to drive sexual selection, and describing comparative studies of interspecific ecomorphs in *Hyalella* that hold promise for shedding light on these issues. We conclude with a discussion of the prospects and challenges that freshwater amphipods offer as empirical models to address current conceptual problems in the evolution of mating behavior.

Our treatment is necessarily selective and often speculative and is offered with the hope that it will stimulate discussion and, better yet, empirical studies that fill the many gaps in our understanding of sexual selection in this group. To a large degree, the current issues that confront advances in our knowledge of mating biology of amphipods, such as the relative roles of sexual conflict and mate preference and the maintenance of genetic variation for sexually selected traits, are precisely the challenges faced by the broader discipline of sexual selection. Thus, empirical advances in amphipod mating behavior have the potential to clarify fundamental issues in behavior and evolution.

Overview of Reproductive Biology and Mating System

Some key features of the reproductive biology of freshwater amphipods are important for understanding evolution of their mating system and behavior. As in many other crustaceans, the opportunity to fertilize eggs is limited to the period just after the female's molt (Sutcliffe 1992). This time-limited opportunity for fertilization, together with external fertilization, favors the evolution of contact mate guarding, a ubiquitous behavior in freshwater amphipods (Ridley 1983). In precopulatory mate guarding (referred to here as "pairing" for simplicity), males are positioned dorsal to the female and grasp the female's integument with prehensile thoracic appendages called gnathopods. Males hold females in this way until the female molts, at which time ova are passed into the female's external brood chamber, where they are fertilized by the guarding male (Sutcliffe 1992). Males discontinue pairing soon after fertilization, and females do not store sperm. Females carry the developing embryos until the embryos hatch and leave the brood chamber, before or coincident with the next female molt. Females are iteroparous and generally capable of producing a new clutch with each molt. Timing of female receptivity is usually asynchronous among females.

Pairs do not remain together after fertilization, and sexes are promiscuous. Shuster and Wade (2003) classify this typical amphipod mating system as coercive polygynandry because male coercion can involve costs to females. As we discuss below, it is very difficult to distinguish between successful male coercion and female selectivity (Cordero and Eberhard 2003), and it is not clear that entirely coerced pairings ever occur in freshwater amphipods. Nonetheless, sexual conflict over precopulatory pairing is likely (see chapter 8).

Sexual Selection

Mating is characterized by intrasexual competition and by complex intersexual interactions reflecting a tension between cooperation and conflict (Trivers 1972). Evolutionary dynamics that drive sexual selection and shape mating behavior derive from multiple mechanistic processes, including sexual conflict, intrasexual competition for access to mates, direct and indirect selection on female preference, and sensory bias. Much effort, both theoretical and empirical, has focused on assessing the relative prevalence of these mechanisms and the circumstances under which they might operate (Andersson 1994, Kokko et al. 2002, Chapman et al. 2003). Most recently, there has been an emerging recognition that these processes are likely to operate simultaneously and interactively in the evolution of mating traits (Kokko et al. 2003, Arnqvist and Rowe 2005, Cordero and Eberhard 2005, Kokko 2005), suggesting that synthetic research approaches are needed. In amphipods, few studies have considered the nature and scope of sexual selection and its ultimate consequences for traits associated with intrasexual and intersexual mating interactions. In this section, we explore components of the amphipod mating sequence, focusing on the roles of each sex, and the variety of ways that sexual selection may shape mating traits in both sexes.

Searching Phase

In freshwater amphipods, time-limited and asynchronous female receptivity give rise to male-biased operational sex ratios (Jormalainen 1998), a condition expected to promote intense competition among males for mating opportunities (Emlen and Oring 1977, Shuster and Wade 2003). One way males may enhance mating success is to increase investment in mate searching behavior, as is common in amphipods (Ridley 1983). Searching may be costly, however, exposing males to higher rates of predation and parasitism. Indirect evidence for these costs comes from adult sex ratios in the field, which are often female biased (Ward 1986, Wellborn and Bartholf 2005), and higher parasite loads in males, which carry an associated reduction in mating success (Ward 1986, Bollache et al. 2001). Searching also may be energetically costly (Sparkes et al. 2002). For example, unpaired *Gammarus pulex* males had lower energy reserves than did paired males in a natural population (Plaistow et al. 2003), and because males had the potential to replenish energy reserves while paired, costs of mate searching behavior may have contributed to low energy reserves in unpaired males.

Mate searching costs in males can influence the evolution of pairing duration (Jormalainen et al. 1994, Yamamura and Jormalainen 1996, Cothran 2004; see also chapter 8). Predation risk (Cothran 2004) and feeding rate (Robinson and Doyle 1985), for example, may differ between searching and pairing phases, altering optimal pairing duration (Yamamura and Jormalainen 1996). Search costs can also play an important role in determining the intensity of intersexual conflict over guarding duration (Parker 1979, Jormalainen 1998). Resolution of guarding conflict should depend on the relative ability of each sex to influence the outcome of the conflict and on each sex's fitness payoff from alternative outcomes (Yamamura and Jormalainen 1996). When male search costs are high, males and females may differ in the relative costs of pairing because, for males, the cost of pairing may be offset by the cessation of searching costs, but because females do not suffer a search cost, any cost of pairing represents a net cost (Jormalainen 1998, Cothran 2004). Such an asymmetry between sexes sets the stage for intersexual conflict over guarding duration (see chapter 8), and traits that are important in determining the outcome of this interaction may be especially prone to evolutionary change as a result of such conflicting interests.

Costs of mate searching will depend on ecological conditions (Andersson 1994, Rowe 1994; see also chapter 10). For example, freshwater habitats vary in the form and intensity of predation, which may lead to variation in the direction and intensity of selection acting on male activity (Strong 1973, Wellborn et al. 1996). Other aspects of an individual's habitat, including resource availability, temperature, sex ratio, and the temporal and spatial distribution of females, are likely to affect searching costs (Jormalainen 1998) and thus foster interpopulation and interspecific variation in mating behavior (Strong 1973, Wellborn 1995, Jormalainen et al. 2000, Wellborn and Bartholf 2005).

Assessment, Resistance, and Pairing

Encounters between males and females may precipitate pair formation, but often they do not (Elwood et al. 1987, Sparkes et al. 2000, Strong 1973). Qualities of individuals that determine whether pairing will occur have been considered primarily from a male perspective. Female characteristics that enhance their value to males include larger size, owing to its correlation with fecundity (Ward 1988, Wellborn 1995), and shorter time until the female molt, which decreases missed opportunity costs for guarding males and lowers costs arising from reduced feeding or increased predation during pairing. Additionally, small female size, relative to male size, may be favored under some conditions due to reduced energetic costs of carrying females (Adams and Greenwood 1987). Traits of males that influence their value to females have received much less attention but are likely to be important considering the growing recognition of the role female behavior plays in mediating pair formation (Borowsky and Borowsky 1987, Sutcliffe 1992, Jormalainen 1998, Sparkes et al. 2000, 2002). In this section, we explore the potential for females to exercise choice in pairing and examine how sexual selection driven by female preference may operate in freshwater amphipods.

The degree of female control in mating is unclear. Most researchers interpret pair formation as being under mutual control of both sexes, although not necessarily with both sexes having equal influence (see Table 2 in Jormalainen 1998). Some researchers interpret pair formation to be controlled strictly by male choice, with females being passive (Jormalainen 1998). These conflicting interpretations probably reflect some degree of behavioral differences among species studied, but female passivity and female control have sometimes been reported in the same species (Birkhead and Clarkson 1980, Ward 1984, Elwood et al. 1987). We use the term "female control" to indicate circumstances in which females have the ability to prevent some or all unwanted pairings, and "female preference" to refer to circumstances in which female traits (resistance behavior, chemical signals, etc.) cause some male phenotypes to have higher mating success than others. Our intuition is that female control and preference are likely to be common in amphipods both because observations of female resistance are common, and because costs and benefits of when and with whom to pair are likely to differ between sexes (Jormalainen 1998, Arnqvist and Rowe 2005), and thus there should be selection on both sexes' ability to control pair formation.

Female physical "resistance" to male mating attempts is common in animals (Parker 1979, Crudgington and Siva-Jothy 2000, Hosken et al. 2003, Arnqvist and Rowe 2005), including amphipods and isopods (Jormalainen 1998, Sparkes et al. 2000). The functional role of female resistance is often attributed to minimizing naturally selected costs of pairing (Jormalainen 1998, Chapman et al. 2003). For example, female amphipods may use resistance behavior to counter the costs of prolonged precopulatory pairing with males (Jormalainen and Merilaita 1995; see also chapter 8). Another, but not mutually exclusive, explanation for resistance behavior is that females exercise active mate preference through selective resistance (Sparkes et al. 2002, Cordero and Eberhard 2003, Kokko 2005). In either case, resistance behavior may cause male mating success to vary across male phenotypes, imposing sexual selection on male traits.

The issue of the extent to which each sex is able to exercise control of mating is important in understanding the evolution of mating behavior (Fincke 1997) but is exceedingly difficult to evaluate in amphipods due in part to the nature of the tactile interaction that occurs at encounter. Is what appears to be "resistance behavior" a general female rejection of all males, or is it sometimes assessment behavior that facilitates female preference? Is resistance an honest signal to males indicating that the female is far from her molt, and thus is to the mutual benefit of both individuals to forgo pairing, or is resistance a sometimes deceptive signal used for female preference? Although studies that manipulate female behavior are valuable (Jormalainen and Merilaita 1995), such studies could be misleading if males use female behavior to assess receptivity or other qualities of females. Ultimately, studies that focus on fitness consequences accruing to females and males exposed to different choice regimes may be particularly informative in elucidating the role of each sex in pairing dynamics (Cordero and Eberhard 2005).

A large male mating advantage is common in freshwater amphipods (Birkhead and Clarkson 1980, Ward 1984, Wellborn 1995, Jormalainen 1998), and variance in other male traits, including gnathopod size and antennal length, may influence pairing success (Conlan 1991, Wellborn 1995, 2000) but have received less attention. Most explanations for a large male mating bias have centered on competition among males for access to mates, with larger males having an advantage in scramble competition or takeover attempts (Ward 1988, Bollache and Cezilly 2004). Although intrasexual competition for receptive females should be intense in freshwater amphipods, direct takeovers of paired females by single males are apparently rare (Strong 1973, Elwood et al. 1987). Scramble competition among males is likely, however, and should select for high rates of searching activity (Ridley 1983). Higher activity will increase encounter rates with females, but its effectiveness in securing mates will also depend on female preference, provided that females can exercise control in pairing (Sparkes et al. 2000). For example, if females control pair formation and discriminate among males based on body size, then we must look to female preference as an important determinant of size-biased pairing in males.

Evolution of mating preference requires that selective mating yield higher fitness than indiscriminate mating (Andersson 1994, Kokko et al. 2003). Variation in mate quality can cause direct selection on female mating biases when nonrandom mating increases female fecundity or viability (Price et al. 1993, Kokko et al. 2003). In amphipods, direct selection on female mating preferences may be common during the precopulatory pairing phase, but we are not aware of studies explicitly evaluating this issue. Precopulatory pairing might influence predation risk (Ward 1986, Cothran 2004), foraging (Robinson and Doyle 1985, Sparkes et al. 1996), habitat use (Sparkes et al. 1996), and mobility (Adams and Greenwood 1987), and it seems likely that the impact of these effects on a female will often depend on her mate's body size or other traits.

A second potential mechanism for the evolution of female preference in amphipods is indirect selection that occurs when mating preferences increase offspring fitness (Kokko 2001). Females gain indirect fitness benefits because preferred male traits are genetically correlated with components of fitness, and offspring inherit these "good" genes. The potential importance of "good genes" indirect selection in driving female mating preference has been debated (Andersson 1994, Gavrilets et al. 2001, Cameron et al. 2003, Cordero and Eberhard 2003), with much attention focused on how genetic variance in male fitness is maintained, and the related issue of whether indirect selection can be sufficiently strong to drive mating preferences when preference is costly (Kokko et al. 2003). Recent studies have at least partially clarified this issue by demonstrating that additive genetic variance in fitness related traits can be considerable (Burt 1995, Pomiankowski and Møller 1995, Houle 1998).

Explanations for how genetic variation in fitness is maintained have focused on the idea that male traits that are targets of selection by female preference are condition-dependent traits that capture genetic variation at the level of the genome (Houle and Kondrashov 2002, Tomkins et al. 2004, Hunt et al. 2004). Traits that are costly to produce, such as body size, may be genetically correlated with condition, or overall quality, of an individual because large size is the product of all of the developmental, physiological, and behavioral processes that underlie size. Because such condition-dependent traits are determined by the collective influence of a large number of genes, the traits are large targets for mutation, allowing substantial genetic variance to be maintained (Rowe and Houle 1996, Houle 1998). Some empirical studies are now beginning to illuminate the issue of indirect selection on female choice (Jones et al. 1998, Iyengar and Eisner 1999, Hine et al. 2002), including a study of crickets demonstrating that female preference for attractive males can provide indirect fitness benefits to females that outweigh direct costs (Head et al. 2005). In an isopod, Sparkes et al. (2002) manipulated male condition by causing some males to expend glycogen reserves and found that males with higher condition had greater pairing success because females resisted more vigorously in encounters with low-condition males.

Male traits for which preference is most likely to evolve are those that reliably indicate fitness, and thus preferred traits should be those having fitness correlations that are relatively unaffected by environmental variance and genotype by environment interactions (Hunt et al. 2004). Body size or gnathopod size, for example, might satisfy this requirement because, although these can vary with resource levels or other environmental conditions, relatively large size might always denote high genetic quality in habitats where larger size is beneficial. That body size and correlated traits may be targets of sexual selection is particularly interesting because amphipods have indeterminate growth, implying that phenotypic variance in size results from variance in both growth rate and age. Provided that some mortality occurs across mature age classes, variation in size due to age differences will also reflect variation in viability. Thus, size-biased mating can impose sexual selection on growth and viability (Trivers 1972, Kokko 1998), presumably mediated through the underlying traits, such as resource acquisition ability, that cause variation in growth and viability (Tomkins et al. 2004). Furthermore, both growth rate and viability are likely to be condition dependent, suggesting a mechanism for maintenance of additive genetic variance in these traits (Houle 1998). As a methodological note, indeterminate growth in amphipods greatly hinders quantification of sexual selection intensity directly from mating biases because mating biases do not necessarily denote variation in lifetime mating success (Arnold and Wade 1984). Nonetheless, mating biases with respect to size imply sexual selection can act to the extent that size variance is due to differential growth and survival (Wellborn 1995).

Ecological Context and Sexual Selection in *Hyalella* Ecomorphs

The causes and phenotypic outcome of sexual selection are mediated by its interaction with natural selection (Partridge and Endler 1987, Andersson 1994, Kokko et al. 2003; see also chapter 10). Because natural selection is driven largely by ecological processes, ecological context may modulate the evolution of mating behavior in important ways (Houde 1997, Shuster and Wade 2003). In this section, we describe comparative studies of interspecific ecomorphs within a group of *Hyalella* amphipods (Fig. 7.1). The ecomorph species experience disparate ecological conditions that determine the benefits and costs of mating activities, which in turn shape evolution

Figure 7.1 Females of large (top) and small (bottom) ecomorphs within the *Hyalella azteca* species complex collected in Oklahoma. These species are undescribed (Wellborn et al. 2005).

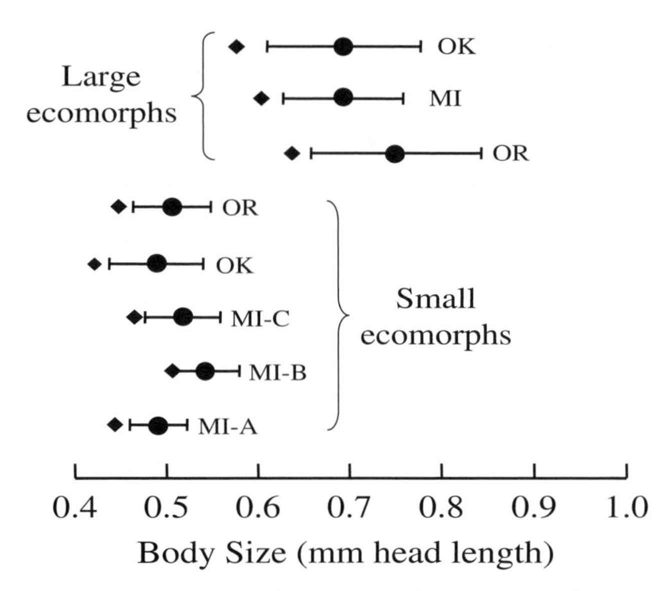

Figure 7.2 Mean (±1 SD) body size and maturation size (diamonds) of large and small regional ecomorphs in the *Hyalella azteca* species complex. Notation indicates collection site (OK, Oklahoma; MI, Michigan; OR, Oregon). Letter designations for Michigan small ecomorphs indicate separate species. Data based on Strong (1972), Wellborn and Cothran (2004), and Wellborn et al. (2005).

of mating behavior. Thus, understanding the role of ecological factors in generating interspecific diversity in mating behavior can illuminate the mechanisms by which sexual and natural selection interact to form mating phenotypes.

Hyalella amphipods are common in permanent freshwater habitats throughout much of the New World. Here, we explore diversity in mating behavior among "interspecific ecomorphs" (i.e., sets of species that differ consistently both in habitat and in adaptive traits that mediate ecological success in the different habitats). These ecomorphs differ in body size and life history (Fig. 7.2) and occur in at least three regions of North America: Oregon (Strong 1972, 1973), Michigan (Wellborn 1994), and Oklahoma (Wellborn et al. 2005). The ecomorphs are undescribed species that fall within a broader array of previously hidden species diversity recently discovered through genetic analyses (Witt and Hebert 2000, Wellborn et al. 2005). Ecomorphs within the same geographic region segregate among local habitats based on the habitats' ecological qualities. "Small ecomorph" species mature early, maintain a small adult body size, and occur in habitats where they are subject to intense predation by fish, particularly sunfish in the genus *Lepomis*. "Large ecomorph" species mature at a large body size and occur in habitats lacking this intense predation by fish but are subject to predatory invertebrates and some fish predation (Wellborn et al. 2005).

Ecomorph phenotypes are adaptive, given the disparate ecological conditions of their habitats (Wellborn 1994). The ecology of regional ecomorphs is best understood for Michigan species, but similar mechanisms are likely to operate in the other regions (Strong 1973, Wellborn et al. 2005). In Michigan, small ecomorph species experience

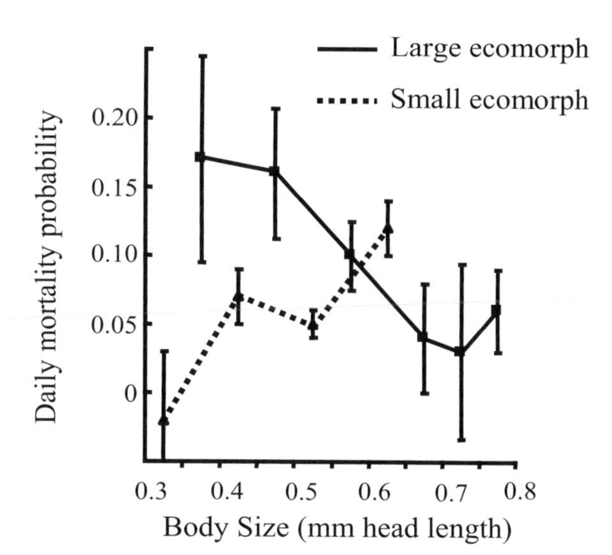

Figure 7.3 Mean (±1 SD) daily mortality in natural habitats for large and small ecomorphs in Michigan replicated in two years. Based on data from Wellborn (1994).

intense size-selective predation by *Lepomis* sunfish, and mortality increases with body size (Fig. 7.3). In contrast, the large ecomorph species is found in fishless habitats where larval dragonflies are the primary predators. These predators impose heaviest mortality on smaller, juvenile individuals, and mortality declines with increasing body size. Ecomorph differences in maturation size and adult body size are consistent with life history adaptation under these disparate mortality regimes (Taylor and Gabriel 1992).

Hyalella *Mating Behavior*

Our observations of mating behavior in Michigan and Oklahoma *Hyalella* species (Wellborn 1995, Wellborn and Bartholf 2005) largely coincide with those of Oregon ecomorphs reported in Strong (1973). Males actively search for females. Movement of males appears haphazard, and there is no evidence of waterborne chemical communication between the sexes (Strong 1973, Wellborn and Cothran, unpublished observations). Because females do not appear to search, males bear costs of mate searching, which may include elevated predation risk. Indeed, female-biased sex ratios are typical for *Hyalella*. For example, large- and small-ecomorph populations in Oklahoma were 20% and 34% male, respectively (Wellborn and Bartholf 2005), and Michigan ecomorph populations were approximately 40% male (Wellborn 1995). The operational sex ratio, however, is male biased, because only about 9% of females are expected to become receptive each day, assuming an 11-day molt cycle that is asynchronous among females (Wellborn 1995, Othman and Pascoe 2001), whereas virtually all unpaired males are able to pair. Operational sex ratios for Oklahoma ecomorph species were approximately 72% and 84% male for large and small species, respectively, and 86% male in Michigan ecomorph species (calculated from Wellborn 1995, Wellborn and Bartholf 2005).

Females, through resistance behavior, appear to control pair formation in *Hyalella* (Strong 1973, Wellborn 1995). While actively searching, males appear to attempt to pair with each female encountered, and females often appear to resist. Resistance behavior in female *Hyalella* is similar to that reported in other amphipods and isopods (Ward 1984, Hunte et al. 1985, Borowsky and Borowsky 1987, Elwood et al. 1987, Caine 1991, Jormalainen and Merilaita 1993, Sparkes et al. 2000) and includes thrashing and tightly curling their body (Strong 1973, Wellborn 1995). Because it is associated with failed pairing attempts, we interpret this behavior to indicate that females can control the onset of pairing (Wellborn 1995), a conclusion shared by Strong (1973). This inference is also supported by ongoing experimental work showing that males pair earlier in the female molt cycle in trials with inactive females, for which resistance behavior is experimentally removed by anesthetization, than with active females (R.D. Cothran, unpublished observations). A detailed study of a freshwater isopod recorded similar behavioral interactions and concluded that females can control pairing by persisting in resistance behaviors (Sparkes et al. 2000).

Timing of the onset of precopulatory pairing is likely to involve intersexual conflict and will depend on ecological conditions that shape costs of pairing (Härdling et al. 2001; see also chapter 8). Pairing is beneficial for both sexes very near the female's molt, but net benefits decline progressively as the time to the female's molt increases because costs of pairing accrue with pairing duration (Jormalainen 1998). Sex differences in costs of pairing generate disparity between sexes in the optimal duration of pairing. How is intersexual conflict in pairing duration resolved in *Hyalella*? Although manipulative experiments are needed, we suggest that current evidence indicates conflict may be largely resolved in favor of females. That males appear indiscriminant in attempting to pair with females, whereas females successfully resist males, offers observational evidence consistent with our interpretation. Additionally, the observational evidence is corroborated by the recently completed study described above, showing that males guard anesthetized females longer than they do active females. These results suggest that optimum guarding duration is greater for males than for females, as may be typical in amphipods (Jormalainen 1998).

Although males may suffer a missed opportunity cost by pairing with a female that is too far from her molt, receptive females (i.e., females that ultimately accept males) may always be sufficiently near their molt that they fall within the male optimal pairing duration. If true, female receptivity acts as a reliable indicator of female quality for males. One difficulty with this interpretation, however, is that males could profit from forced pairings with females whose molt stage falls within the male optimum guarding duration, but not yet within the female optimum. Because of the nature of the tactile interaction that precedes pairing, it is very difficult to differentiate male force from female preference enacted by selective resistance (Cordero and Eberhard 2003). If forced pairing does occur, this behavior raises the question of how males assess female molt stage.

Costs of forced pairing for males may be high, especially if males cannot determine the time remaining until a female molts and thus cannot assess the intensity of missed opportunity costs (Jormalainen 1998). Given that information about the female's internal physiological state "belongs" to females, it is not clear why females would reveal this information to males when its use precipitates a fitness cost to females. One possibility is that females cannot control information about their molt stage, for example, if molting hormones unavoidably leak from females. Indeed, in some amphipod species, waterborne or contact chemicals released by females influence pairing (Lyes 1979, Borowsky and Borowsky 1987, Borowsky 1991, Sparkes et al. 2000), but it is not clear whether females control display of chemical cues in these species, and it is not clear whether females suffer a direct fitness cost as a result of the cues, as would be expected if males use cues in coercive pairings. An alternative possibility for male assessment of female molt stage is that males judge the time to female molt based on intensity of female resistance (Jormalainen 1998), but females may be able to deceive males by altering resistance. We know of no a priori reason why females *must* honestly signal their time to molt during resistance behaviors. Clearly, better empirical understanding of the dynamics of coercive pairing and information transfer is needed to resolve these issues.

Ecological factors may shape pairing duration by influencing the costs of guarding, and to the extent that these ecologically mediated costs differ between sexes, they may shift levels of intersexual conflict (Cothran 2004; see also chapter 8). In *Hyalella*, increased susceptibility to predation is a potentially important cost of pairing, and because predation regime differs between habitat types, the two *Hyalella* ecomorphs are likely to be differentially affected by predation. Strong (1973), studying ecomorphs in Oregon, found that mean pairing duration in the large ecomorph species was 4.8 days, whereas the small species pairs for only 2.6 days. In Oklahoma, large and small ecomorph species pair for an average of 3.0 and 1.4 days, respectively (Wellborn and Bartholf 2005). Strong (1973) hypothesized that size-selective predation by *Lepomis* and other fish in the small species' habitat causes elevated predation risk for paired individuals, and this risk drives the evolution of short pairing duration.

This hypothesis was tested experimentally in Oklahoma ecomorph species (Cothran 2004). In fish predation treatments, pairs of the small species were consumed about twice as frequently as unpaired individuals, suggesting a substantial cost of pairing in fish habitats. Although fish always consumed both individuals in pairs, the cost of pairing was greater for females because mortality of single males in the experiment tended to be higher than that of single females, perhaps because male searching behavior makes them more conspicuous to fish. In large ecomorph habitats, predatory invertebrates, especially dragonfly larvae, are common. These predators had an effect opposite to that of fish, consuming single individuals more frequently than paired individuals. Thus, at least in terms of predation risk, pairing may be beneficial for large ecomorph individuals, and this benefit was similar between sexes. Overall, the experiment points to habitat differences in predation as a likely cause of the ecomorph differences in pairing duration.

A further implication of the study by Cothran (2004) is that small-species females, given their high cost of pairing, may be especially likely to delay pairing until very close to their molt and that intersexual conflict over pairing duration may be high. Furthermore, by delaying pairing until she is very near her molt, a small-species female obtains the added benefit of reduced risk to her developing offspring because offspring often leave the brood pouch a day or two before the female's molt. In a study of Oklahoma ecomorphs, few small-species females in pairs carried embryos or offspring in the brood pouch, compared with single females and with paired females of the large species (Fig. 7.4). Thus, elevated predation risk may drive short pairing duration in the small species due to the twofold benefit of increased survival for the female and increased survival of her current offspring.

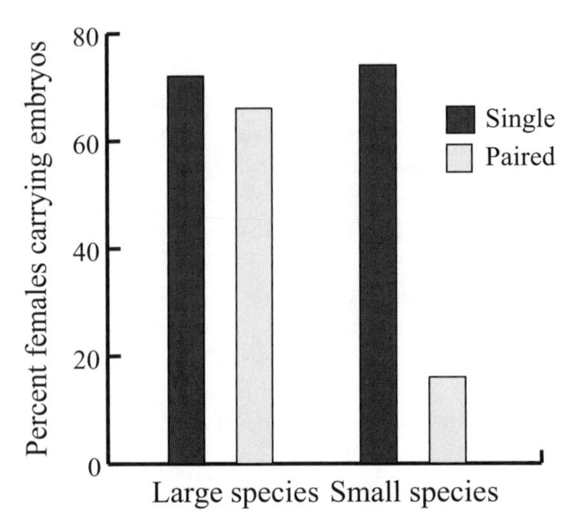

Figure 7.4 Percentage of unpaired (dark bars) and paired (light bars) females with embryos in the marsupium for the small and large *Hyalella* species in Oklahoma. Because the small ecomorph species delays pairing until late in its molt cycle, most females have released offspring before pairing begins. Based on unpublished data collected for Wellborn and Bartholf (2005).

Mating Biases and Sexual Selection

Sexual selection is influenced by natural selection in multiple ways. Expression of traits subject to directional sexual selection is balanced by countervailing natural selection (Kokko et al. 2002). Form or intensity of direct benefits that shape evolution of mate choice may depend on ecological conditions (Andersson 1994). Also, mate preferences driven by indirect benefits can arise because sexual selection favors females that mate with males possessing traits favored by natural selection (Kokko et al. 2002), and thus it follows that preferences will depend on the regime of natural selection operating in a system. Variation among populations or species in the action of natural selection, therefore, can produce variation in the action or outcome of sexual selection (Partridge and Endler 1987, Houde 1997). In this section, we explore how differences in natural selection may interact with sexual selection to give rise to mating biases observed in *Hyalella* ecomorphs. A full understanding of selection mechanisms operating in this system will require extensive experimental work, some of which is underway. At present, our approach is necessarily limited to documenting ecomorph differences in mating biases and attempting to make sense of these differences in the context of the ecomorphs' disparate ecological contexts.

Mating success of male *Hyalella* is often dependent on male body size (Wellborn 1995, Wellborn and Bartholf 2005). Although other traits may also be under sexual selection (Wellborn 1995, 2000), we focus here on body size because we understand much about how size affects ecological success of these species (Wellborn 1994, 2002) and thus how ecological factors may influence sexual selection. In Oklahoma ecomorphs, field studies demonstrated a large male mating advantage for both ecomorphs, but ecomorphs appeared to differ in the form of the relationship between body size and pairing success (Wellborn and Bartholf 2005). In the large ecomorph,

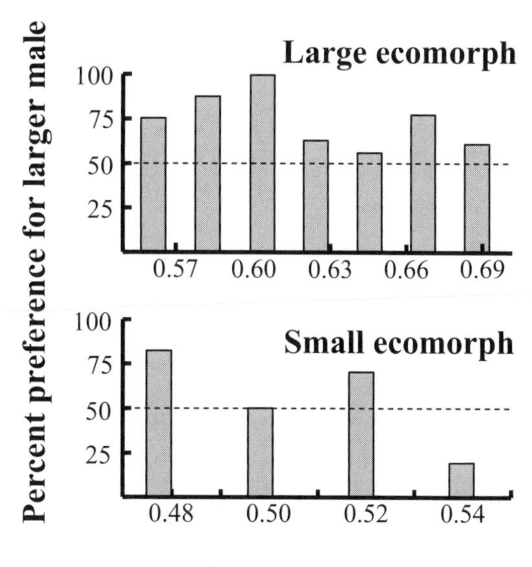

Figure 7.5 Results of mate choice trials for large and small ecomorph species in Oklahoma. Females chose between two males that differed in size. Shown is percentage of trials in which the larger male mated with the female, categorized by size class of the smaller male in the trial. In the large species, females preferred larger males independent of the absolute size of males, but preference in the small species depended significantly on the size of the smaller male in choice trials. Small females had an aversion to males in the smallest size class but were nonselective when choosing between intermediate and larger males. Based on Wellborn and Bartholf (2005).

pairing success increased continuously across male size. Pairing success in the small ecomorph, however, exhibited a threshold relationship in which smaller males had low pairing success, but pairing success did not differ across intermediate- to large-sized males. This field pattern was also evident in a mate choice experiment showing that, in the large ecomorph species, a female tended to mate with the larger male regardless of the absolute size of the males in the trial (Fig. 7.5). Size-biased pairing in the small species, however, depended significantly on the absolute size of the smaller male in the trial. Males of the smallest mature size class had low pairing success, but above this smallest size class, larger male size did not enhance pairing success. This pattern of mating biases is also seen in Michigan ecomorphs (Wellborn 1995), suggesting that similar ecological processes cause similar mating behavior.

Mating biases suggest that sexual selection in large ecomorphs pushes males toward ever larger size. In contrast, sexual selection in small ecomorphs imposes only a minimum threshold size at the lower extent of the adult size range, pushing male size over the threshold, but not larger. The disparate forms of these mating biases have an obvious correspondence to the disparate ecological value of body size for the two ecomorphs. Females prefer larger males in environments where larger size equates to greater ecological success (Wellborn 1994, 2002), but in environments where large size entails a mortality cost, females do not prefer larger males above a threshold size. Our challenge is to understand the mechanisms of sexual selection operating in *Hyalella* ecomorphs and how these drive observed mating biases.

Our understanding of the *Hyalella* mating system points to two possible forms of intrasexual selection among males. First, direct male–male interactions might occur if a single male is able to displace a guarding male. Such takeovers, however, appear to be rare in *Hyalella* and unlikely to play a significant role in sexual selection (Wen 1993). For example, Strong (1973) reported that takeovers did not occur in laboratory trials, even when paired males were smaller than unpaired males. A second mechanism of intrasexual selection in *Hyalella* is scramble mate competition, in which males attempt to find and guard receptive females before females are guarded by competing males. This mechanism is made more likely by the rarity of takeovers because males must find receptive, but unguarded, females. Sexual selection arising from scramble mate competition favors male traits, such as high activity rate, that increase encounters with receptive females. Although male *Hyalella* actively search for mates, currently there is little evidence available to assess the operation of sexual selection arising from mate searching.

Because females appear to influence pair formation, female preference may contribute to variance in male mating success and impose sexual selection on male traits. Female mate preferences may evolve when females gain direct fitness benefits as a result of mate choice. As in other amphipods, opportunities for direct benefits in *Hyalella* may occur as a consequence of precopulatory pairing, and effects arising from predation seem especially plausible because pairing influences predation risk (Cothran 2004). Given that individuals of both ecomorphs are subject to size-dependent mortality from predation (Fig. 7.3), a female's mortality risk may covary with the size of her mate because apparent size of the pair would depend in part on male size. Furthermore, because the direction of size-biased predation differs between ecomorphs, one could expect female preferences to be divergent between ecomorphs, with females of the large species preferring larger males, and females of the small species preferring smaller males. Although large-ecomorph females do prefer larger males, small-ecomorph females do not prefer smaller males, suggesting that this explanation may contribute to observed mating biases but cannot fully explain them.

Indirect selection, mediated through a correlation between preferred traits and offspring fitness (Kokko et al. 2003), may also contribute to evolution of mating biases in the ecomorphs. The qualitatively disparate mortality regimes experienced by large and small species suggests that mate size may correlate with offspring fitness. Because mortality declines with size in the large species, while competitive ability and foraging rate increase, as does mating success (Wellborn 1994, 1995, 2002), larger body size may indicate male genetic quality, provided size is heritable. In the small species, however, small or moderate size in males may indicate highest genetic quality due to the high mortality cost of large size. Alternatively, selection on preference for larger size may simply be relaxed in the small species because larger size is not associated with higher fitness. The disparity in mating biases observed between *Hyalella* ecomorphs is roughly consistent with evolution of mate preferences by indirect sexual selection acting across different regimes of natural selection, suggesting that further research is warranted.

Sexual Size Dimorphism

Ecomorph differences in the form of sexual selection acting on male body size can leave an evolutionary footprint in patterns of sexual size dimorphism (Andersson 1994).

Patterns of dimorphism in *Hyalella* are consistent with these disparate regimes of sexual selection. Large and small ecomorph species generally differ in the direction of sexual size dimorphism, with males being the larger sex in large ecomorph species, and the smaller sex in most small ecomorph species (Wellborn 1995, Wellborn and Cothran 2004, Wellborn et al. 2005). In Oklahoma, for example, large-species males are about 34% larger than females in mass, but males of the small species are 10% smaller than females (Wellborn and Bartholf 2005). Furthermore, as juveniles, males of the large ecomorph grow faster than females, contributing to species differences in adult sexual size dimorphism and suggesting strong selection for rapid growth in large-ecomorph males (Wellborn and Bartholf 2005). This interspecific variation in sexual size dimorphism can be driven by species disparity in female mate preference and thus may ultimately derive from the evolutionary effects of ecological constraints on sexual selection.

Future Directions

While it is evident that many characteristics of freshwater amphipods make them well suited as research models for understanding evolution of mating behavior, several challenges must be overcome before we can realize their full potential for informing current conceptual issues confronting the study of mating behavior and sexual selection. These hurdles largely parallel those facing the science as a whole (Kokko et al. 2003, Cordero and Eberhard 2003, Arnqvist and Rowe 2005). First, it is essential that we better understand the roles of males and females in pair formation by disentangling female mate preference from male coercion, a difficult problem in many species (Bisazza et al. 2001, Cordero and Eberhard 2003, Kokko 2005). Such information is required because mechanisms of sexual selection, and selected traits will depend on the extent to which each sex exercises control over mating (Fincke 1997). Detailed behavioral studies and manipulative experiments can contribute substantially to our understanding of this issue (Jormalainen and Merilaita 1995, Sparkes et al. 2000), and studies empirically evaluating direct and indirect selection on female preference hold much potential for illuminating this difficult issue (Head et al. 2005). The related issue of confidently assessing paternity also deserves further attention, and molecular markers will be especially beneficial for evaluating this subject. Although takeovers are generally considered rare based on visual observations during the guarding phase, takeovers might be most likely during the narrow time widow of the female molt and copulation, especially if multiple copulation bouts are needed for successful fertilization (Hume et al. 2005).

Second, evaluating sexual selection on body size, and traits that covary with size, is hindered in amphipods by the natural correlation between size and age that is characteristic of indeterminate growth (Wellborn 1995). The correlation between size and age implies that observed mating biases do not directly estimate biases in lifetime mating success but does imply sexual selection is acting on body size to the extent that variation in male body size reflects variation in age-specific size. At minimum, we need to employ laboratory rearing studies to assess levels of variation in age-specific body size. A better method is to assess mating success across the lifetime of individuals, rather than during "snapshot" samples, but measurement of lifetime reproductive success is a seemingly impossible task in these small organisms living in complex environments. Molecular genetic methods, however, may hold much promise for allowing measurement of reproductive success and sexual selection under natural conditions (Ritland 1996, Gibbs and Weatherhead 2001). Such molecular methods will likely be essential for achieving substantial strides toward a broader understanding of sexual selection under natural conditions.

Finally, it is imperative that we examine the nature of direct and indirect benefits that can shape female mating preferences and male traits. Some progress will be relatively straightforward, such as assessing effects of male mate size on direct benefits accruing to females through changes in the female's predation risk, feeding rate, or energetic costs, for example. Progress in understanding the nature and importance of indirect benefits of mate choice will be more challenging (Kokko et al. 2003), but laboratory studies assessing associations between female preference and offspring fitness are feasible (Hine et al. 2002, Head et al. 2005), and molecular genetic methods may allow such assessments under more natural conditions (Gibbs and Weatherhead 2001).

Summary and Conclusions

Although freshwater amphipods have been a research model for studies of male time-investment strategies during mating, much less is known about female mating preferences, the nature of benefits that drive female choice, and the influence of sexual conflict on female preference. Females may exercise mate choice by altering the form or intensity of resistance behavior. Direct selection on female mating preferences is likely to occur during the contact pairing phase because costs incurred or benefits gained by females during pairing are likely to depend on traits of guarding males. Indirect benefits from mate choice may also influence evolution of female preferences and male traits, but this has not been addressed empirically. Comparative studies of mating biases in interspecific ecomorphs in the genus *Hyalella* help to shed light on the evolution of female preference and preferred male traits. In species that occupy habitats where large body size is favored by ecological processes, females prefer larger males throughout the size range of males. For species subject to intense fish predation, however, mortality selection favors small body size, and female preference for larger males is weak, exhibiting a bias against the smallest males but indifference across intermediate to large male size. These divergent mating biases are consistent with direct and indirect selection on female preference acting under the disparate regimes of natural selection faced by the ecomorphs. Future advances in understanding the evolution of mating behavior in amphipods will require better integration of the joint roles of males and females in pairing, and mechanisms of sexual selection that drive mating biases.

Acknowledgments We thank Ola Fincke, Trish Schwagmeyer, and two anonymous reviewers for insightful comments on the manuscript. This research was supported by the National Science Foundation (DEB 98–15059).

References

Adams, J., and P.J. Greenwood. 1987. Loading constraints sexual selection and assortative mating in peracarid Crustacea. Journal of Zoology 211:35–46.

Andersson, M. 1994. Sexual selection. Princeton University Press, Princeton, N.J.

Arnold, S.J., and M.J. Wade. 1984. On the measurement of natural and sexual selection: applications. Evolution 38:720–734.

Arnqvist, G., and L. Rowe. 2005. Sexual conflict. Princeton University Press, Princeton, N.J.

Birkhead, T.R., and K. Clarkson. 1980. Mate selection and precopulatory guarding in *Gammarus pulex*. Zeitschrift für Tierpsychologie 52:365–380.

Bisazza, A., G. Vaccari, and A. Pilastro. 2001. Female mate choice in a mating system dominated by male sexual coercion. Behavioral Ecology 12:59–64.

Bollache, L., and F. Cezilly. 2004. Sexual selection on male body size and assortative pairing in *Gammarus pulex* (Crustacea: Amphipoda): field surveys and laboratory experiments. Journal of Zoology 264:135–141.

Bollache, L., G. Gambade, and F. Cezilly. 2001. The effects of two acanthocephalan parasites, *Pomphorhynchus laevis* and *Polymorphus minutus*, on pairing success in male *Gammarus pulex* (Crustacea: Amphipoda). Behavioral Ecology and Sociobiology 49:296–203.

Borowsky, B. 1991. Patterns of reproduction of some amphipod crustaceans and insights into the nature of their stimuli. Pages 33–49 *in*: R.T. Bauer and J.W. Martin, editors. Crustacean sexual biology. Columbia University Press, New York.

Borowsky, B., and R. Borowsky. 1987. The reproductive behaviors of the amphipod crustacean *Gammarus palustris* (Bousfield) and some insights into the nature of their stimuli. Journal of Experimental Marine Biology and Ecology 107:131–144.

Burt, A. 1995. The evolution of fitness. Evolution 49:1–8.

Caine, E.A. 1991. Reproductive behavior and sexual dimorphism of a caprellid amphipod. Journal of Crustacean Biology 11:56–63.

Cameron, E., T. Day, and L. Rowe. 2003. Sexual conflict and indirect benefits. Journal of Evolutionary Biology 16:1055–1060.

Chapman, T., G. Arnqvist, J. Bangham, and L. Rowe. 2003. Sexual conflict. Trends in Ecology and Evolution 18:41–47.

Conlan, K.E. 1991. Precopulatory mating behavior and sexual dimorphism in the amphipod Crustacea. Hydrobiologia 223:255–282.

Cordero, C., and W.G. Eberhard. 2003. Female choice of sexually antagonistic male adaptations: a critical review of some current research. Journal of Evolutionary Biology 16:1–6.

Cordero, C., and W.G. Eberhard. 2005. Interaction between sexually antagonistic selection and mate choice in the evolution of female responses to male traits. Evolutionary Ecology 19:111–122.

Cothran, R.D. 2004. Precopulatory mate guarding affects predation risk in two freshwater amphipod species. Animal Behaviour 68:1133–1138.

Crudgington, H.S., and M.T. Siva-Jothy. 2000. Genital damage, kicking and early death. Nature 407:855–856.

Elwood, R.W., J. Gibson, and S. Neil. 1987. The amorous *Gammarus*: size assortative mating in *G. pulex*. Animal Behaviour 35:1–6.

Emlen, S.T., and L.W. Oring, 1977. Ecology, sexual selection, and the evolution of mating systems. Science 197:215–223.

Fincke, O.M. 1997. Conflict resolution in the Odonata: implications for understanding female mating patterns and female choice. Biological Journal of the Linnean Society 60:201–220.

Gavrilets, S., G. Arnqvist, and U. Friberg. 2001. The evolution of female mate choice by sexual selection. Proceedings of the Royal Society of London, Series B 268:531–539.

Gibbs, H.L., and P.L. Weatherhead. 2001. Insights into population ecology and sexual selection in snakes through the application of DNA-based genetic markers. Journal of Heredity 92:173–179.

Härdling, R., H.G. Smith, V. Jormalainen, and J. Tuomi. 2001. Resolution of evolutionary conflicts: costly behaviors enforce the evolution of cost-free competition. Evolutionary Ecology Research 3:829–844.

Head, M.L., J. Hunt, M.D. Jennions, and R. Brooks. 2005. The indirect benefits of mating with attractive males outweigh the direct costs. PLOS Biology 3:289–294.

Hine, E., S. Lachish, M. Higgie, and M.W. Blows. 2002. Positive genetic correlation between female preference and offspring fitness. Proceedings of the Royal Society of London, Series B 269:2215–2219.

Hosken, D.J., O.Y. Martin, J. Born, and F. Huber. 2003. Sexual conflict in *Sepsis cynipsea*: female reluctance, fertility and mate choice. Journal of Evolutionary Biology 16:485–490.

Houde, A.E. 1997. Sex, color, and mate choice in guppies. Princeton University Press, Princeton, N.J.

Houle, D. 1998. How should we explain variation in the genetic variance of traits? Genetica 103:241–253.

Houle, D., and A.S. Kondrashov. 2002. Coevolution of costly mate choice and condition-dependent display of good genes. Proceedings of the Royal Society of London, Series B 269:97–104.

Hume, K.D., R.W. Elwood, J.T.A. Dick, and J. Morrison. 2005. Sexual dimorphism in amphipods: the role of male posterior gnathopods revealed in *Gammarus pulex*. Behavioral Ecology and Sociobiology 58:264–269.

Hunt, J., L.F. Bussieere, M.D. Jennions, and R. Brooks. 2004. What is genetic quality? Trends in Ecology and Evolution 19:329–333.

Hunte, W., R.A. Myers, and R.W. Doyle. 1985. Bayesian mating decisions in an amphipod, *Gammarus lawrencianus* Bousfield. Animal Behaviour 33:366–372.

Iyengar, V.K., and T. Eisner. 1999. Female choice increases offspring fitness in an arctiid moth (*Utetheisa ornatrix*). Proceedings of the National Academy of Sciences, USA 96:15013–15016.

Jones, T.M., R.J. Quinnell, and A. Balmford. 1998. Fisherian flies: benefits of female choice in a lekking sandfly. Proceedings of the Royal Society of London, Series B 265: 1651–1657.

Jormalainen, V. 1998. Precopulatory mate guarding in crustaceans: male competitive strategy and intersexual conflict. Quarterly Review of Biology 73:275–304.

Jormalainen, V., and S. Merilaita. 1993. Female resistance and precopulatory guarding in the isopod *Idotea baltica* (Pallas). Behaviour 125:219–231.

Jormalainen, V., and S. Merilaita. 1995. Female resistance and duration of mate-guarding in three aquatic peracarids (Crustacea). Behavioral Ecology and Sociobiology 36:43–48.

Jormalainen, V., J. Tuomi, and N. Yamamura. 1994. Intersexual conflict over precopula duration in mate guarding Crustacea. Behavioral Processes 32:265–283.

Jormalainen, V., S. Merilaita, and R. Härdling. 2000. Dynamics of intersexual conflict over precopulatory mate guarding in two populations of the isopod *Idotea baltica*. Animal Behaviour 60:85–93.

Kokko, H. 1998. Good genes, old age and life-history trade-offs. Evolutionary Ecology 12:739–750.

Kokko, H. 2001. Fisherian and "good genes" benefits of mate choice: how (not) to distinguish between them. Ecology Letters 4:322–326.

Kokko, H. 2005. Treat 'em mean, keep 'em (sometimes) keen: evolution of female preferences for dominant and coercive males. Evolutionary Ecology 19:123–135.

Kokko, H., R. Brooks, J.M. McNamara, and A.I. Houston. 2002. The sexual selection continuum. Proceedings of the Royal Society of London, Series B 269:1331–1340.

Kokko, H., R. Brooks, M.D. Jennions, and J. Morley. 2003. The evolution of mate choice and mating biases. Proceedings of the Royal Society of London, Series B 270:653–664.

Lyes, M.C. 1979. The reproductive behaviour of *Gammarus duebeni* (Lilljeborg), and the inhibitory effect of a surface active agent. Marine Behavior and Physiology 6:47–55.

Othman, M.S., and D. Pascoe. 2001. Growth, development and reproduction of *Hyalella azteca* (Saussure, 1858) in laboratory culture. Crustaceana 74:171–181.

Parker, G.A. 1979. Sexual selection and sexual conflict. Pages 123–166 *in*: M.S. Blum and N.A. Blum, editors. Sexual selection and reproductive competition in insects. Academic Press, New York.

Partridge, L., and J.A. Endler. 1987. Life history constraints on sexual selection. Pages 265–277 *in*: J.W. Bradbury and M.B. Andersson, editors. Sexual selection: testing the alternatives. John Wiley and Sons, New York.

Plaistow, S.J., L. Bollache, and F. Cezilly. 2003. Energetically costly precopulatory mate guarding in the amphipod *Gammarus pulex*: causes and consequences. Animal Behaviour 64:683–691.

Pomiankowski, A., and A.P. Møller. 1995. A resolution of the lek paradox. Proceedings of the Royal Society of London, Series B 260:21–29.

Price, T., D. Schluter, and N.E. Heckman. 1993. Sexual selection when the female directly benefits. Biological Journal of the Linnean Society 48:187–211.

Ridley, M. 1983. The explanation of organic diversity: the comparative method and adaptations for mating. Clarendon Press, Oxford.

Ritland, K. 1996. Marker-based method for inferences about quantitative inheritance in natural populations. Evolution 50:1062–1073.

Robinson, B.W., and R.W. Doyle. 1985. Trade-off between male reproduction (amplexus) and growth in the amphipod *Gammarus lawrencianus*. Biological Bulletin 168:482–488.

Rowe, L. 1994. The costs of mating and mate choice in water striders. Animal Behaviour 48:1049–1056.

Rowe, L., and D. Houle. 1996. The lek paradox and the capture of genetic variance by condition dependent traits. Proceeding of the Royal Society of London, Series B 263:1214–1421.

Shuster, S.M., and M.J. Wade. 2003. Mating systems and strategies. Princeton University Press, Princeton, N.J.

Sparkes, T.C., D.P. Keogh, and R.A. Pary. 1996. Energetic costs of mate guarding behavior in male stream-dwelling isopods. Oecologia 106:166–171.

Sparkes, T.C., D.P. Keogh, and K. Haskins. 2000. Female resistance and male preference in a stream-dwelling isopod: effects of female molt characteristics. Behavioral Ecology and Sociobiology 47:145–155.

Sparkes, T.C., D.P. Keogh, and T.J. Orsburn. 2002. Female resistance and mating outcomes in a stream-dwelling isopod: effects of male energy reserves and mating history. Behaviour 139:875–895.

Strong, D.R., Jr. 1972. Life history variation among populations of an amphipod (*Hyalella azteca*). Ecology 53:1103–1111.

Strong, D.R., Jr. 1973. Amphipod amplexus, the significance of ecotypic variation. Ecology 54:1383–1388.

Sutcliffe, D.W. 1992. Reproduction in *Gammarus* (Crustacea, Amphipoda): basic processes. Freshwater Forum 2:102–129.

Taylor, B.E., and W. Gabriel. 1992. To grow or not to grow: optimal resource allocation for *Daphnia*. American Naturalist 139:248–266.

Tomkins, J.L., J. Radwan, J.S. Kotiaho, and T. Tregenza. 2004. Genic capture and resolving the lek paradox. Trends in Ecology and Evolution 19:323–328.

Trivers, R.L. 1972. Parental investment and sexual selection. Pages 136–179 *in*: B. Campbell, editor. Sexual selection and the descent of man: 1871–1971. Aldine, Chicago, Ill.

Ward, P.I. 1984. The effects of size on the mating decisions of *Gammarus pulex* (Crustacea, Amphipoda). Zeitschrift für Tierpsychologie 64:174–184.

Ward, P.I. 1986. A comparative field study of the breeding behavior of a stream and a pond population of *Gammarus pulex* (Amphipoda). Oikos 46:29–36.

Ward, P.I. 1988. Sexual selection, natural selection, and body size in *Gammarus pulex* (Amphipoda). American Naturalist 131:348–359.

Wellborn, G.A. 1994. Size-biased predation and prey life histories: a comparative study of freshwater amphipod populations. Ecology 75:2104–2117.

Wellborn, G.A. 1995. Determinants of reproductive success in freshwater amphipod species differing in body size and life history. Animal Behaviour 50:353–363.

Wellborn, G.A. 2000. Selection on a sexually dimorphic trait in ecotypes within the *Hyalella azteca* species complex (Amphipoda: Hyalellidae). American Midland Naturalist 143:212–225.

Wellborn, G.A. 2002. Tradeoff between competitive ability and antipredator adaptation in a freshwater amphipod species complex. Ecology 83:129–136.

Wellborn, G.A., and S. Bartholf. 2005. Ecological context and the interaction of natural and sexual selection in two amphipod species. Oecologia 143:308–316.

Wellborn, G.A., and R.D. Cothran. 2004. Phenotypic similarity and differentiation among sympatric cryptic species in a freshwater amphipod species complex. Freshwater Biology 49:1–13.

Wellborn, G.A., D.K. Skelly, and E.E. Werner. 1996. Mechanisms creating community structure across a freshwater habitat gradient. Annual Review of Ecology and Systematics 27:337–363.

Wellborn, G.A., R.D. Cothran, and S. Bartholf. 2005. Life history and allozyme diversification in regional ecomorphs of the *Hyalella azteca* (Crustacea: Amphipoda) species complex. Biological Journal of the Linnean Society 84:161–175.

Wen, Y.H. 1993. Sexual dimorphism and mate choice in *Hyalella azteca* (Amphipoda). American Midland Naturalist 129:153–160.

Witt, J.D.S., and P.D.N. Hebert. 2000. Cryptic species diversity and evolution in the amphipod genus *Hyalella* within central glaciated North America: a molecular phylogenetic approach. Canadian Journal of Fisheries and Aquatic Sciences 57:687–698.

Yamamura, N., and V. Jormalainen. 1996. Compromised strategy resolves intersexual conflict over pre-copulatory guarding duration. Evolutionary Ecology 10:661–680.

Mating Strategies in Isopods

From Mate Monopolization to Conflicts

Veijo Jormalainen

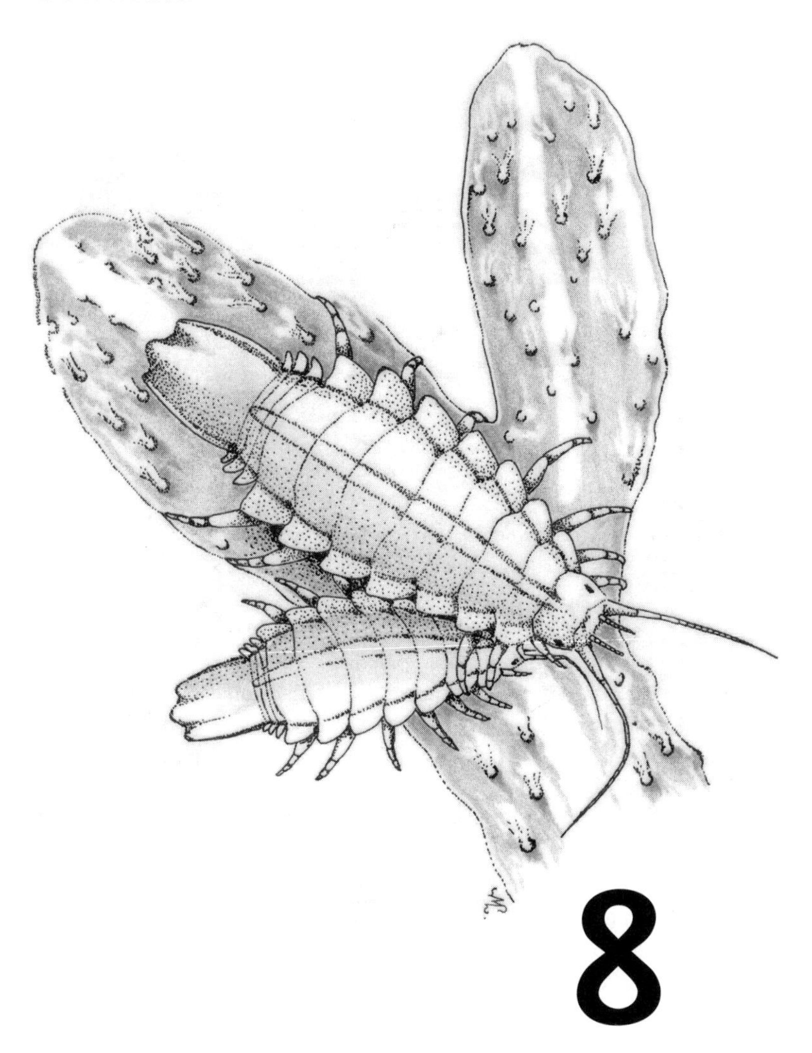

8

Sexual reproduction requires teamwork in bringing together the male and female gametes, but rather than harmonious cooperation, this often involves conflicts between the sexes. The mating system reflects the pattern of mate monopolization in space and time (Shuster and Wade 2003). It consists of sex-specific strategies for producing offspring; these evolve to maximize fitness in each environment. In addition to selection imposed by the environment, mating strategies have to be successful when faced not only with the behaviors of other individuals of the same sex but also with those of the opposite sex, suggesting that the game between the sexes may largely drive their evolution (Arnqvist and Rowe 2005).

Isopods have adapted successfully to a wide variety of environments. Currently, there are about 9,500 known isopod species (Schotte et al. 2005), 5,300 of them living in marine and freshwater habitats, 4,200 in terrestrial habitats, and 1,300 species having adopted a parasitic lifestyle. It is thus no wonder that isopod mating strategies are widely variable, ranging from sequential hermaphroditism to gonochorism, from promiscuity to strict monogamy, and from serial mate guarding of free-ranging iteroparous females to semelparous harem polygynandry (W.S. Johnson et al. 2001; see also chapter 2). The great majority of isopod species are gonochoristic, that is, reproducing in the role of a male or a female their whole life (Brook et al. 1994). Reproduction follows the basic peracaridan sequence (W.S. Johnson et al. 2001): females grow for several molt cycles, develop ovaries, and lay eggs at the so-called parturial molt, when a ventral marsupium appears. The eggs are incubated in the marsupium, from which the live juveniles, called mancae, crawl out after a few weeks. There may be only one brood, after which the female dies, or there may be several broods either in successive molts or alternating with a normal molt between reproductive molts. Males achieve maturity with the development of the penes and the appendages used in sperm transfer, the appendices masculinae, often at a relatively large size after several molt cycles (Wilson 1991). Thus, reproduction is closely tied to the molting cycle, but patterns of timing of receptivity for copulation and oviposition, sperm storage capability, and the number of mates vary among species.

One common behavior of male isopods is to guard females, either by physical attachment to her or by attending in close proximity, before fertilization and egg laying. This precopulatory guarding, also known as "passive phase," "amplexus," or "precopula," has been described in most isopod suborders, with the notable exception of the terrestrial suborder Oniscidea. My focus here is on the causes and consequences of this male reproductive strategy in isopods. I briefly review the theory of mate guarding as a male competitive strategy and the empirical knowledge available on variation in isopod guarding strategies, including the mating behavior of terrestrial isopods. I then review empirical findings concerning the various consequences of guarding. As has become evident during the last two decades, sexual selection, in addition to other forms of natural selection, is a fueling force in evolutionary change. It not only contributes to the evolution of secondary sexual characteristics in morphology, physiology, and behavior (Andersson 1994) but also may drive diversification within species and boost speciation (Partridge and Parker 1999). Mate-guarding behavior in isopods generates sexual selection in several ways: it involves competition among males and male mate choice; perhaps most interestingly, it also augments intersexual selection by giving rise to conflicts of interests between the sexes.

To Guard or Not to Guard?

Time-Limited Female Receptivity and Optimizing Guarding Duration

Male mate guarding as a competitive strategy has been studied extensively as an example of adaptive, optimal behavior. The obvious benefit of guarding for a male is mate monopolization when the probability of encountering receptive mates is low (Parker 1970). The probability of finding mates depends on the availability of mates that are ready for mating, that is, the operational sex ratio (Parker 1974). In peracarids, two characteristics of the female reproductive cycle make the operational sex ratio highly male biased. First, female receptivity to copulation usually—though not always—starts at the parturial molt, when the exoskeleton is still soft, and second, receptivity ends with immediate oviposition following copulation when the oostegites, the plates forming the external egg-pouch, and/or developing eggs close access to oviducts (W.S. Johnson et al. 2001). Thus, the functional significance of precopulatory guarding for males in isopods, and in peracarids in general (see, e.g., chapters 2, 7), lies in ensuring presence at the moment of receptivity for copulation. Temporally limited receptivity has long been recognized as a prerequisite for the evolution of precopulatory mate-guarding both in theoretical models and empirically. The first to conceptualize and model precopulatory guarding as a competitive time investment strategy was Parker (1974). He realized that guarding is beneficial when the resulting fitness gain is greater than that of continuing the search for females, and that the time invested in guarding should thus increase with a decrease in the encounter rate of the sexes. Ridley (1983) was the first to find the taxonomic association between precopulatory guarding and time-limited opportunity for fertilization in Crustacea. Since Parker's initial work, several models have explored the factors affecting optimal mate-guarding duration from the male perspective. I have previously summarized these models in detail (Jormalainen 1998); here, I merely identify the factors determining guarding duration.

The Evolutionary Stable Strategy model by Grafen and Ridley (1983) was the first to show mathematically the crucial importance of the population sex ratio and encounter frequency of males and females: optimal guarding duration increases with an increasing male bias in the sex ratio and with a lower rate of male encounter of females. In fact, continuous guarding in male-biased sex ratios was predicted. Encounter rate is affected by the availability of females, which varies among other things with the synchrony of the female reproductive cycle. Thus, the level of synchrony in the female population in approaching the reproductive molt modifies optimal guarding duration, as shown by the model of Yamamura and Jormalainen (1996). Grafen and Ridley (1983) also modeled a case where they allowed male–male competition in the form of takeovers; they found that the optimal guarding duration for males capable of takeovers is shortened due to the apparent female bias in their favor in the sex ratio. Yamamura (1987) generalized the model further by including guarding and searching costs for males and variation in the duration of receptivity for females. Guarding duration now decreased with increasing guarding costs; consequently, temporally limited guarding was commonly found in male-biased sex ratios, as well. Costs of guarding versus costs of searching, in terms of survival probability, were also included in an optimization model by Jormalainen et al. (1994b), with a similar effect on guarding duration. Most interestingly, momentary receptivity to copulation was not a necessary condition for the evolution of guarding (Yamamura 1987). Short receptivity favors guarding, especially under certain specific conditions, but guarding was able to evolve over a range of receptivity durations. Elwood and Dick (1990) took into account size-specific energetic guarding costs and fecundity-related variation in fitness benefits with female size; their graphic model suggested that guarding duration was positively related to both male and female size.

In conclusion, optimal guarding duration for the male is expected to vary with the sex ratio, the encounter rate of the sexes, the occurrence of takeovers, the synchrony of the female molt cycle, guarding versus searching costs, the duration of receptivity, and the size of both sexes in a pair. The value of these models is that they predict the evolution of male guarding behavior with respect to the above factors, although—with the exception of Jormalainen et al. (1994b) and Yamamura and Jormalainen (1996)—they unrealistically assume the start of guarding to be a solely male decision, with the female playing an entirely passive role in pair formation.

Variation in Guarding Strategies

The models of guarding summarized above can be interpreted as predicting that male decisions to start guarding will vary with the factors listed; in other words, rather than fixed guarding criteria, what evolves is adaptive phenotypic plasticity. Since guarding duration is predicted to be critical to male fitness, males are expected to evolve the sensory capability to assess the time remaining before female receptivity as well as ways of acquiring information on the availability of females, and to adjust their guarding criteria plastically according to the mating environment. This prediction has been tested a number of times, and plasticity of decision making has been found in both isopods and amphipods (reviewed in Jormalainen 1998). The problem with most of these experiments is that, owing to experimental design, it is often impossible to determine whether the observed variation in guarding duration is due to male or female decision making. Jormalainen and Shuster (1999) compared mate-guarding duration in *Thermosphaeroma milleri* and *T. thermophilum* in triplets of one male to two females versus two males to one female. Males were found to adjust guarding duration to the sex ratio as predicted: guarding lasted longer with a male-biased sex ratio in both species, and the difference in guarding duration was due to males adjusting their guarding criteria rather than to female responses to the level of male harassment (Jormalainen and Shuster 1999). In the marine isopod *Idotea baltica*, males adjust their guarding criteria depending on their experience of the male population (Jormalainen et al. 2000). Interestingly, this adjustment occurs in a size-specific manner: large males with experience of the occurrence of small males in the mating population postpone the start of guarding when put together with a female, compared to small males with experience of the occurrence of large males. This implies that males are able to collect information on the composition of the mating population and that they respond to it in a size-specific way. Since large *I. baltica* males are capable of taking over females from small ones (Jormalainen et al. 1994a), the sex ratio is functionally more female biased for them, and they may therefore postpone the start of guarding as predicted by the model (Grafen and Ridley 1983).

One interesting male mate-guarding strategy, found in certain Janiridae species, is the guarding of newborn mancae (Franke 1993; see also Thiel 2002 and references therein). In *Iais pubescens*, a commensal isopod on the shore-living isopod *Exosphaeroma gigas*, males manipulate the marsupium of females to obtain virgin mancae (Thiel 2002; Fig. 8.1a,b). They carry the manca for about a week until the first postmarsupial molt, copulate, and release it. The same behavior had previously been described as a form of paternal care; since, however, guarding ends with copulation, it rather represents a mating strategy. At the population level, there is a high frequency of guarding, and males compete for access to newborn mancae. Males even have a specific adaptation for carrying the mancae: their fourth pereopod is reduced and can be used to grasp a manca tightly underneath the male. Thiel (2002) suggests that the guarding of mancae may have evolved in response to an unpredictable encounter probability between the sexes.

Jaera hopeana, a commensal species on the isopod *Sphaeroma serratum*, has a similar kind of guarding behavior (Franke 1993): males guard the mancae up to 12 days until the first postmarsupial molt, at which point copulation takes place. The immature mancae then store the sperm until they reach maturity but remain continuously sexually receptive and may engage in en passant copulations (Franke 1993). Thus, guarding cannot be explained in this case solely by temporally restricted receptivity, and there may be other selective agents for the evolution of guarding. Franke (1993) suggests two possibilities. First, there may be sperm precedence of the first copulatory male. Second, mating with the first manca stage may be beneficial for males because of the tendency of females to refuse additional copulations. This suggests that female behavior may drive males to early guarding. Franke (1993) further suggests that the possibility of colonizing new hosts after copulation may have selected for early receptivity to mating in females. Thus, both the sperm storage capability and early receptivity to mating may be adaptations of the female mating strategy to enhance colonization success in a species with a commensal lifestyle, and the guarding of mancae may be the counteradaptation of males to this female strategy.

Precopulatory guarding and copulation may sometimes be followed by short postcopulatory guarding. *I. baltica* and *T. thermophilum* males routinely guard females until the anterior cuticle is shed and the female has oviposited, even though copulation is possible already after the shedding of the posterior cuticle (Jormalainen and Shuster 1999; V. Jormalainen and S.M. Shuster, personal observations). Such guarding probably functions as paternity assurance when there is a risk of sperm competition. The stakes of the male in a sperm competition situation are set by the structure and function of female genitalia, which determine the patterns of sperm mixing and precedence. Postcopulatory guarding is selected for when there is high degree of sperm mixing or sperm precedence of the last copulatory male (Simmons 2001 and references therein). First-male-sperm precedence selects for precopulatory guarding, which is the likely explanation for the occurrence of precopulatory guarding in species with continuous receptivity.

The above cases were ones where males physically attach to females while guarding, but guarding may also take the form of cohabitation of the pair. Conlan (1991), in a review of guarding behavior in amphipods, called this kind of guarding an "attending" strategy. Guarding by cohabitation is common in isopods that inhabit cavities in various materials. In the wood-boring isopod *Limnoria tripunctata*, the male

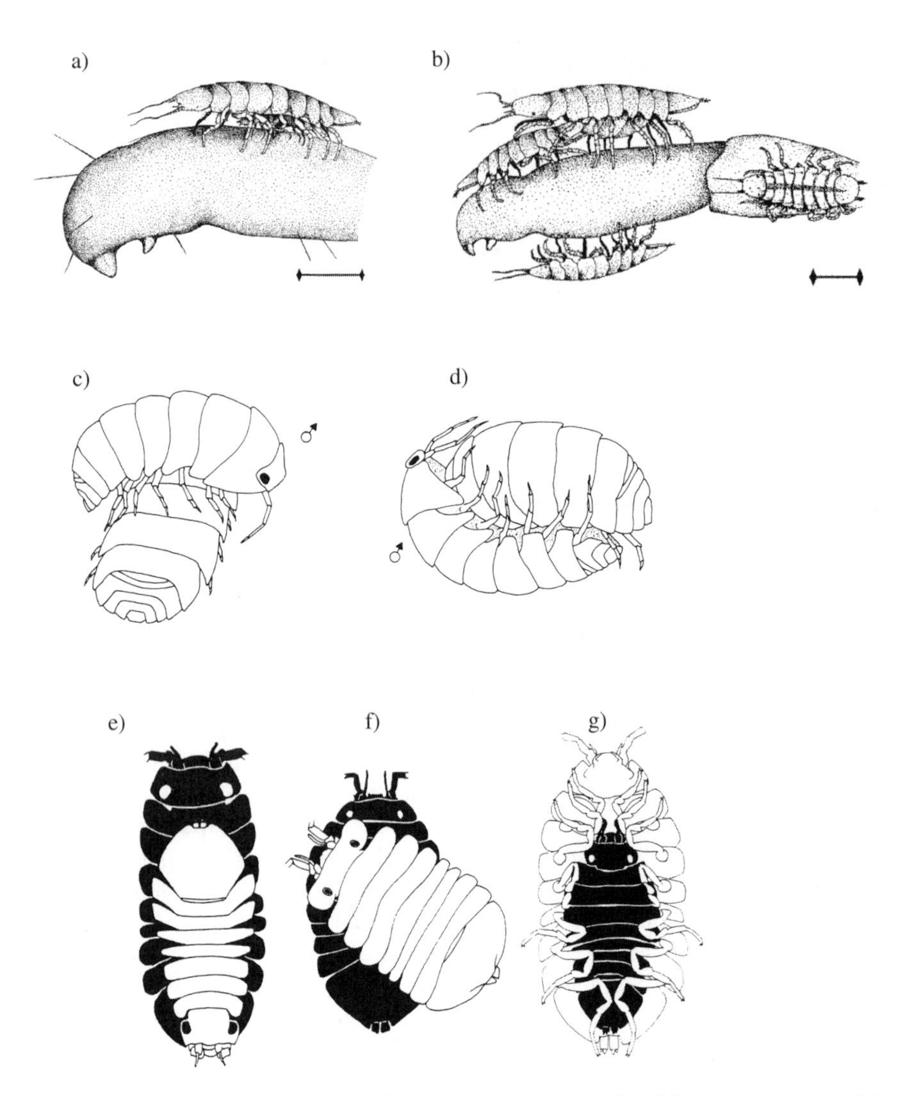

Figure 8.1 Variation in isopod mating behaviors. Top row: a male of the symbiotic isopod *Iais pubescens* (Janiroidea) on its host guarding a manca below him (a) and several males surrounding a female with juveniles ready to leave the marsupium (b). Males manipulate the female to obtain newborn mancae for guarding (from Thiel 2002; reprinted by permission of Springer-Verlag Gmbh). Scale bar 0.5 mm. Middle row: the terrestrial isopod *Venezillo evergladensis* in the short nuptial ride preceding copulation (c) and in copulatory position (d; modified from C. Johnson 1985; reprinted by permission of the University of Notre Dame). Bottom row: sexual size dimorphism and precopulatory guarding positions of the male (white) and female (black) in *Jaera albifrons* (e), *J. istri* (f), and *J. nordmanni* (g) (from Veuille 1980; reprinted by permission of the Linnean Society of London).

and female may spend extended periods in a burrow, during which they produce multiple broods (Menzies 1954, as cited in Wilson 1991). In the wood-boring isopod *Sphaeroma terebrans*, the males cohabit in the females' burrows, usually with a female that has ripe ovaries, and leave the burrow after egg laying (Thiel 1999), which suggests temporally restricted female receptivity and copulation taking place at molt. In this species, females share the burrow with the offspring; this has been interpreted as parental care, since the burrow provides both shelter and an osmotically stable environment for the development of the juveniles. In the sandstone-boring *S. wadai*, heterosexual pairs inhabit burrows (Murata and Wada 2002). The female is in the inner part and the male in the outer part of the burrow, implying that pair formation occurs when the male enters the female's burrow. Murata and Wada (2002) consider cohabitation as mate guarding that lasts up to five months. The males leave the burrows after the females become ovigerous, probably right after copulation. The cohabitation of males and females has also been found in leaf-mining isopods (see van Tussenbroek and Brearley 1998 and references therein). Thus, the data available suggest that, functionally, the cohabitation of the sexes is male mate guarding and that this guarding strategy is also associated with the temporally restricted female receptivity.

Terrestrial Isopods: An Exception Supporting the Rule

An exception to guarding behavior in free-living isopods is the terrestrial Oniscidea suborder (Zimmer 2001). Oniscidean isopods lack prolonged guarding, although they may have a short, approximately one-minute "nuptial ride" before copulation (C. Johnson 1985; Fig. 8.1c,d). The only known exceptions to this are the supralittoral species of the genus *Ligia* and the terrestrial *Helleria brevicornis*, both of which have guarding and momentary female receptivity for copulation during the parturial molt (Zimmer 2001), and possibly some burrow-digging *Porcellio* species (Linsenmair 1989). A short receptivity for copulation during parturial molt is the rule in aquatic isopods (reviewed in Jormalainen 1998), with the exception of the deep-sea Janiridae, in which low population densities may have selected for prolonged receptivity for copulation (see Wilson 1991 and references therein). In terrestrial isopods, the rule is long receptivity for copulation, lasting for several days during the intermolt, the whole preparturial intermolt, or over several molts excluding incubation periods (reviewed in Zimmer 2001). This led Zimmer (2001) to suggest that the evolutionary loss of guarding is probably due to the loss of temporal restrictions on female receptivity. Terrestrial isopods can also store sperm from previous copulations (see Moreau et al. 2002 and references therein), which means that monopolization of a female is ineffective if she is already carrying sperm from previous copulations. For example, *Venezillo evergladensis* females are continuously receptive and store sperm for extended periods, using it for up to eight successive broods (C. Johnson 1982). Multiple males may copulate, but the sperm are utilized sequentially: first the sperm of the first male, and while his sperm deplete over several broods, sperm from later copulations replaces that of the first male (C. Johnson 1982). Thus, there is a clear first-male advantage, and virgin females will give males the greatest return. However, both the absent need to ensure presence during the short receptivity and the long-term sperm storage probably select against guarding in terrestrial isopods.

In addition to factors arising directly from female characteristics, the terrestrial environment as such may have had an influence on the loss of guarding. Carrying females for prolonged periods may be too costly in a terrestrial environment (Zimmer 2001). There may be an increased energetic cost of carrying, or the maneuverability of precopulatory pairs may be poor in typical habitats in narrow crevices and under logs and stones (Zimmer 2001). Terrestrialization may well have contributed to the loss of guarding, assuming guarding to have been the ancestral condition as indicated by its prevalence in aquatic species and in the semiterrestrial *Ligia*. Mate guarding, on the other hand, is common in several groups of terrestrial insects in variable habitats, for example, damselflies, crickets, water striders and beetles (Choe and Crespi 1997), suggesting that environmental constraints on the evolution of contact guarding are not absolute.

In Oniscidea, promiscuous mating systems are common (Zimmer 2001). Moreau et al. (2002) studied the mating system of *Armadillidium vulgare* and found multiple female matings and a high level of sperm mixing. Interestingly, they found a low remating rate of females during the same intermolt period (Moreau et al. 2002). This was not due to the mating ability of males, because they can fertilize more than 10 females in a short time (Caubet et al. 1998). Moreau et al. (2002) hypothesized that males may induce refractory mating behavior in females to ensure their own paternity. Such a strategy could have evolved after the loss of guarding behavior, as an alternative to guarding (Moreau et al. 2002). While such a male strategy is possible, evidence is needed to confirm it.

An exception to promiscuity as the rule is the desert isopod *Hemilepistus reaumuri*. This species has strict monogamy: the male and female form a cooperative pair and inhabit and guard a burrow together (Linsenmair 1989; see also chapter 16). There is intraspecific competition for the burrows, which are a prerequisite for reproduction and survival in a warm and dry environment. Females are semelparous; the offspring cohabit in the burrow, and the male participates in brood care. Leaving the burrow is necessary for foraging; this may be why both the male and the female guard the burrow. Thus, environmental constraints may make monogamy the best option for the male, as well. Circumstantial evidence for the importance of burrow building and defense for monogamy comes from the close relative *H. elongatus*, whose mating system is highly promiscuous. In the habitat of *H. elongatus*, natural shelters are abundant; therefore, there is no competition for them, and cooperative guarding of shelters does not occur (Röder and Linsenmair 1999).

Intersexual Stimulus of Vitellogenesis

Terrestrial isopods lack guarding behavior, but intersexual stimulus has been found to affect the reproductive development of females. This is especially interesting because it may indicate that males may have evolved means—other than contact guarding—to increase their mating success. A case where the rate of reproductive development of females varies with the presence of males may indicate that males can manipulate females to lay eggs earlier. The onset of reproduction is usually induced by an environmental stimulus, for example, changes in temperature, in photoperiod, or in the availability of food (e.g., Caubet et al. 1998, W.S. Johnson et al. 2001). Jassem et al. (1991) were the first to show that the presence of males can stimulate the female

Armadillidium vulgare to accelerate vitellogenesis and, consequently, to shorten the preparturial intermolt and lay eggs earlier. The phenomenon has since been confirmed repeatedly; female *A. vulgare* reared with males have been found to come to the parturial molt about 40% more quickly than those reared alone (Caubet et al. 1998, Lefebvre and Caubet 1999). The females are only stimulated when they are in direct contact with males (Jassem et al. 1991). Lefebvre and Caubet (1999), in a series of ingenious manipulations of various tactile organs, showed further that the male copulatory organs, the appendices masculinae, are responsible for the effect. Moreover, there is geographical variability in the degree of male-induced stimulation: the higher the latitude, the weaker the effect (Souty-Grosset et al. 1994).

Ovarian maturation takes place as a consequence of vitellogenesis and the accumulation of vitellogenin (Lefebvre and Caubet 1999). The period sensitive to the stimulus, weeks before egg laying around the fourth week of the preparturial intermolt, corresponds to the beginning of vitellogenin synthesis in females (Caubet et al. 1998). This led Caubet et al. (1998) to suggest that the presence of a male could either speed up synthesis or increase the release of vitellogenin in females. While this provides a proximal explanation for the stimulus, the fundamental reason remains obscure. Is the speeding up of vitellogenesis a female response to the availability of a mating partner or her response to male harassment, or does it represent successful manipulation of the female by the male? In their discussion, Lefebvre and Caubet (1999) suggest that delaying oviposition when there are no males present would be a real advantage for females. The mature eggs will be laid anyway, but if they are not fertilized they will be aborted. For virgin females, and for those which have depleted their sperm storage, egg laying would be a waste of resources. Consequently, although the intersexual stimulus of vitellogenesis has been called a "male effect" (Lefebvre and Caubet 1999), it may rather represent an adaptation of females to the varying availability of mates and, if so, should rather be called "delayed oviposition" in the absence of a mating partner. Male availability may sometimes be low, especially because many terrestrial isopods—including *A. vulgare*—commonly have female-biased sex ratios due to endosymbiotic sex-ratio distorters (Bouchon et al. 1998).

How generally the rate of reproductive development varies with the availability of mates in isopods is not known. In addition to *A. vulgare*, the phenomenon has been found in only one species, the desert isopod *Hemilepistus reaumuri* (Nasri et al. 1996, as cited in Caubet et al. 1998). But what looks like rarity may reflect a lack of research effort. Such intersexual stimulus has not been reported in mate-guarding isopod species. In *I. baltica*, it does not exist (V. Jormalainen, unpublished data), except in the sense that copulation induces oviposition that is otherwise delayed for up to about three days. Clearly, further studies are needed to recognize the generality and consequences of the plasticity in the rate of vitellogenesis as well as to identify the selective agents for it.

Sexual Conflict of Interests

In sexual reproduction both sexes try to maximize their own fitness, which may occur at a cost to the other sex (Arnqvist and Rowe 2005 and references therein). Differences in the fitness maximizing strategies of the sexes follow from fundamental differences in the allocation of resources to gametes, and conflicts of interests concerning mating decisions are consequently unavoidable. Such intersexual conflicts are probably widespread, although current evidence is biased toward insect taxa (reviewed in Choe and Crespi 1997, Chapman et al. 2003, Arnqvist and Rowe 2005). At the heart of the conflict are the costs incurred by one sex due to an action by the other. Parker (1979), in conceptualizing the conflict, used male mate guarding in dung flies as an example and suggested that guarding may impose some energetic costs for females. These would generate a conflict, because male fitness can be maximized by behaviors that are costly for females. In isopods, conditions favoring long guarding for males are common (see above), while for the female copulation(s) with a male at the onset of receptivity is enough to fertilize the clutch. Thus, the optimum guarding duration is longer for males than for females, and there is a conflict over the decision as to when to start guarding (see chapter 10 for a possible conflict over the start of pairing in fiddler crabs). Depending on the female optimum, this conflict may concern either guarding duration or whether or not to form a guarding pair at all. There is both theoretical (Jormalainen et al. 1994b, Yamamura and Jormalainen 1996) and experimental (see below; see also chapter 7) evidence on sexual differences in optimal guarding duration in mate-guarding peracarids.

Conflicting interests regarding mating decisions are often, though not always, revealed by intersexual contests. In mate-guarding isopods, female resistance to male mating attempts is common: the female may assume a scrolled or hooked position, making firm attachment by the male difficult, or she may vigorously kick the male who is attempting guarding (reviewed in Jormalainen 1998). There is also some evidence in terrestrial isopods that the female, by rolling up her body, by exhibiting jerky body maneuvers, or by escaping from the situation (see Zimmer 2001 and references therein), may make the decision whether or not to copulate. The existence of such costly behaviors implies that they are used to avoid mating costs, which indicates conflicting interests. However, fitness payoffs or costs of resistance cannot be inferred from behavioral contests alone; quantification of the costs arising from behaviors of the opposite sex is needed to demonstrate the conflict (see, e.g., Pizzari and Snook 2003). While guarding costs for males are often clear, for example, in terms of energy and lost mating opportunities, these costs are assumed to be offset by the benefit of ensuring a mate; thus, they merely limit the optimal guarding duration. Crucial to the conflict are the costs of guarding for females. These costs to the female, arising from restricted feeding opportunities or from the increased risk of predation, of intersexual cannibalism, or of being flushed by currents, have been suggested repeatedly, and some empirical support for them exists (reviewed in Jormalainen 1998, Cothran 2004). There are three species of aquatic mate-guarding isopods—*Lirceus fontinalis*, *Thermosphaeroma thermophilum*, and *Idotea baltica*—where mate guarding has been studied from the point of view of intersexual conflict; in the following, I review these cases briefly.

Lirceus fontinalis is a freshwater detritivore found in North American stream habitats. At encounters between the male and the female, there is a struggle, during which the male attempts to overpower the female (Styron and Burbanck 1967, as cited in Sparkes et al. 1996). If the male is successful, precopulatory guarding of one to three days will follow, and copulation and egg laying take place immediately following the parturial molt (Sparkes et al. 1996). Males are able to evaluate the molting status of the female based on variation in the amount of the arthropod molting hormone (Sparkes et al. 2000). Sparkes et al. (1996) measured the energetic costs of guarding for males by measuring their glycogen reserves. Glycogen is the energy-rich storage compound in muscles and other tissue that is first used to fuel energy-demanding tasks. Guarding was found to be costly for males. Furthermore, the glycogen cost during guarding equalled that of having starved for the same period, suggesting that the cost arose from being unable to feed during guarding. Since predation pressure on these isopods is high, Sparkes et al. (1996) hypothesized that precopulatory pairs stay at predation refuges where food is unavailable. If their hypothesis is correct and the pair cannot feed during guarding, guarding should be costly also for females, which would lead to conflict over the duration of guarding.

Female behavior in *L. fontinalis* is crucial for the outcomes of contests (Sparkes et al. 2000). At the beginning of the encounter, females resist actively and often succeed in escaping. Later, active struggling ceases, but the female can still decide whether pair formation will occur: she either remains coiled, so that the male cannot move freely and has to return to active struggling, or uncoils and thereby allows guarding (Sparkes et al. 2000). Sparkes et al. (2000) suggested that the function of female resistance may be to shorten the period of precopulatory guarding; they further hypothesized that resistance may function as a mate choice mechanism, because it may bias pairings toward those males that can overcome the resistance. In further experiments, where male condition was manipulated by depleting glycogen reserves, the last stage of pair formation occurred most often with the male with higher glycogen reserves (Sparkes et al. 2002). Thus, female resistance biased matings toward males with higher glycogen reserves.

Thermosphaeroma isopods are a group of endemic species living on bottom sediments and vegetation in thermal springs in southwestern United States and Mexico. In *T. thermophilum*, the sex ratio is male biased throughout the season (Shuster 1981), which favors long guarding duration for males. Shuster (1981) observed that females resist guarding attempts by rolling into a ball, often making males lose their grip. Because such resistance indicated conflict over the starting of guarding, Jormalainen and Shuster (1999) used a manipulative experiment to test which sex determines the start of guarding. They manipulated the females' ability to resist guarding attempts by anesthetizing them with alcohol for a few hours daily. The females gradually recovered from the anesthesia, but their ability to resist guarding was markedly reduced. The experiment also involved reducing the ability of males to hold the resisting female, by cutting off dactyls from the pereopods. Because males use all their pereopods to hold the female, they assumed that this amputation would effectively reduce the male's mate-guarding ability. The manipulation of the female's ability to resist guarding increased guarding duration by 57% compared to the control group, but the manipulation of the male's ability to guard had no effect. The result showed that females effectively shorten guarding duration by resistance and that females are able to decide the time when guarding will start. Thus, the experiment provided strong evidence for the existence of conflicting male and female interests over the start of guarding.

Idotea baltica is a widespread generalist herbivore inhabiting marine littoral vegetation. The mating behavior of this species (Fig. 8.2) closely resembles that of the two species described above (Jormalainen and Merilaita 1993). There is a period of

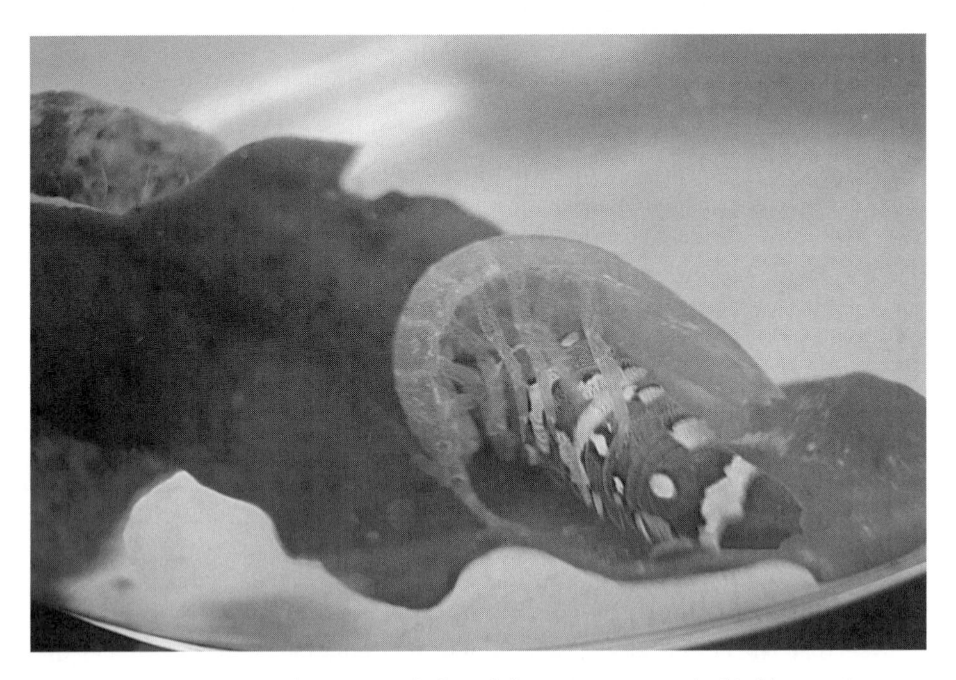

Figure 8.2 (top) A precopulatory pair of *Idotea baltica* swimming in the bladder wrack habitat, with the large dark male carrying the small, passive female. The guarding phase ends in the parturial molt, during which copulation takes place. (bottom) A copulating pair, with the uniformly colored male coiled up on the side of the white-spotted female. Photographs by V. Jormalainen.

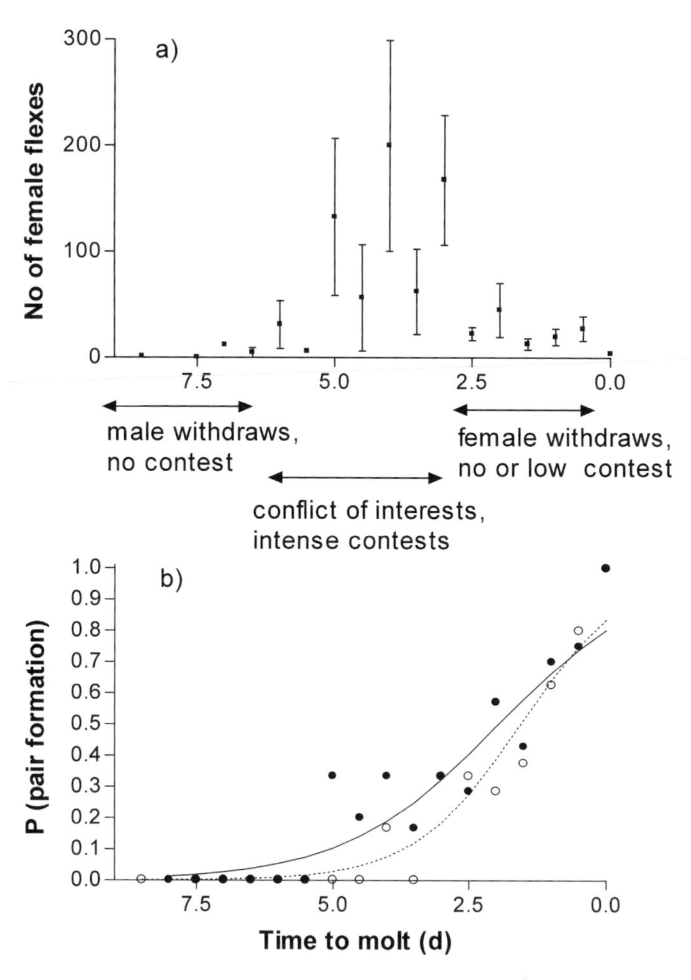

Figure 8.3 Dynamics of conflict behavior and consequent formation of guarding pairs in relation to the time left before the parturial molt in a population of *I. baltica* in the northern Baltic Sea. Data come from a laboratory experiment in which single males were presented to individual females at 12-hour intervals until the parturial molt, and the responses of the females and possible pair formation were observed. In encounters with females more than six days from their molt, the males usually withdraw, and there is no contest behavior (a). In encounters with more mature females, males attempt to initiate guarding, but females resist intensely (a) and often manage to escape; that is, pair formation fails (b). Around two days before the molt, female resistance ceases (a), and the probability of pair formation consequently increases. In (b), solid circles represent large males and open circles small ones, but in this population the two groups do not differ statistically significantly. From Jormalainen et al. (2000), by permission of Elsevier.

several days preceding the start of the guarding phase, during which males continuously try to initiate guarding and females resist these attempts (Fig. 8.3). During this period, females reared with a male escape from the male dozens of times a day, revealing the intensity both of male-guarding attempts and of female resistance. There is mutual aggression: males respond to female resistance by forcefully kicking back or by quickly bending their body back and forth while keeping the female in a solid grip. Some time before the parturial molt, female resistance ceases and a guarding period of one to more than six days begins (Fig. 8.3). Because such intense mutual aggression certainly indicates conflict, Jormalainen and Merilaita (1995) conducted a manipulation of male and female behavior, similar to the one described for *T. thermophilum*, to explore the interests of the sexes. They conducted two experiments with *I. baltica*, in which they manipulated the females' ability to resist by either predisposing them to osmotic stress or by treating them with a neuromuscular blocking agent. Both experiments also included manipulating the male's ability to overcome female resistance, either by osmotic stress or by removing dactyls (Jormalainen and Merilaita 1995). The results were similar in both experiments: in the group with reduced female resistance, guarding duration was about twice that of the control, and the manipulation of males had no effect on guarding duration. Similarly to *T. thermophilum*, conflict over guarding duration in *I. baltica* was thus resolved according to the female interest.

Costs that generate conflicts have seldom been demonstrated. Jormalainen et al. (2001) confirmed the existence of conflict in *I. baltica* by showing that mating is costly for females. Females depleted their glycogen reserves in resisting male mating attempts, and females that were guarded by males laid smaller eggs than did females that were not allowed to be guarded (Fig. 8.4). Thus, females faced both an energetic cost due to resistance behavior and a fecundity cost due to male mate guarding. The latter cost is probably related to restricted feeding or suboptimal diet during guarding, because the female cannot control where to feed and what to feed on while being guarded. The fecundity cost is also shared by the male, suggesting that such a cost to some extent restrains the prolongation of guarding duration and, furthermore, that intersexual conflict leads to decreased fitness of both sexes.

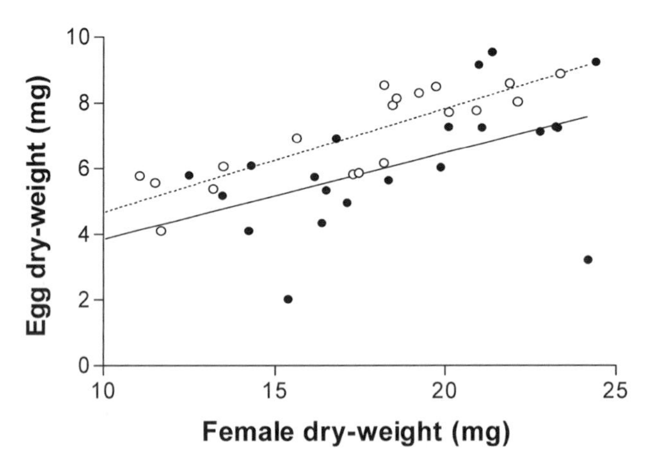

Figure 8.4 Costs of guarding in terms of pooled weight of eggs, shown in relation to maternal weight. During the intermolt preceding oviposition, the females were maintained either alone (open symbols, dashed regression line) or with a male (solid symbols, solid regression line); the latter group of females thus experienced precopulatory guarding. Egg weight differs statistically significantly (analysis of covariance, $F_{1,137} = 8.52$, $P < 0.01$) between the two female groups. From Jormalainen et al. (2001), by permission from Blackwell Science Ltd.

Consequences of Conflict

Compromised Guarding Duration

When the interests of the sexes are in conflict, how is guarding duration determined? Models for situations where both male and female fitness depends on the strategy of the opposite sex offer some insight into the resolution of sexual conflicts. In such models, a conflict is predicted to occur if the optimal strategies of the sexes differ. According to the models, such conditions are common (reviewed in Arnqvist and Rowe 2005; for mate-guarding models, see Jormalainen 1998). In a conflict, the relative power of the contestants as well as the costs of the behavioral conflict become important for the resolution; since fighting is costly, both contestants are expected to withdraw before the conflict costs exceed the fitness payoff for the winner (Yamamura and Jormalainen 1996, Härdling et al. 2001). Thus, a compromise guarding duration should arise, shorter than the male optimum but longer than the female one (Yamamura and Jormalainen 1996). This compromise strategy is mainly determined by the population sex ratio, which in turn determines the payoffs of winning for the male, and by the relative physical powers of the contestants.

In order to have a behavioral conflict, that is, mutual conflict behaviors of the sexes, both sexes must have a chance to win the conflict; otherwise, conflict behavior is suppressed, and the resolution follows the interest of the more powerful party (Härdling et al. 1999). The evolutionary dynamics of the conflict behaviors can in theory be variable (for details on various outcomes, see Yamamura and Jormalainen 1996, Härdling et al. 1999, 2001). First, the level of aggressiveness in encounters may reach an evolutionarily stable level for both sexes. Second, levels of male and female aggression may evolve in a cyclic, arms-race-like dynamics. Third, the extent of the conflict may be suppressed, and conflict behaviors are not realized except in cases where the contestants have inaccurate information concerning the environment and each other. Thus, the models tell us that the conflict over the start of guarding is likely to reach a compromise solution and that the conflict may exist even if overt behavioral contests are rare or nonexistent.

Guarding and Conflict as Selective Agents

Since males are the active sex in finding a mate, sexual selection is expected to favor sensory abilities that enhance mate finding and allow assessment of the female reproductive status. Detection of a potential mate from a distance is based on hormonal signals (see Jormalainen 1998 and references therein). In peracarids, the main chemoreception organs are the antennae (see Bertin and Cezilly 2003 and references therein). The role of the antennae in sexual selection has been studied in *Asellus aquaticus*, in which males not only are larger than females but also have longer antennae relative to body size (Bertin and Cezilly 2003). Bertin and Cezilly (2003) found that body size was a better predictor of male pairing success than antenna length. However, when they manipulated antenna length in the laboratory, they found that males with short antennae were less likely to detect, orient toward, and pair with females than were males with long antennae, implying sexual selection for antenna length. Sexual dimorphism in antenna length of terrestrial, promiscuous species has also been suggested to have evolved due to scramble competition among males to be the first to find receptive females (Lefebvre et al. 2000).

Molt constrains the male's ability to guard, since he will lose the female if he molts during guarding (Franke 1993; V. Jormalainen, personal observation). In a gammarid species, the time left before molt affects the male's propensity to start guarding (Bollache and Cezilly 2004), and in *I. baltica*, males delay the start of guarding until after their own molt (Borowsky 1987). In some amphipod species, the duration of the male molt cycle is longer than that of females (Wickler and Seibt 1981, Borowsky 1986). Thus, it is possible that in mature males, guarding generates selection for long intermolt duration (Wickler and Seibt 1981); so far, however, there are too few data for generalizations.

Direct male–male competition in the form of taking over females from other males selects for large male size, and possibly for size and form of the appendages used in carrying females (Schmalfuss 2003, reviewed in Jormalainen 1998). Such selection contributes to sexual size dimorphism, males commonly being the larger sex in mate-guarding crustaceans (Conlan 1991, Jormalainen 1998). The species in the genus *Jaera* provide a nice example where the direction of sexual dimorphism in size is related to the intensity of intrasexual competition among males (Veuille 1980): in species showing a high level of male–male competition, males are larger than females (Fig. 8.1e,f,g). In addition to male–male competition, also female resistance against male mating attempts selects for large male size: in *I. baltica*, large males are better able to overcome female resistance and thus have better chances in mate monopolization (Jormalainen and Merilaita 1993, 1995). The idea of contact mate-guarding behavior selecting for large male size gets circumstantial support from a comparison between free-living and burrow-living sphaeromatids. The limited evidence suggests that in free-living species contact guarding is common, while the males of burrow-living species guard by attending female crevices (see Murata and Wada 2002 and references therein). Interestingly, sexual dimorphism seems to be related to living habit; in burrow-living species, males are smaller than females, while in free-living species, the reverse is true (Thiel 1999, Murata and Wada 2002). This suggests that in free-living species guarding by physical contact selects for large male size, while guarding by cohabitation does not.

Thus, large male size is beneficial in mate-guarding species in both intra- and intersexual interactions. Interestingly, this sex-specific size dependency in mating success may select for a completely different reproductive strategy, namely, protogynous hermaphroditism (Brook et al. 1994, Abe and Fukuhara 1996). Hermaphroditism is a rare reproductive strategy among peracarids (W.S. Johnson et al. 2001), but in isopods it occurs in two different forms (Brook et al. 1994). Protandry, starting the life cycle as a male and subsequently changing sex to a female, is the more common strategy, found mostly in parasitic species (Brook et al. 1994) but also in some terrestrial ones, where females are bigger than males (M. Zimmer, personal communication). Protogyny, starting the life cycle as a female and changing sex to a male later on with larger size, has been described in fewer than 10 aquatic, free-living species with guarding males larger than females. The evolution of sequential hermaphroditism in general has been attributed to situations where it is difficult to find a mate, where size-specific reproductive returns for males and females are different, and where populations are small and genetically isolated (Ghiselin 1969). More specifically, the sex role is expected to change with the size of the individual following the size-specific reproductive returns for the sexes (Ghiselin 1969). Brook et al. (1994) noted an evident association between protogyny and precopulatory guarding in isopods. This led them to suggest that benefits for large males due to precopulatory mate-guarding behavior may be the primary selective agent favoring protogyny. Precopulatory guarding has not been observed in protandrous hermaphrodites (Brook et al. 1994), although dwarf males may live attached to a female in some parasitic epicarids (Wilson 1991).

Conflicts are suggested to lead to antagonistic coevolution of male and female traits (Parker 1979, Rice and Holland 1997, Chapman et al. 2003, Arnqvist and Rowe 2005), although the role of conflicts compared to other mechanisms is still somewhat controversial. In the front line of the conflict are the behaviors used in male–female interactions. If we assume the initial condition to be a population without sexual conflict, aggressive behavior in intersexual encounters will first emerge as a female strategy for resisting and escaping costly guarding by males (see Härdling et al. 1999 and references therein). Female resistance in turn will select for male persistence in fights, that is, behaviors attempting to overcome the resistance. Male aggressions, such as quick flexing of the body in response to female kicking, are often involved in behavioral conflicts over the start of guarding, as described above.

Sexual conflicts may generate selection for timing of reproductive events and for reproductive anatomy, including sperm storage capability and morphology of genitalia. In *T. thermophilum*, reproductive molt and oviposition are temporally separated (Jormalainen and Shuster 1999). Males guard females before the reproductive molt, during which copulation takes place, for about a week. Guarding partly overlaps with the period of ovarian maturation, which continues until oviposition about two weeks after the reproductive molt. The females store sperm in their oviducts from copulation to oviposition (Jormalainen et al. 1999). Guarding during ovarian development is probably costly for females because it may interfere with their ability to provision the developing ova. This led Jormalainen et al. (1999) to suggest that separating guarding from oviposition by delaying ovary development may be a female trait, evolved in the context of sexual conflict. Sperm storage can be viewed as part of such female adaptation, because it is a precondition for the temporal separation of copulation and fertilization. In *Jaera hopeana*, in which males guard the mancae and copulate with them at the first postmarsupial molt, the females refuse copulation after the manca stage (Franke 1993). Avoidance of guarding costs during ovary maturation, that is, sexual conflict, could also have been the selective agent for the evolution of such an early receptivity to copulation.

Female Resistance Versus Female Choice

Female discrimination among male candidates, female choice, is the classic mechanism of sexual selection. Female choice assumes that, by biasing mating toward certain males, females will achieve either direct benefits, for example, better fertilization ability or resources provided by the male, or indirect benefits, such as genes providing either better survival and viability or traits in sons that will increase their mating success (Andersson 1994). The direct benefits are probably more important (see Cameron et al. 2003 and references therein), although the importance of direct versus indirect effects in selecting for female preferences has remained controversial due to a lack of studies with careful quantification of both fitness components in the same system (Pizzari and Snook 2003, Cordero and Eberhard 2003). Intersexual conflicts provide an alternative, though not mutually exclusive, mechanism of sexual selection. Conflicts generate sexual selection due to the costs generated by the fitness maximization strategy of the opposite sex and through antagonistic evolution of the traits important for conflict resolution (Parker 1979, Härdling et al. 1999, Arnqvist and Rowe 2005). Thus, the direct benefits of avoiding costs due to behaviors or other manipulative traits of the opposite sex are the causal agent of selection. In the case of mate-guarding crustaceans, female resistance to male mating attempts will thus evolve because it reduces the direct costs arising out of being guarded. The alternative female choice interpretation would be that resistance evolved as a mechanism for choosing among male candidates that differ in their quality in providing direct or indirect benefits.

Demarcation between these two mechanisms is not easy, for two reasons. First, resistance, just like choice, may lead to nonrandom mating; second, the inclusive quantification of direct and especially indirect benefits with respect to male traits, as well as measurement of the direct costs of mating, including the costs of antagonistic behaviors, is very difficult. Arnqvist and Rowe (2005) suggest that the most promising approach would be to look at the evolutionary forces acting on female behavior (resistance or preference). In the case of mate-guarding crustaceans, the idea of sexual selection by female choice has been put forward (Ridley and Thompson 1979, Sparkes et al. 2002; see also chapter 7), but I am not aware of any evidence for the existence of indirect benefits associated with mating with a particular male phenotype. Ridley and Thompson (1979) first suggested that female resistance in *A. aquaticus* could be a female mate choice strategy for large males. Similarly, Sparkes et al. (2002) suggested that female resistance in *L. fontinalis* may act as a mechanism of female choice for good-condition males, that is, those with high energy reserves and good fertilization ability. In *I. baltica*, female resistance leads to better pairing success of large males and to discrimination against males in poor condition (Jormalainen and Merilaita 1993, 1995). Thus, female resistance can clearly lead to nonrandom mating among males. However, the proximate consequence of resistance, a pairing bias toward large or high-condition males, cannot be used as evidence of sexual selection through female choice, since it may follow independently of how the resistance evolved. The study by Sparkes et al. (2002) identified one possible direct benefit of choice for females, namely, the higher fertilization success of males that have not mated recently and thus depleted their energy reserves. However, before further evidence is available as to the direct and indirect benefits of female mate choice, and a comparison of the magnitude of these with the direct costs of guarding for females, the most parsimonious explanation of intersexual selection in mate-guarding isopods is that nonrandom mating arises as a side effect of the evolution of resistance in females to avoid direct guarding costs. Such evolution has been shown to be plausible over a range of conditions (Gavrilets et al. 2001).

Female preference for large males may occur in terrestrial isopods. In *Porcellio linsenmairi*, females allow copulation within a minute in encounters with large males, while in encounters with small males they withdraw and achieving copulation takes more time (Linsenmair 1989). In the field, such withdrawals are likely to lead to escape of the female. Linsenmair (1989) further found that, in some *Porcellio* species in the Canary Islands, females search for large males that dig specific copulation burrows and join them to copulate. While this may represent a preference for sheltered copulations, mating will be biased toward large males since small males do not dig such burrows. To confirm female choice for male size, however, experimental data are needed on female responses to different-sized males, in setups where male age, maturity, and burrow-digging behavior are controlled.

Future Directions

Although sexual selection for various traits related to the mating interactions of isopods has been suggested, the mechanisms of selection are not always clear. Also more generally, there is no consensus as to the relative importance in sexual selection of the indirect benefits of female choice and of antagonistic selection due to avoidance of direct mating costs. More case studies are therefore needed. Mate-guarding isopods and other crustaceans whose mating system includes intersexual conflict have the potential to act as model systems, where the direct and indirect costs and benefits of mating can be quantified and the nature of intersexual coevolution clarified.

The mechanisms of the behavioral optimization of mating decisions are insufficiently understood. This concerns not only the neural levels of sending, receiving, and processing signals (see chapter 4), but also behavioral processes. For example, there are no studies on how males obtain the information on population composition needed in adjusting the guarding criterion. The nature of the chemical cues of receptivity, especially their evolutionary origin, is poorly understood. When males detect receptive females on the basis of hormonal signals, is this a matter of active signaling of receptivity by females or male exploitation of female cues not originally evolved for signaling purposes? Knowing the exact mechanism may help understand the evolution of the trait.

The role of sexual selection, and especially of sexual conflicts, in the evolution of insect genitalia has recently been emphasized (see Hosken and Stockley 2004 and references therein). Genital morphology mediates conflicts and is likely to evolve in that context. For example, in *A. vulgare* structural changes in the female genitalia occur during the molt cycle, as genitalia of "reproductive" and "nonreproductive" type are successively produced after maturation (Suzuki 2001). During the preparturial intermolt, there are copulatory openings present, but they are transformed into oopores within hours after the parturial molt. By means of such genital transformations, females are able to influence the timing of copulation and the period during which males are interested in them. The role of various mechanisms of sexual selection in explaining the variation in structure and function of isopod genitalia (reviewed in Wilson 1991) is an unstudied area, with the potential to provide new, more generally interesting insights into the evolution of the genitalia.

Sexual selection in general has been proposed to act as an engine of speciation (see, e.g., Higashi et al. 1999 and references therein). Isopod mating systems can be used to study the role of mating behavior in diversification and speciation. Because these systems are widely variable, the mechanisms of sexual selection are likely to differ considerably among species. Antagonistic sexual selection through sexual conflicts may be especially important here, because it may create runaway coevolution between the sexes, which may promote rapid behavioral divergence between populations and enforce sympatric speciation (see Martin and Hosken 2003 and references therein), but the empirical data are rudimentary. In isopods, there is some evidence for the contribution of sexual selection to speciation. In the *Jaera albifrons* species complex found on North Atlantic coasts, five species can be distinguished only by the male secondary sexual characters (Solignac 1981). Where species occur sympatrically, their microhabitats overlap, but hybridization is rare and sexual isolation is due to male mating behavior and female responses to it (Solignac 1981): the male mounts the female in a head-to-tail position (Fig. 8.1e) and emits a highly specific signal, consisting of brushing and touching specific regions of the female. Nonreceptive females resist intensely. This suggests that male manipulation of females and female responses to it are species specific. In another example, sexual conflict generates diversification of mating behavior. Jormalainen et al. (2000) studied the dynamics of behavioral conflict over the start of guarding in two different populations of *I. baltica*. They found that the intensity of conflict as well as the conflict behaviors had diverged between the populations. Although it is not known how this may affect pairings between individuals from different populations, the coevolution of male and female traits suggests that divergent conflict behaviors may potentially act as an isolation mechanism.

Summary

Isopods have adapted to a wide variety of habitats and consequently show highly variable mating systems. Mate guarding by males before copulation, however, occurs commonly in most free-living taxa, with the notable exception of the terrestrial Oniscidea. Such guarding prior to copulation is a male mate monopolization strategy, evolved as a response to short female receptivity to copulation. Males are able to assess female maturity and to adjust guarding duration accordingly; relatively long guarding duration is often optimal for males. Guarding has no known benefits for females. Moreover, guarding is likely to impose costs for females, thus leading to a sexual conflict. Behaviors suggesting the existence of such conflict have been found in several isopod species and have been studied in detail in *Lirceus fontinalis*, *Thermosphaeroma thermophilum*, and *Idotea baltica*. These studies suggest that conflicts over the start of guarding generate sexual selection for traits related to obtaining or resisting mates, such as female resistance, male persistence, and large male size. Furthermore, conflicts may select for female traits related to timing of receptivity for copulation, timing of ovarian development, and the capability to store sperm. Isopod mating systems can be used to clarify the roles played by traditional female choice and intersexual conflicts in sexual selection.

Acknowledgments This work was conducted while I enjoyed the position of Academy Research Fellow with the Academy of Finland. I am grateful to Roger Härdling, Sami Merilaita, Steve Shuster, Juha Tuomi, and Norio Yamamura for their collaboration and for discussions on isopod mating behavior and sexual conflicts, which have deeply contributed to an understanding of the role of conflicts in isopod mating systems. Martin Zimmer and an anonymous referee provided valuable suggestions for an earlier version of the manuscript.

References

Abe, M., and H. Fukuhara. 1996. Protogynous hermaphroditism in the brackish and freshwater isopod, *Gnorimosphaeroma naktongense* (Crustacea: Isopoda, Sphaeromatidae). Zoological Science 13:325–329.

Andersson, M. 1994. Sexual selection. Princeton University Press, Princeton, N.J.

Arnqvist, G., and L. Rowe. 2005. Sexual conflict. Princeton University Press, Princeton, N.J.

Bertin, A., and F. Cezilly. 2003. Sexual selection, antennae length and the mating advantage of large males in *Asellus aquaticus*. Journal of Evolutionary Biology 16:491–500.

Bollache, L., and F. Cezilly. 2004. State-dependent pairing behaviour in male *Gammarus pulex* (L.) (Crustacea, Amphipoda): effects of time left to moult and prior pairing status. Behavioural Processes 66:131–137.

Borowsky, B. 1986. Laboratory observations of the pattern of reproduction of *Elasmopus levis* (Crustacea: Amphipoda). Marine Behaviour and Physiology 12:245–256.

Borowsky, B. 1987. Laboratory studies of the pattern of reproduction of the Isopod Crustacean *Idotea baltica*. Fishery Bulletin 85:377–380.

Bouchon, D., T. Rigaud, and P. Juchault. 1998. Evidence for widespread *Wolbachia* infection in isopod crustaceans: molecular identification and host feminization. Proceedings of the Royal Society of London, Series B 265:1081–1090.

Brook, H.J., T.A. Rawlings, and R.W. Davies. 1994. Protogynous sex change in the intertidal isopod *Gnorimosphaeroma oregonense* (Crustacea: Isopoda). Biological Bulletin 187:99–111.

Cameron, E., T. Day, and L. Rowe. 2003. Sexual conflict and indirect benefits. Journal of Evolutionary Biology 16:1055–1060.

Caubet, Y., P. Juchault, and J.P. Mocquard. 1998. Biotic triggers of female reproduction in the terrestrial isopod *Armadillidium vulgare* Latr. (Crustacea Oniscidea). Ethology Ecology and Evolution 10:209–226.

Chapman, T., G. Arnqvist, J. Bangham, and L. Rowe. 2003. Sexual conflict. Trends in Ecology and Evolution 18:41–47.

Choe, J.C., and B.J. Crespi. 1997. The evolution of mating systems in insects and arachnids. Cambridge University Press, Cambridge.

Conlan, K.E. 1991. Precopulatory mating behavior and sexual dimorphism in the amphipod Crustacea. Hydrobiologia 223:255–282.

Cordero, C., and W.G. Eberhard. 2003. Female choice of sexually antagonistic male adaptations: a critical review of some current research. Journal of Evolutionary Biology 16:1–6.

Cothran, R.D. 2004. Precopulatory mate guarding affects predation risk in two freshwater amphipod species. Animal Behaviour 68:1133–1138.

Elwood, R.W., and J.T.A. Dick. 1990. The amorous *Gammarus*: the relationship between precopula duration and size-assortative mating in *G. pulex*. Animal Behaviour 39:828–833.

Franke, H.D. 1993. Mating system of the commensal marine isopod *Jaera hopeana* (Crustacea). 1. The male-manca(I) amplexus. Marine Biology 115:65–73.

Gavrilets, S., G. Arnqvist, and U. Friberg. 2001. The evolution of female mate choice by sexual conflict. Proceedings of the Royal Society of London, Series B 268:531–539.

Ghiselin, M.T. 1969. Evolution of hermaphroditism among animals. Quarterly Review of Biology 44:189–208.

Grafen, A., and M. Ridley. 1983. A model of mate guarding. Journal of Theoretical Biology 102:549–567.

Härdling, R., V. Jormalainen, and J. Tuomi. 1999. Fighting costs stabilize aggressive behavior in intersexual conflicts. Evolutionary Ecology 13:245–265.

Härdling, R., H.G. Smith, V. Jormalainen, and J. Tuomi. 2001. Resolution of evolutionary conflicts: costly behaviors enforce the evolution of cost-free compromises. Evolutionary Ecology Research 3:829–844.

Higashi, M., G. Takimoto, and N. Yamamura. 1999. Sympatric speciation by sexual selection. Nature 402:523–526.

Hosken, D.J., and P. Stockley. 2004. Sexual selection and genital evolution. Trends in Ecology and Evolution 19:87–93.

Jassem, W., P. Juchault, C. Soutygrosset, and J.P. Mocquard. 1991. Male-induced stimulation of the initiation of female reproduction in the terrestrial isopod *Armadillidium vulgare* Latr. (Crustacea, Oniscidea). Acta Oecologica 12:643–653.

Johnson, C. 1982. Multiple insemination and sperm storage in the isopod, *Venezillo evergladensis* Schultz, 1963. Crustaceana 42:225–232.

Johnson, C. 1985. Mating behavior of the terrestrial isopod, *Venezillo evergladensis* (Oniscoidea, Armadillidae). American Midland Naturalist 114:216–224.

Johnson, W.S., M. Stevens, and L. Watling. 2001. Reproduction and development of marine peracaridans. Advances in Marine Biology 39: 105–260.

Jormalainen, V. 1998. Precopulatory mate guarding in crustaceans—male competitive strategy and intersexual conflict. Quarterly Review of Biology 73:275–304.

Jormalainen, V., and S. Merilaita. 1993. Female resistance and precopulatory guarding in the isopod *Idotea baltica* (Pallas). Behaviour 125:219–231.

Jormalainen, V., and S. Merilaita. 1995. Female resistance and duration of mate-guarding in three aquatic peracarids (Crustacea). Behavioral Ecology and Sociobiology 36:43–48.

Jormalainen, V., and S.M. Shuster. 1999. Female reproductive cycle and sexual conflict over precopulatory mate-guarding in *Thermosphaeroma* isopods. Ethology 105:233–246.

Jormalainen, V., S. Merilaita, and J. Tuomi. 1994a. Male choice and male-male competition in *Idotea baltica* (Crustacea, Isopoda). Ethology 96:46–57.

Jormalainen, V., J. Tuomi, and N. Yamamura. 1994b. Intersexual conflict over precopula duration in mate guarding Crustacea. Behavioural Processes 32:265–283.

Jormalainen, V., S.M. Shuster, and H. Wildey. 1999. Reproductive anatomy, precopulatory mate guarding, and paternity in the Socorro isopod, *Thermosphaeroma thermophilum*. Marine and Freshwater Behaviour and Physiology 32:39–56.

Jormalainen, V., S. Merilaita, and R. Härdling. 2000. Dynamics of intersexual conflict over precopulatory mate guarding in two populations of the isopod *Idotea baltica*. Animal Behaviour 60:85–93.

Jormalainen, V., S. Merilaita, and J. Riihimäki. 2001. Costs of intersexual conflict in the isopod *Idotea baltica*. Journal of Evolutionary Biology 14:763–772.

Lefebvre, F., and Y. Caubet. 1999. On the male-effect in the terrestrial Crustacean *Armadillidium vulgare* (Latreille, 1804). Invertebrate Reproduction and Development 35:55–64.

Lefebvre, F., M. Limousin, and Y. Caubet. 2000. Sexual dimorphism in the antennae of terrestrial isopods: a result of male contests or scramble competition? Canadian Journal of Zoology 78:1987–1993.

Linsenmair, K.E. 1989. Sex-specific reproductive patterns in some terrestrial isopods. Pages 19–47 *in*: A. Rasa, editor. The sociobiology of sexual and reproductive strategies. Chapman and Hall, London.

Martin, O.Y., and D.J. Hosken. 2003. The evolution of reproductive isolation through sexual conflict. Nature 423:979–982.

Moreau, J., S. Seguin, Y. Caubet, and T. Rigaud. 2002. Female remating and sperm competition patterns in a terrestrial crustacean. Animal Behaviour 64:569–577.

Murata, Y., and K. Wada. 2002. Population and reproductive biology of an intertidal sandstone-boring isopod, *Sphaeroma wadai* Nunomura, 1994. Journal of Natural History 36:25–35.

Parker, G.A. 1970. Sperm competition and its evolutionary consequences in the insects. Biological Reviews 45:525–567.

Parker, G.A. 1974. Courtship persistence and female guarding as male time investment strategies. Behaviour 48:157–184.

Parker, G.A. 1979. Sexual selection and sexual conflict. Pages 123–166 *in*: M.S. Blum and N.A. Blum, editors. Sexual selection and reproductive competition in insects. Academic Press, New York.

Partridge, L., and G.A. Parker. 1999. Sexual conflict and speciation. Pages 130–159 *in*: A.E. Magurran, and R.M. May, editors. Evolution of biological diversity. Oxford University Press, New York.

Pizzari, T., and R.R. Snook. 2003. Perspective: sexual conflict and sexual selection: chasing away paradigm shifts. Evolution 57:1223–1236.

Rice, W.R., and B. Holland. 1997. The enemies within: intergenomic conflict, interlocus contest evolution (ICE), and the intraspecific Red Queen. Behavioral Ecology and Sociobiology 41:1–10.

Ridley, M. 1983. The explanation of organic diversity: the comparative method and adaptations for mating. Clarendon Press, Oxford.

Ridley, M., and D. Thompson. 1979. Size and mating in *Asellus aquaticus* (Crustacea: Isopoda). Zeitschrift für Tierpsychologie 51:380–397.

Röder, G., and K.E. Linsenmair. 1999. The mating system in the subsocial desert woodlouse, *Hemilepistus elongatus* (Isopoda Oniscidea): a model of an evolutionary step towards monogamy in the genus *Hemilepistus* sensu stricto? Ethology Ecology and Evolution 11:349–369.

Schmalfuss, H. 2003. Leg structure and mate-guarding in the Ligiidae (Isopoda, Oniscidea). Crustaceana Monographs 2:53–68.

Shuster, S.M. 1981. Sexual selection in the Socorro isopod *Thermosphaeroma thermophilum* (Cole) (Crustacea: Peracarida). Animal Behaviour 29:698–707.

Schotte, M., B.F. Kensley, and S. Shilling. (1995 onwards). World list of marine, freshwater and terrestrial crustacea isopoda. National Museum of Natural History Smithsonian Institution, Washington D.C.. Available at *http://www.nmnh.si.edu/iz/isopod/*

Shuster, S.M., and M.J. Wade. 2003. Mating systems and strategies. Princeton University Press, Princeton, N.J.

Simmons, L.W. 2001. Sperm competition and its evolutionary consequences in the insects. Princeton University Press, Princeton, N.J.

Solignac, M. 1981. Isolating mechanisms and modalities of speciation in the *Jaera albifrons* species complex (Crustacea, Isopoda). Systematic Zoology 30:387–405.

Souty-Grosset, C., D. Bouchon, J.P. Mocquard, and P. Juchault. 1994. Interpopulation variability of the seasonal reproduction in the terrestrial isopod *Armadillidium vulgare* Latr (Crustacea, Oniscidea)—a review. Acta Oecologica 15:79–91.

Sparkes, T.C., D.P. Keogh, and R.A. Pary. 1996. Energetic costs of mate guarding behavior in male stream-dwelling isopods. Oecologia 106:166–171.

Sparkes, T.C., D.P. Keogh, and K.E. Haskins. 2000. Female resistance and male preference in a stream-dwelling isopod: effects of female molt characteristics. Behavioral Ecology and Sociobiology 47:145–155.

Sparkes, T.C., D.P. Keogh, and T.H. Orsburn. 2002. Female resistance and mating outcomes in a stream-dwelling isopod: effects of male energy reserves and mating history. Behaviour 139:875–895.

Suzuki, S. 2001. Structural changes of the female genitalia during a reproductive cycle in the isopod crustacean, *Armadillidium vulgare*. Invertebrate Reproduction and Development 40:9–15.

Thiel, M. 1999. Reproductive biology of a wood-boring isopod, *Sphaeroma terebrans*, with extended parental care. Marine Biology 135:321–333.

Thiel, M. 2002. Reproductive biology of a small isopod symbiont living on a large isopod host: from the maternal marsupium to the protective grip of guarding males. Marine Biology 141:175–183.

van Tussenbroek, B.I., and A. Brearley. 1998. Isopod burrowing in leaves of turtle grass, *Thalassia testudinum*, in a Mexican Caribbean reef lagoon. Marine and Freshwater Research 49:525–531.

Veuille, M. 1980. Sexual behaviour and evolution of sexual dimorphism in body size in *Jaera* (Isopoda: Asellota). Biological Journal of the Linnean Society 13:89–100.

Wickler, W., and U. Seibt. 1981. Monogamy in Crustacea and man. Zeitschrift für Tierpsychologie 57:215–234.

Wilson, G.D.F. 1991. Functional morphology and evolution of isopod genitalia. Pages 228–245 *in*: R.T. Bauer and J.M. Martin, editors. Crustacean sexual biology. Columbia University Press, New York.

Yamamura, N. 1987. A model on correlation between precopulatory guarding and short receptivity to copulation. Journal of Theoretical Biology 127:171–180.

Yamamura, N., and V. Jormalainen. 1996. Compromised strategy resolves intersexual conflict over precopulatory guarding duration. Evolutionary Ecology 10:661–680.

Zimmer, M. 2001. Why do male terrestrial isopods (Isopoda: Oniscidea) not guard females? Animal Behaviour 62:815–821.

Sperm Demand and Allocation in Decapod Crustaceans

Bernard Sainte-Marie

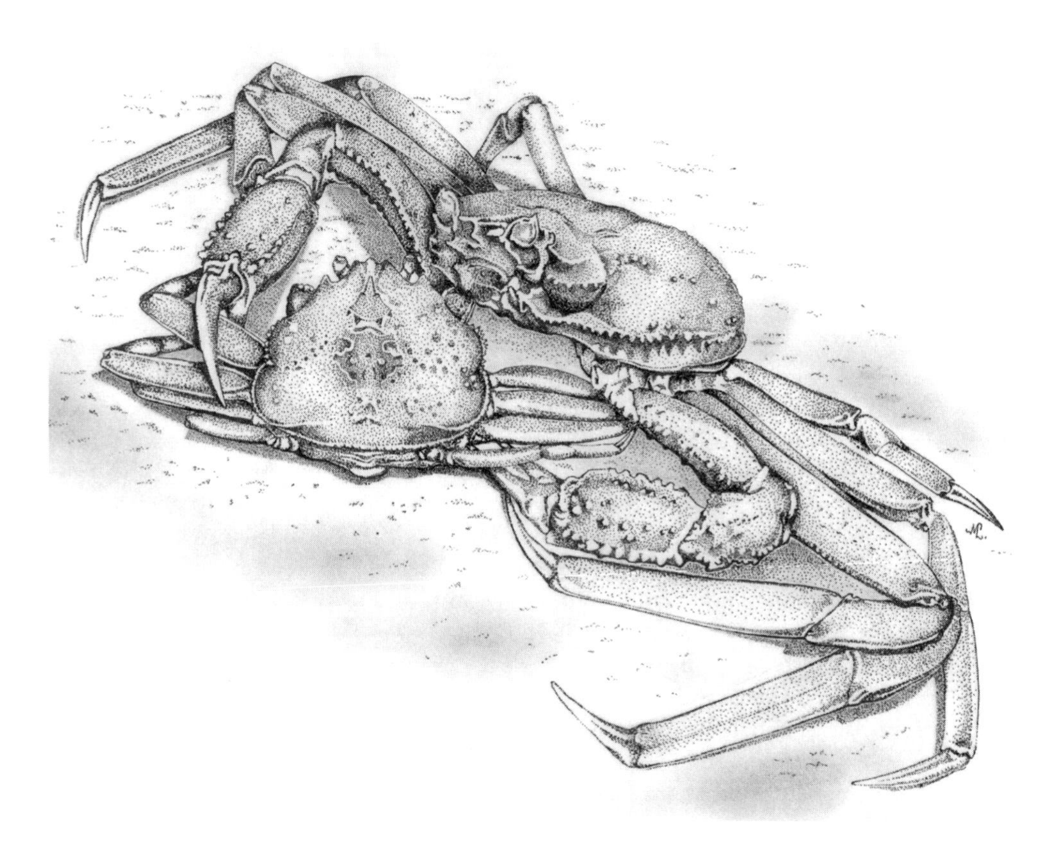

9

In most animal species, females produce a limited number of large nutrient-rich ova, whereas males produce numerous small sperm (Trivers 1972). Anisogamy is often associated with very different sex roles and maternal care, and it leads to the prediction that female reproductive success is maximized by choice of mates that confer material or genetic benefits, whereas male reproductive success is maximized by mating with as many females as possible (Trivers 1972, Clutton-Brock and Parker 1992). Competition for access to mates and sexual conflict arising from discordant female and male reproductive interests foster sexual selection that shapes many aspects of morphology, physiology, and behavior and that can lead to rapid speciation (Stockley 1997, Chapman et al. 2003, Shuster and Wade 2003).

Gametic strategies describe how much energy is invested into individual ova and sperm, how many ova and sperm are produced over a lifetime, and how those ova and sperm are allocated over time and among mates (Pitnick and Markow 1994). Female and male gametic strategies are composed of a suite of characters that coevolve within and between the sexes. Female gametic strategies include the number and size of eggs and the partitioning of egg production among reproductive episodes (Stearns 1992). Female sperm storage, when extant, is an integral part of the female's broader mating strategy that can modify her mating frequency and favor the control/manipulation of male gametes (Neubaum and Wolfner 1999). Males were traditionally assumed to have an almost unlimited supply of sperm and a ubiquitous gametic strategy of dispensing sperm "recklessly" and in excess of female requirements (Trivers 1972). However, this traditional view of male behavior was changed by the recognition that sperm, and the accessory fluids and/or spermatophores that accompany them, may be expensive to produce and can be depleted more quickly than they are renewed (Dewsbury 1982). Subsequent research that focused mainly on insects, birds, and fishes demonstrated tradeoffs between size and number of sperm and judicious sperm allocation patterns—which may even be suboptimal from the female perspective—in response to female traits, number of mating opportunities, and the risk of sexual competition at the organism or gamete level (Parker 1970, Smith 1984, Gage 1994, Stockley et al. 1997, Birkhead and Møller 1998, Wedell et al. 2002). Males of insect and vertebrate species with promiscuous females may also have evolved larger testes and/or sperm production rates when sperm competition follows a "raffle" mode (Svard and Wiklund 1989, Pierce et al. 1990, Gage 1994, Stockley et al. 1997), whereby the share of offspring fathered by one male depends wholly or partly on the number of sperm he contributes relative to other males (Parker 1993).

Female and male gametic strategies can vary considerably among species and higher taxonomic groupings, in reflection of phylogenetic, life history, and ecological/environmental constraints and adaptations. Comparisons of gametic strategies across different taxa can therefore challenge the universality of responses to sexual competition and conflict and provide a broader insight into the evolution of mating behaviors and systems (e.g., Pitnick and Markow 1994, Stockley et al. 1997). To date, models for the study of gametic strategies have included mainly insects and birds (I refer to these as "traditional models" below) and, to a lesser degree, reptiles, mammals, and fishes. Here, I explore gametic strategies within the crustacean order Decapoda (hereafter, decapods), which includes the familiar crayfishes, shrimps, lobsters, and crabs. This is a very diverse group, with more than 10,000 described species representing about one-quarter of known crustaceans. The range of maximum size at maturity for decapods is probably uniquely high among orders of the Animal Kingdom: the leg span can vary from scarcely a couple of millimeters to as much as 3.5–4 meters (Hartnoll 1983, Manning and Felder 1996). Decapods can be found in terrestrial coastal habitats, in freshwater habitats, and in the oceans from the Arctic to the Antarctic and from the intertidal to the deep sea. Some species live more than 10–15 years.

The basic aspects of decapod reproduction are the following. Decapod sperm are aflagellate and nonmotile, and they are packaged by groups of tens to millions into spermatophores; both sperm and spermatophores may differ considerably in shape and size among species (Krol et al. 1992). The spermatophores are transferred directly on or into the female, accompanied by or containing varying but sometimes copious amounts of seminal fluids (Subramoniam 1993). Eggs are brooded on the pleopods beneath the female's abdomen, except in penaeid shrimp, where females broadcast them directly into the environment (Bauer 1986). Development is usually indirect with free-living planktotrophic or lecithotrophic larvae hatching from the eggs; however, a few species exhibit almost direct development and extended maternal care. Because of the diversity of decapod morphology, life history, habit, and habitat, this order undoubtedly features a very high yet insufficiently appreciated number of mating systems/strategies (Christy 1987, Correa and Thiel 2003; see also chapter 2) even at the family or genus level (Orensanz et al. 1995, Brockerhoff and McLay 2005a).

This chapter focuses on sperm supply and allocation (male gametic strategies) as they relate to female sperm demand in gonochoristic (i.e., separate-sex) decapods. I begin by reviewing how female life history and sperm storage complexity interact to determine the potential mating frequency and immediate and lifetime sperm requirements, and how these factors are reflected in female sperm storage capacity. Aspects of male fertility, sperm allocation, and sperm competition are then discussed in relation to female requirements and population intrinsic and extrinsic factors. My review emphasizes the very few species for which most of the fundamental aspects of sperm transfer, female sperm storage, and sperm competition have been quantified. However, they alone provide an interesting insight into the scope of possible gametic strategies that have evolved among the decapods.

Sperm Demand

Female Life History Pattern and Sperm Storage Mode

Life history patterns divide female decapods into two categories (Hartnoll 1983). Females in the first category have indeterminate growth and continue to molt after first maturity. In these species, the time between molts and the number of spawns produced per intermolt period generally increase as females become larger (older). Females in the second category have determinate growth. The puberty molt during which they acquire external traits of maturity is their last molt and is followed by the production of one or more spawns until the onset of senescence and death.

Sperm storage structures of decapod females rival those of traditional insect and vertebrate models in diversity, in complexity of form (and perhaps function), and in potential for long-term sperm storage (e.g., reviewed by Neubaum and Wolfner 1999). Female decapods exhibit a phylogenetic trend of sperm storage complexity

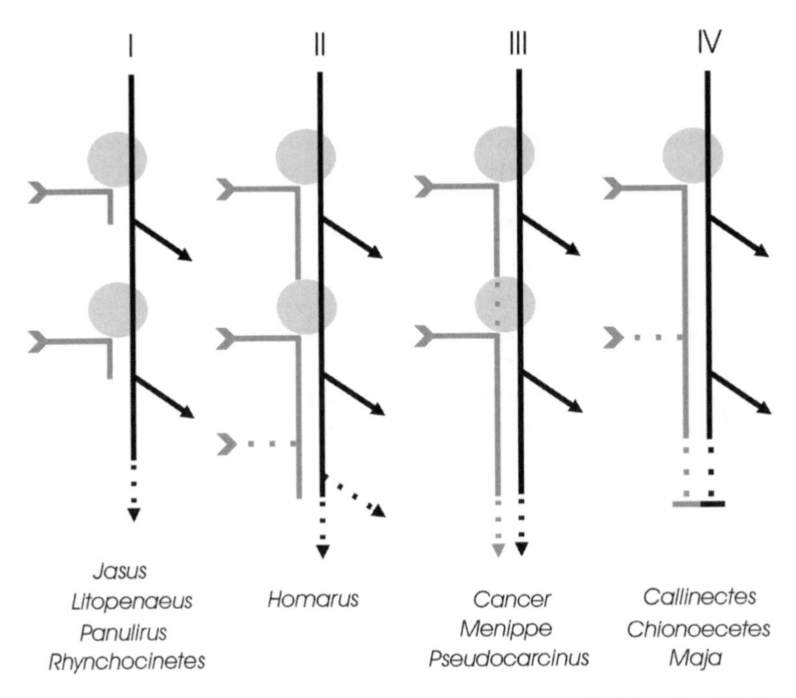

Figure 9.1 Decapod female reproductive types (with genus examples) based on life history and sperm storage mode. Type I, II, and III females have indeterminate growth combined with ephemeral, short-term, or long-term sperm storage, respectively. Type IV females have determinate growth and long-term sperm storage. The pale gray circle represents a molt; the dark gray arrow tail and line represent insemination and subsequent sperm storage on or in the female; the vertical black line represents ovaries, and the diagonal black lines indicate ovipositions. Dashed segments indicate optional or within-type variable events. The time (distance) between insemination and oviposition varies among species within each type.

(Bauer 1986, Subramoniam 1993) that seemingly correlates with potential duration for sperm storage (Fig. 9.1). In many shrimps, some crayfishes, some lobsters, and all anomuran crabs, females lack internalized sperm storage organs. Males plaster or attach their spermatophore(s) onto the female's abdominal sternites, which may be provided with protuberances or recesses forming specialized receptor areas that facilitate species recognition, mate assessment, and positioning and adherence of spermatophores (George 2005). This is "ephemeral" storage because sperm is used at the next oviposition and the period of time between insemination and sperm use/disposal is short relative to the female's potential reproductive lifespan, usually hours to weeks and exceptionally up to several months (MacDiarmid and Butler 1999). In other shrimps, crayfishes, and lobsters, and in primitive brachyuran crabs, females have partially or completely internalized chitinous storage areas. These structures have little extensibility and are shed along with their contents at each molt. Sperm may be stored between molts but usually serves to fertilize only a small subset of the female's lifetime spawns over a period of time—usually a few weeks/months but exceptionally up to 2–3 years (Waddy and Aiken 1986)—that varies from short to moderately long relative to female reproductive lifespan.

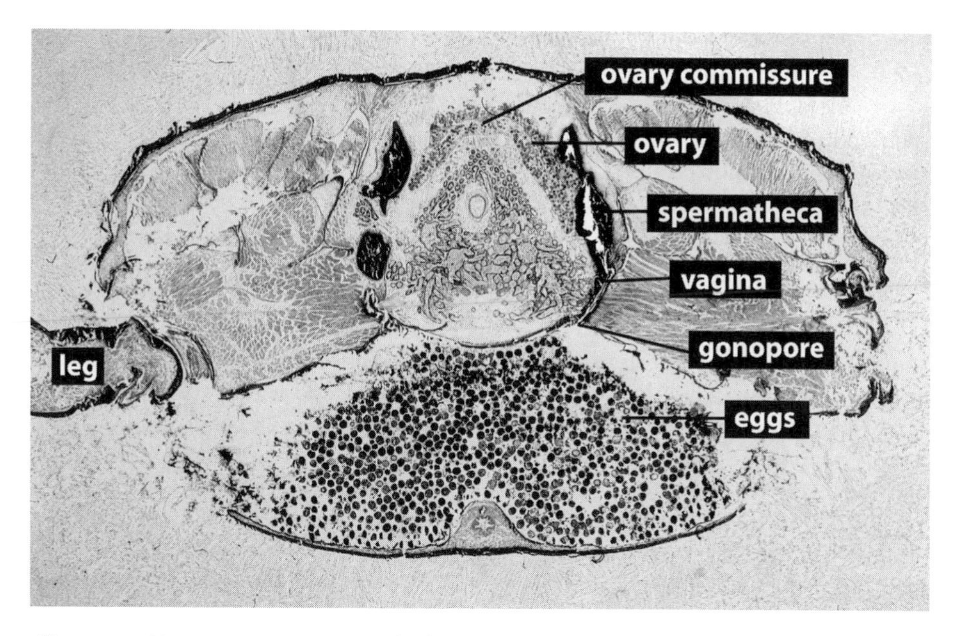

Figure 9.2 Transverse cross section of a female snow crab (*Chionoecetes opilio*) showing the arrangement of paired ovaries, spermathecae, vaginas and gonopores, and the egg incubating chamber formed by the abdomen.

The most elaborate female sperm storage organs are found in the higher brachyuran crabs (Krol et al. 1992). Each of the paired spermathecae is composed of an extensible dorsal reservoir with a glandular epithelium, derived from and connecting to the oviduct, which opens into a ventral chitinous area leading to the exterior via a vagina (Fig. 9.2). Accessory or transitory sperm storage chambers may also exist (Jensen et al. 1996). Chitinous parts of the female reproductive tract are shed at molting, but sperm may be retained across molts within the dorsal reservoir in some but not all species (e.g., Swartz 1978, Orensanz et al. 1995). Sperm acquired during one mating period may remain viable and be used over the female's full reproductive lifespan, which may last up to at least 4–6 years (Kon and Sinoda 1992, Gardner and Williams 2002, B. Sainte-Marie, unpublished data), and so this represents "long-term" storage. Gamete fusion or fertilization is thought to occur internally in the higher brachyurans (but see discussion in Lee and Yamazaki 1990), whereas it is external in all other decapods (Bauer 1986, Krol et al. 1992).

I recognize four basic female reproductive configurations based on life history and sperm storage mode. These are ranked in Figure 9.1 from left to right by order of potentially decreasing frequency of mating and of potentially increasing sperm requirements at individual mating episodes. Type I females, represented by the spiny lobster (*Jasus edwardsii*), grow indeterminately, have ephemeral sperm storage, and hence must mate each time before extruding eggs. Type II females, represented by American lobster (*Homarus americanus*), grow indeterminately but can retain sperm between molts for multiple (two to three) spawnings, and thus mating episodes may be reduced to once after each molt. Type III females, represented by rock crab (*Cancer irroratus*), grow indeterminately and may retain sperm across molts, and therefore may be relieved from mating for substantial periods of time. Type IV females, represented by blue crab (*Callinectes sapidus*), have determinate growth and long-term sperm storage. Females can further be divided into those that mate only when soft shelled (i.e., postmolt), those that mate first when soft shelled but can (optionally) also mate later when hard shelled, and those that mate only when hard shelled (e.g., Hartnoll 1969).

Female Sperm Requirements

Females will need enough sperm to fertilize their immediate spawn, and depending on the interaction of life history, sperm storage mode, and ecology, they may need excess sperm for future use. Spawn weight or volume is usually positively correlated with female mean size across species of similar body configuration and with female individual size for any given species. In brachyurans and other decapods with rigid body walls, spawn size may be limited by internal body volume available for ovary development, and there may be a tradeoff between egg number and egg size (Hines 1982). The number of eggs in a spawn is highly variable among decapod species and spans five orders of magnitude, ranging from a few eggs in the minute pinnotherid crab *Nannotheres moorei* (Manning and Felder 1996) to a few million in some cancrid and portunid crabs (Shields 1991, Hines et al. 2003). Moreover, within the same species, fecundity can vary by up to two or three orders of magnitude between the smallest and the largest female (Shields 1991, Gardner 1997), and this feature should promote male choosiness. Such extensive variability in female fecundity is generally not found among the traditional insect or vertebrate models used for the study of gametic strategies. Female immediate sperm requirements will depend on the number of eggs to be fertilized in the next spawn. Excess sperm requirements will depend on the product of eggs per spawn by the number of spawns that might or must be fertilized "autonomously" after one mating episode. This product may reach up to 54×10^6 eggs, partitioned among 18 spawns, in the blue crab (Hines et al. 2003).

Little is known of the efficiency of sperm use in decapods. Calculated decreases in number of stored sperm through one or more autonomous fertilizations resulted in rather similar expenses of about 25 and 70 sperm per ovum, respectively, for the blue crab (Hines et al. 2003) and the snow crab *Chionoecetes opilio* (B. Sainte-Marie and Lovrich 1994), which are both thought to mix gametes or to fertilize internally. Hines et al. (2003) indicated that these ratios are lower than in most animals, and considering that decapod sperm are nonmotile, this can suggest selection for improved efficiency in use of sperm. Although intuitively it would seem that decapods with external gamete fusion and fertilization should be less efficient at using sperm, Heisig (2002) found that the operational sperm-to-egg ratio ranged only from 21 to 37 in the spiny lobster *Panulirus argus*.

It is likely that females of many decapods can perceive how much sperm (or ejaculate) they have stored, as appears to be the case in some insects (e.g., Sakurai 1998, Wedell 2005). In the snow crab, females do not extrude eggs when the ratio of sperm to ova is less than 7:1 (B. Sainte-Marie and Lovrich 1994). Moreover, genetic analysis shows that recently but unilaterally inseminated female snow crabs may fertilize their eggs only with sperm from the freshly provisioned spermatheca when the other spermatheca contains few or no sperm (M. Carpentier, B. Sainte-Marie, and J.-M. Sévigny, unpublished data). This is possible because a commissure between the two ovaries (Fig. 9.2) allows females to discharge (and fertilize) eggs via one side of the reproductive tract only (Diesel 1989). Females with external sperm storage may perceive the number and size of accumulated spermatophores by tactile means, whereas perception of sperm reserves in species with partially or completely internalized sperm storage might involve tensors, muscles associated with the dorsal reservoir of spermathecae, or even chemosensory means.

Advantages and Capacity of Female Sperm Storage

Sperm storage is advantageous to female decapods in the same ways that it may be to other animals (Neubaum and Wolfner 1999). Ephemeral or short-term storage is essential in decapods where female receptivity and mating are temporally dissociated from ovary maturity and fertilization. Females need long-term storage to realize their full reproductive potential when they are limited to one mating period but produce multiple broods in a lifetime, for example, the gall crab *Hapalocarcinus marsupialis* (Kotb and Hartnoll 2002) and the blue crab (Jivoff 2003). Long-term sperm storage may protect other females against variable and sometimes unfavorable mating contexts, for example, a female-biased sex ratio, a scarcity of males, and sperm-depleted or infertile males (Elner and Beninger 1995). The genus *Chionoecetes* provides a striking example of predictable ontogenetic changes in mating context. Primiparous (first-time spawning) females mate in loose mesoscale aggregations over a protracted period of time in the winter, while multiparous (repeat-spawning) females become receptive in high-density aggregations over a short period of time in the spring (Stevens et al. 1993, Elner and Beninger 1995, Duluc 2004). These two contrasting spatial and temporal patterns of receptivity may determine different variances of mating success (Shuster and Wade 2003; see also chapter 2) for the two female ontogenetic stages: primiparous females are always mated, whereas some multiparous females in need of sperm may go unmated (B. Sainte-Marie et al. 2002, Duluc 2004).

Sperm storage may reduce female mating frequency and thus decrease her exposure to physical stresses (e.g., desiccation and heat for intertidal species) or predators when advertising or searching for mates (Morgan et al. 1983, Koga et al. 1998; see also chapter 10), to male inflicted injuries or microbe introduction (Elner and Beninger 1995), and likely, by analogy with insects, to sexually transmitted diseases (Knell and Webberley 2004). Finally, sperm storage, whether ephemeral, short, or long term, may afford females the opportunity to accumulate sperm from several males (e.g., type I females, MacDiarmid and Butler 1999; type II females, Gosselin et al. 2005; type III females, Jensen et al. 1996; type IV females, Diesel 1991), resulting in genetically more diverse progeny and/or the possibility of "cryptic" (postcopulatory) mate choice (Eberhard and Cordero 1995). For the time being, evidence to support cryptic mate choice in decapods is scarce and conjectural. In the snow crab, multiparous females preparing to fertilize a spawn using sperm stored over from a previous mating period may express glandular spermathecal activity, which seems implicated in the mobilization of sperm, vis-à-vis only one of many stored ejaculates (G. Sainte-Marie et al. 2000). Females of the rock shrimp *Rhynchocinetes typus* manipulate spermatophores attached to their sternites and selectively discard some prior to fertilizing (Thiel and Hinojosa

2003). Whether these two behaviors have any adaptive value—material or genetic—is still unclear, as are the criteria for and the mechanisms of sperm selection.

The extent to which short- and long-term sperm storage can potentially release females from the obligation to remate depends on storage capacity relative to female intermolt or lifetime sperm requirements and survivorship of stored sperm. The capacity of female sperm storage areas or organs has been investigated in very few decapods. For comparative purposes, I have calculated a female sperm reserve index (SRI) as the ratio of storage area loading (i.e., weight of attached spermatophores) or of storage organ (content or total) weight to female body weight. The range of mean SRI values is very large among the various female types, and there appears to be a gradient of increasing relative capacity from type I and II to type IV females (Table 9.1). The two extremes of SRI represented by American lobster and blue crab are separated by about one to two orders of magnitude, although the SRI value for lobster is somewhat deflated by inclusion of the exceptionally large chelae in body weight (chelae account for about 30% of total body weight of mature females). The number of sperm and the volume or weight of stored ejaculate are positively correlated in recently inseminated female decapods (Heisig 2002, B. Sainte-Marie et al. 2002, Hines et al. 2003), but over time that relationship may break down owing to volumetric attrition of seminal fluids (matrix) and sperm mortality or losses (B. Sainte-Marie 1993, González-Gurriarán et al. 1997, Hines et al. 2003, Wolcott et al. 2005).

The average values of SRI in Table 9.1 mask the fact that some type III and IV female decapods have highly extensible spermathecae whose limit capacity is determined by unoccupied internal body space (e.g., G. Sainte-Marie and Sainte-Marie 1998). The maximum recorded value of relative ejaculate capacity among female snow crabs is 7%, which is similar to the average 8–10% relative weight of ovaries or recently extruded eggs (B. Sainte-Marie, unpublished data). The extreme extensibility of some spermathecae has two possible implications. First, sperm storage may be antagonistic to ovary development because the two processes compete for space in the female body cavity. This problem is skirted when female receptivity is temporally dissociated from ovary maturity and the initially large volume of stored ejaculate is progressively reduced as ovaries develop, as in the blue crab (Hines et al. 2003, Wolcott et al. 2005). Alternatively, in the snow crab, primiparous females can accumulate very large sperm reserves by mating several times after oviposition (in addition to mating, usually once, immediately before oviposition) when the spent ovaries do not hinder expansion of spermathecae, and then—as in the blue crab—the volume of the spermathecae decreases gradually as ovaries mature again (B. Sainte-Marie 1993, Urbani et al. 1998). Attrition of spermathecal content may be largely caused by loss of seminal fluids (matrix). The functions of seminal fluids in decapods are poorly understood: demonstrated roles in sperm competition are obstruction of female sperm receptor areas or their accesses and blocking or displacement of rival sperm; hypothetical roles are sperm nourishment, sperm stabilization, and bacterial/microbe control (Diesel 1991, Subramoniam 1993, Beninger and Larocque 1998, Wolcott et al. 2005). Additional roles that should be considered because they exist in the insects are provision of nutrients to the female and male control of female receptivity by saturation of sperm receptor areas or chemical control (Chapman and Davies 2004, Wedell 2005). Second, highly extensible spermathecae may allow the orderly storage of large numbers of ejaculates, thereby favoring last male sperm precedence or female

Table 9.1. Female reproductive configuration (♀ type; refer to Fig. 9.1), eggs per spawn and sperm reserve capacity, and male sperm reserve and allocation in decapod crustaceans.

Species	♀Type	Eggs	SRI (%)[a]	♀N_{sperm}	VSI (%)	EJI (%)	EJ N_{sperm}	Reference
Rock lobster *Jasus edwardsii*	I	10^5	0.07	?	0.15	15	?	MacDiarmid 1989 Mauger 2001
American lobster *Homarus americanus*[b]	II	10^3	0.03	?	0.17	16	?	Gosselin et al. 2003
Rock crab *Cancer irroratus*	III	10^5	0.19[c]	?	0.74	<7[d]	?	B. Sainte-Marie, unpublished data
Stone crab *Menippe* sp.	III	10^5	?	6×10^6	?	8	5×10^6	Wilber 1989
Giant crab *Pseudocarcinus gigas*[b]	III	10^6	0.26[c]	?	0.39	<27[d]		Gardner and Williams 2002
Spider crab *Libinia emarginata*	IV	?	?	?	1.60[e]	?	?	Homola et al. 1991
Snow crab *Chionoecetes opilio*	IV	10^4	0.82	5×10^7	2.03[f]	2	5×10^6	Sainte-Marie and Lovrich 1994 Sainte-Marie et al. 2002, unpublished data
Blue crab *Callinectes sapidus*	IV	10^6	2.21	8×10^8	3.27	47	8×10^8	Hines et al. 2003 Jivoff 2003

Abbreviations: SRI, female sperm reserve index—wet weight (WW) of sperm reserve as a percentage of female body WW; ♀ N_{sperm}, number of sperm in storage; VSI, vasosomatic index—WW of vas deferens as a percentage of male body WW; EJI, ejaculate size index—WW of ejaculate or number of sperm (for *Menippe* sp. only) as a percentage of vas deferens WW or sperm count; EJ N_{sperm}, number of sperm transferred to the female by one male.

[a]When not provided by the author, I calculated ♀ SRI and VSI from length-weight relationships published elsewhere (available from the author upon request); [b]species with disproportionately large chelae; [c]weight of spermatheca included; [d]calculated assuming weight of spermathecal contents were transferred in one mating; [e]value for old-shell (abraded) males; [f]value for intermediate-shell males.

cryptic choice. In some majid crabs, there is no visual (by histology) evidence of sperm mixing even when females accumulate five to eight ejaculates in each spermatheca, and sperm precedence mechanisms may be effective even when females are intensely promiscuous (Diesel 1991, G. Sainte-Marie et al. 2000). By contrast, multiple ejaculates apparently mix extensively in the spermathecae of the ocypodid crab *Macrophthalmus hirtipes* (Jennings et al. 2000), and sperm precedence mechanisms in other animals may become ineffective when females mate with more than two males (e.g., Zeh and Zeh 1994).

Sperm Supply

Male Sperm Reserves

There is for the decapods a stunning sexual asymmetry in the information available on fertility. While it has been routine practice to measure female reproductive output, there is little information on most aspects of male fertility except for the size at onset of sperm production. The issue of sperm quality—functionally speaking—has barely been touched upon, although it is becoming of interest for species with potential for aquaculture. In penaeid shrimps, for example, larger and older males may have a greater proportion of live sperm and a smaller proportion of deformed sperm than smaller or younger males (Ceballos-Vázquez et al. 2003). There is a suggestion that male snow crabs may, perhaps uniquely, pass different types of sperm and spermatophores adapted either for short- or long-term storage (Elner and Beninger 1995, Moriyasu and Benhalima 1998, G. Sainte-Marie et al. 2000). In other animal taxa, the sperm of individual males may differ in fertilization success, and this may be related to environment, age, reproductive experience, phenotype, and/or genotype (e.g., Jones and Elgar 2004, Pizarri et al. 2004).

The better documented aspect of male fertility in decapods is the size of reproductive organs. In decapods, sperm develop and mature completely or partly in the testes, while spermatophores are formed and stored with seminal fluids in the so-called vas deferens and associated structures (Krol et al. 1992). Under normal circumstances, mature males seem to always have sperm in reserve. Depending on the species, the testes may be as large as the vas deferens (MacDiarmid 1989, Sato et al. 2005) or much smaller (Homola et al. 1991, B. Sainte-Marie et al. 1995), and this difference may relate to a shifting balance between sperm production rate and sperm reserve capacity that reflect different mating systems/strategies. There are striking differences among species in the vasosomatic index (VSI), which is the ratio of vas deferens weight to male body weight. Overall, there seems to be a trend of increasing VSI from species with type I females to species with type IV females (Table 9.1). The VSI is slightly more than one order of magnitude greater in the blue crab, in which female sperm requirements at mating are very high, than in the lobsters *Jasus edwardsii* and *Homarus americanus* in which female immediate needs are comparatively small.

There is considerable variability in the size of sperm reserves among individual males within decapod species. One reason for this is that size at maturity can be as plastic in males as it is in females, and this can determine steep size-dependent gradients of male reproductive potential. Indeed, the vas deferens usually increases in size with growing male body size (e.g., MacDiarmid 1989, Homola et al. 1991,

Gosselin et al. 2003, Jivoff 2003), and where examined, the number of stored sperm is positively correlated to vas deferens or male size and can vary by up to two orders of magnitude from the smallest to the largest male (MacDiarmid 1989, Wilber 1989, Sato et al. 2005). Also, the VSI increases with time elapsed since last molt, so relatively old males have larger sperm stores than recently molted males of the same size (Homola et al. 1991, B. Sainte-Marie et al. 1995). Therefore, larger and "older" males can potentially inseminate individual females more generously or can equally inseminate more females than can smaller males (MacDiarmid and Butler 1999, Jivoff 2003, Sato et al. 2005).

However, it is increasingly evident for decapods that male ability to service females can be limited by sperm reserve and regeneration rate. The time necessary for males to fully recover their sperm reserve after sexual activity may be substantial: 9–21 days after only one mating in some subtropical or temperate crabs (Ryan 1967, Kendall and Wolcott 1999) and from more than 28 days to up to one year or more after one season of intensive mating in species from colder waters (Mauger 2001, Rondeau and Sainte-Marie 2001, Sato et al. 2005). In populations subject to intense male-only fisheries, surviving large males may become sperm depleted or never achieve sperm reserve levels recorded in prefishery times or in currently unfished populations (B. Sainte-Marie et al. 1995, Hines et al. 2003, Sato et al. 2005). The male's ability to recuperate sperm reserves may be affected by various factors, notably, population abundance and availability of food resources (B. Sainte-Marie, personal observations). Sperm supply at the population level is therefore a composite function of male number, size and age structure, and physiological condition.

Sperm Allocation

Sperm allocation by male decapods has been poorly studied, and there is a problem of currency for interspecies comparisons. Most often, studies have reported only the duration of coitus (e.g., Hartnoll 1969), which says nothing of the quantity of sperm transferred, although the two variables may in some cases be correlated (e.g., B. Sainte-Marie et al. 1997). Only a small number of studies have documented the number of sperm and/or the volume or weight of ejaculate passed to females and their relationship to male sperm reserves. Nevertheless, it is clear from these few studies that sperm allocation varies widely among species and that males of a given species allocate differently depending on their characteristics and the sociosexual context.

Differences in the potentially accessible share of female lifetime egg production and in the risk, intensity, and outcome of sperm competition probably explain some striking patterns of sperm allocation among decapods (Table 9.1). The largest absolute and relative values of sperm allocation belong to the blue crab, in which a sole brief period of female receptivity and male postcopulatory guarding make female promiscuity an uncommon event (<10% of females) of low intensity (two mates at most). Males that succeed in mating a female are therefore likely to father a large part or all of the female's prolific lifetime production of eggs (Hines et al. 2003, Jivoff 2003). The smallest relative value (about 2%) belongs to the snow crab *Chionoecetes opilio* in which females are frequently (70–100% of females, depending on sociosexual context) and intensely promiscuous (two to four mates each on average) during their first breeding season and may mate again later as multiparous females (Urbani et al. 1998,

B. Sainte-Marie, N. Roy, and J.-M. Sévigny, unpublished data). Sperm competition is almost certain to occur in snow crab, and the outcome may be independent of male contribution because of effective sperm precedence mechanisms and the possibility of cryptic female choice. Under some circumstances, the first father's sperm may never serve again. The scarce other decapod species examined to date fall somewhere between these two extremes (Table 9.1). In the American lobster, female promiscuity is uncommon and of low intensity (0–20% of females depending on population, maximum of two to three mates per female), and when sperm competition occurs, the outcome is mixed paternity seemingly conforming to a raffle (Gosselin et al. 2005). However, an individual male lobster will never fertilize more than one, two, or perhaps three spawns out of one female's lifetime production, because his sperm will be shed when the female molts, and therefore he should not invest excessively into any single mating event.

Within decapod species, male characteristics may influence the quantity and rate of sperm transfer. In a noncompetitive context, the quantity of ejaculate passed to each female is positively correlated with male body size in the white shrimp *Litopenaeus vannamei* (Ceballos-Vázquez et al. 2003), the spiny lobster *Panulirus argus* (MacDiarmid and Butler 1999), and the blue crab (Jivoff 2003). In the American lobster, large males pass more ejaculate than do small males, but it represents a smaller share of their total sperm reserves (Gosselin et al. 2003). Female American lobsters prefer large males and may queue up in front of their dens and stagger their molts (i.e., control their receptivity) to mate with them (Cowan and Atema 1990), a preference that may be explained in part by the greater allocation of sperm (Gosselin et al. 2003). However, the amount of sperm transferred to individual females is independent of male body size in crabs of the genus *Chionoecetes* (Adams and Paul 1983, B. Sainte-Marie and Lovrich 1994) and the species *Hemigrapsus sexdentatus* (Brockerhoff and McLay 2005b). Additionally, in the snow crab, large adult males pass their sperm during one or two long copulations, whereas small adult males partition their allocation among several brief copulations. These two behaviors represent alternative mating strategies reflecting the fact that large males are not easily displaced from mating pairs while small males are (B. Sainte-Marie et al. 1997). Size-related sperm allocation patterns may be subordinated to VSI. In both the blue and snow crabs, males with a large VSI pass more ejaculate than do males with a small VSI (Kendall and Wolcott 1999, Rondeau and Sainte-Marie 2001, Jivoff 2003).

Sociosexual context encompasses population characteristics such as the quality and abundance of reproductive females and competing males, and the distribution of female receptivity in time and space. Female attributes that modify number of eggs per spawn or lifetime egg production—that is, the potential return on the male's investment—may change a male's motivation to mate and his sperm allocation. Female size clearly influences sperm allocation in some species, when it is a good index of fecundity. Males in three lobster species increase the weight of ejaculate they allocate by a factor of 2–6 over a female carapace length range of about 40 mm (MacDiarmid and Butler 1999, Mauger 2001, Gosselin et al. 2003). In the rock shrimp, dominant males double the number of spermatophores they pass to individual females over a carapace length range of about 8 mm (Hinojosa and Thiel 2003). In the intertidal crab *Hemigrapsus sexdentatus*, males increase the amount of ejaculate about 10-fold over a female carapace width range of 23 millimeters (Brockerhoff and McLay

2005b). However, males do not scale sperm allocation to female size in the blue crab, where (exceptionally) it is not a particularly good index of fecundity (Hines 1982), or in the snow crab (B. Sainte-Marie and Lovrich 1994), where it is a good index but the risk of sperm competition is very high.

Female mated status may also affect decapod sperm allocation as in other taxa (Wedell et al. 2002). In the snow crab, successive male mates may inject increasingly larger ejaculates that more effectively displace and isolate previously inserted ejaculates away from the oviduct (Rondeau and Sainte-Marie 2001), thereby reducing the probability of immediate paternity for previous rivals and perhaps facilitating female cryptic choice later. Other easily detected attributes of type IV females that might signal a potentially low return on male investment and cause sperm allocation to be reduced are physical handicaps, for example, missing limbs that hinder foraging or predator avoidance, and an old shell indicating the female has little residual reproductive output because senescence is imminent (Carrière 1995).

The abundance of potential mates and the intensity of male competition may strongly influence male sperm allocation. Laboratory studies have shown that male decapods guard longer and pass more sperm at larger than at smaller male-to-female ratios, and this is considered to be a mate monopolization strategy and a response to increased risk of sexual/sperm competition (Wilber 1989, Jivoff 2003; see also chapter 8). In the snow crab, dominant males also scale guard time and sperm allocation inversely to number of mating opportunities (Rondeau and Sainte-Marie 2001). However, rather than reflecting a response to inevitable sperm competition, this pattern may reflect a balance between a male bet-hedging strategy for partitioning a limited sperm reserve among a variable number of mates (Pitnick and Markow 1994) and a female sperm extortion strategy that operates by delaying oviposition (i.e., increasing male mating costs by forcing more extensive guarding) until some desired amount of sperm is obtained (B. Sainte-Marie and Lovrich 1994).

Sociosexual context may be subject to ecological or environmental forcing. In the snow crab, populations may experience large decadal swings in the sex ratio of mature individuals that are caused by the interaction of pronounced sexual size/age dimorphism with intrinsic or environmentally mediated multiyear patterns of autocorrelated recruitment (B. Sainte-Marie et al. 2002). As a consequence of this variability, primiparous females are in some years an extremely limited resource and are all intensely promiscuous (by choice or by force) and accumulate large sperm stores that represent more than their lifetime requirements; in other years, they outnumber all classes of mature males, may be sperm limited due to parsimonious allocation by dominant males, exhibit relatively low frequency and intensity of promiscuity, and accumulate small sperm reserves. The difference in mean quantity of ejaculate or number of sperm stored by primiparous females in different years may reach up to one order of magnitude (Fig. 9.3). In many decapods, the latitudinal and depth distributions of species are sufficiently ample to produce steep temperature (or other ecological and environmental) and demographic gradients that modify female size at maturity, intermolt duration, lifetime spawning frequency, and ova size and number per spawn and result in different "ecomorphs" (chapter 7). For example, females reproduce annually in warmer and biennially in colder parts of the range of the Dungeness crab, *Cancer magister* (Swiney et al. 2003), or produce many more broods and eggs per year and lifetime in the southern than in the northern part of the blue

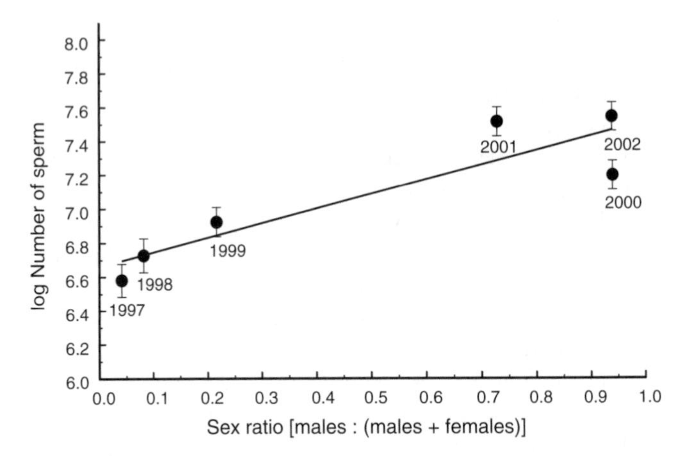

Fig. 9.3 Mean (±95% confidence interval) number of sperm ($\log_{10}$-transformed) stored by primiparous female snow crabs (*Chionoecetes opilio*) at different sex ratios in a wild population. The mean is based on a sample of 38–41 females having matured in the year indicated beside the data point. The least squares regression line is significant ($r^2 = 0.816$, $P = 0.014$). Data from B. Sainte-Marie et al. (2002).

crab range (Hines et al. 2003). Therefore, female lifetime sperm requirements, intensity of male competition and costs of mating may differ among environments, and this is expected to be reflected in among-population differences in reproductive behavior (see chapter 7).

Future Directions

This review has illustrated the surprising paucity of information on male gametic strategies in decapods, which is particularly distressing because many species are subject to intense male-only fisheries. With very few exceptions, biological research and management practices for decapods continue to reflect the archaic assumption that males, once sexually mature, have virtually unlimited resources for inseminating females. It follows from this review that future investigations of decapods should put as much emphasis on deciphering gametic strategies for males as for females. More fundamentally, because of the very high female fecundity and seemingly attenuated anisogamy, the decapods apparently offer fertile ground for the exploration of relationships among (1) female lifetime fecundity, (2) female temporal partitioning of receptivity and fecundity, (3) female sperm storage, (4) male lifetime fecundity, (5) potential for and outcome of sperm competition, and (6) sperm allocation patterns in variable ecological and environmental contexts. These aspects of reproduction ultimately determine the variance and covariance of reproductive success in groups of individuals, which is the raw material for operation of sexual selection (Shuster and Wade 2003; see also chapter 2).

Decapod researchers should strive for consistency in the way they report information on male (and the case arising, female) sperm reserves and ejaculate size (as weight and number of sperm) and provide appropriate reference information (body weight or, better still, body claw-free weight) for scaling these traits to facilitate interspecies comparisons. There is also a need to quantify sperm competition in nature, in terms of both its frequency/intensity of occurrence and its outcome, because laboratory studies may be biased (Shuster and Wade 2003). The existence, the criteria, mechanisms, and adaptive value of cryptic mate choice require attention. In species with internal and external sperm storage, the possibility that components of seminal fluids offer nutritional benefits or serve to control female mating behavior or receptivity (Neubaum and Wolfner 1999, Wedell 2005) still remains to be explored. The potential conflict between female internal sperm storage and ovary development, and how this relates to temporal patterns of mating and oviposition, is an intriguing issue. The relationships of sperm size and quality and of spermatophore size and properties to the duration of storage, the efficiency of sperm use, and the potential for sperm competition require attention (Elner and Beninger 1995, G. Sainte-Marie et al. 2000). Finally, decapod species with small size at maturity and low female fecundity may provide an interesting contrast to the larger, usually highly fecund species that have been studied so far.

Summary

This review examined trends in male gametic strategies in relation to female sperm demand among gonochoristic decapod crustaceans. Female lifetime fecundity and number of eggs per spawn can be considerably variable within and among decapod species and usually much more so than in other taxa in which gametic strategies have been explored (insects, birds, mammals). Female anatomy and life history determine a gradient of complexity and potential duration for sperm storage, from ephemeral external attachment of spermatophores to fully internalized sperm storage, lasting from a few hours to several years. The spermathecae of some decapods have enormous capacity, and filling may be antagonistic to ovary development. Sperm requirements associated with high female fecundity can be met by large sperm investments from individual males, promiscuity, and very effective use of sperm. As in many other taxa, male decapods may allocate sperm strategically as a function of female size (fecundity) and mated status, number of mating opportunities, and risk and intensity of sexual (sperm) competition. Vas deferens size may be tuned more to reflect female sperm requirements than it is to the risks of sperm competition. The species extremes of sperm allocation strategies are illustrated by the blue crab and the snow crab. In the former, males allocate a very generous share (about 50%) of their large sperm reserves to virgin females, which are never receptive again, have enormous potential lifetime fecundity, and are little inclined to promiscuity. In the snow crab, sperm allocation by dominant males to virgin females is extremely parsimonious (about 2% of their reserve), reflecting a bet-hedging strategy adapted to periodic episodes of high female relative abundance, a very high frequency and intensity of female promiscuity after first oviposition, and an outcome of sperm competition (or hypothetically of sperm cryptic selection) that favors a single male even when females have mated several times. I conclude that crustacean decapods are a most interesting group for the study of animal gametic strategies, albeit largely unappreciated to date.

Acknowledgments I acknowledge past or recent discussions with P.G. Beninger, M. Carpentier, C. Duluc, R.W. Elner, T. Gosselin, A.H. Hines, P.R. Jivoff, G.A. Lovrich, A.B. MacDiarmid, J.M. (Lobo) Orensanz, A.J. Paul, A. Rondeau, J.-M. Sévigny, and N. Urbani. Thanks to H. Dionne for preparing Fig. 9.1 and to C. Rouleau and T. Gosselin for Fig. 9.2. This chapter was improved by the comments of E. Duffy, M. Thiel, and two anonymous reviewers.

References

Adams, A.E., and A.J. Paul. 1983. Male parent size, sperm storage and egg production in the crab *Chionoecetes bairdi* (Decapoda, Majidae). International Journal of Invertebrate Reproduction 6:181–187.

Bauer, R.T. 1986. Phylogenetic trends in sperm transfer and storage complexity in decapod crustaceans. Journal of Crustacean Biology 3:313–325.

Beninger, P.G., and R. Larocque. 1998. Gonopod tegumental glands: a new accessory sex gland in the Brachyura. Marine Biology 132:435–444.

Birkhead, T.R., and A.P. Møller, editors. 1998. Sperm competition and sexual selection. Academic Press, San Diego, Calif.

Brockerhoff, A.M., and C.L. McLay. 2005a. Comparative analysis of the mating strategies in grapsid crabs with special references to the intertidal crabs *Cyclograpsus lavauxi* and *Helice crassa* (Decapoda: Grapsidae) from New Zealand. Journal of Crustacean Biology 25:507–520.

Brockerhoff, A.M., and C.L. McLay. 2005b. Mating behaviour, female receptivity and male-male competition in the intertidal crab *Hemigrapsus sexdentatus* (Brachyura: Grapsidae). Marine Ecology Progress Series 290:179–191.

Carrière, C. 1995. Insémination et fécondité chez la femelle du crabe des neiges *Chionoecetes opilio* de l'estuaire maritime du Saint-Laurent. Unpublished M.Sc. thesis, Université du Québec à Rimouski, Rimouski, Québec.

Ceballos-Vázquez, B.P., C. Rosas, and I.S. Racotta. 2003. Sperm quality in relation to age and weight of white shrimp *Litopenaeus vannamei*. Aquaculture 228:141–151.

Chapman, T., and S.J. Davies. 2004. Functions and analysis of the seminal fluid proteins of male *Drosophila melanogaster* fruit flies. Peptides 25:1477–1490.

Chapman, T., G. Arnqvist, J. Bangham, and L. Rowe. 2003. Sexual conflict. Trends in Ecology and Evolution 18:41–47.

Christy, J.H. 1987. Competitive mating, mate choice and mating associations of brachyuran crabs. Bulletin of Marine Science 41:177–191.

Clutton-Brock, T.H., and G.A. Parker. 1992. Potential reproductive rates and the operation of sexual selection. Quarterly Reviews of Biology 67:437–456.

Correa, C., and M. Thiel. 2003. Mating systems in caridean shrimp (Decapoda: Caridea) and their evolutionary consequences for sexual dimorphism and reproductive biology. Revista Chilena de Historia Natural 76:187–203.

Cowan, D.F., and J. Atema. 1990. Moult staggering and serial monogamy in American lobsters, *Homarus americanus*. Animal Behaviour 39:1199–1206.

Dewsbury, D.A. 1982. Ejaculate cost and male choice. American Naturalist 119:601–610.

Diesel, R. 1989. Structure and function of the reproductive system of the symbiotic spider crab *Inachus phalangium* (Decapoda: Majidae): observations on sperm transfer, sperm storage, and spawning. Journal of Crustacean Biology 9:266–277.

Diesel, R. 1991. Sperm competition and the evolution of mating behavior in Brachyura, with special reference to spider crabs (Decapoda, Majidae). Pages 145–163 *in*: R.T. Bauer and J.W. Martin, editors. Crustacean sexual biology. Columbia Press, New York.

Duluc, C. 2004. Les femelles multipares du crabe des neiges, *Chionoecetes opilio*, s'accouplent-elles pour compenser des réserves de sperme insuffisantes? Unpublished M.Sc. thesis, Université du Québec à Rimouski, Rimouski, Québec.

Eberhard, W.G., and C. Cordero. 1995. Sexual selection by cryptic female choice on male seminal products—a new bridge between sexual selection and reproductive physiology. Trends in Ecology and Evolution 10:493–496.

Elner, R.W., and P.G. Beninger. 1995. Multiple reproductive strategies in snow crab, *Chionoecetes opilio*: physiological pathways and behavioral plasticity. Journal of Experimental Marine Biology and Ecology 193:93–112.

Gage, M.J.G. 1994. Associations between body size, mating pattern and sperm lengths across butterflies. Proceedings of the Royal Society of London, Series B 258:247–254.

Gardner, C. 1997. Effect of size on reproductive output of giant crabs *Pseudocarcinus gigas* (Lamarck): Oziidae. Marine Freshwater Research 48:581–587.

Gardner, C., and H. Williams. 2002. Maturation in the male giant crab, *Pseudocarcinus gigas*, and the potential for sperm limitation in the Tasmanian fishery. Marine Freshwater Research 53:661–667.

George, R.W. 2005. Comparative morphology and evolution of the reproductive structures in spiny lobsters, *Panulirus*. New Zealand Journal of Marine and Freshwater Research 39:493–501.

González-Gurriarán, E., L. Fernández, J. Freire, and R. Muiño. 1997. Mating and role of seminal receptacles in the reproductive biology of the spider crab *Maja squinado* (Decapoda, Majidae). Journal of Experimental Biology and Ecology 220:269–285.

Gosselin, T., B. Sainte-Marie, and L. Bernatchez. 2003. Patterns of sexual cohabitation and female ejaculate storage in the American lobster (*Homarus americanus*). Behavioral Ecology and Sociobiology 55:151–160.

Gosselin, T., B. Sainte-Marie, and L. Bernatchez. 2005. Geographic variation of multiple paternity in wild American lobster, *Homarus americanus*. Molecular Ecology 14:1517–1525.

Hartnoll, R.G. 1969. Mating in the Brachyura. Crustaceana 16:161–181.

Hartnoll, R.G. 1983. Strategies of crustacean growth. Memoirs of the Australian Museum 18:121–131.

Heisig, J.S. 2002. Male reproductive dynamics in the Caribbean spiny lobster *Panulirus argus*. Unpublished M.Sc. thesis, Old Dominion University, Norfolk, Va.

Hines, A.H. 1982. Allometric constraints and variables of reproductive effort in brachyuran crabs. Marine Biology 69:309–320.

Hines, A.H., P.R. Jivoff, P.J. Bushmann, J. van Montfrans, S.A. Reed, D.L. Wolcott, and T.G. Wolcott. 2003. Evidence for sperm limitation in the blue crab, *Callinectes sapidus*. Bulletin of Marine Science 72:287–310.

Hinojosa, I., and M. Thiel. 2003. Somatic and gametic resources in male rock shrimp, *Rhynchocinetes typus*: effect of mating potential and ontogenetic male stage. Animal Behaviour 66:449–458.

Homola, E., A. Sagi, and H. Laufer. 1991. Relationship of claw form and exoskeleton condition to reproductive system size and methyl farnesoate in the male spider crab, *Libinia emarginata*. Invertebrate Reproduction and Development 20:219–225.

Jennings, A.C., C.L. McLay, and A.M. Brockerhoff. 2000. Mating behaviour of *Macrophthalmus hirtipes* (Brachyura: Ocypodidae). Marine Biology 137:267–278.

Jensen, P.C., J.M. Orensanz, and D.A. Armstrong. 1996. Structure of the female reproductive tract in the Dungeness crab (*Cancer magister*) and implications for the mating system. Biological Bulletin 190:336–349.

Jivoff, P.R. 2003. A review of male mating success in the blue crab, *Callinectes sapidus*, in reference to the potential for fisheries-induced sperm limitation. Bulletin of Marine Science 72:273–286.

Jones, T.M., and M.A. Elgar. 2004. The role of male age, sperm age and mating history on fecundity and fertilization success in the hide beetle. Proceedings of the Royal Society of London, Series B 271:1311–1318.

Kendall, M.S., and T.G. Wolcott. 1999. The influence of male mating history on male-male competition and female choice in mating associations in the blue crab, *Callinectes sapidus* (Rathbun). Journal of Experimental Marine Biology and Ecology 239:23–32.

Knell, R.J., and K.M. Webberley. 2004. Sexually transmitted diseases of insects: distribution, evolution, ecology and host behaviour. Biological Reviews 79:557–581.

Koga, T., P.R.Y. Backwell, M.D. Jennions, and J.H. Christy. 1998. Elevated predation risk changes mating behaviour and courtship in a fiddler crab. Proceedings of the Royal Society of London, Series B 265:1385–1390.

Kon, T., and M. Sinoda. 1992. Zuwai crab population. Marine Behaviour Physiology 21:185–226.

Kotb, M.M.A., and R.G. Hartnoll. 2002. Aspects of the growth and reproduction of the coral gall crab *Hapalocarcinus marsupialis*. Journal of Crustacean Biology 22:558–566.

Krol, R.M., W.E. Hawkins, and R.M. Overstreet. 1992. Reproductive components. Pages 295–343 *in*: F.W. Harrison and A.G. Humes, editors. Microscopic anatomy of invertebrates. Wiley-Liss, New York.

Lee, T.-H., and F. Yamazaki. 1990. Structure and function of a special tissue in the female genital ducts of the Chinese freshwater crab *Eriocheir sinensis*. Biological Bulletin, Woods Hole 178:94–100.

MacDiarmid, A.B. 1989. Size at onset of maturity and size-dependent reproductive output of female and male spiny lobsters *Jasus edwardsii* (Hutton) (Decapoda, Palinuridae) in northern New Zealand. Journal of Experimental Marine Biology and Ecology 127:229–243.

MacDiarmid, A.B., and M.J. Butler IV 1999. Sperm economy and limitation in spiny lobsters. Behavioral Ecology and Sociobiology 46:14–24.

Manning, R.B., and D.L. Felder. 1996. *Nannotheres moorei*, a new genus and species of minute pinnotherid crab from Belize, Carribean Sea (Crustacea: Decapoda: Pinnotheridae). Proceedings of the Biological Society of Washington 109:311–317.

Mauger, J.W. 2001. Sperm depletion and regeneration in the spiny lobster *Jasus edwardsii*. Unpublished M.Sc. thesis, University of Auckland, Auckland, New Zealand.

Morgan, S.G., J.W. Goy, and J.D. Costlow, Jr. 1983. Multiple ovipositions from single matings in the mud crab *Rhithropanopeus harrisii*. Journal of Crustacean Biology 3:542–547.

Moriyasu, M., and K. Behalima. 1998. Snow crabs, *Chionoecetes opilio* (O. Fabricius, 1788) (Crustacea: Majidae) have two types of spermatophore: hypotheses on the mechanism of fertilization and population reproductive dynamics in the southern Gulf of St. Lawrence, Canada. Journal of Natural History 32:1651–1665.

Neubaum, D.M., and M.F. Wolfner. 1999. Wise, winsome, or weird? Mechanisms of sperm storage in female animals. Current Topics in Developmental Biology 41:67–97.

Orensanz, J.M., A.M. Parma, D.A. Armstrong, and P. Wardrup. 1995. The breeding ecology of *Cancer gracilis* (Crustacea: Decapoda: Cancridae) and the mating system of cancrid crabs. Journal of Zoology 235:411–437.

Parker, G.A. 1970. Sperm competition and its evolutionary consequences in the insects. Biological Reviews 45:525–567.

Parker, G.A. 1993. Sperm competition games—sperm size and sperm number under adult control. Proceedings of the Royal Society of London, Series B 253:245–254.

Pierce, J.D., Jr., B. Ferguson, A.L. Salo, D.K. Sawrey, L.E. Shapiro, S.A. Taylor, and D.A. Dewsbury. 1990. Patterns of sperm allocation across successive ejaculates in four species of voles (*Microtus*). Journal of Reproduction and Fertility 88:141–149.

Pitnick, S., and T.A. Markow. 1994. Male gametic strategies: sperm size, testes size, and the allocation of ejaculate among successive mates by the sperm-limited fly *Drosophila pachea* and its relatives. American Naturalist 143:785–819.

Pizarri, T., P. Jensen, and C.K. Cornwallis. 2004. A novel test of the phenotype-linked fertility hypothesis reveals independent components of fertility. Proceedings of the Royal Society of London, Series B 271:51–58.

Rondeau, A., and B. Sainte-Marie. 2001. Variable mate-guarding time and sperm allocation by male snow crabs (*Chionoecetes opilio*) in response to sexual competition, and their impact on the mating success of females. Biological Bulletin 201:204–217.

Ryan, E.P. 1967. Structure and function of the reproductive system of the crab *Portunus sanguinolentus* (Herbst) (Brachyura: Portunidae). I. The male system. Pages 506–521 *in*: Proceedings of the Symposium on Crustacea, 12–15 January 1965, Ernakulum, India. Marine Biological Association of India, Symposium Series. Bangalore Press, Bangalore, India.

Sainte-Marie, B. 1993. Reproductive cycle and fecundity of primiparous and multiparous female snow crab, *Chionoecetes opilio*, in the northwest Gulf of Saint Lawrence. Canadian Journal of Fisheries and Aquatic Sciences 50:2147–2156.

Sainte-Marie, B., and G.A. Lovrich. 1994. Delivery and storage of sperm at first mating of female *Chionoecetes opilio* (Brachyura: Majidae) in relation to size and morphometric maturity of male parent. Journal of Crustacean Biology 14:508–521.

Sainte-Marie, B., S. Raymond, and J.-C. Brêthes. 1995. Growth and maturation of the benthic stages of male snow crab, *Chionoecetes opilio* (Brachyura: Majidae). Canadian Journal of Fisheries and Aquatic Sciences 52:903–924.

Sainte-Marie, B., J.-M. Sévigny, and Y. Gauthier. 1997. Laboratory behavior of adolescent and adult males of the snow crab (*Chionoecetes opilio*) (Brachyura: Majidae) mated noncompetitively and competitively with primiparous females. Canadian Journal of Fisheries and Aquatic Sciences 54:239–248.

Sainte-Marie, B., J.-M. Sévigny, and M. Carpentier. 2002. Interannual variability of sperm reserves and fecundity of primiparous females of the snow crab (*Chionoecetes opilio*) in relation to sex ratio. Canadian Journal of Fisheries and Aquatic Sciences 59:1932–1940.

Sainte-Marie, G., and B. Sainte-Marie. 1998. Morphology of the spermatheca, oviduct, intermediate chamber and vagina of the adult snow crab (*Chionoecetes opilio*). Canadian Journal of Zoology 76:1589–1604.

Sainte-Marie, G., B. Sainte-Marie, and J.-M. Sévigny. 2000. Ejaculate-storage patterns and the site of fertilization in female snow crabs (*Chionoecetes opilio*; Brachyura, Majidae). Canadian Journal of Zoology 78:1902–1917.

Sakurai, T. 1998. Receptivity of female remating and sperm number in the sperm storage organ in the bean bug, *Riptortus clavatus* (Heteroptera: Alydidae). Researches on Population Ecology 40:167–172.

Sato, T., M. Ashidate, S. Wada, and S. Goshima. 2005. Effects of male mating frequency and male size on ejaculate size and reproductive success of female spiny crab *Paralithodes brevipes*. Marine Ecology Progress Series 296:251–262.

Shields, J.D. 1991. The reproductive ecology and fecundity of *Cancer* crabs. Pages 193–213 *in*: A. Wenner and A. Kuris, editors. Crustacean egg production. A.A. Balkema, Rotterdam, The Netherlands.

Shuster, S.M., and M.J. Wade. 2003. Mating systems and strategies. Princeton University Press, Princeton, N.J.

Smith, R.L., editor. 1984. Sperm competition and the evolution of animal mating system. Academic Press, San Diego, Calif.

Stearns, S.C. 1992. The evolution of life histories. Oxford University Press, New York.

Stevens, B.G., W.E. Donaldson, J.A. Haaga, and J.E. Munk. 1993. Morphometry and maturity of paired tanner crabs, *Chionoecetes bairdi*, from shallow- and deepwater environments. Canadian Journal of Fisheries and Aquatic Sciences 50:1504–1516.

Stockley, P. 1997. Sexual conflict resulting from adaptations to sperm competition. Trends in Ecology and Evolution 12:154–159.

Stockley, P., M.J.G. Gage, G.A. Parker, and A.P. Møller. 1997. Sperm competition in fishes: the evolution of testis size and ejaculate characteristics. American Naturalist 149:933–954.

Subramoniam, T. 1993. Spermatophores and sperm transfer in marine crustaceans. Pages 129–214 *in*: J.H.S. Blaxter and A.J. Southward, editors. Advances in Marine Biology. Academic Press, New York.

Svard, L., and C. Wiklund. 1989. Mass and production rate of ejaculates in relation to monandry/polyandry in butterflies. Behavioral Ecology and Sociobiology 24:395–402.

Swartz, R.C. 1978. Reproductive and molt cycles in the xanthid crab, *Neopanope sayi* (Smith, 1869). Crustaceana 34:15–32.

Swiney, K.M., T.C. Shirley, S.J. Taggart, and C.E. O'Clair. 2003. Dungeness crab, *Cancer magister*, do not extrude eggs annually in southeastern Alaska: an *in situ* study. Journal of Crustacean Biology 23:280–288.

Thiel, M., and I.A. Hinojosa. 2003. Mating behavior of female rock shrimp *Rhynchocinetes typus* (Decapoda: Caridea)—indication for convenience polandry and cryptic female choice. Behavioral Ecology and Sociobiology 55:113–121.

Trivers, R.L. 1972. Parental investment and sexual selection. Pages 136–179 *in*: B. Campbell, editor. Sexual selection and the descent of man. Heineman, London.

Urbani, N., B. Sainte-Marie, J.-M. Sévigny, D. Zadworny, and U. Kuhnlein. 1998. Sperm competition and paternity assurance during the first breeding period of female snow crab *Chionoecetes opilio* (Brachyura: Majidae). Canadian Journal of Fisheries and Aquatic Sciences 55:1104–1113.

Waddy, S.L., and D.E. Aiken. 1986. Multiple fertilization and consecutive spawning in large American lobsters, *Homarus americanus*. Canadian Journal of Fisheries and Aquatic Sciences 43:2291–2294.

Wedell, N. 2005. Female receptivity in butterflies and moths. Journal of Experimental Biology 208:3433–3440.

Wedell, N., M.J.G. Gage, and G.A. Parker. 2002. Sperm competition, male prudence and sperm-limited females. Trends in Ecology and Evolution 17:313–320.

Wilber, D.H. 1989. The influence of sexual selection and predation on the mating and postcopulatory guarding behavior of stone crabs (Xanthidae, *Menippe*). Behavioral Ecology and Sociobiology 24:445–451.

Wolcott, D.L., C.W.B. Hopkins, and T.G. Wolcott. 2005. Early events in seminal fluid and sperm storage in the female blue crab *Callinectes sapidus* Rathbun: effects of male mating history, male size, and season. Journal of Experimental Marine Biology and Ecology 319:43–55.

Zeh, J.A., and D.W. Zeh. 1994. Last-male sperm precedence breaks down when females mate with 3 males. Proceedings of the Royal Society of London, Series B 257:287–292.

Predation and the Reproductive Behavior of Fiddler Crabs (Genus *Uca*)

John H. Christy

10

Failure to avoid predators results in death, while failure to meet other challenges seldom is fatal. This may explain why predator avoidance behavior often compromises an animal's performance while engaged in nearly all other functional categories of behavior (Lima and Dill 1990, Magnhagen 1991, 1993, Sih 1994, Endler 1995). Behavioral responses to predation can be direct, such as escape elicited by the presence of a predator, or they can be indirect, such as habitat choice based on the presence of refuges. In the former, an interaction occurs in ecological time and is governed by the detection of a predator and decisions about the threat it poses (Helfman 1989). In the latter, the behavior is the result of differences in fitness between ancestral phenotypes that were predictably exposed to predation in past environments. Such antipredator adaptations often do not depend on predator detection and risk assessment. They can even result from selection on another life stage, such as parental behavior that reduces predation of young. Choice of habitat and activity period determine not only the rates and kinds of encounters between predator and prey (Endler 1995) but also the environments in which animals are active and hence the context-dependent expression of a broad range of behavior.

Purpose and Scope of Discussion

Here I discuss some of the many ways that two kinds of behavioral responses to predation affect the reproductive behavior of fiddler crabs (genus *Uca*, about 97 species; Rosenberg 2001). My intent is to illustrate how research on this group has provided new insights or a different perspective on problems of general interest. Both sets of examples illustrate the multiple effects of adaptations to predation that do not depend on predator detection and risk assessment. First I show how variation in performance at one life stage can profoundly affect the behavioral phenotype of another (Podolsky 2003). I review the evidence for a functional link between avoidance of predators by newly hatched larvae and the timing of larval release by females relative to the tidal amplitude, tide height and diurnal light cycles, which together govern predator activity and hence predation risk (see Yamahira 2004 for an applicable fitness model). Cycles of larval release are produced by cycles in courtship and mating and are under endogenous control. Predation on larvae selects for precise timing of the reproductive behavior of both sexes, and it does so by affecting the phase of the crabs' endogenous rhythms. In contrast, in many other organisms, including crustaceans, temporal elements of reproductive behavior result not from endogenous clocks but from responses by individuals to current, variable, social, and ecological conditions, including their effect on an animal's internal state (e.g., anurans, McCauley et al. 2000; amphipods, Jormalainen 1998 [see also chapter 8]; lobsters, chapter 6).

In the second example, I show how the ecology of courtship and searching for a mate, and adaptations for predator avoidance can drive the evolution of courtship signals (Endler 1992). Adult fiddler crabs are relatively small, nontoxic, poorly defended morphologically, and highly conspicuous to their predators, primarily shorebirds, as they interact socially on the surface of open intertidal mud and sand flats. Research on fiddler crab visual and nonvisual systems for predator detection and avoidance and on male courtship signals and mechanisms of mate choice suggests that characteristics of these systems have strongly biased how females choose mates and which male signals are attractive. This conclusion has important general implications for understanding sexual selection and signal evolution (Christy 1995, Dawkins and Guilford 1996, Endler and Basolo 1998).

There are other studies of how predation has shaped fiddler crab behavior that have general implications, including studies of predator detection and avoidance by hiding in refuges (Koga et al. 2001, Hugie 2004, Jennions et al. 2003, Pratt et al. 2005, Wong et al. 2005) and studies of risk reduction by aggregating in a "selfish herd" when individuals are far from safe sites (Viscido and Wethey 2002). For lack of space, I do not review these topics here. I concentrate on how predation influences reproductive ecology and behavior.

Before I discuss (1) the effects of reproductive timing for predator avoidance by larvae on courtship and mating behavior and (2) how antipredator adaptations have biased courtship signal evolution, it will be useful to give a brief overview of the diversity, distribution, and general patterns of reproduction in the genus *Uca*.

Systematics and Mating Systems

Rosenberg's (2001) comprehensive review and synthesis of the systematics and taxonomy of the genus *Uca* (see also Rosenberg 2006) produced a detailed and credible phylogeny for the group based on adult morphology, with tree branching constraints suggested by a coarser phylogeny based on 16S ribosomal DNA sequence (Sturmbauer et al. 1996). There are now 97 recognized species, compared to the 67 species and numerous subspecies described by Crane (1975), the previous authority. Especially pertinent for this chapter, *Uca musica*, the name I used previously for a species that has figured prominently in my studies of courtship and mate choice, becomes *U. terpsichores*. There are three major branches in the genus: (1) a basal clade of nine species, which includes one species, *Uca tangeri*, in the eastern Atlantic, the only *Uca* species in that region, and eight species in the Americas: six in the tropical eastern Pacific and two in the western Atlantic and Caribbean; (2) a "crown" American clade of 57 species, 21 in the western Atlantic and Caribbean and 36 in the eastern Pacific (nearly all tropical); and (3) an Indo-Pacific clade with 39 species (again, mostly tropical).

DeRivera and Vehrencamp (2001) summarized the ecological and social correlates of mating systems in fiddler crabs within a phylogenetic perspective. There are two general modes of mating in the genus. Males of many American species (but not those in the basal clade) court from and defend burrows to which they attract females for mating and in which females breed. Females search for mates and breeding burrows by leaving their own burrows, walking on the surface, and sequentially visiting from a few males (one or two minimum, *Uca pugilator*; Christy 1983) to many (up to 106! *U. crenulata*; deRivera 2005). This can take an hour or more, and the female may move many meters in short punctuated steps as she stops at one male's burrow after another (repeat visits are extremely rare). Mate choice is indicated when the female stays in a male's burrow, and he plugs the entrance, sealing himself and the female below. After mating, the female ovulates, fertilizes, extrudes, and attaches her eggs to her abdominal appendages ("oviposits") in an enlarged terminal chamber of the burrow. She typically will stay in this burrow for about two weeks until she releases her planktonic

Table 10.1. Characteristics of the species of fiddler crab (genus *Uca*) discussed in this chapter.

Clade	Species	Predation[a]	Mating Location	Male-Built Structure
American	*beebei*	++	Male's burrow, surface	Pillar
	crenulata	−	Male's burrow	None
	pugilator	++	Male's burrow	Semidome
	pugnax	+	Male's burrow	Semidome (rare)
	stenodactylus	++	Male's burrow, surface	None
	terpsichores	++	Male's burrow	Hood
Basal	*stylifera*	+	Surface	None
	tangeri	+	Male's burrow	None
Indo-Pacific	*annulipes*	− −	Male's burrow, surface	Semidome (rare)
	lactea	−	Male's burrow, surface	Semidome
	paradussumieri	−	Female's burrow	None
	perplexa	− −	Male's burrow, surface	Lip, pillar (rare)
	rosea	−	Male's burrow, female's burrow, surface	None
	tetragonon	−	Male's burrow, surface	None

[a]++, +, −, − −: very frequent to very infrequent predation of adults based on personal observations and the opinions of other field researchers.

larvae. Most of the species showing this mode of mating are small to medium in size (typically <2 cm carapace width), and they live at high densities on well-drained sediment in the upper intertidal or supratidal zone. In contrast to this pattern of female searching, in many of the Indo-Pacific species and those in the basal clade in the Americas (with notable exceptions), it is the males that search for mates among nearby female burrow residents. Courtship may not include male claw waving but always includes (poorly known) tactile stimulation and possibly seismic signals transmitted through the substratum. The pair usually mates on the surface at the entrance of the female's burrow, and the female subsequently breeds alone. Many of the species with this male-searching behavior are relatively large (>2 cm carapace width) and live at lower densities in muddier sediment in the mid to upper intertidal zones. Several species in the Americas and in the Indo-Pacific mate both on the surface and in males' burrows (Table 10.1; *Uca beebei*, Christy 1987; *U. lactea* and its relatives, Backwell and Passmore 1996, Murai et al. 1987, Nakasone and Murai 1998; *U. tetragonon*, Murai et al. 1995, Goshima et al. 1996; *U. stenodactylus*, J.H. Christy, personal observation; and a few others, Murai et al. 1996).

Temporal Patterns of Reproductive Behavior

Fiddler crabs have been the subjects of extensive research on physiological and behavioral rhythms, their endogenous properties and control, and their adaptive significance (for a review, see Thurman 2004). Because fiddlers live in the intertidal zone and emerge from their burrows primarily when they are exposed to the air during the day, the diurnal and tidal cycles together determine when they are active. The phase relationship between these physical cycles varies in time and across tidal regions but predictably so. It is not surprising, then, that the daily activity and biweekly or monthly reproductive rhythms of fiddler crabs are governed by endogenous clocks. There is one clock with a 24 hour circadian period and a continuing debate whether there is but one other clock with a 12.4 hour circatidal period or two other clocks with 24.8 hour circalunidian periods that are 180° out of phase, each tracking one of the two semidiurnal tides. Thurman (2004) suggests that the two circalunidian clock model is more likely the correct one. The period of reinforcement of the circadian and one or the other of the two circalunidian clocks is about every two weeks. Hence, the interaction between these three clocks, set to local conditions, can flexibly govern the timing of reproductive cycles with respect to the spring-neap tidal amplitude cycle (14.8 day average period) on coasts with semidiurnal tides (e.g., western Atlantic and tropical eastern Pacific) and with respect to the tropic-equatorial tidal amplitude cycle (13.7 day average period) that dominates the tidal pattern on some coasts with mixed semidiurnal tides (e.g., eastern Gulf of Mexico).

Timing of Larval Release

Biweekly or monthly cycles of larval release have been described for 10 species of fiddler crabs (family Ocypodidae), and numerous species of intertidal crabs in several other families (e.g., Grapsidae, Gecarcinidae, Xanthidae; for reviews, see Morgan 1995, Morgan and Christy 1995, Christy 2003, and Thurman 2004). Intra- and interspecific comparative studies indicate that these cycles result from the hierarchical expression of diel (circadian clock), tidal (circalunidian clock), and tidal amplitude (circadian × circalunidian clock interaction) rhythms (Morgan and Christy 1994, Morgan 1996, Kellmeyer and Salmon 2001). Most intertidal crabs release larvae at night, close to the time of high tide on the days of the month with larger amplitude tides, typically during the spring or tropic tides. As a result, newly hatched larvae move rapidly on nocturnal ebb currents from shallow habitats where they are released, to the deeper coastal ocean where they grow and develop. Since small planktivorous fish are numerically most abundant in shallow water, and most feed primarily during the day (Morgan and Christy 1995), emigration, a consequence of the timing of larval release, allows crab larvae to escape many predators. Several weeks later fiddler crab megalopae (postlarvae) immigrate into estuaries, primarily on nocturnal flood tides, and settle in adult habitats (Christy and Morgan 1998). The predator avoidance hypothesis as an explanation for the reproductive timing of intertidal crabs has received strong support from comparative studies; larvae that are protected by spines or cryptic colors do not emigrate from shallow waters and adults of the species with these larvae lack the strong temporal patterns of reproduction that characterize species with vulnerable larvae (Christy 1986, Morgan and Christy 1995).

Timing of Courtship and Mating

Precise timing of larval release, a female behavioral trait that promotes larval survival, is known in several species to be preceded by cycles of female sexual receptivity and male investment in courtship (e.g., *Uca pugilator*, Christy 1978, 1983; *Uca terpsichores*, Zucker 1976, Christy et al. 2001, Christy 2003; *Uca perplexa*, Murai and Backwell 2005). Little is known about the hormonal control of crab reproductive behavior or the role of endogenous rhythms in hormone production. However, since larval release ends a sequence of reproductive events that begins with courtship and mate choice, and since larval release is under endogenous control, it is likely that the earlier stages of breeding are, as well.

Temporal patterns of investment in courtship by male *U. terpsichores* additionally suggest that endogenous rhythms govern the early stages of the reproductive sequence. Courting males of this species sometimes build sand hoods at the entrances of their burrows; these structures attract mate-searching females (Christy et al. 2002; see also below). Hoods are destroyed by the tide and must be built anew daily. On the hourly scale, during the tidal activity period, most males build their hoods before females begin to search for mates (Fig. 10.1); hence, males do not adjust their investment in courtship signaling according to their perception of the number of females that are receptive on a given day, for example, through encounter rates with mate-searching females. On the biweekly temporal scale, the mating rate of hood builders is about 9% per day and does not vary with the number of hoods built each day (Christy et al. 2001). This constant proportional relationship between the male hood building cycle and the female cycle of receptivity and mate choice indicates that these two cycles

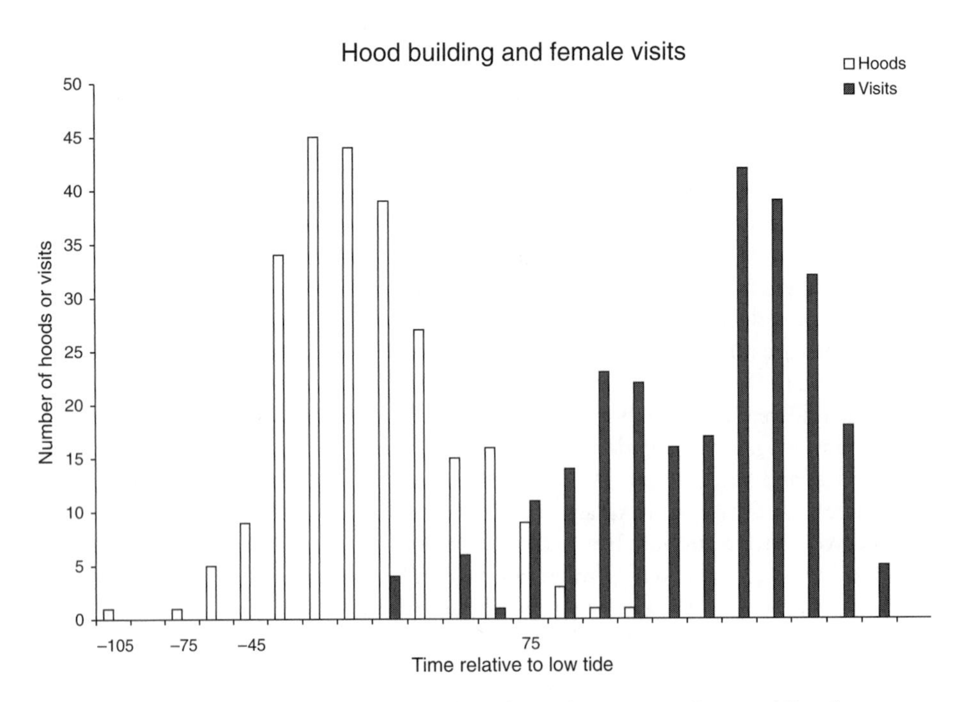

Figure 10.1 Timing of hood building (open bars) by male *Uca terpsichores* and female visits to courting males (solid bars) during the tidal activity period. The counts of newly appearing hoods and visits each 15 = minute interval were summed across three consecutive days at the beginning of a biweekly reproductive cycle.

are coincident. Expression of a behavior in anticipation of or coincident with favorable conditions are the classic characteristics of endogenous rhythms. The daily number of hoods is an accurate proxy (by a factor of 0.09) for the daily number of females seeking mates, for the number of females that begin incubation about 1.5 days later, and for the number that release larvae about 15 days later. If males adjusted their tendency to build hoods according to the number of mate-seeking females the preceding day, the number of hoods would have proportionately lagged and overshot the number of receptive females early and late in the reproductive cycle, respectively.

Male *Uca terpsichores* evidently invest in courtship exactly according to the expected payoff. This may be true generally for fiddler crabs (*U. pugilator*, Christy 1978; *U. crenulata*, deRivera 2003; *U. pugnax*, Greenspan 1982; *U. perplexa*, Murai and Backwell 2005) because the sexes share the same endogenous mechanism governing biweekly reproductive cycles (but see Kim and Choe 2003). Biweekly (semilunar) reproductive rhythms, whether endogenous or not, are common among shallow-water marine organisms presumably because selection affecting adults or young varies predictably with the light and tidal cycles across the geographic range of a species (e.g., Warner 1997, Yamahira 2004). In contrast, in terrestrial animals, with the exception of seasonal cycles, temporal patterns in sexual competition usually are thought to reflect day-to-day assessments and strategic decision making rather than the expression of underlying rhythms. Hence, the terrestrial and marine realms may differ in the predictability of predation and, conversely, safe periods on longer (>24 hours) but subseasonal time scales. This may account for a fundamental difference in the mechanisms that regulate the expression of antipredator behaviors and their diverse effects on the rest of the behavioral phenotype.

Timing of Aggressive and Coercive Elements in Courtship

Cycles of mate choice by females may not perfectly anticipate ovulation and fertilization cycles (Christy 1978, deRivera 2003, Murai and Backwell 2005), and this may affect male courtship tactics. It is not known whether such tactics are endogenously controlled or reflect male assessment of the reproductive state of individual females (e.g., Goshima et al. 1996, Murai et al. 2002). Early in a breeding cycle, after choosing a mate, some females take longer to ovulate compared to females that mate at or after the peak of the cycle. Since males usually guard their mates until they oviposit, a male who attracts a mate early in a cycle will guard longer, feed less, and miss more future mating opportunities than a male who attracts a mate later. This lag between female behavioral receptivity and ovulation may create a conflict between the sexes over whether to commit to pairing in a plugged burrow (Jormalainen 1998; see also chapter 8). On the one hand, due to the male-based sex ratio, mating opportunities per male are rare, and males should accept any sexually receptive female. Guarding time is lost feeding time for the female, too, so the lag between mate choice and oviposition should never be large, even when a female chooses a mate early. In addition, one might expect "early" females to be well fed and to produce relatively large clutches, which would additionally favor their acceptance and guarding by the male. On the other hand, sexual selection should always favor males that increase their mating rate by decreasing the time they spend with each mate. Male *U. pugilator* can mate with up to three females in a single breeding cycle while at the same burrow (each in a separate incubation chamber; see Christy 1982), and the record is four for *U. crenulata* (deRivera 2003). Males always should prefer receptive females who are ready to ovulate so that they can return quickly to the surface and attract additional mates. On balance, a conflict seems likely; how is it resolved?

Males of several other mate-guarding crustaceans that mate when the female molts distinguish among potential mates on the basis of time to molting, which they probably assess after contact and reception of chemical cues (e.g., Jormalainen 1998). Although a male fiddler crab may chemically assess a female's readiness to ovulate (Murai et al. 2002) after she has entered his burrow, it would be more efficient for the male to screen the female at an earlier stage in courtship. One possible method may be to include threats in the courtship sequence. Threats (e.g., chela flicks, lunges, short chases) are commonly given by males during courtship in species in which males attract females to their burrows. In *U. pugilator* (Christy 1983), *U. beebei* (Christy 1987), *U. terpsichores* (Christy et al. 2002), *U. lactea* (Murai et al. 1987), *U. tetragonon* (Goshima et al. 1996, Koga et al. 2000), and *U. rosea* (Murai et al. 1996), females that are not receptive nevertheless leave their burrows and move on the surface, perhaps in search of a better burrow or feeding area or to escape aggression from neighboring males (Murai et al. 1987). Males court these females and females visit these males, perhaps as a way to reduce their predation risk while changing burrow locations. Clearly, it is not in a male's interest to court a female who will not mate. As a female nears ovulation, her tendency to approach courting males may increase (decreasing response threshold; Murai and Backwell 2005) while her tendency to feed may decline. Nonreceptive female *U. beebei* (Christy 1988b) and *U. terpsichores* (Christy et al. 2002) feed, sometimes extensively, and they do not respond as quickly or directly to courting males as do receptive females. Males who readily switch from an attracting wave to a threat upon the slightest hesitation from the female may most effectively repel nonreceptive females and receptive females that are several days from ovulation. If true, males should less often threaten females as the breeding cycle advances and proportionately more wandering females become ready to ovulate. At the same time, the costs of failing to exclude a female that requires a longer mate guarding time will diminish because the chance of attracting an additional mate after emerging from guarding will decline quickly as the cycle wanes. This will additionally favor eliminating threats from courtship on and after peak mating days.

The diminishing cost to males of committing to mate guarding as a cycle peaks and then wanes also should favor more frequent use of coercive behavior later in each breeding period. Males of some species (Crane 1975) cut off, push, herd, capture, carry, and otherwise directly and aggressively attempt to force females into their burrows (Zucker and Denny 1979). Males of other species in which the female usually follows the male into his burrow sometimes use tactics that get the female to enter his burrow first (Murai et al. 1996). A female always has ultimate control over mating because she must lower her abdominal flap to allow the male access to her gonopores for intromission and sperm transfer. However, by forcing a female into a burrow and keeping her there shortly before she will ovulate, the male may leave the female no choice but to mate with him if she is to produce a clutch and begin incubation on time. Hence, as the reproductive cycle peaks and wanes, aggressive elements may become less frequent in courtship sequences that function to attract females into males' burrows, and coercive interactions may become more frequent.

Staying on Time by Choice of Breeding Site

In species that breed in male-defended burrows, selection for precise timing of larval release to promote larval escape from predators may affect not only when but also where females choose to breed. Comparisons of features of the burrows females enter and leave with those in which they stay and mate indicate that burrow structure affects mate choice in *U. pugilator* (deep, stable burrows; Christy 1983), *U. annulipes* (deep but compact burrows of moderate size; Backwell and Passmore 1996), and *U. crenulata* (longer burrows of a specific shaft diameter; deRivera 2005). Indeed, burrow features are more important than male size as criteria for mate choice in these species, at least during mating peaks. There are several possible ways that burrows may affect female reproductive success (Christy 1983, Christy and Salmon 1984), but until recently, there has been no experimental demonstration of such effects. DeRivera (2005) has published an exciting study of *U. crenulata* showing experimentally that the diameter of the burrow relative to the size of the female critically affects (how is not known) the timing of larval release and that females choose burrows that allow them to release larvae on time. Thus, predation on newly hatched larvae may select for female choice of breeding sites where they can incubate their eggs and release their larvae during safe periods. Choice of breeding sites should affect male–male competition for the burrows that females prefer. Linkage between mortality of young, female breeding site preference, and male–male competition has been demonstrated in *Uca pugilator* (Christy 1983).

Antipredator Behavior Biases Signal Evolution

Predation on adult fiddler crabs appears to be far more common in the Americas (Backwell et al. 1998, Iribarne and Martinez 1999, Ribeiro et al. 2003) than in the Indo-Pacific (P. Backwell, M. Murai, and others, personal communication). Hence, most of my examples of how predation may bias signal evolution are of American species, but I include a few Indo-Pacific species to illustrate behavioral patterns where predation evidently is relatively infrequent. At best, these comparisons are only suggestive because species in the two regions diverged long ago (Rosenberg 2001), making it difficult to separate the effects of history and predation.

Theory

Female mating preferences are thought to evolve primarily due to selection that is a consequence of mate choice (Kokko et al. 2003, Fuller et al. 2005). In this nearly universal view, preferences are adaptations for choice of a mate that will contribute the most to female or offspring fitness. However, it has become increasingly apparent that preferences are based on features of female sensory-response systems that evolve for a variety of reasons unrelated to mate choice (Endler and Basolo 1998). Research on fiddler crab courtship, mechanisms of mate choice, and mating preferences has contributed to this view as expressed in the sensory trap model of signal evolution (West-Eberhard 1984, Christy 1995). A sensory trap occurs during courtship when a signal elicits a response that has an ecological or social function other than mate choice. The receiver responds to the signal because it mimics stimuli that elicit the response for the other function. The words "mimicry" and "trap" are sometimes thought to imply that sensory trap preferences must be costly and maladaptive (e.g., Marcías and Ramirez 2005, Stuart-Fox 2005). This implication does not follow from the sensory trap model (Christy 1995, 1997, Dawkins and Guilford 1996); the preference may or may not be costly. Most important, the net effect of a sensory trap response on fitness, across all the contexts in which the response is made, must be positive or the response will be eliminated (Christy 1995).

In courtship, sensory trap preferences may often be beneficial because they increase the efficiency (by decreasing time or energy expenditure) of mate localization and because they reduce predation risk during mate searching. Research on the function of the behavioral mechanism that governs the female preference for sand hoods built by courting male *Uca terpsichores* has provided a detailed example of a sensory trap preference that is based on an antipredator behavior. I begin this example with a brief overview of fiddler crab orientation and antipredator behavior and the contexts in which they function. This will help show how a behavior that reduces predation risk has come to play a role in mate choice.

Predator-Escape Behavior, Vision, and Orientation Mechanisms

When a fiddler crab detects a predator, it runs quickly back to the burrow it most recently left (Zeil and Layne 2002). Crabs can do this at night, if they are experimentally blinded, and if the burrow entrance is covered; they do not need to see their burrow entrance to find it. Indeed, if a crab is more than about seven to eight body lengths from its burrow (Ribeiro et al. 2006), the opening becomes invisible due to perspective foreshortening and the relatively poor resolution of the ventral portion of the crab's eye that "looks" at the ground. With vision of limited use, crabs rely on a remarkable mechanism, probably based on leg odometry, to construct a path map to their burrow and safety (Layne et al. 2003a, 2003b). The operation of this mechanism is revealed by the near alignment between a crab's transverse axis and the bearing back to its burrow.

Crabs that leave a burrow eventually abandon their path map to it and use other means to reduce their predation risk when a predator comes near. Perhaps most often, the crab will visually move to where a nearby resident disappeared and thereby gain access to a burrow. If for any reason these cues to safety are unavailable, crabs will run to a nearby stone, shell, piece of wood, plant part, or even a lump of sediment and remain motionless at the base of the object. Although critical studies have not been done, this presumably reduces the crab's risk of predation because it makes the crab more difficult for the predator to detect. Landmark orientation, this tendency to orient to objects projecting from the surface, seems to be widespread in the genus *Uca* (Herrnkind 1983, Christy 1995) and is also expressed in other semiterrestrial crabs (Diaz et al. 1995).

Fiddler crab visual systems have three features that are well designed for predator detection and avoidance (Zeil and Hemmi 2006): (1) they have a zone of high vertical resolution in a band around the equator of their eyes (Zeil and Al-Mutairi 1996), (2) they keep this zone aligned with the horizon through fine muscular control of all three axes of rotation of their long eyestalks (Nalbach et al. 1989), and (3) they classify as threatening nearly all moving objects they see above their visual horizon (Layne et al. 1997). Crabs look for, detect, classify, and avoid even small predators that are many meters away. These features of crab visual systems and orientation mechanisms operate during mate search producing preferences that shape male courtship signals and signaling behavior.

Structure Building by Courting Males

Courting males of 18 species of fiddler crabs are known to build structures at the openings of their burrows to which they attract females for mating (listed in Christy 1988a, Christy et al. 2001, plus *Uca uruguayensis*, P. Ribeiro, personal communication). Most structure builders are in the crown American clade, but *U. lactea* and its relatives in the Indo-Pacific also sometimes build structures. Structure building and structure size and shape vary considerably. Male *U. terpsichores* build sand hoods (Fig. 10.2a), the largest structures relative to male size in the genus, and they do so following a biweekly or monthly (upper intertidal populations only) cycle. Male *U. beebei*, a species that is sympatric with *U. terpsichores* (but with little spatial overlap) and is about the same size, build narrow mud pillars (Fig. 10.2b) that are as tall as hoods. Both structures attract females to males' burrows for mating (Christy et al. 2003a).

We studied how hoods attract female *Uca terpsichores* by recording the responses of females to courting hood-building and nonbuilding *U. terpsichores* males with and without natural and replica hoods. These experiments controlled for differences between builders and nonbuilders that might affect their attractiveness (Christy et al. 2002). Female *U. terpsichores* significantly more often approached the males that courted them if they had hoods (or replicas) at their burrows. Hoods did not attract females from a distance, nor did they affect female mating decisions after they reached males' burrows. We placed hood replicas about 3 centimeters to the side of males' burrow entrances and found that females sometimes moved to the offset replicas, not to the courting males as the males led the females to their burrows.

Why do females preferentially move to hoods when they leave one burrow and go to the next? Sexually receptive and nonreceptive female *U. terpsichores* both preferentially approach males with hoods (Christy et al. 2002), suggesting that they use the same orientation mechanisms. As they leave one male's burrow and move to the next, females must abandon their path maps and move on. This sometimes is clearly evident in a "break" in the transverse orientation of the female toward the burrow she is leaving. At this point, the female quickly moves to the next burrow by either following a male or moving to a hood. The results of several experiments with and without model predators support the hypothesis that hoods elicit landmark orientation (Christy 1995, Christy et al. 2003a, 2003b). The most telling experiments showed that receptive and nonreceptive females, when not given burrows, moved spontaneously to hoods (replicas), shells, stones, and pieces of wood. We added these objects to courting males' burrows and found that receptive females did not prefer hoods over the other natural objects that are common on the beach (Christy et al. 2003b).

Comparative studies also support the idea that the female preference for hoods is based on landmark orientation. Males and females of species that do not build structures run to hoods and pillars when they are chased by a model predator (Christy 1995, Christy et al. 2003a). Using cast replicas, we switched structure types between

Figure 10.2 Male fiddler crabs and their structures: (a) *U. terpsichores*, hood, Rodman, Panama; (b) *U. beebei*, pillar, Rodman, Panama; (c) *U. pugilator*, Cayo Pelau, Florida; (d) massive semidome of *U. pugilator*, Cayo Pelau, Florida; (e) *U. lactea*, semidome, Aitsu, Japan; (f) *U. perplexa*, lip, Okinawa, Japan. All photographs by the author.

U. beebei and *U. terpsichores* and found that mate-searching females of both species preferred hoods; structure preferences are not species specific as they should be if they have evolved for mate choice. Female *U. stenodactylus*, a species that does not build structures, show a significant but relatively weak tendency to move to hoods, pillars, and other objects. Perhaps the tendency for females to orient to structures is enhanced in structure-building species. In these species, male-built structures may be the most abundant objects for landmark orientation (at least during courtship peaks), and they may indicate especially safe sites because, unlike other objects, they nearly always lead to an open burrow.

I know of no other example of a sexual signal that is attractive because it elicits an antipredator orientation response. Females of many species are at an increased risk of predation when searching for a mate and many modify their behavior to reduce this risk (Jennions and Petrie 1997, Hazlett and Rittschof 2000), suggesting that signals allowing safe searching should be common. Warner and Dill (1999) proposed that the bright sexual colors and vigorous displays of some male reef fish indicate to attentive females that it is safe to visit these males. And it seems likely that properties of signals that improve efficacy are favored in part because they facilitate detection (Fleishman 2000) and localization of the courting male, perhaps allowing the females to approach directly and with less risk. Either signals that elicit predator escape behavior are truly rare, or they have gone unnoticed.

Origins of Structure Building

We have demonstrated current sexual selection by a female preference for structure building in two species that build tall structures. Recent observations and experiments indicate that hoods have a second function. Male *U. terpsichores* that had hood replicas placed beside their burrow entrance sometimes led females to these objects instead of to their burrows. Experiments (Ribeiro et al. 2006) have shown that male *U. terpsichores* that have had errors introduced experimentally in their path maps significantly more often and more quickly relocate their burrows if they have hoods at the entrance. Courting males without hoods sometimes fail to relocate their burrows and become wandering rogues (Christy et al. 2002). Thus, structure building in *U. terpsichores* may currently be under sexual selection by indirect male–male competition for resource-holding ability and by a female preference, and both may be mediated by landmark orientation, an antipredator behavior.

Some species build structures that are too low to be imaged in the acute zone of the eyes of adults, so they may not elicit landmark orientation. Male *U. pugilator* make massive but low semidomes (Fig. 10.2c,d; Christy 1982). *U. lactea* make more delicate low semidomes (Fig. 10.2e; Kim et al. 2004), and *U. perplexa*, a close relative, makes low asymmetrical lips on their burrows (Fig. 10.2f). Presumably, structure building evolved from sediment manipulation behavior that produced at first only simple low ridges or lips around the burrow entrance. How do males benefit from making low structures? Semidomes and even lips tilt the image of the burrow opening up from the horizontal. This will reduce perspective foreshortening and should make the opening visible from a greater distance, at least when seen from the front. Discriminating this oblate image from the visual background noise would be a formidable task (Zeil and Layne 2002) that might be made easier if the crab knew where to expect the image. Recent research has shown that fiddler crabs can use their nonvisual path map to project the location of their burrow in visual space (Hemmi and Zeil 2003). They use this ability to determine when an approaching crab is on a trajectory toward their burrow so they can defend their ownership. They could also use it to identify the area in their visual field where the image of their burrow opening should be if it were large enough to be seen. This would be useful to a male if spatially complex courtship interactions far from his burrow produce errors in his path map causing him to lead a female to the wrong place or to err when running back to his burrow to escape a predator. This function for lips and semidomes is entirely conjectural, but it is consistent with the dual function of hoods in *U. terpsichores* that we are just beginning to understand.

Finally, it also is possible that both lips and semidomes allow females to see and more easily orient to the burrow entrance when they are close to it at the end of an approach. When approached by a female, courting male *U. perplexa* (lips) sometimes do not enter their burrows first. Instead, the male briefly and partially enters, sometimes repeatedly (J.H. Christy, personal observations), and then steps to the other side of the opening and waits for the female to go into the burrow first (Nakasone and Murai 1998). Perhaps this stimulates the female to find the burrow by orienting to where the male appears to enter the ground. The tilted image of the opening would only enhance her ability to locate the entrance by presenting an additional visual cue. Avian predation on this species and *U. lactea* is very infrequent. Landmark orientation may not bias sexual selection for tall structures in these species, leaving their architecture to be shaped by other orientation mechanisms used by both sexes to find the male's burrow.

Claw Waving and Other Visual Signals

Fiddler crabs are best known for their extreme sexual dimorphism, which is most apparent when males wave their single greatly enlarged cheliped, usually toward passing females, at least in American species (Pope 2000). Claw waving would seem to be a classic example of signaling behavior that is selected by female choice. Remarkably, there is no experimental evidence that females are attracted by claw waving (but see Oliveira and Custódio 1998). The best evidence, based on an analysis of videotapes of courtship (Backwell et al. 1999), is correlative, making it difficult to isolate which features of the wave or waving context are most important. Murai et al. (1996) noted that females of several species mate on the surface with males that do not direct claw waving toward them. Male *U. lactea* (Yamaguchi 1971) and *U. beebei* (J.H. Christy, personal observations) without large claws nevertheless can mate on the surface. Christy and Salmon (1991) suggested that the vertical components of the claw wave may be especially stimulatory given the female's highly structured perceptual field. Perhaps claw waving serves simply and primarily to attract the female's attention and reveal the location (and perhaps the species) of the male (Land and Layne 1995). Just as structures may elicit landmark orientation, claw waving and other male visual signals, such as the raised-carpus display of *U. beebei* (Christy 1985), may play to the tendency for crabs to follow residents to their burrows.

Future Directions

Predictable selection favoring precise timing of larval release for predator avoidance and equally precise timing of female receptivity and male investment in courtship have apparently led to their control by endogenous timers, one with a circadian period and perhaps two with circalunidian periods running in antiphase. Definitive experimental demonstration of the roles of these timers in the biweekly behavioral cycles that are so apparent in the field is unlikely to be forthcoming for obvious practical reasons. However, indirect evidence may be sought in long-term monitoring studies with an eye toward the predictive, rather than the reactive, nature of variation in these temporal patterns. For example, species that live in the upper intertidal zone experience an approximately seven-month cycle in which one of the two spring tides is higher and larger in amplitude. This cycle is a consequence of the changing phase relationship between the perigee and apogee cycle (varying distance of the moon from the earth, due to the moon's elliptical orbit) and the syzygies (approximate alignment of the sun, earth, and moon). If only one of the two spring tides is suitable for larval release, courtship and mating should be relatively more intense about two weeks prior to those tides, during the lunar phase with relatively low tides. Where in an estuarine or associated coastal system a species lives may also affect reproductive timing and synchrony. We have found that *Uca terpsichores* releases larvae on the two to three morning high spring tides just before dawn, skips a day, and then releases larvae on two to three subsequent evening tides, producing a relatively drawn out six to eight day hatching period (J. Christy and P. Backwell, unpublished observations). This species often lives on medium-energy sand beaches near or just inside the mouth of an estuary. Because oceanic waters are nearby, nighttime seaward migration of newly hatched larvae may be less important, permitting larval release on spring tides just before dawn, an atypical pattern for estuarine species. Differences in the temporal variance of reproduction will affect the operational sex ratio, perhaps also mate guarding tactics, and as suggested here, the use of aggressive and coercive elements in courtship. Finally, the tidal regimes on the coasts where fiddler crabs live vary considerably (see Thurman 2004); so, too, do the times best for larval release (Morgan and Christy 1994), yet the effects of this variation on the timing of courtship and mating, mating modes, and courtship behavior have hardly been explored.

We are just beginning to understand the structural and functional organization of the fiddler crab visual system and how it interacts with the nonvisual orientation mechanism to keep crabs safe from predation as they feed, defend their burrows, change locations, and court or be courted. I have argued here for a dominant role for predator detection and escape behavior in the courtship (signals and responses) of smaller fiddlers that live in open habitats in the Americas where predation is frequently seen. Some of these ideas could be tested by comparing the visual systems, orientation mechanisms, courtship signals, and responses of these species with those that experience less frequent predation, including larger species in the Americas (less preferred prey; Backwell et al. 1998), and species in the Indo-Pacific, where predation appears to be infrequent. For example, coercive courtship tactics that stimulate predator detectors and elicit escape responses would not be expected in species that are not frequently startled to their burrows by predators, while coercive behavior that elicits burrow defense mechanisms would be. Systematic comparisons of fiddler crab courtship signals and displays that may operate as sensory traps based on antipredation and burrow defense behavior would be informative.

Summary and Conclusions

Predation affects fiddler crab reproductive timing, male competition for females, visual systems and orientation mechanisms, mechanisms and patterns of female choice, and the kinds of signals males use to attract females for mating. Predation is pervasive in the lives of these animals as both larvae and adults; so, too, are adaptations to avoid being eaten, and these adaptations strongly influence how crabs communicate and compete for mates. The bulk of theoretical and empirical research on courtship signal evolution considers predation to be a constraint on signaling behavior, signal elaboration, mate sampling, and choice (e.g., Koga et al. 1998, Acharya and McNeil 1998, Jones et al. 2002). Fiddler crab courtship shows how predation can also be a creative force in signal evolution. Male-built structures and even the well-known waving display of these animals may be selected by preferences that allow females to better detect, locate, and visit males safely. The number of examples of sensory traps in animal courtship is increasing (Christy 1995, Sakaluk 2000, Córdoba-Aguilar 2002, Fleishman 2000, Rodd et al. 2002, Stålhandske 2002, Zimmer et al. 2003, Marcías and Ramirez 2005). These studies show that it is not sufficient to ask what a courtship signal may indicate about the benefits a female and her young receive as a consequence of mating with males that use that particular signal. To understand the origin and maintenance of a mating preference and how it shapes signal evolution, it is necessary to ask how the sensory and behavioral mechanism on which it is based evolves, particularly how the mechanism functions and is selected in other social and ecological contexts (Stuart-Fox 2005).

Acknowledgments I am especially grateful to Pat Backwell for her invaluable contributions to the studies we did together during our six-year association and for her thoughtful discussion of many of the ideas in this chapter. I thank Ira Rubinoff, Director of the Smithsonian Tropical Research Institute, for providing sufficient annual support so that I can spend my time in the field doing research rather than in my office writing grant applications. Finally, I thank Emmett Duffy and Martin Thiel for their editorial guidance and, especially, for their patience.

References

Acharya, L., and J.N. McNeil. 1998. Predation risk and mating behavior: the responses of moths to bat-like ultrasound. Behavioral Ecology 9:552–558.

Backwell, P.R.Y., and N.I. Passmore. 1996. Time constraints and multiple choice criteria in the sampling behaviour and mate choice of the fiddler crab, *Uca annulipes*. Behavioral Ecology and Sociobiology 38:407–416.

Backwell, P.R.Y., P.D. O'Hara, and J.H. Christy. 1998. Prey availability and selective foraging in shorebirds. Animal Behaviour 55:1659–1667.

Backwell, P.R.Y., M.D. Jennions, J.H. Christy, and N.I. Passmore. 1999. Female choice in the synchronously waving fiddler crab *Uca annulipes*. Ethology 105:415–421.

Christy, J.H. 1978. Adaptive significance of reproductive cycles in the fiddler crab *Uca pugilator*: a hypothesis. Science 199:453–455.

Christy, J.H. 1982. Burrow structure and use in the sand fiddler crab, *Uca pugilator* (Bosc). Animal Behaviour 30:487–494.

Christy, J.H. 1983. Female choice in the resource-defense mating system of the sand fiddler crab, *Uca pugilator*. Behavioral Ecology and Sociobiology 12:169–180.

Christy, J.H. 1985. Iconography in the courtship behavior of the fiddler crab *Uca beebei*. American Zoologist 25:60A.

Christy, J.H. 1986. Timing of larval release by intertidal crabs on an exposed shore. Bulletin of Marine Science 39:176–191.

Christy, J.H. 1987. Female choice and breeding behavior of the fiddler crab *Uca beebei*. Journal of Crustacean Biology 7:624–635.

Christy, J.H. 1988a. Pillar function in the fiddler crab *Uca beebei* (I): effects on male spacing and aggression. Ethology 78:53–71.

Christy, J.H. 1988b. Pillar function in the fiddler crab *Uca beebei* (II): competitive courtship signaling. Ethology 78:113–128.

Christy, J.H. 1995. Mimicry, mate choice and the sensory trap hypothesis. American Naturalist 146:171–181.

Christy, J.H. 1997. Deception: the correct path to enlightenment? Trends in Ecology and Evolution 12:160.

Christy, J.H. 2003. Reproductive timing and larval dispersal of intertidal crabs: the predator avoidance hypothesis. Revista Chilena de Historia Natural 76:177–185.

Christy, J.H., and S.G. Morgan. 1998. Estuarine immigration by crab postlarvae: mechanisms, reliability and adaptive significance. Marine Ecology Progress Series 174:51–65.

Christy, J.H., and M. Salmon. 1984. Ecology and evolution of mating systems of fiddler crabs (genus *Uca*). Biological Reviews 59:483–509.

Christy, J.H., and M. Salmon. 1991. Comparative studies of reproductive behavior in mantis shrimps and fiddler crabs. American Zoologist 31:329–337.

Christy, J.H., P.R.Y. Backwell, and S. Goshima. 2001. The design and production of a sexual signal: hoods and hood building by male fiddler crabs *Uca musica*. Behaviour 138:1065–1083.

Christy, J.H., P.R.Y. Backwell, S. Goshima, and T.J. Kreuter. 2002. Sexual selection for structure building by courting male fiddler crabs: an experimental study of behavioral mechanisms. Behavioral Ecology 13:366–374.

Christy, J.H., P.R.Y. Backwell, and U.M. Schober. 2003a. Interspecific attractiveness of structures built by courting male fiddler crabs: experimental evidence of a sensory trap. Behavioral Ecology and Sociobiology 53:84–91.

Christy, J.H., J.K. Baum, and P.R.Y. Backwell. 2003b. Attractiveness of sand hoods built by courting male fiddler crabs, *Uca musica*: test of a sensory trap hypothesis. Animal Behaviour 66:89–94.

Córdoba-Aguilar, A. 2002. Sensory trap as the mechanism of sexual selection in a damselfly genitalic trait (Insecta: Calopterygidae). American Naturalist 160: 594–601.

Crane, J. 1975. Fiddler crabs of the world: Ocypodidae: genus *Uca*. Princeton University Press, Princeton, N.J.

Dawkins, M.S., and T. Guilford. 1996. Sensory bias and the adaptiveness of female choice. American Naturalist 148:937–942.

deRivera, C.E. 2003. Causes of a male-biased operational sex ratio in the fiddler crab *Uca crenulata*. Journal of Ethology 21:137–144.

deRivera, C.E. 2005. Long searches for male-defended breeding burrows allow female fiddler crabs, *Uca crenulata*, to release larvae on time. Animal Behaviour 70:289–297.

deRivera, C.E., and S.L. Vehrencamp. 2001. Male versus female mate searching in fiddler crabs: a comparative analysis. Behavioral Ecology 12:182–191.

Diaz, H., B. Orihuela, and R.B. Forward, Jr. 1995. Visual orientation of postlarval and juvenile mangrove crabs. Journal of Crustacean Biology 15:671–678.

Endler, J. 1992. Signals, signal conditions and the direction of evolution. American Naturalist 139:S125–S153.

Endler, J. 1995. Multiple-trait coevolution and environmental gradients in guppies. Trends in Ecology and Evolution 10:22–29.

Endler, J.A., and A.L. Basolo. 1998. Sensory ecology, receiver bias and sexual selection. Trends in Ecology and Evolution 13:415–420.

Fleishman, L.J. 2000. Signal function, signal efficacy and the evolution of anoline lizard dewlap color. Pages 209–257 *in*: Y. Espmark, T. Admusen, and R. Rosenqvist, editors. Animal signals. Tapir Academic Press, Trondheim, Norway.

Fuller, R.C., D. Houle, and J. Travis. 2005. Sensory bias as an explanation for the evolution of mate preferences. American Naturalist 166:437–446.

Goshima, S., T. Koga, and M. Murai. 1996. Mate acceptance and guarding by male fiddler crabs *Uca tetragonon* (Herbst). Journal of Experimental Marine Biology and Ecology 196:131–143.

Greenspan, B.N. 1982. Semi-monthly reproductive cycles in male and female fiddler crabs, *Uca pugnax*. Animal Behaviour 30:1084–1092.

Hazlett, B.A., and D. Rittschof. 2000. Predation-reproduction conflict resolution in the hermit crab *Clibanarius vittatus*. Ethology 106:811–818.

Helfman, G.S. 1989. Threat-sensitive predator avoidance in damselfish-trumpetfish interactions. Behavioral Ecology and Sociobiology 24:47–58.

Hemmi, J.M., and J. Zeil. 2003. Robust judgment of inter-object distance by an arthropod. Nature 421:160–163.

Herrnkind, W.F. 1983. Movement patterns and orientation. Pages 41–105 *in*: F.J. Vernberg, and W.B. Vernberg, editors. The biology of Crustacea, Volume 7. Behavior and ecology. Academic Press, New York.

Hugie, D.M. 2004. A waiting game between the black-bellied plover and its fiddler crab prey. Animal Behaviour 67:823–831.

Iribarne, O.O., and M.M. Martinez. 1999. Predation on the southwestern Atlantic fiddler crab (*Uca uruguayensis*) by migratory shorebirds (*Pluvialis dominica, P. squatarola, Arenaria interpres*, and *Numenius phaeopus*). Estuaries 22:47–54.

Jennions, M.D., and M. Petrie. 1997. Variation in mate choice and mating preferences: a review of causes and consequences. Biological Reviews 72:283–327.

Jennions, M.D., P.R.Y. Backwell, M. Murai, and J.H. Christy. 2003. Hiding behaviour in fiddler crabs: how long should prey hide in response to a potential predator? Animal Behaviour 66:251–257.

Jones, G., A. Barabas, W. Elliot, and S. Parsons. 2002. Female greater wax moths reduce sexual display behavior in relation to the potential risk of predation by echolocating bats. Behavioral Ecology 13:375–380.

Jormalainen, V. 1998. Precopulatory mate guarding in crustaceans: male competitive strategy and intersexual conflict. Quarterly Review of Biology 73:275–304.

Kellmeyer, K., and M. Salmon. 2001. Hatching rhythms of *Uca thayeri* Rathbun: timing in semidiurnal and mixed tidal regimes. Journal of Experimental Marine Biology and Ecology 260:169–183.

Kim, T.W., and J.C. Choe. 2003. The effect of food availability on the semilunar courtship rhythm in the fiddler crab *Uca lactea* (de Haan) (Brachyura: Ocypodidae). Behavioral Ecology and Sociobiology 54:210–217.

Kim, T.W., J.H. Christy, and J.C. Choe. 2004. Semidome building as sexual signaling in the fiddler crab *Uca lactea* (Brachyura: Ocypodidae). Journal of Crustacean Biology 24:673–679.

Koga, T., P.R.Y. Backwell, M.D. Jennions, and J.H. Christy. 1998. Elevated predation risk changes mating behaviour and courtship in a fiddler crab. Proceeding of the Royal Society of London, Series B 265:1385–1390.

Koga, T., M. Murai, S. Goshima, and S. Poovachiranon. 2000. Underground mating in the fiddler crab *Uca tetragonon*: the association between female life history traits and male mating tactics. Journal of Experimental Marine Biology and Ecology 248:35–52.

Koga, T., P.R.Y. Backwell, J.H. Christy, M. Murai, and E. Kasuya. 2001. Male-biased predation of a fiddler crab. Animal Behaviour 62:201–207.

Kokko, H., R. Brooks, M.D. Jennions, and J. Morley. 2003. The evolution of mate choice and mating biases. Proceeding of the Royal Society of London, Series B 270:653–664.

Land, M., and J. Layne. 1995. The visual control of behaviour in fiddler crabs. I. Resolution, thresholds and the role of the horizon. Journal of Comparative Physiology A 177:81–90.

Layne, J.E., M. Land, and J. Zeil. 1997. Fiddler crabs use the visual horizon to distinguish predators from conspecifics: a review of the evidence. Journal of the Marine Biological Association of the United Kingdom 77:43–54.

Layne, J.E., W.J. P. Barnes, and L.M.J. Duncan. 2003a. Mechanisms of homing in the fiddler crab *Uca rapax*. 1. Spatial and temporal characteristics of a system of small-scale navigation. Journal of Experimental Biology 206:4413–4423.

Layne, J.E., W.J.P. Barnes, and L.M.J. Duncan. 2003b. Mechanisms of homing in the fiddler crab *Uca rapax*. 2. Information sources and frame of reference for a path integration system. Journal of Experimental Biology 206:4425–4442.

Lima, A.L., and L.M. Dill. 1990. Behavioral decisions made under the risk of predation: a review and prospectus. Canadian Journal of Zoology 68:619–640.

Magnhagen, C. 1991. Predation risk as a cost of reproduction. Trends in Ecology and Evolution 6:183–186.

Magnhagen, C. 1993. Conflicting demands in gobies: when to eat, reproduce, and avoid predators. Marine Behavior and Physiology 23:79–90.

Marcías, G.C., and E. Ramirez. 2005. Evidence that sensory traps can evolve into honest signals. Nature 434:501–505.

McCauley, S.J., S.S. Bouchard. B.J. Farina, K. Isvaran, S. Quader, D.W. Word, and C.M. St. Mary. 2000. Energetic dynamics and anuran breeding phenology: insights from a dynamic game. Behavioral Ecology 11:429–436.

Morgan, S.G. 1995. The timing of larval release. Pages 157–191 *in*: L. McEdwards, editor. Ecology of marine invertebrate larvae. CRC Press, Boca Raton, Fla.

Morgan, S.G. 1996. Plasticity in reproductive timing by crabs in adjacent tidal regimes. Marine Ecology Progress Series 139:105–118.

Morgan, S.G., and J.H. Christy. 1994. Plasticity, constraint and optimality in reproductive timing. Ecology 75:2185–2203.

Morgan, S.G., and J.H. Christy. 1995. Adaptive significance of the timing of larval release by crabs. American Naturalist 145:457–479.

Murai, M., and P.R.Y. Backwell. 2005. More signaling for earlier mating: conspicuous male claw waving in the fiddler crab *Uca perplexa*. Animal Behaviour 70:1093–1097.

Murai, M., S. Goshima, and Y. Henmi. 1987. Analysis of the mating system of the fiddler crab, *Uca lactea*. Animal Behaviour 35:1334–1342.

Murai, M., T. Koga, S. Goshima, and S. Poovachiranon. 1995. Courtship and the evolution of underground mating in *Uca tetragonon* (Decapoda: Ocypodidae). Journal of Crustacean Biology 15:655–658.

Murai, M., S. Goshima, K. Kawai, and H.-S. Yong. 1996. Pair formation in the burrows of the fiddler crab *Uca rosea* (Decapoda: Ocypodidae). Journal of Crustacean Biology 16:522–528.

Murai, M., T. Koga, and H.-S. Yong. 2002. The assessment of female reproductive state during courtship and scramble competition in the fiddler crab, *Uca paradussumieri*. Behavioral Ecology and Sociobiology 52:137–142.

Nakasone, Y., and M. Murai. 1998. Mating behavior of *Uca lactea perplexa* (Decapoda: Ocypodidae). Journal of Crustacean Biology 18:70–77.

Nalbach, H.-O., G. Nalbach, and L. Forzin. 1989. Visual control of eyestalk orientation in crabs: vertical optokinetics, visual fixation of the horizon and eye design. Journal of Comparative Physiology A 165:577–587.

Oliveira, R.F., and M.R. Custódio. 1998. Claw size, waving display and female choice in the European fiddler crab, *Uca tangeri*. Ethology, Ecology and Evolution 10:241–251.

Podolsky, R.D. 2003. Integrating development and environment to model reproductive performance in natural populations of an intertidal gastropod. Integrative and Comparative Biology 43:450–458.

Pope, D.S. 2000. Testing function of fiddler crab claw waving by manipulating social context. Behavioral Ecology and Sociobiology 47:432–437.

Pratt, A.E., D.K. McLain, and A.S. Berry. 2005. Variation in the boldness of courting sand fiddler crabs (*Uca pugilator*). Ethology 111:63–76.

Ribeiro, P.D., O.O. Iribarne, L. Jaureguy, D. Navarro, and E. Bogazzi. 2003. Variable sex-specific mortality due to shorebird predation on a fiddler crab. Canadian Journal of Zoology 81:1209–1221.

Ribeiro, P.D., J.H., Christy, J.R. Rissanen, and T.W. Kim. 2006 A male courtship signal is attractive to both sender and receiver. Behavioral Ecology and Sociobiology. 61:81–89.

Rodd, F.H., K.A. Hughes, G.F. Grether, and C.T. Baril. 2002. A possible non-sexual origin of a mate preference: are male guppies mimicking fruit? Proceedings of the Royal Society of London, Series B 269:475–481.

Rosenberg, M.S. 2001. The systematics and taxonomy of fiddler crabs: a phylogeny of the genus *Uca*. Journal of Crustacean Biology 21:839–869.

Rosenberg, M.S. 2006. Fiddler Crab Species. Available at *http://www.fiddlercrab.info/uca_species.html* Website owned by Michael S. Rosenberg. Last updated August 16, 2005.

Sakaluk, S.K. 2000. Sensory exploitation as an evolutionary origin to nuptial food gifts in insects. Proceedings of the Royal Society of London, Series B 267:339–343.

Sih, A. 1994. Predation risk and the evolutionary ecology of reproductive behaviour. Journal of Fish Biology 45(Supplement A):111–130.

Stålhandske, P. 2002. Nuptial gifts of male spiders function as sensory traps. Proceedings of the Royal Society of London, Series B 269:905–908.

Stuart-Fox, D. 2005. Deception and the origins of honest signals. Trends in Ecology and Evolution 20:521–523.

Sturmbauer, C., J.S. Levinton, and J.H. Christy. 1996. Molecular phylogeny analysis of fiddler crabs: test of the hypothesis of increasing behavioral complexity in evolution. Proceedings of the National Academy of Sciences, USA 93:10855–10857.

Thurman, C.L. 2004. Unraveling the ecological significance of endogenous rhythms in intertidal crabs. Biological Rhythms Research 35:43–67.

Viscido, S.V., and D.S. Wethey. 2002. Quantitative analysis of fiddler crab flock movement: evidence for "selfish herd" behaviour. Animal Behaviour 63:735–741.

Warner, R.R. 1997. Evolutionary ecology: how to reconcile pelagic dispersal with local adaptation. Pages 75–80 *in*: H.A. Lessios and I.G. Macintyre, editors. Proceedings of the 8th international coral reef symposium volume 1, Smithsonian Tropical Research Institute, Balboa, Republic of Panama.

Warner, R.R., and L.M. Dill. 1999. Courtship displays and coloration as indicators of safety rather than of male quality: the safety assurance hypothesis. Behavioral Ecology 112:444–451.

West-Eberhard, M.J. 1984. Sexual selection, competitive communication and species-specific signals in insects. Pages 283–324 *in*: T. Lewis, editor. Insect communication. Academic Press, London.

Wong, B.B.M., C. Bibeau, K.A. Bishop, and G.G. Rosenthal. 2005. Responses to perceived predation threat in fiddler crabs: trust thy neighbor as thyself? Behavioral Ecology and Sociobiology 58:345–350.

Yamaguchi, T. 1971. Courtship behavior of a fiddler crab, *Uca lactea*. Kumamoto Journal of Science: Biology 10:13–37.

Yamahira, K. 2004. How do multiple environmental cycles in combination determine reproductive timing in marine organisms? A model and test. Functional Ecology 18:4–15.

Zeil, J., and M.M. Al-Mutairi. 1996. The variation of resolution and of ommatidial dimensions in the compound eyes of the fiddler crab *Uca lactea annulipes* (Ocypodidae, Brachyura, Decapoda). Journal of Experimental Biology 199:1569–1577.

Zeil, J., and J.M. Hemmi. 2006. The visual ecology of fiddler crabs. Journal of Comparative Physiology A 192:1–25.

Zeil, J., and J.E. Layne. 2002. Path integration in fiddler crabs and its relation to habitat and social life. Pages 227–246 *in*: K. Wiese, editor. Crustacean experimental systems in neurobiology. Springer, Berlin, Germany.

Zimmer, M., O. Diestelhorst, and K. Lunau. 2003. Courtship in long-legged flies (Diptera: Dolichopodidae): function and evolution of signals. Behavioral Ecology 14:526–530.

Zucker, N. 1976. Behavioral rhythms in the fiddler crab *Uca terpsichores*. Pages 145–159 *in*: P.J. DeCoursey, editor. Biological rhythms in the marine environment. University of South Carolina Press, Columbia, S.C.

Zucker, N., and R. Denny. 1979. Interspecific communication in fiddler crabs: preliminary report of a female rejection display directed toward courting heterospecific males. Zeitschrift für Tierpsychologie 50:9–17.

Hermaphroditism in Caridean Shrimps

Mating Systems, Sociobiology, and Evolution, with Special Reference to *Lysmata*

Raymond T. Bauer

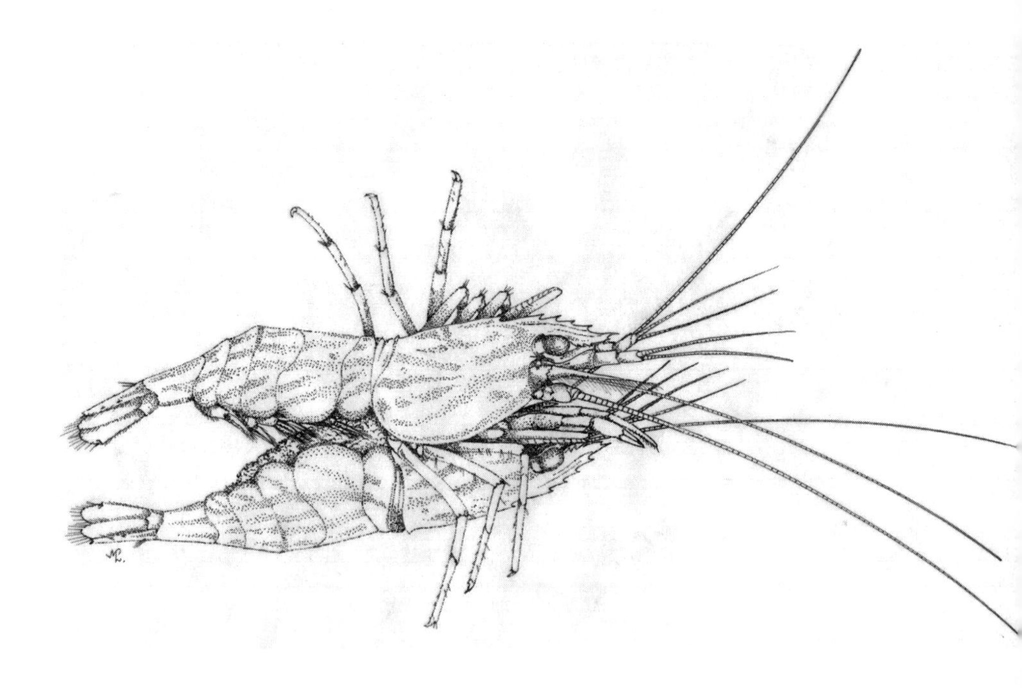

11

The distribution of hermaphroditism among species of plants and animals is often puzzling (Policansky 1982). This is certainly true of the decapod Crustacea (e.g., shrimps, lobsters, crabs), in which hermaphroditism is relatively rare, mainly occurring in caridean shrimps (Bauer 2000). Although caridean species are usually gonochoristic (separate sexes) (Fig. 11.1a), protandry (sequential sex change from male to female) has been described in more than 40 caridean species from at least 15 (of 350+) genera and 7 (of 28+) families (Bauer 2000, 2004). Recently, a variation of protandry (protandric simultaneous hermaphroditism [PSH]) has been discovered in the genus *Lysmata* (Bauer and Holt 1998, Fiedler 1998, Bauer and Newman 2004). Sex-changed "females" of *Lysmata* species, although with a primarily female phenotype, are capable of mating as both male and female. Both protandry with complete sex change and PSH appear quite adaptive to individuals of species with these sexual systems. Yet hermaphroditism has not evolved in many other caridean species closely related to hermaphroditic ones or in unrelated species with

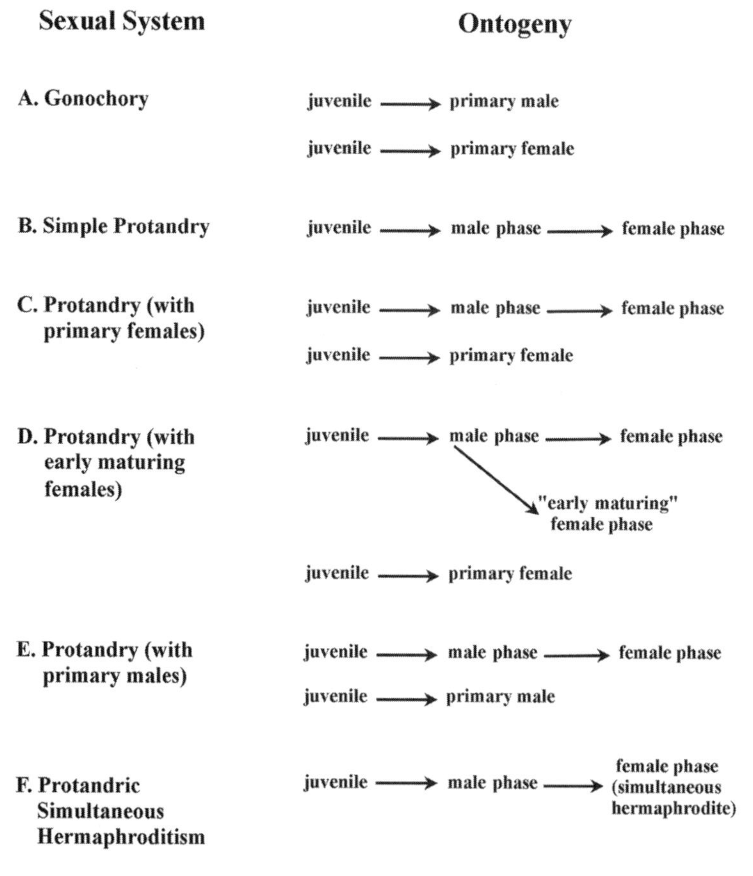

Figure 11.1 Ontogeny of individuals in different sexual systems of caridean shrimps. Examples: (a) *Heptacarpus sitchensis*, (b) *Pandalus goniurus*, (c) *Crangon crangon*, *Processa edulis*, (d) *Pandalus borealis*, (e) *Thor manningi*, (f) *Lysmata wurdemanni*. Figure from Bauer (2004) with permission, University of Oklahoma Press, after Bauer (2000).

apparently similar socioecological characteristics (e.g., social organization, mating systems, sexual dimorphism, habitat). A major question posed in this chapter is, given its considerable adaptive value, why has hermaphroditism not evolved more often in carideans or other decapod crustaceans?

The effect of social environment on the occurrence and timing of sex change has been studied and documented in a variety of taxa (Charnov 1982), especially in prosobranch mollusks (Warner et al. 1996, Collin et al. 2005) and bony fishes (e.g., Warner 1975, 1984, Fricke and Fricke 1977, Ross 1990). Given the lability of sex determination in a number of caridean species (Charnov 1982, Charniaux-Cotton and Payen 1985), an effect of social environment on sex allocation (occurrence, size, or age of sex change) is a hypothesis that can be tested using *Lysmata* and other caridean shrimps as model species. Additionally, the unique sexual system of *Lysmata* shrimps, coupled with a sharp dichotomy in the sociobiology of species in the genus (Bauer 2000), suggests that historical contingency (unique past occurrence of particular selective pressures: Gould 1989) might explain the evolution of hermaphroditism in *Lysmata* and potentially in other taxa, as well.

Protandry in Caridean Shrimps

Protandric carideans first mature as male-phase individuals (MP), phenotypically similar to males of gonochoristic carideans (Bauer 2004). Typical male sexual characters include the appendices masculinae on the second pleopods and coxal gonopores on the last thoracic legs. Initially, the gonads are dominated by testicular tissues and ducts, and the individual functions only as a male. However, rudimentary ovarian components are present. As the MP approaches sex change, the ovarian portions develop further, with the oocytes often beginning vitellogenesis, as testicular components regress (Charniaux-Cotton and Payen 1985, Bauer 2000). Female-phase individuals (FPs) have a purely female phenotype, without male gonopores and sexual appendices but with the female "breeding dress" associated with spawning and incubation (Bauer 2004). The change from male (MP) to female (FP) characters takes place over a variable number of molts (Bauer 1986, Bergström 2000). After sex change, gonads are completely ovarian. In a purely protandric caridean species, all individuals undergo male-to-female sex change (Fig. 11.1b). However, in many "protandric" species, the population is composed of protandric sex changers and individuals that do not change sex (primary females or males) (Fig. 11.1c–e).

Sociobiology and Mating Systems of Protandric Shrimps

One generalization that can be made about most protandric species is that individuals occur in mobile, high-density aggregations. The commercially important caridean shrimps, including protandric species in the genera *Pandalus* and *Crangon* (Bauer 2004), are found in schools over sandy-mud bottoms, facilitating fishing by trawls. Protandric seagrass shrimps (e.g., *Hippolyte inermis*, *Thor manningi*) often occur at high density. Individuals of these protandric species have not been reported to establish territories or occupy specific home ranges. The population structure, as one would expect from species composed completely or mainly of protandric individuals, is composed or dominated by small MPs and larger FPs.

Most protandric carideans have a polygynous mating system with a male mating tactic of "pure searching" (Correa and Thiel 2003, Bauer 2004). In pure searching, a male does not guard the female for any length of time before or after mating. "Parturial" females (with ovaries full of vitellogenic oocytes) become receptive to mating just after molting. Males initiate copulatory behavior upon touching such females with the long chemotactile antennal flagella.

Males are highly mobile, and therefore, in aggregated species, contact with females is frequent. Males briefly contact encountered individuals, and if the individuals are newly molted females, the males will attempt to seize and copulate with them. Immediately after mating, the male leaves the female with no postcopulatory mate guarding. Males have the possibility and are capable of mating with other females within a relatively short time (immediately, within hours, or a few days), whereas females will spawn and brood embryos without being able to mate again until the next spawning cycle (weeks to months). "Pure searching" is characteristic of caridean species in which males are small and females are larger (Bauer and Abdalla 2001). In such species, there is little sexual dimorphism in cheliped or other weaponry (Bauer 2004).

Although "pure searching" mating systems are typical of most protandric species, it seems apparent that this mating system is not a selective pressure promoting protandry but simply is a consequence of their population structure (small MPs, larger FPs), high density, and high mobility. The same attributes coincide in many caridean species without protandry, for example, many hippolytids (Bauer 2004). Palaemonidae is a large caridean family with many "small male, larger female" species, but no hermaphroditism has been reported (Correa and Thiel 2003). Likewise, protandry has rarely been reported (Bauer 2000) and never verified in the penaeid shrimps, another decapod taxon with aggregated populations, small males and larger females, and weak cheliped weaponry (Bauer 1996).

Application of Hermaphroditism Models to Protandric Species

The size advantage model explains well the direction of sex change (male to female) in shrimps (Ghiselin 1969, Warner 1975, Charnov 1982), as it does for sex change, protandrous or protogynous, in other taxa, such as fishes (Warner 1975). Sex change may evolve when reproductive success is positively correlated with larger size or age in one sex but not in the other. In protandric species, male mating success does not appear to increase with male size or age, whereas fecundity is greater in larger females (Bauer 2004).

In my view, protandric carideans evolved from ancestors with small males and larger females. As discussed above, both in caridean and penaeid shrimps, such a population structure is generally associated with high density (aggregated dispersion) and a polygynous "pure searching" mating system. Large male size is not selected for in such a mating system, and smaller males invest less energy in growth and may have lower mortality rates from predators. The mix of sex changers and primary males or females in a population, as well as the timing of sex change, may be explainable by sex allocation theory (Charnov 1982). However, experimental evidence on the proximate factors involved is still scanty (Bauer 2002, Baeza and Bauer 2004).

An alternative explanation to the size advantage hypothesis for explaining sequential hermaphroditism is the "inbreeding version" of the gene dispersal hypothesis, also proposed by Ghiselin (1969). Protandry or protogyny might evolve to prevent inbreeding among siblings in organisms with limited dispersal. However, this clearly does not apply to carideans shrimps, which are usually quite mobile with planktonic larvae for dispersal (for exceptions, see Bauer 2004).

The low-density model predicts simultaneous hermaphroditism rather than gonochory or sequential hermaphroditism when low population density limits contact among the sexes. High predation rates act in the same way by reducing effective mobility in organisms, such as shrimps, that are able to walk and swim about. Pure simultaneous hermaphroditism is not yet known in caridean or penaeid shrimps. Instead, perhaps the social monogamy (living in male–female pairs) of many caridean species (e.g., many alpheid and pontoniine shrimp: Bauer 2004) has evolved in response to low density or mobility. However, sequential/simultaneous hermaphroditism does occur at least in one caridean taxon, the hippolytid *Lysmata*. This special case is explored in depth below.

Protandric Simultaneous Hermaphroditism in *Lysmata*

Shrimps of the caridean genus *Lysmata* (Fig. 11.2), with at least 30 species, are found in subtropical and tropical waters in rocky-bottom and coral-reef habitats. *Lysmata* shrimps have received much attention from aquarium enthusiasts and divers because of the bright coloration and striking fish-cleaning behaviors of some species. Recently, a novel sexual system for decapod crustaceans, protandric simultaneous hermaphroditism (PSH; Bauer 2000; Fig. 11.1f) has been confirmed by mating studies (Bauer and Holt 1998, Fiedler 1998, Bauer and Newman 2004). As in protandry, an individual first matures as a male (MP), but after a series of molts resulting in an increase in size, the MP molts into an individual with the external phenotype of breeding caridean females. These female-phase individuals (FPs) lack male appendices and show the female "breeding dress" (Bauer 2004), including female gonopores. FPs spawn eggs and brood the resulting embryos. Externally, the only male character is a pair of gonopores on the last thoracic segment, typical of male decapods. The gonads of small MPs are ovotestes dominated by testicular tissue and male ducts but with an anterior ovarian portion containing immature oocytes and rudimentary oviducts. As the MPs become larger, the ovarian portion of the gonads becomes proportionately larger. In the last intermolt before external change from MP to FP, the oocytes in the ovarian portion become large and vitellogenic and are spawned soon after the sex change molt. However, unlike other protandric shrimps, the testes and male ducts (vasa deferentia, ejaculatory ducts) do not disappear or become vestigial after the sex change molt. Although relatively small in size, the testes continue to produce sperm that is stored in typical male ejaculatory ducts. MPs can copulate and inseminate only as males, whereas FPs are fully capable of functioning as male or female; that is, they function sexually as simultaneous hermaphrodite.

Figure 11.2 Two species of *Lysmata*, protandric simultaneous hermaphrodites. Above, *L. debelius*, a "pairs" species; below, *L. californica*, a "crowd" species (adapted from Bauer 2004 with permission, University of Oklahoma Press).

As do many caridean females, FPs of *Lysmata* species undergo successive spawnings. As the brood of embryos from one spawning develops, the gonads fill again with vitellogenic oocytes. Within one to two days after the brooded embryos hatch as planktonic larvae, the FP molts and is receptive to mating, which occurs within seconds to minutes if an MP or another FP is present. In the usually single copulation, a spermatophore mass is attached by the male mating partner (MP or FP) to the underside of the molting FP. The latter spawns within a few hours, utilizing the recently attached spermatophores (decapod "type 1" reproductive pattern—see chapter 9), and the fertilized eggs are attached below the abdomen for incubation. FPs are capable of mating as a male any time during the molting, spawning, and incubation cycle

(Bauer and Holt 1998). In an FP × FP copulation, one FP (newly molted, prespawning) copulates only as a female, while the other FP copulates only as a male.

Spawned eggs must pass from the female gonopores to the underside of the abdomen for attachment. In transit, the spawned eggs pass the FPs own male gonopores. However, although it appears mechanically possible, FPs do not fertilize themselves. Isolated FPs undergo the parturial molt but do not produce successful (fertilized) broods (Bauer and Holt 1998, Fiedler 1998, Bauer and Newman 2004).

Thus, FPs of *Lysmata* are nonreciprocally outcrossing simultaneous hermaphrodites. Although other terminology might be used to describe the simultaneous hermaphrodite FPs of *Lysmata* species, I prefer "FP" because the female phase of *Lysmata* has the external phenotype of a caridean female and incubates the embryos. It is developmentally analogous and probably homologous with the FPs of purely protandric species (Fig. 11.1; Bauer 2000). Additionally, the term "FP" has been used in the majority of reports dealing with the sexual system of *Lysmata* (Bauer and Holt 1998, Fiedler 1998, Bauer 2000, 2002, Baldwin and Bauer 2003, Baeza and Bauer 2004, Bauer and Newman 2004).

Sex Allocation: Age (Size) of Sex Change in Hermaphroditic Shrimps

The partitioning of reproductive output or effort into male and female components is sex allocation (Charnov 1982). In sequential hermaphrodites, time or resources spent during in each sexual phase of the life history may be unequal and variable. The life history and sexual biology of hermaphroditic shrimps provide interesting systems with which to test hypotheses about sex allocation (e.g., occurrence of sex change in some individuals of a population but not others, timing of sex change, influence of social environment vs. abiotic environmental factors). In protandric pandalids, the age (size) at which MPs change to FPs is variable in different spatial and temporal populations (Charnov 1982, Bergström 2000). Some MPs change so early in their life history that, although male sexual characters begin to develop, they never breed as males but rather first as females. These "early maturing females" (Fig. 11.1d) are thus females (FPs) of small size. Other individuals develop directly into females from juveniles without any sign of maleness ("primary females"; Fig. 11.1d).

Size frequency distributions of samples taken from commercially fished species have shown that the frequency and size of MP-to-FP change are correlated to the relative abundance of MPs and FPs in the population. Fishing mortality is greater on older FPs because of their larger size, and their abundance relative to MPs decreases in fished populations. In *Pandalus jordani* (Charnov et al. 1978) and *P. borealis* (Charnov 1981), the size (age) of sex change was related to the proportion of FPs in the population. In heavily fished populations, the larger and older FPs were less abundant and the age (size) of development as female (FP) was much younger (smaller) than in populations with less fishing pressure and thus more abundant older (larger) FPs.

Thus, the age (size) of sex change may be a phenotypically labile trait influenced by the demographic environment experienced by a maturing individual (Charnov 1982). In this view ("environmental sex determination"), the developing individual can somehow measure the proportion of MPs and FPs in the population and adjust the timing (or even occurrence) of sex change accordingly. Charnov et al. (1978) and

Charnov (1982) did not venture hypotheses on "how" this might be done (the proximate mechanisms). Presumably, individuals would measure sex ratio by the number and kind of interactions with surrounding conspecifics ("socially mediated sex change in different demographic environments"; Bauer 2002).

According to sex allocation theory, the rarer sex (sexual morph) has a greater reproductive fitness than the more common one (Charnov 1982). This is an assumption that is not often confirmed empirically. However, a good argument that the rarer sex has a higher reproductive fitness can be made for protandric shrimps (Bauer 2000, 2002). In populations of *Pandalus* with abundant FPs, an MP might gain additional inseminations by delaying sex change because an MP can potentially inseminate several FPs during the breeding season. In contrast, an individual of *Pandalus* species reproducing as female can produce only one brood per year. If labile sex change is possible, an individual might, when FPs are low in abundance, benefit by early sex change to FP because insemination of a female's brood of eggs is assured. If remaining as an MP, the individual would have considerable competition from the other numerous males to inseminate the rare FPs in such a population.

The only experiments testing the effect of variable demographic environments on sex change in pandalids were performed by Marliave et al. (1993). Working with *Pandalus danae*, these investigators reared large numbers of young individuals in laboratory populations together with variable proportions of adults (breeding MPs, larger FPs). Marliave et al. (1993) concluded that the seemingly adaptive changes in the proportions of sexual morphs are caused by between-generation selection on sexual genotypes instead of the within-generation, demographically influenced control proposed by Charnov et al. (1978) and Charnov (1981, 1982). Likewise, demographic evidence from *P. borealis* populations indicates that timing of sex change in *P. borealis* may be best explained by frequency-dependent selection on genotypes for sex change at different ages (sizes) (Bergström 2000).

In *Lysmata wurdemanni*, the size of sex change from the smaller MP to the larger simultaneous hermaphrodite FP is variable (Baldwin and Bauer 2003). Below a certain minimum size (Baldwin and Bauer 2003), individuals are MPs and cannot function as females presumably because they do not have the energy resources to produce the large yolky eggs typical of caridean shrimps. However, most MPs do not change until they are larger, often much larger than the minimum size of sex change. Bauer (2002) proposed that a delay in sex change might be favored in populations with high abundances of FPs. In such a population, an MP has more opportunity to mate as a male. However, in *L. wurdemanni*, FPs can also mate as male and thus are potential competitors for insemination of prespawning FPs. MPs would benefit from delayed sex change in a population with abundant FPs only if they were *better* at obtaining copulations with prespawning FPs than are FPs mating as male. This is not the case. Competitive mating experiments showed that FPs are just as successful as MPs in inseminating prespawning FPs (Fig. 11.3). Thus, a superior male mating ability of MPs over the simultaneous hermaphrodite FPs is not a likely selective pressure promoting delayed sex change in *L. wurdemanni*.

However, mating experiments do not directly address the question whether there is socially mediated sex change in *Lysmata wurdemanni*. Baeza and Bauer (2004) did test this hypothesis by rearing sex-change "candidates," that is, MPs larger enough to change sex, in different demographic environments. Sex-change candidates reared in

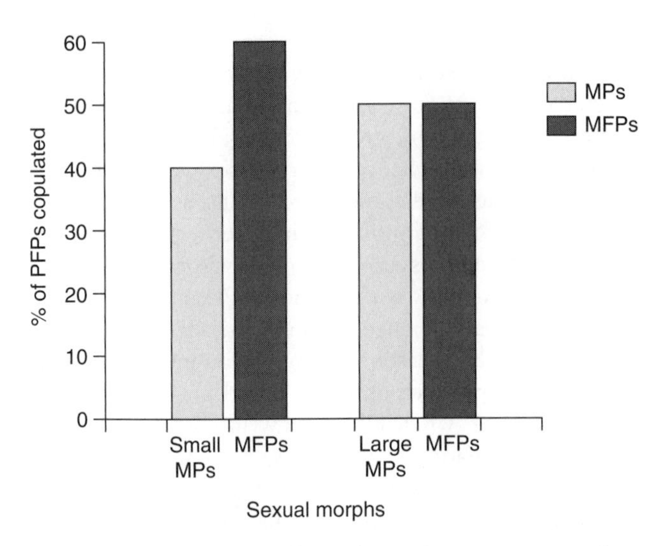

Figure 11.3 Comparative mating success of sexual morphs in *Lysmata wurdemanni*. In two separate experiments, male-phase individuals (MPs) smaller or larger than the minimum size of sex change competed with female-phase individuals (FPs) mating as males (MFPs) for copulations with prespawning FPs (mating as females = PFPs). Based on data from Bauer (2002).

the absence of FPs changed to FP most rapidly, while those maintained with "abundant" FPs changed most slowly (Fig. 11.4). The adaptive value of rapid change in populations without FPs or with low FP abundance is obvious. In such environments, there is little or no chance for an MP to reproduce, because it can mate only as a male, and there are no or few FPs to mate with. By changing as soon as possible to FP, the individual can immediately mate and spawn as a female while still having the ability to mate as a male if the abundance of FPs increases. In environments with FPs, especially abundant FPs, MPs delay sex change. The adaptive benefit is not the increased male mating opportunities afforded by abundant FPs, because FPs can mate as males just as well as the MPs and thus are competitors of MPs for prespawning FPs (Bauer 2002) (Fig. 11.3).

What might be the proximate factors stimulating or delaying sex change in populations of varying sex and age composition? I believe that we are still at the point indicated by Charnov (1982), when discussing labile sex change in *Pandalus jordani*: "We can only wonder at what cues the animals may be using." Presumably, the number of contacts a sex-change candidate makes with different sexual morphs would induce hormonal changes initiating or delaying sex change depending on the current mix of sexual types in the population.

Socially mediated sex change, based on the current demographic environment, may be unlikely in pandalid shrimps because, as Bergström (2000) has pointed out, MPs of *Pandalus borealis* change sex in spring and summer, but breeding takes place in the autumn. Thus, if sex change is a response to the size and age composition of the population, the potential sex changers would have to "forecast" what the breeding population will be like some months in the future. In *Lysmata wurdemanni*, sex change may occur throughout the year, and the breeding season is an extended one.

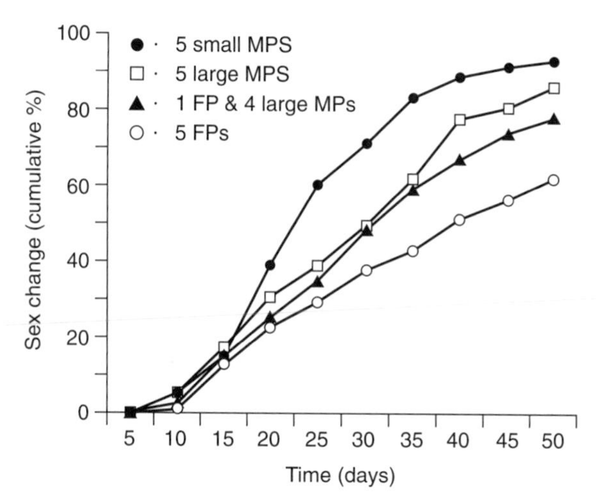

Figure 11.4 Influence of demographic environment on the speed of sex change from the male phase (MP) to the simultaneous-hermaphrodite female phase (FP) in *Lysmata wurdemanni*. The cumulative percentage of large MPs ("sex-change candidates") molting to FP is compared among four treatments in which five sex change candidates were maintained with five small MPs (solid circles), five large MPs (open squares), one FP and four large MPs (solid triangles), or five FPs (open circles). Note that sex change occurred more rapidly in all-MP treatments than in those with FPs. Based on data from Baeza and Bauer (2004).

For an MP large enough to change sex, a maximum of one molt cycle (2–3 weeks) would be necessary to fill the gonad with vitellogenic oocytes and prepare the "female" exoskeleton that will emerge at the sex change molt. Thus, an MP sex-change candidate could respond to changes in its demographic environment with a delay of only a few weeks, which seems reasonable for socially mediated sex change to be adaptive.

The influence of the abiotic environment on adaptive sex change should not be ignored. Zupo (2000) showed that seasonal differences in natural diets explained the sexual maturation of individuals into either primary females or protandric males. In *Lysmata wurdemanni*, change from MP to the FP occurred more rapidly (at a smaller size) in the late spring/early summer than in the autumn (Bauer 2002, Baldwin and Bauer 2003). Sex-change candidates experimentally exposed to spring/early summer conditions of long photoperiod and high water temperatures changed more rapidly than those maintained under autumn conditions of shorter photoperiod and falling temperatures (Bauer 2002). Thus, individuals recruited in the early spring grow rapidly, changing sex quickly to FP, with production of embryos at a time favorable to larval survival and subsequent recruitment. Individuals recruited in later summer and autumn, on the other hand, did not change sex when arriving at the size of possible sex change. Instead, sex change was delayed until the following spring, when these now large MPs changed to FPs of large size. Thus, instead of producing broods in the autumn or winter, time periods presumably not favorable for larval development, summer/fall-recruited MPs put their energy into growth for successful female breeding at the start of the next year's breeding season. This is a pattern common in seasonally breeding gonochoristic shrimps; that is, spring-recruited females mature rapidly and breed all summer, while fall-recruited ones delay reproduction to the next breeding season (Bauer 2004). Thus, sex change in *Lysmata wurdemanni* may be mediated by both the social/demographic environment (Baeza and Bauer 2004) or by seasonally related abiotic factors (Bauer 2002) acting on breeding physiology.

Evolution of Protandric Simultaneous Hermaphroditism

The sexual system of *Lysmata* species (PSH) appears to be extremely advantageous to the individual. A major puzzle is why this sexual system has not evolved more frequently in the Caridea, given the taxon's propensity to protandry, the likely precursor to PSH (Bauer 2004). The most difficult step in the evolution of a protandrous sexual system, the building of the opposite sexual system in the male or female of gonochoristic species (Charnov 1982), has been achieved in a variety of unrelated caridean genera. The retention of the male system after change into the female phase, nearly attained in some *Pandalus* species (Charniaux-Cotton and Payen 1985), seems a small evolutionary step compared to the evolution of protandry directly from gonochory.

Perhaps the costs of changing sex or the costs of maintaining a functional dual system represent a barrier that has prevented PSH from evolving from protandry more frequently. Hoffman et al. (1985) found that, in three species of labrid fishes with female-to-male (protogynous) sex change, deferred reproduction was the price paid for sex change. In *Pandalus* shrimps, the change from MP to FP takes place in a series of molts in which there is a gradual loss of the male sexual appendices as female external characters are attained. However, at least in *P. borealis*, sex change takes place outside of the breeding season (Bergström 2000), so there is no deferred reproduction. In *Lysmata* species, the mating success of transitional MPs has not been specifically investigated. However, it is likely that these MPs have full male mating abilities, as do the FPs, which have no male appendices at all.

Is there a cost of maintaining the male system in the simultaneous hermaphrodite female phase of *Lysmata* shrimps? The energetic costs of the relatively small testicular part of the ovotestes and male ducts in *Lysmata* FPs have not been measured, but certainly the mass of sperm is very small compared to the huge mass of ovarian tissue composed of energy-rich vitellogenic oocytes found in a prespawning FP. However, Bauer (2005) has measured a cost of maleness on brood production in *L. wurdemanni*. In two treatments, focal FPs were maintained with other FPs or a mixture of MPs and FPs; focal FPs could reproduce as both male and female. In another treatment, FPs were maintained solely with MPs, so that they could only reproduce as females. Those FPs that could exercise both sexual functions had smaller brood sizes but more frequent broods than did FPs reproducing only as females. In spite of somewhat more frequent spawning, FPs that could reproduce as both male and female had a lower projected fecundity than those FPs reproducing only as female. Thus, there is some cost of maleness to female fecundity in FPs of *L. wurdemanni*. However, taking into account the broods an FP might fertilize mating as a male, the total reproductive fitness of an FP of a species with PSH still seems to far outweigh that of an FP of strictly protandric species. This cost of maleness is not sufficient, in my view, to explain the low incidence of PSH in caridean shrimps or other decapods.

The comparative method is a powerful tool for identifying the selective pressures responsible for a particular adaptive trait. One looks for an association across species between a particular trait and some set of environmental factors (selection pressures). Bauer (2000) looked for some unique feature of social organization or ecology in *Lysmata* species that might explain the occurrence of PSH in that taxon and not in others. Instead of some unique feature, a wide variation was encountered in the socioecological traits of those *Lysmata* species for which there is at least some good information. *Lysmata wurdemanni*, *L. californica* (Fig. 11.2), and *L. seticaudata* have been termed "crowd" species (Bauer 2000). They occur at high density in aggregations on rocky substrata in subtropical to warm temperate areas of the world. Individuals have a subdued red-striped coloration, are nocturnal, and are facultative "imperfect" cleaners of fishes and other organisms (Bauer 2004). In contrast, in "pairs" species (e.g., *L. grabhami*, *L. amboinensis*, and *L. debelius*; Fig. 11.2), which occur in coral reefs in tropical waters, population density is low, FPs occur in breeding pairs, coloration is conspicuous, and diurnal behavioral displays are performed to attract fish "clients" by these specialized fish cleaners. Thus, there is no obvious selective pressure common to all or most *Lysmata* species that might explain their unique sexual system.

Instead of considering current selective pressures, we might look at past selective pressures to understand the origin of PSH in *Lysmata* species. Charnov (1982) and Warner (1984) conceded that the history of a group might be important in understanding the evolution of hermaphroditism. Species may show hermaphroditism that evolved in ancestors that were subject selective pressures no longer in operation. The hermaphroditic system is retained because it fortuitously has adaptive value even under a more recent and different selective regime. Additionally, the co-occurrence of the different selective pressures whose combination gave rise to PSH may have taken place rarely or uniquely (historical contingency hypotheses; Gould 1989), perhaps explaining the rarity of PSH in the Caridea.

I have proposed a possible evolutionary scenario for the evolution of PSH (Bauer 2000). Given the number of caridean species with protandry, the similarity of PSH to protandry, and the difficulty of the evolution from gonochory to hermaphroditism (Charnov 1982) as opposed to from protandry to PSH (see above), the assumption is made that the ancestor of *Lysmata* was protandric (Fig. 11.5). As in most recent protandric carideans, the *Lysmata* ancestor occurred at high density in aggregations or schools in which encounters between potential mating partners were frequent. Upon invasion of a new habitat or by environmental change in the existing habitat, low effective population density was imposed by some combination of low available resources (e.g., lower food supply, competition) and/or low mobility (e.g., response to high predation pressure). Low encounter rates between potential mates favored the evolutionary step from protandry to PSH and, simultaneously, the evolution of pair living. Breeding pairs of extant "pairs" species are composed of two FPs, each of which can serve as the other's male or female mating partner.

This scenario (hypothesis) explains PSH in "pairs" species, which do occur at low density and low mobility. However, PSH is also the sexual system of "crowd" species in which MPs and simultaneous hermaphrodite FPs live together in aggregations. There is no shortage of encounters among individuals or potential mating partners. If PSH evolved under a low-density selective regime, what explains its presence in "crowd" species? I proposed (Bauer 2000) that "crowd" species are descendants of

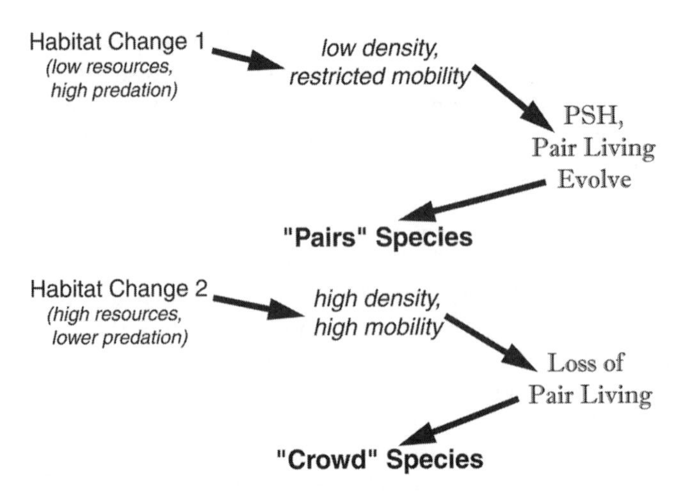

Figure 11.5 A proposed scenario for the evolution of protandric simultaneous hermaphroditism (PSH) and pair living ("pairs" species) in *Lysmata* from a protandric ancestor, with the subsequent loss of pair living but not PSH in "crowd" species.

"pairs" species that encountered a new selective regime (Fig. 11.5). "Pairs" species later experiencing a habitat with greater resources and/or lower predation pressure, either by invasion or by in situ habitat change, could occur again at high densities and move more freely within the habitat. In such an environment, allowing frequent contact among potential mating partners, pair living might be selected against. Living with a conspecific has a variety of costs, such as agonistic encounters with the pair partner and limitation of contacts with other mating partners. However, PSH would not be selected against in "crowd" species because it bestows the simultaneous hermaphrodite FP with a great reproductive advantage (over a purely female FP) at apparently little cost. PSH in "crowd" species would thus be a fortuitous adaptation inherited from an ancestor.

Although this "historical contingency" hypothesis is complex, it can be generally tested by analyzing the distribution of "pairs" and "crowd" species in the phylogeny of *Lysmata* species. "Pairs" species should be ancestral to "crowd" species. "Crowd" species could have evolved independently from various "pairs" species or as a single clade from a single "pairs" species. If "crowd" species are uniformly ancestral to "pairs" species, the scenario given above would be rejected.

Hermaphroditism in Shrimps, Fishes, and Other Selected Groups

Mating system and population density, the latter depending in part on the availability of resources, greatly influence the evolution of hermaphroditic sexual systems (Ghiselin 1969, Warner 1984). Thus, the many protogynous fish species tend to have mating systems in which large dominant males have the greatest reproductive success (size advantage hypothesis; Warner 1975). The degree of protogyny (percentage of protogynous individuals) may vary with population density (Warner 1984). Simultaneous hermaphroditism is not uncommon in the animal kingdom (Ghiselin 1969, Michiels 1998). It has evolved, often together with reciprocal egg trading, in some fish species that live or breed in pairs and/or occur a low density (Warner 1984). A similar sexual system has evolved in a polychaete worm (*Ophryotrocha diadema*) that lives in breeding pairs (Sella 1990). Bidirectional sex change occurs in a related species (*O. puerilis*) that may occur in pairs in nature (Berglund 1986). In both polychaete species, the larger more successful simultaneous hermaphrodite (*O. diadema*) or sex-changing (*O. puerilis*) adult stage is preceded by an adolescent male phase. In fishes of the genus *Lythrypnus* (Gobiidae) (St. Mary 1996), which breed in "male–female" nesting pairs, individuals have both male and female gonadal tissue simultaneously. However, they only exhibit one gender externally and behaviorally at a time. Therefore, they are not functionally simultaneous hermaphrodites but rather bidirectional sex-changers because some individuals are capable of changing back to the sex in which they first reproduced. In contrast to most fishes, hermaphroditism in shrimps is protandric, a function of the interaction of a mating system ("pure searching" polygyny) and aggregated dispersion (high density), two characteristics found in many shrimp species. Although hermaphroditism in somewhat rare in peracarid crustaceans, both protandry and protogyny are found in the isopods, apparently a consequence of the variable life histories (e.g., parasitic, free-living) and mating systems found in the group (see chapter 8).

Protandry has evolved from a different route in the anemone fishes (*Amphiprion* spp.; Fricke and Fricke 1977) in response to selective pressures (low density and predator-enforced low mobility) that favor simultaneous hermaphroditism in other fishes. In protandric *Amphiprion* species, a monogamous breeding pair lives within the protective shelter of a sea anemone. Several subadult males and juveniles may be associated with the breeding pair and host. When the larger female is removed, the breeding male changes to female and the next largest subadult male moves up to breeding status. There are many species of caridean shrimps that live in pairs in association with sea anemones, sea urchins, and other invertebrates (Bauer 2004). In some species, juveniles and subadults live suboptimally around the host with the superior position occupied by the breeding pair. It is not unconceivable that an *Amphiprion*-like protandry has evolved in some of those species.

Unanswered Questions

The major question about hermaphroditism in shrimps is one that has continually puzzled students of hermaphroditism of many animal groups: why hasn't hermaphroditism evolved more frequently? Protandry and its more specialized form, PSH, certainly seem adaptive and fairly well explained by the size advantage hypothesis (Ghiselin 1969, Charnov 1982). An argument that genetic, developmental, and morphological barriers ("evolutionary constraints") limit selection for this advantageous system is facile but not tenable. Protandry is scattered among unrelated caridean genera. Even within genera with protandric species, species with a similar natural history may be gonochoristic. Phylogenetic analyses of such "protandric" genera would reveal much about the evolution of protandry in the Caridea.

Evolutionary hypotheses on the origin of PSH in *Lysmata* species have been proposed above. Just as important as phylogenetic analysis of the genus are sociobiological studies on the many *Lysmata* species whose sexual biology and social organization are still unknown. I suspect that the "pairs" versus "crowd" dichotomy in socioecological traits may be modified to a more complex scheme when the full variation in sexual systems, breeding biology, and social organization is known. Knowledge of these traits is needed in many species to analyze their distribution on a phylogenetic tree, yet another task that needs to be done. The costs and benefits of living in pairs versus group living must be analyzed. Likewise, a cost–benefit analysis of male function versus female function should be done to test whether or not the costs of maleness to FPs is a major barrier to the evolution of simultaneous hermaphroditism in decapod shrimps.

Other major questions about hermaphroditism in carideans and other organisms have been treated in this chapter. The factors that influence the timing of sex change, such as abiotic factors influencing seasonality of reproduction and life history, as well as the effect of social environment on sex change, need further clarification. In particular, the actual social stimuli that may influence sex change, such as the number and nature of contacts among different sexual morphs, need to be elucidated. Certainly, the study of hermaphroditism in shrimps can provide many scientific lifetimes of work. Much theory has been generated on hermaphroditism in shrimps and other organisms. What is really needed is an emphasis on empirical observation and testing. There are too few facts available to test the voluminous theory, a problem that interesting and fulfilling field and laboratory work can resolve.

Summary and Conclusions

Although hermaphroditism is rare in the decapod Crustacea, protandry (male-to-female sex change) is not uncommon in the caridean shrimps. Protandric shrimps are generally mobile, living in high-density aggregations, with pronounced sexual dimorphism in body size (small males, larger females). As in many gonochoristic species, males have no well-developed weaponry for defending or guarding females. The mating system is polygynous, with males using a "pure searching" tactic to maximize mating success. Protandry is fairly well explained by the size advantage model, in which male mating success is not positively correlated with size but female fecundity is. In the genus *Lysmata* ("cleaner shrimps"), male-phase (MP) individuals change to a female phase (FP) as in protandric shrimps but, unlike them, retain male gonadal tissues. Although *Lysmata* FPs have an external female phenotype, they function as nonreciprocal outcrossing simultaneous hermaphrodites. Labile sex change, in which individuals make the physiological "decision" if and when to change sex, has been suggested in the commercially important species of protandric pandalids and *Lysmata* species. Experiments on *Lysmata wurdemanni* confirm that the timing of sex change is modified by an individual's demographic environment. However, field observations and other experiments show that abiotic factors related to seasonality of breeding are important in delayed sex change in this species. Major questions about hermaphroditism in caridean are asked: What explains the occurrence of protandry in some caridean but not in their close relatives? What are the important factors influencing the time of sex change in carideans? Given the propensity of carideans to protandry, why has not the very advantageous PSH found in *Lysmata* species evolved more often? A dichotomy in the sociobiology of *Lysmata* species ("pairs" and "crowd" species) suggests a historical contingency hypothesis. Phylogenetic analyses (chapter 3) of protandric species, *Lysmata* species, and related species are needed to test this hypothesis about the evolution of hermaphroditism in caridean shrimps.

Acknowledgments Studies on *Lysmata* and other hermaphroditic shrimps were supported by grants from the National Science (grant OCE-9982466) and Neptune Foundations. My thanks to J. Antonio Baeza for his stimulating interest and work in *Lysmata* shrimps (and for help with Fig. 11.4). Many thanks to the several undergraduate and graduate research assistants who worked with me on this topic. This is Contribution 102 of the Laboratory for Crustacean Research, University of Louisiana at Lafayette.

References

Baeza, J.A., and R.T. Bauer. 2004. Experimental test of socially mediated sex change in a protandric simultaneous hermaphrodite, the marine shrimp *Lysmata wurdemanni* (Caridea: Hippolytidae). Behavioral Ecology and Sociobiology 5:544–550.

Baldwin, A.P., and R.T. Bauer. 2003. Growth, survivorship, life span, and sex change in the hermaphroditic shrimp *Lysmata wurdemanni* (Decapoda: Caridea: Hippolytidae). Marine Biology 143:157–166.

Bauer, R.T. 1986. Sex change and life history pattern in the shrimp *Thor manningi* (Decapoda: Caridea): a novel case of protandric hermaphroditism. Biological Bulletin 170:11–31.

Bauer, R.T. 1996. A test of hypotheses on male mating systems and female molting in decapod shrimp, using *Sicyonia dorsalis* (Decapoda: Penaeoidea). Journal of Crustacean Biology 16:429–436.

Bauer, R.T. 2000. Simultaneous hermaphroditism in caridean shrimps: a unique and puzzling sexual system in the Decapoda. Journal of Crustacean Biology 20(Special Number 2):116–128.

Bauer, R.T. 2002. Test of hypotheses on the adaptive value of an extended male phase in the hermaphroditic shrimp *Lysmata wurdemanni* (Caridea: Hippolytidae). Biological Bulletin 203:347–357.

Bauer, R.T. 2004. Remarkable shrimps: adaptations and natural history of the carideans. University of Oklahoma Press, Norman, Okla.

Bauer, R.T. 2005. Cost of maleness on brood production in the shrimp *Lysmata wurdemanni* (Decapoda: Caridea: Hippolytidae), a protandric simultaneous hermaphrodite. Journal of the Marine Biological Association of the United Kingdom 85:101–106

Bauer, R.T., and J.A. Abdalla. 2001. Male mating tactics in the shrimp *Palaemonetes pugio* (Decapoda, Caridea): precopulatory mate guarding vs. pure searching. Ethology 107:185–199.

Bauer, R.T., and G.J. Holt. 1998. Simultaneous hermaphroditism in the marine shrimp *Lysmata wurdemanni* (Caridea: Hippolytidae): an undescribed sexual system in the decapod Crustacea. Marine Biology 132:223–235.

Bauer, R.T., and W.A. Newman. 2004. Protandric simultaneous hermaphroditism in the marine shrimp *Lysmata californica* (Caridea: Hippolytidae). Journal of Crustacean Biology 24:131–139.

Berglund, A. 1986. Sex change by a polychaete: effects of social and reproductive costs. Ecology 67:837–845.

Bergström, B. 2000. The biology of *Pandalus*. Advances in Marine Biology 38:1–245.

Charniaux-Cotton, H., and G. Payen. 1985. Sexual differentiation. Pages 217–299 *in*: D.E. Bliss and H. Mantel, editors. The biology of crustacea, Volume 9. Integument, pigments, and hormonal processes. Academic Press, New York.

Charnov, E.L. 1981. Sex reversal in *Pandalus borealis*: effect of a shrimp fishery? Marine Biology Letters 2:53–57.

Charnov, E.L. 1982. The theory of sex allocation. Princeton University Press, Princeton, N.J.

Charnov, E.L., D.W. Gotshall, and J.G. Robinson. 1978. Sex ratio: adaptive response to population fluctuations in pandalid shrimps. Science 200:201–206.

Collin, R., M. McLellan, K. Gruber, and C. Bailey-Jourdain. 2005. Effects of conspecific associations on size at sex change in three species of calyptraeid gastropods. Marine Ecology Progress Series 293:89–97.

Correa, C., and M. Thiel. 2003. Mating systems in caridean shrimp (Decapoda: Caridea) and their evolutionary consequences for sexual dimorphism and reproductive biology. Revista Chilena de Historia Natural 76:187–203.

Fiedler, G.C. 1998. Functional, simultaneous hermaphroditism in female-phase *Lysmata amboinensis* (Decapoda: Caridea: Hippolytidae). Pacific Science 52:161–169.

Fricke, H., and S. Fricke. 1977. Monogamy and sex change by aggressive dominance in coral reef fish. Nature 266:830–832.

Ghiselin, M.T. 1969. The evolution of hermaphroditism among animals. Quarterly Review of Biology 44:189–208.

Gould, S.J. 1989. Wonderful life. The Burgess Shale and the nature of history. W.W. Norton, New York.

Hoffman, S.G., M.P. Schildauer, and R.R. Warner. 1985. The costs of changing sex and the ontogeny of males under contest competition for mates. Evolution 39:915–927.

Marliave, J.B., W.F. Gergits, and S. Aota. 1993. F_{10} pandalid shrimp: sex determination; DNA and dopamine as indicators of domestication; and outcrossing for wild pigment pattern. Zoo Biology 12:435–451.

Michiels, N.K. 1998. Mating conflicts and sperm competition in simultaneous hermaphrodites. Pages 219–254 *in*: T.R. Birkhead and A.P. Møller, editors. Sperm competition and sexual selection. Academic Press, New York.

Policansky, D. 1982. Sex change in plants and animals. Annual Reviews of Ecology and Systematics 13:417–495.

Ross, R.M. 1990. The evolution of sex-change mechanisms in fishes. Environmental Biology of Fishes 29:81–93.

Sella, G. 1990. Sex allocation in the simultaneously hermaphroditic polychaete worm *Ophryotrocha diadema*. Ecology 7:27–32.

St. Mary, C. 1996. Sex allocation in a simultaneous hermaphrodite, the zebra goby *Lythrypnus zebra*: insights gained through a comparison with its sympatric congener, *Lythrypnus dalli*. Environmental Biology of Fishes 45:177–190.

Warner, R.R. 1975. The adaptive significance of sequential hermaphroditism in animals. American Naturalist 109:61–82.

Warner, R.R. 1984. Mating behavior and hermaphroditism in coral reef fishes. American Scientist 72:128–136.

Warner, R.R., D.L. Fitch, and J.D. Standish. 1996. Social control of sex change in the shelf limpet, *Crepidula norrisiarum*: size-specific responses to local group composition. Journal of Experimental Marine Biology and Ecology 204:155–167.

Zupo, V. 2000. Effect of microalgal food on the sex reversal of *Hippolyte inermis* (Crustacea: Decapoda). Marine Ecology Progress Series 201:251–259.

The Mating System of Symbiotic Crustaceans
A Conceptual Model Based on Optimality and Ecological Constraints

J. Antonio Baeza

Martin Thiel

12

During the last decades, several conceptual, graphical, and mathematical models have been proposed to explain particular or more general aspects of the breeding ecology of both marine and terrestrial organisms (e.g., Parker 1970, Jarman 1974, Bradbury and Vehrencamp 1977, Wickler and Seibt 1981, Grafen and Ridley 1983). The pioneering and influential work by Emlen and Oring (1977), and most recently that of Shuster and Wade (2003), are among the most comprehensive contributions. Most models predict particular mating systems depending on a limited set of ecological conditions, mostly the abundance and distribution of resources and receptive females in space and time, which ultimately determines the "environmental potential for polygamy" (Emlen and Oring 1977). Also, Shuster and Wade (2003) have recently proposed a method for quantifying the source and intensity of selection with which to improve our understanding about the evolution of mating systems.

Among marine crustaceans, mating systems are diverse (see reviews on hermit crabs, Hazlett 1975; brachyuran crabs, Christy 1987; caridean shrimps, Correa and Thiel 2003; isopods and amphipods, Jormalainen 1998), and several studies have attempted to identify the factors determining their mating systems (i.e., Christy 1987, Correa and Thiel 2003, Bauer 2004). Experimental and descriptive studies of crustaceans have largely corroborated Emlen and Oring's model but have also drawn attention to the importance of additional factors such as sperm competition (Parker 1970) and pre- and postcopulatory mate choice (Eberhard 1996) in shaping mating strategies. Although our understanding of mating behaviors in several groups of

Figure 12.1 (*Contd.*)

Figure 12.1 (A) The porcelain crab *Allopetrolisthes spinifrons* that dwells solitarily on the column and occasionally among the tentacles of its host, the sea anemone *Phymactis papillosa* (= *clematis*) in intertidal coastal waters of Chile. (B) The porcelain crab *Liopetrolisthes mitra* that lives in unstructured multi-individual aggregations among the spines and beneath the oral disk of its host, the sea urchin *Tetrapygus niger* in subtidal coastal waters of Chile.

crustaceans has improved, no study so far has attempted to explain the variety of mating systems of symbiotic crustaceans. There is also increasing appreciation that environmental factors, such as habitat characteristics, exert an influence on mating systems (Knowlton 1980, Thiel and Baeza 2001). Promising taxa in which to explore the effects of the environment are those inhabiting well-defined microenvironments, namely, species that have assumed a symbiotic lifestyle.

Crustaceans are among the most diverse marine invertebrates, and many (including some isopods, amphipods, shrimps, crabs) have developed complex interrelations with other marine invertebrates from many different taxa (i.e., sponges, sea anemones, corals, among others) (Ross 1983, Thiel and Baeza 2001). These hosts differ widely in body plan, size, morphology, and general ecology (i.e., abundance, dispersion pattern) (Thiel and Baeza 2001, Baeza and Thiel 2003). In turn, symbiotic crustaceans feature a wide array of population or deme structures on their hosts (Thiel and Baeza 2001). Some dwell in/on hosts as solitary individuals (Diesel 1986; Fig. 12.1A), some are found as heterosexual pairs (Knowlton 1980), and others as dense unstructured aggregations of individuals (Baeza and Thiel 2000; Fig. 12.1B). The wide range of hosts used by symbiotic crustaceans and the diversity of their population or deme structures represent an opportunity to explore and understand those environmental conditions constraining or promoting particular mating associations in marine and terrestrial organisms exploiting small discontinuous habitats (including parasitoid insects, parasite helminthes).

Our aim in this chapter is to propose and evaluate a conceptual model that predicts the mating system of symbiotic crustaceans depending on particular host characteristics and environmental conditions (e.g., predation pressure).

Model Structure and Assumptions

Our model assumes that males and females have different optimal mating strategies, which they attempt to realize by defending (i.e., host guarding) and moving (host switching) between hosts, and that the environment limits the behavioral options of symbiotic individuals.

Different Optimal Mating Strategies of the Sexes

We assume that polygyny is the optimal male mating strategy whereas monogamy coupled with mate choice is the optimal female mating strategy (Emlen and Oring 1977). Males are usually considered as "eager" to mate because their fitness is maximized by reproducing with as many females as possible. In contrast, females are perceived as "reluctant" to mate because their reproductive potential is lower and fitness is generally maximized by choosing a male of "high quality" (but see chapter 2). Females may obtain direct nongenetic benefits (i.e., resources, parental care) or indirect genetic benefits (i.e., "good genes"; Hunt et al. 2004) by choosing particular males. In species where females obtain only sperm from males, such as in most marine crustaceans with dispersing larvae, females are thought to derive indirect genetic benefits (good genes) by choosing a male of high quality because offspring inherit both the genes underlying choice and the genes for quality (Hunt et al. 2004). In turn, direct nongenetic benefits such as food, protection against natural enemies, and a dwelling in which to brood their embryos may also be important for females. Because of the close interdependence between symbionts and their hosts, these direct benefits may more often outweigh genetic benefits in symbiotic crustaceans than in free-living species. Overall, this perception about the different optimal mating strategies of males and females is based on classical sexual selection and parental investment theories (Darwin 1871, Bateman 1948, Trivers 1972) and still constitutes the classic Darwinian paradigm on which most theoretical models about the evolution of animal breeding systems are based.

Host Guarding and Host Switching

The host organism constitutes a critical resource for a symbiotic organism. Males may increase their chances to mate by monopolizing hosts or by roaming among them if these harbor females. Similarly, females may be able to choose among mates only if they are capable of visiting hosts harboring males of dissimilar quality. Host guarding is any activity performed to secure a particular host against intruders (see Wilson 1975). Host switching is the movement of individuals among hosts. These are completely different behavioral traits. For instance, a symbiont may change hosts frequently and exclude all other conspecifics each time it occupies a new host or may be sedentary and share space on the host. Resource defense and site tenacity have been recognized previously as important elements in crustacean mating behavior (e.g., Christy 1987; see also chapter 10). For instance, males of the shrimp *Alpheus armatus* have been shown to increase their reproductive success by increasing the rate of host switching, which in turn provides more opportunities to find receptive females (Knowlton 1980). Similarly, monopolization of hosts allows males of symbiotic species to defend their refuges and potential mating partners on them (Nakashima 1987). Thus, we expect both host guarding and host switching to have important effects on individual reproductive success because both behaviors affect the rate and number of interactions between potential mates.

Environmental Constraints

Emlen and Oring (1977) considered the temporal and spatial distribution of receptive females as important elements shaping mating systems. We also consider the spatial distribution of receptive females as of primary importance, but we focus on host characteristics because we believe they ultimately control female abundance and distribution in symbiotic organisms. Three host characteristics—relative size, abundance, and morphological complexity—together with predation risk, are proposed as primary ecological factors affecting host guarding and host switching (Fig. 12.2). These host characteristics were chosen based on our review of effects of ecological factors on intraspecific associations in symbiotic crustaceans. For instance, the hosts of solitary symbiotic crustaceans are, on average, smaller and less abundant and have simpler morphological complexity than the hosts of crustaceans that live in large unstructured groups (Thiel and Baeza 2001, Baeza and Thiel 2003). Predation is known to affect many reproductive traits of marine crustaceans, including mate-searching behavior (chapter 10). Therefore, we expect that predation risk, host size, abundance, and morphological complexity will have important effects on both host guarding and host switching, ultimately determining the rate and number of interactions between potential competitors and mates.

The Optimality Approach

We assume that individuals attempt to respond "optimally" to their environment (Maynard-Smith 1978). This optimal response by symbionts is determined by costs and benefits associated with specific behaviors (i.e., host guarding and host switching) that vary with particular environmental conditions (i.e., host size, abundance, morphological complexity, predation risk). The optimal behavior of symbionts is that

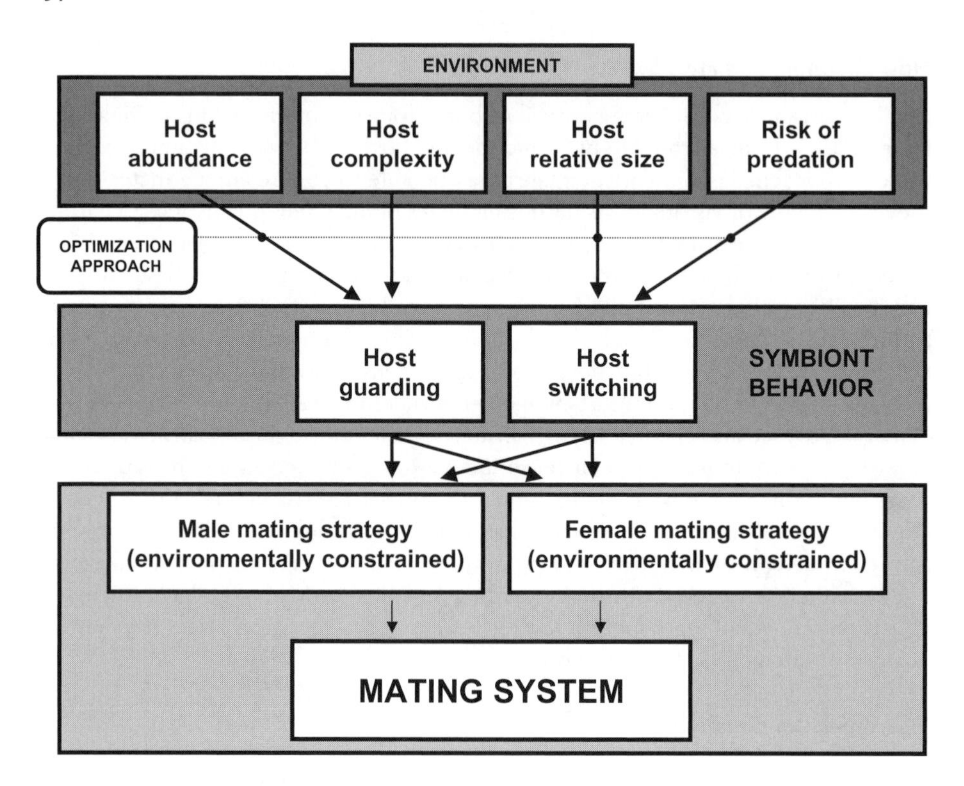

Figure 12.2 Proposed model of mating systems of symbiotic crustaceans. Risk of predation and host characteristics (i.e., relative size, abundance, morphological complexity) constrain and/or promote host monopolization and host switching, behavioral traits used by individuals to achieve their optimal mating strategy. The decision whether or not to guard and roam among individual hosts is an optimality problem crucial in determining the behavioral strategies of both males and females and therefore the mating system of symbiotic crustaceans.

where the net benefit or the benefit-to-cost ratio is the largest under prevailing conditions (Maynard-Smith 1978). Increases in survivorship, growth, or ultimately, mating opportunities are potential benefits of host guarding and host switching. In turn, the amount of energy an individual must expend on resource-guarding activities and the risks of falling victim to predators when away from hosts are potential costs. These benefits and costs ultimately influence lifetime reproductive success, the currency being optimized in this model.

By following this optimality approach, we make the following predictions. All else being equal, with increasing host abundance, the probability of symbionts monopolizing hosts should decrease, while the probability of symbionts moving among host individuals (i.e., host switching) should increase. This is because increased host abundance decreases the benefits of monopolizing hosts and the costs of roaming around them. Second, with increasing morphological complexity, the probability of symbionts monopolizing hosts should decrease, while the probability of movement among hosts should remain constant. This is because the host's morphological complexity renders host guarding too costly but affects neither the cost nor the benefits of roaming among them. Third, increasing host size (relative to symbiont size) should not affect the

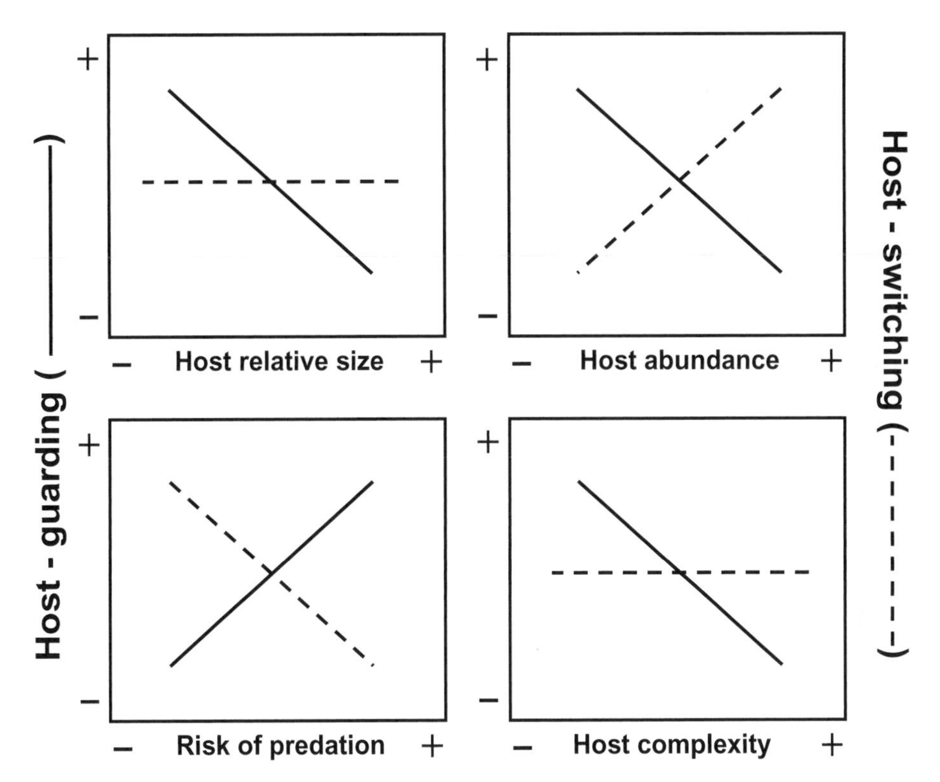

Figure 12.3 Predicted changes in host guarding and host switching of symbiotic crustaceans depending upon host characteristics and predation risk.

frequency of movements among hosts (i.e., host switching), but should decrease host guarding. This is because neither the costs nor the benefits of roaming among hosts change with host relative body size, but increases in host relative body size should render host monopolization by symbionts too costly. Finally, with increasing predation pressure, host guarding should increase but host switching should decrease. This is because the host's value in offering protection against predators increases with predation risk but renders movements among hosts more risky (Fig. 12.3).

Sexual Dimorphism and Alternative Mating Tactics

Theory predicts that intrasexual selection favors weapons (i.e., chelae, gnathopods, and maxillipeds in crustaceans) or other traits (e.g., large body size) that improve the potential of males to monopolize females. In turn, intersexual selection promotes ornaments in males that increase their chance to be chosen by and mate with females (Darwin 1871). Both intra- and intersexual selection, together with the action of natural selection (e.g., fecundity selection on females, cost of ornaments in terms of growth and survival on males), should ultimately determine how different in terms of morphology, coloration, and/or behavior males should be from females. Sexual selection theory also predicts that intense intrasexual selection favors alternative reproductive tactics in subdominant males (e.g., sneaking or sex change; Andersson 1994). The consequences in terms of sexual dimorphism and alternative mating tactics for symbionts adopting a particular mating system are discussed below.

Mating Systems

We propose five mating system categories that should occur in particular environments, and we discuss the behavioral tactics adopted by individuals for each. These mating categories are named in terms of number and variability of copulations experienced by both males and females (*sensu* Shuster and Wade 2003). Thus, monogamy refers to the mating system in which both males and females have a single mate for life. In polygyny, females mate with a single male in their lives, but males may mate with more than one female. Polygynandry occurs when both sexes are variable in their mate numbers, but males are more variable than females (see Shuster and Wade 2003). Herein, we distinguish two categories of polygyny and two categories of polygynandry. These mating categories are envisioned as occurring within a three-dimensional space resulting from the interaction among predation risk, host relative size/host complexity, and host abundance (Fig. 12.4, Table 12.1). Even though

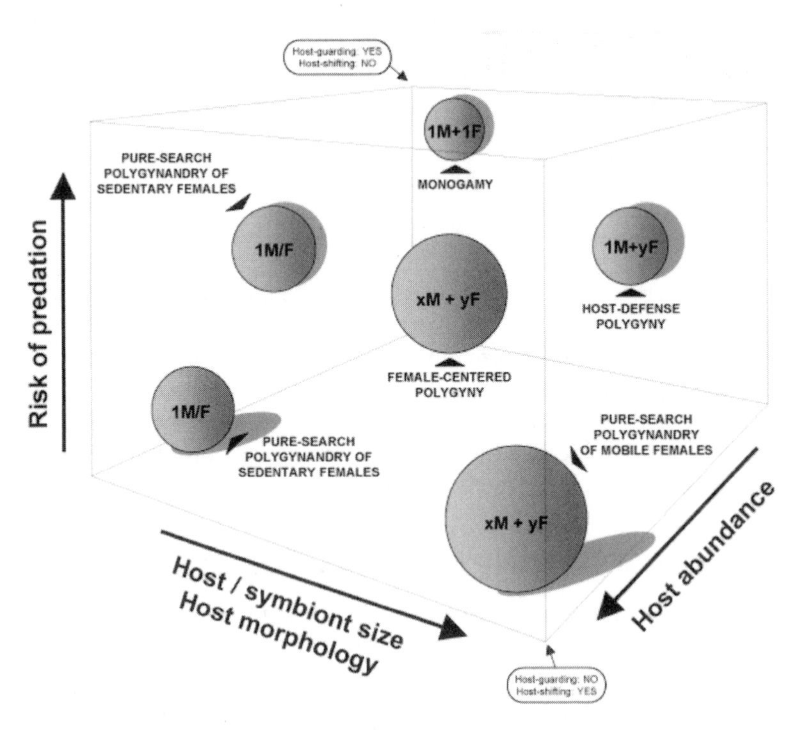

Figure 12.4 Mating systems predicted for symbiotic crustaceans according to a specific set of environmental conditions. The interaction among predation risk, host relative size, and abundance is envisioned as a three-dimensional space in which different mating systems occur. x and y, number of males (M) and females (F), respectively, found in a host under the different mating systems.

Table 12.1. The five mating system categories occurring in symbiotic crustaceans proposed in the present model.

Mating System and Selected Examples	Environmental Conditions				Sexual selection intensity[e]	Consequences	
	Host size[a]	Host complexity[b]	Host abundance[c]	Risk of predation[d]		Sexual dimorphism[f]	Alternative mating tactics[g]
1. Monogamy	S	S, M	S	L	S	M = F	N
Alpheus armatus (Knowlton 1980)							
Periclimenes ornatus (Omori et al. 1994)							
2. Host-defense polygyny	M	S, M	S, M	L	L	M > F	Y
Anamixis hanseni (Thomas and Barnard 1983)[h]							
Paracerceis sculpta (Shuster and Wade 1991)							
3. Pure-search polygynandry of mobile females	L	L, M	L	S	L, M	M < F	Y
Liopetrolisthes mitra (Baeza and Thiel 2000)[h]							
4. Pure-search polygynandry of sedentary females	S	S, M, L	L, M	M, L	L, M	M < F	Y
Allopetrolisthes spinifrons (Baeza and Thiel 2003)[h]							
Zaops ostreum (Christensen and McDermott 1958)							
Zebrida adamsii (Yanagisawa and Hamaishi 1986)							
5. Female-guarding polygyny	M	M	M	M	L, M	M > or ≈F	Y
Inachus phalangium (Diesel 1986)							
Athanas kominatoensis (Nakashima 1987)							

Environmental conditions determining each mating system, the intensity of sexual selection, and consequences in terms of sexual dimorphism and the presence of alternative mating tactics are shown. Selected empirical examples supporting the proposed model in each mating system category are listed. Additional studies may be found in Thiel and Baeza (2001). [a]S, small; M, intermediate; L, large. [b]S, simple; M, moderate; L, complex. [c]S, scarce; M, intermediate; L, abundant. [d]S, low; M, moderate; L, high. [e]S, low; M, moderate; L, high. [f]M, male; F, female. [g]Y, yes; N, no. [h]Additional information on the mating behavior and reproductive biology of this species is required.

these mating systems are introduced as discrete categories, they are better considered "variations of a single theme." After introducing each mating category, examples that support the predictions of this model are discussed.

Monogamy

Monogamy should evolve when hosts are morphologically simple, small enough to support few individuals, and relatively rare and when predation risk away from hosts is high. Under these circumstances, movements among hosts are severely constrained. Host guarding is favored in males and females due to host scarcity and because of the host's value in offering protection against predators. The small body size of hosts makes it possible for individuals to monopolize the host by guarding. Small host body size and scarcity also result in females not being clumped on a single host, but rather distributed more uniformly among them. It is therefore difficult either for a male to monopolize more than a single female at the same time on the same host or to visit additional females on different hosts. Because spatial constraints allow only few adult symbiotic individuals to cohabit on the same host, both males and females maximize their reproductive success by sharing "their" dwelling with a member of the opposite sex during their lifetime.

Temporal synchrony of receptivity among females should reinforce this mating system while low synchrony should provide an opportunity for males to seek extrapair copulations. No or low sexual dimorphism in body size and weaponry is expected due to the low intensity of sexual selection in this mating system (i.e., low variance in mating success among both males and females) (see Shuster and Wade 2003) (Table 12.1). However, if these monogamous species experience food limitation, natural selection might favor sexual dimorphism in body size. Small body size in males could be favored because that would make more food available for their female partners for offspring production (i.e., a form of indirect paternal care; Andersson 1994, Shuster and Wade 2003). As a result, monogamous symbiotic crustaceans with a limited food supply (e.g., parasites) should feature monogamy characterized by dwarf males.

In agreement with our prediction, many species of symbiotic shrimps, crabs, and isopods have been reported living in male–female pairs, in relatively small and scarce hosts inhabiting tropical environments where predation risk off hosts is presumed to be high (see references in Thiel and Baeza 2001). A monogamous mating system with reduced or no dimorphism in body size has been confirmed in at least two of these species (*Alpheus immaculatus*, Knowlton 1980, Knowlton and Keller 1983; *Periclimenes ornatus*, Omori et al. 1994). In other monogamous species that live trapped within host body cavities and that strongly depend on their hosts for—most probably—limited food (e.g., a parasitic pinnotherid crab *Pinnotheres halingi*, Hamel et al. 1999), males are dwarfs, in agreement with the prediction from this model.

Host-Defense Polygyny

Host-defense polygyny should evolve when hosts are morphologically simple, of a size that supports more than two but not too many symbiotic individuals, and are relatively rare and when predation risk off hosts is high or moderate. Under these circumstances, movement among hosts is constrained, but their intermediate carrying capacity and morphological simplicity allow males to defend relatively stable aggregations of females and to efficiently exclude intruders of the same sex. That is, males have the opportunity to mate with more than a single female in a short period of time in or on a guarded host. Host guarding by males is favored because of host scarcity and high predation risk and is further enforced by the clumped distribution of females. Thus, one should expect a correlation between host size, male size, and the number of females per host, with large hosts harboring large males and numerous females and small hosts harboring small males and few females. While males are predicted to maximize their reproductive success by monopolizing host individuals and reproducing with all females present on "their" hosts, females may select large hosts guarded by large "high-quality" males in order to increase their fitness under either a "good genes" of direct-benefits scenario.

Breeding asynchrony of females should reinforce this mating pattern, but synchrony of female receptivity should make this mating system drift toward pure-search polygynandry (see below). Because males defend hosts, sexual dimorphism in body size and in structures potentially usable as weapons (chelae, gnathopods, and/or maxillipeds) should be favored by intrasexual competition. Female mate choice may additionally select for male ornaments (e.g., colorful stripes). Alternative mating tactics such as sneaking, satellite, and/or female-mimicking males could be expected in species featuring host-defense polygyny because host characteristics and predation risk promote variance in mating success among males.

Species featuring host-defense polygyny include the ascidian-dwelling amphipod *Anamixis hanseni* (Thomas and Barnard 1983) and the sponge-dwelling isopod *Paracerceis sculpta* (Shuster and Wade 1991). In these species, a single male guards (the entrance of) a relatively large host that allows the cohabitation of several females. In *P. sculpta*, information on temporal and spatial receptivity of females and on male and female reproductive behavior and sexual dimorphism is extensive. The dominant males (alpha males) are larger than females and female receptive synchrony is relatively low, supporting our predictions (Shuster and Wade 2003). However, information on host abundance and predation risk is lacking in both species. Also in *P. sculpta*, three genetically determined male morphs featuring dissimilar mating strategies coexist in the same population (Shuster and Wade 1991), supporting our prediction.

Female-Centered Polygyny

Female-centered polygyny should evolve when host body size supports a moderate number of symbiotic individuals at the same time and when host abundance and predation pressure are moderate. This combination of factors imposes constraints (albeit not severe) on host switching. Host guarding is constrained due to the morphological complexity and intermediate body size of hosts and the consequent high costs of defense. Under these circumstances, females should be randomly distributed among hosts, and at moderate densities. As a result of the moderate abundance of females within and among hosts and their dispersion pattern, guarding of females by males becomes more profitable than host guarding or pure-searching strategies (see below). Males are expected to roam within hosts and among hosts (when possible) in search of receptive females. When females are found, males should guard and protect them against other males until they become receptive, at which time they should be mated.

After mating, males should leave females in search of additional females in the same or other hosts. Females may resist copulations, and even select high-quality males, to increase their reproductive success under a good-genes scenario because moderate predation risk allows at least some host-switching behavior.

Breeding asynchrony should reinforce this mating pattern because it allows males to search, find, guard, and mate multiple females in sequence. On the other hand, if females become receptive synchronously this mating system should shift toward monogamy since there are no additional females with whom to reproduce, particularly in species where males are tied up in mate guarding activities for days or weeks. Increases in host abundance or decreases in predation risk (facilitating roaming among hosts) should make this mating system drift toward pure-search polygynandry (see below). Because males defend females, sexual dimorphism in body size and structures potentially usable as weapons should be favored by intrasexual competition. Female mate choice should additionally select for male ornaments. Alternative mating tactics may also be expected here (see Shuster and Wade 2003) (Table 12.1).

Female-centered polygyny as described here has been reported for the anemone-dwelling crab *Inachus phalangium* (Diesel 1986) and the urchin-dwelling shrimp *Athanas kominatoensis* (Nakashima 1987). The hosts inhabited by these crustaceans are intermediate in body size (relative to their symbionts) and occur at moderate abundance in shallow subtidal habitats (Diesel 1986). Unfortunately, information on the intensity of predation off hosts of both symbionts is lacking, but its occurrence should be moderate since these species inhabit subtropical or warm temperate environments where predation pressure should be neither as high as in the tropics nor as low as in cold temperate zones (Diesel 1986, Nakashima 1987). It must be noted that the environment inhabited by *I. phalangium* is also similar to that in which pure-search polygynandry is expected. Different populations or individuals in the same population of *I. phalangium* may be experiencing conditions promoting both this mating system and pure-search polygynandry at the same time. Comparative studies that explicitly address this possibility would be very interesting. Slight shifts in environmental characteristics could easily cause the individuals to switch from one strategy to another. In general, in these species, chelae are well developed as weapons (see, e.g., Nakashima 1987). In *A. kominatoensis*, small males change sex to females if they have not been successful in copulating with females early during the reproductive season, while small males of *I. phalangium* may attempt sneaking copulations with females even if these are guarded by large dominant males (Diesel 1986, Nakashima 1987). As predicted, these two species exhibit marked sexual dimorphism and alternative mating tactics.

Pure-Search Polygynandry of Mobile Females

Pure-search polygynandry should evolve when hosts are morphologically complex, large enough to support many symbionts at the same time and highly abundant and when predation risk off hosts is low. Under these circumstances, both males and females should move freely and frequently among and within hosts. Females are expected to be uniformly dispersed among hosts. Host guarding is constrained by the invasion rate and consequent high cost of defense. High host morphological complexity and large size also render their monopolization ineffective. Under these circumstances, males should maximize their reproductive success by roaming within and among hosts in search of females. As soon as receptive females are found, males should mate and abandon them immediately to continue searching for other receptive females. On the other hand, females may reject male advances and attempt to choose and mate high-quality males among those visiting them. Multiple mating by females may occur due to female choice or male coercion. Female reproductive concealment (if allowed by female physiology) may evolve as a response to male coercive tactics.

Low breeding synchrony among females should reinforce this mating system while increased breeding synchrony should drive this mating system toward female-centered polygyny, with males guarding females after finding them (see above). Sexual dimorphism in body size should be found in species featuring this mating system. Because males do not invest in defense of females against other males, sexual selection should not favor the development of weapons. Small body size in males should be favored because that leads to an increase in agility and in the encounter rate with potential mating partners and because small body size renders them less conspicuous to predators (see chapter 7).

No studies on the mating system of symbiotic crustaceans inhabiting abundant, large, morphologically complex hosts have been reported, where pure-search polygynandry of mobile females should occur. However, observations on the porcellanid crab *Liopetrolisthes mitra* that inhabit an abundant, large, and morphologically complex sea urchin suggest that this species may feature this mating system (Baeza and Thiel 2000, Thiel et al. 2003; Fig. 12.1). This and other symbiotic crustaceans may perceive abundant, large, and morphologically complex hosts not as individuals but rather as patchy microhabitats.

Pure-Search Polygynandry of Sedentary Females

Pure-search polygynandry of sedentary females should evolve when hosts are extremely small (supporting no more than a single symbiont) and (1) when host abundance is low/moderate and predation risk off hosts is high/moderate, or alternatively, (2) when predation risk is low while host abundance is high/moderate (Fig. 12.4). Under these circumstances, host switching is constrained, but not severely, with both males and females moving among hosts infrequently. Host guarding is favored in males and females. Host monopolization is efficient regardless of their morphological complexity because of small host body size. This combination of factors results in females being uniformly dispersed among hosts at low abundance (i.e., only one female per host) that makes host guarding by males less profitable than roaming in search for females. Therefore, males maximize their reproductive success by infrequently roaming among hosts in search of receptive females. As soon as a female is found, the male should mate and abandon her immediately to continue searching for other receptive females. In turn, females should choose to mate with high-quality males among those visiting them. Sperm limitation may occur under these circumstances, and multiple mating by females may be favored. This mating system may drift toward monogamy with increases in predation pressure or increases in host size. As in the previous mating system, sexual dimorphism in body size should be found in these species, with males being smaller than females.

Species such as the urchin-dwelling brachyuran crab *Zebrida adamsii* (Yanagisawa and Hamaishi 1986) and various species of parasitic pea-crabs (e.g., *Pinnotheres pisum*,

Haines et al. 1994; *Zaops ostreum*, Christensen and McDermott 1958) have been reported living as solitary individuals in/on small hosts that form relatively dense aggregations in temperate or subtropical temperate environments, where predation risk off hosts is moderate. In agreement with our predictions, studies suggest that males roam among host in search of sedentary females that appear to be mated as soon as they are found. After mating, males do not appear to guard females but leave them immediately to continue searching for others (Christensen and McDermott 1958, Yanagisawa and Hamaishi 1986). Males in all these species are significantly smaller (in body size) than females (Christensen and McDermott 1958, Yanagisawa and Hamaishi 1986).

Discussion

Our model predicts different mating systems in symbiotic crustaceans depending on the interaction among host characteristics and predation risk. These mating systems are determined largely by how males compete for females. In this sense, this model is similar to others previously proposed (Jarman 1974, Bradbury and Vehrencamp 1977, Emlen and Oring 1977). On the other hand, this model differs from previous ones in its inclusion of the behavior of females when predicting particular mating systems. Females were assumed to optimize their reproductive success by choosing a male with "good genes" and/or "good resources." Several recent studies have shown that, in various species, females do not appear to maximize their fitness by choosing a single "high-quality" mate, but rather by being polyandrous (see Zeh and Zeh 2003). Females are assumed to acquire genetic benefits by mating with several males per reproductive event, either by biasing paternity toward males with good genes or by increasing the genetic diversity of their offspring (Andersson 1994, Jennions and Petrie 2000). Whether polyandry has a selective advantage or is simply the result of male coercive tactics is at present a topic of heated debate in behavioral ecology (Zeh and Zeh 2003). The manner in which females maximize their fitness, that is, by mate choice or by preference for polyandry, should be considered more thoroughly in future theoretical and empirical studies of mating systems of symbiotic crustaceans. Another important difference between this model and previous ones is that the risk of predation was explicitly considered as a major determinant of the behavior of these small and vulnerable symbiotic crustaceans. Predation pressure has rarely been considered as a parameter of relevance in previous classical models even though its impact on life history traits in many groups of organisms is widely recognized (see chapters 7, 10). Finally, our model relies on the idea of "economic defensibility of resources" (hosts) to predict the strategy played by individuals to acquire mating partners. This concept is pivotal to most previously proposed models too (Emlen and Oring 1977).

Overall, the limited available empirical studies support the association we predicted among mating system, host characteristics, and risk of predation (Table 12.1). Limited space does not allow discussing a large number of less detailed studies that also support the predictions of our model (see Thiel and Baeza 2001). Nonetheless, we propose that an attempt to falsify this model should be considered the next step in order to improve our understanding of mating systems in symbiotic crustaceans. Here, two different approaches are proposed: (1) manipulative experimental and (2) comparative.

The manipulative experimental approach exploits the notion that different mating systems are variations of a single theme, as indicated above. This means that different populations of the same species or different species with a common ancestor should feature different mating systems whenever individuals in each population or species are experiencing dissimilar ecological conditions. Shifts from one to another mating system should depend not only on environmental conditions but also on how flexible a species is in terms of behavior, physiology, and additional anatomical attributes (e.g., presence or absence of weapons in males, sperm receptacles in females). For instance, it should be relatively easy for species such as *I. phalangium* to switch from female-centered polygyny to pure-search polygynandry with slight shifts in host abundance or predation pressure that facilitates roaming among hosts. Because most of the changes that males and females require to attain their new optimal mating strategy are behavioral, this shift should be easily accomplished. Overall, it should be possible to manipulate specific environmental characteristics one at a time (i.e., host abundance, distribution) in the field or in the laboratory and examine whether or not the behavioral strategies of males and females shift according to the predictions of this model. In this respect, some observational studies agree with the assumptions and predictions here raised, although manipulative experiments are lacking. For instance, the shrimp *Alpheus immaculatus* features a monogamous mating system in environments where the risk of predation off hosts is relatively high (Knowlton 1980). In contrast, in its sibling species *A. armatus*, which inhabits environments with a lower risk of predation off hosts, males roam more frequently among host individuals in search of extrapair copulations (see Knowlton 1980, Knowlton and Keller 1983). In general, decreased risk of predation appears to result in an increase in host-switching behavior and the rate of extrapair copulation by males, and a subsequent shift from monogamy to polygyny, as predicted by this model. Another example is the porcelain crab *Allopetrolisthes spinifrons*, in which both males and females shift more frequently among hosts in the intertidal than in the subtidal zone. The risk of predation appears to be higher and hosts are less abundant in the subtidal than in the intertidal zone (Thiel et al. 2003).

The second, comparative, approach examines whether particular ecological conditions and behavioral traits are related in a group of closely related species. However, because phylogenetic relationships are known to bias the strength of environment–trait correlations, this approach needs to take the phylogenetic relationship among the studied species into consideration (chapter 3). Although various recent studies have elucidated the natural relationships of some groups of symbiotic crustaceans (e.g., Morrison et al. 2004) and have described in detail the mating systems of some others (e.g., Knowlton 1980), the information on a particular group is far from complete to test this model within a comparative framework (for comparative approaches in other groups of arthropods, viz., insects and arachnids, see Choe and Crespi 1997).

An interesting feature of our model is that it focuses on organisms using discrete (i.e., small and discontinuous) refuges (i.e., hosts) as habitat. The fact that these crustaceans depend heavily on their hosts means that these hosts strongly affect the spatial distribution of their associates. In particular, the relative body size of hosts, as well as their abundance and morphological complexity, is proposed to affect the way males search for and monopolize females and the mate choice behavior of females (Knowlton 1980). Not only symbiotic crustaceans but also various other groups of marine and terrestrial vertebrates and invertebrates inhabit discrete refuges, and predator avoidance is an important issue during their lifetime. Examples include parasitic helminths, parasitoid insects, litter-associated amphibians, insects, and arachnids in temperate and tropical forests, as well as several other arthropods associated with plants and algae in the terrestrial and marine environment, respectively (see Price 1980, Godfray 1994). The predictions of our model should also apply to them. In agreement with this model's predictions, monogamy has been reported for symbiotic fish inhabiting refuges with characteristics similar to those of hosts whose crustacean associates feature monogamy (i.e., various clownfish species; Hirose 1995). Also, host-resource polygyny has been described for tree-dwelling lizards where the tree represents a discrete host, large enough to allow the cohabitation of several females but small enough to allow its monopolization by a single dominant male (Manzur and Fuentes 1979). Pure-search polygynandry of mobile females appears to be common in "free-living" shrimps associated with seagrasses and seaweeds (e.g., *Palaeomonetes pugio*, Bauer and Abdalla 2001). These seagrasses and seaweeds constitute habitats with characteristics similar to that of large, abundant, morphologically heterogeneous hosts. At last, pure-search polygynandry also appears to occur in various minute insect parasitoids of animal eggs or plants with a relatively large body size (i.e., various species of fig wasps; Godfray 1994). Limited space precludes discussing many other examples that support the predictions of this model. Whether or not the predictions of this model apply to groups of organisms other than symbiotic crustaceans that inhabit discrete habitats remains to be explored experimentally.

Future Directions

In the model here introduced, various other conditions not necessarily related to the "symbiotic environment" but known to affect animal breeding systems were not included. For instance, female reproductive biology was not addressed here, even though this trait is now recognized as setting the stage for the evolution of male reproductive strategies, because it largely determines the reproductive success of males (Eberhard 1996; see also chapter 9). Similarly, the effects of other important life history traits such as the existence of direct development were not included, even though direct development is recognized as a preadaptation for the evolution of extended parental care (chapters 14, 16) and advanced social behaviors (including eusociality; Duffy 1996; see also chapter 18). In general, by considering the elements above and others in future, more sophisticated versions of the present model, it should be possible to increase the diversity of and accuracy with which the mating systems of symbiotic crustaceans are predicted. Phylogenetic constraints should also be taken into consideration because they may limit the diversity of mating strategies in a particular group of species (see chapter 11). For instance, the fact that sperm storage structures are absent in caridean shrimps (Knowlton 1980) but present in many crabs (e.g., Diesel 1986) indicates that sperm competition and first or last male precedence may be important influences on the mating systems of crustaceans (see chapter 9).

Summary

We propose a conceptual model predicting the mating system of symbiotic crustaceans. It assumes that males and females have different optimal mating strategies that they attempt to attain by defending and moving between hosts, and that host characteristics and predation risk limit the behavioral options of symbiotic individuals. Males are assumed to maximize their reproductive success by mating with as many females as possible and females by choosing a male of high quality among those available. Males and females attempt to maximize their reproductive success mostly by modifying two behavioral traits: host guarding and host switching. Predation risk, host abundance, relative size, and morphological complexity are assumed to affect monopolization of and movement among hosts by symbionts, thereby imposing constraints on their mating strategy. Five mating systems are predicted: (1) monogamy when predation risk off hosts is high and hosts are scarce, morphologically simple, and small in body size; (2) host-defense polygyny when predation risk is high and hosts are relatively scarce, morphologically simple, and intermediate in body size; (3) pure-search polygynandry of mobile females when predation risk is low and hosts are abundant, morphologically complex, and large in body size; (4) pure-search polygynandry of sedentary females when predation risk is moderate to high and hosts are moderately scarce and extremely small in body size; and (5) female-centered polygyny when predation risk is moderate and hosts are neither abundant nor scarce and intermediate in body size. Limited empirical evidence available for symbiotic crustaceans appears largely in agreement with the model's predictions.

Acknowledgments We deeply appreciate the comments of two anonymous reviewers and J. Emmett Duffy that substantially improved previous versions of the manuscript.

References

Andersson, M. 1994. Sexual selection. Princeton University Press, Princeton, N.J.

Baeza, J.A., and M. Thiel. 2000. Host use pattern and life history of *Liopetrolisthes mitra*, a crab associate of the black sea urchin *Tetrapygus niger*. Journal of the Marine Biological Association of the United Kingdom 80:639–645.

Baeza, J.A., and M. Thiel. 2003. Predicting territorial behavior in symbiotic crabs using host characteristics: a comparative study and proposal of a model. Marine Biology 142:93–100.

Bateman, A.J. 1948. Intra-sexual selection in *Drosophila*. Heredity 2:349–368.

Bauer, R.T. 2004. Remarkable shrimps. Oklahoma University Press, Norman, Okla.

Bauer, R.T., and J.H. Abdalla. 2001. Male mating tactics in the shrimp *Palaemonetes pugio* (Decapoda, Caridea): precopulatory mate guarding vs. pure search. Ethology 107:185–199.

Bradbury, J.W., and S.L. Vehrencamp. 1977. Social organization and foraging in emballonurid bats. III. Mating systems. Behavioral Ecology and Sociobiology 2:1–17.

Choe, J.C., and B.J. Crespi, editors. 1997. The evolution of mating systems in insects and arachnids. Cambridge University Press, Cambridge.

Christensen, A.M., and J.J. McDermott. 1958. Life history and biology of the oyster crab, *Pinnotheres ostreum* Say. Biological Bulletin 114:146–179.

Christy, J.H. 1987. Competitive mating, mate choice and mating associations of brachyuran crabs. Bulletin of Marine Science 41:177–191.

Correa, C., and M. Thiel. 2003. Mating systems in caridean shrimp (Decapoda: Caridea) and their evolutionary consequences for sexual dimorphism and reproductive biology. Revista Chilena de Historia Natural 76:187–203.

Darwin, C. 1871. The descent of man, and selection in relation to sex. John Murray, London.

Diesel, R. 1986. Optimal mate searching strategy in the symbiotic spider crab *Inachus phalangium* (Decapoda). Ethology 72:311–328.

Duffy, J.E. 1996. Eusociality in a coral-reef shrimp. Nature 381:512–514.

Eberhard, W.G. 1996. Female control: sexual selection by cryptic female choice. Princeton University Press, Princeton, N.J.

Emlen, S.T., and L.W. Oring. 1977. Ecology, sexual selection and the evolution of mating systems. Science 197:215–223.

Godfray, H.C.J. 1994. Parasitoids. Princeton University Press, Princeton, N.J.

Grafen, A., and M. Ridley. 1983. A model of mate guarding. Journal of Theoretical Biology 102:549–567.

Haines, C.M.C., M. Edmunds, and A.R. Pewsey. 1994. The pea crab *Pinnotheres pisum*, (Linnaeus, 1767), and its association with the common mussel, *Mytilus edulis* (Linnaeus, 1758), in the Solent (UK). Journal of Shellfish Research 13:5–10.

Hamel, J.F., P.K.L. Ng, and A. Mercier. 1999. Life cycle of the pea crab *Pinnotheres halingi* sp. nov., an obligate symbiont of the sea cucumber *Holoturia scabra* Jaeger. Ophelia 50:149–175.

Hazlett, B.A. 1975. Ethological analyses of reproductive behavior in marine Crustacea. Pubblicazione della Statione Zoologica di Napoli, Supplement 39:677–695.

Hirose, Y. 1995. Patterns of pair formation in protandrous anemonefishes, *Amphiprion clarkii*, *A. frenatus* and *A. perideraion*, on coral reefs of Okinawa, Japan. Environmental Biology of Fishes 43:153–161.

Hunt, J., L.F. Bussière, M.D. Jennions, and R. Brooks. 2004. What is genetic quality? Trends in Ecology and Evolution 19:329–333.

Jarman, P.J. 1974. The social organization of antelope in relation to their ecology. Behavior 48:215–267.

Jennions, M.D., and M. Petrie. 2000. Why do females mate multiply? A review of the genetic benefits. Biological Reviews 75:21–64.

Jormalainen, V. 1998. Precopulatory mate guarding in crustaceans: male competitive strategy and intersexual conflict. Quarterly Review of Biology 73:275–304.

Knowlton, N. 1980. Sexual selection and dimorphism in two demes of a symbiotic, pair-bonding snapping shrimp. Evolution 34:161–173.

Knowlton, N., and B.D. Keller. 1983. A new, sibling species of snapping shrimp associated with the Caribbean sea anemone *Bartholomea annulata*. Bulletin of Marine Science 33:353–362.

Manzur, M.I., and E.R. Fuentes. 1979. Polygyny and agonistic behavior in the tree-dwelling lizard *Liolaemus tenuis* (Iguanidae). Behavioral Ecology and Sociobiology 6:23–28.

Maynard-Smith, J. 1978. Optimization theory in evolution. Annual Review of Ecology and Systematics 9:31–56.

Morrison, C.L., R. Ríos, and J.E. Duffy. 2004. Phylogenetic evidence for an ancient rapid radiation of Caribbean sponge-dwelling snapping shrimps (*Synalpheus*). Molecular Phylogenetics and Evolution 30:563–581.

Nakashima, Y. 1987. Reproductive strategies in a partially protandrous shrimp, *Athanas kominatoensis* (Decapoda, Alpheidae)—sex change as the best of a bad situation for subordinates. Journal of Ethology 5:145–159.

Omori, K., Y. Yanagisawa, and N. Hori. 1994. Life history of the caridean shrimp *Periclimenes ornatus* Bruce associated with a sea-anemone in southwest Japan. Journal of Crustacean Biology 14:132–145.

Parker, G.A. 1970. Sperm competition and its evolutionary consequences in the insects. Biological Reviews 45:525–567.

Price, P.W. 1980. Evolutionary biology of parasites. Princeton University Press, Princeton, N.J.

Ross, D.M. 1983. Symbiotic relations. Pages 163–212 *in*: L.G. Abele, editor. The biology of Crustacea, Volume 7. Academic Press, New York.

Shuster, S.M., and M.J. Wade. 1991. Equal mating success among male reproductive strategies in a marine isopod. Nature 350:608–610.

Shuster, S.M., and M.J. Wade. 2003. Mating systems and strategies. Princeton University Press, Princeton, N.J.

Thiel, M., and J.A. Baeza. 2001. Factors affecting the social behaviour of crustaceans living symbiotically with other marine invertebrates: a modelling approach. Symbiosis 30:163–190.

Thiel, M., Zander, A., and J.A. Baeza, 2003. Movements of the symbiotic crab *Liopetrolisthes mitra* between its host sea urchin *Tetrapygus niger*. Bulletin of Marine Science 72: 89–101.

Thomas, J.D., and J.L. Barnard. 1983. Transformation of the *Leucothoides* morph to the *Anamixis* morph (Amphipoda). Journal of Crustacean Biology 3:154–157.

Trivers, R.L. 1972. Parental investment and sexual selection. Pages 136–179 *in*: B. Campbell, editor. Sexual selection and the descent of man. Aldine Press, Chicago, Ill.

Wickler, W., and U. Seibt. 1981. Monogamy in Crustacea and man. Zeitschrift für Tierpsychologie 57:215–234.

Wilson, E.O. 1975. Sociobiology: the new synthesis. Belknap/Harvard University Press, Cambridge, Mass.

Yanagisawa, Y., and A. Hamaishi. 1986. Mate acquisition by a solitary crab *Zebrida adamsii*, a symbiont of the sea urchin. Journal of Ethology 4:153–162.

Zeh, J.A., and D.W. Zeh. 2003. Toward a new sexual selection paradigm: polyandry, conflict and incompatibility. Ethology 109:929–950.

PART IV

Social Systems

Comparative Sociobiology of Spiny Lobsters

Michael J. Childress

13

Understanding how animal social systems evolve remains one of the greatest challenges in evolutionary biology. Why have some species developed close and complex interactions with conspecifics, while others succeed and thrive as solitary individuals? Social aggregation may be favored whenever the benefits (reduced predation risk, increased foraging efficiency, increased reproductive opportunity, increased care of offspring, etc.) exceed the costs of aggregation (increased disease transmission, increased competition, increased aggression, inbreeding, etc.) (Allee 1931, Alexander 1974). Once living in groups, more elaborate social behaviors that serve to mediate conflicts over resources or facilitate mutually beneficial acts of cooperation are likely to evolve (Wilson 1975, Trivers 1985). However, the identification of specific environmental or life history characteristics that favor this transition from solitary to group living remains of primary interest to behavioral ecologists today (Strassmann and Queller 1989, Crespi and Choe 1997, Whitehouse and Lubin 2005). One method used to address this difficult puzzle is to compare closely related species that differ in social structure or behaviors and identify those ecological and life history characteristics most closely correlated with sociality (Avilés 1997, Caro et al. 2004). Such phylogenetic comparative analyses can examine the correlation of social systems with environmental or life history traits while accounting for the nonindependent evolutionary history of closely related species (Harvey and Pagel 1991, Martins 1996).

Crustaceans are clearly model organisms in the study of social evolution. They have extraordinary diversity in their ecology, morphology, and life history and show an equally impressive range of social and sexual systems. Crustacean social systems tend to fall into one of two categories: highly structured dominance hierarchies with intense competition for resources, or cooperative groups of individuals that more efficiently exploit resources. For example, in chapter 5, Moore describes a complex, ritualized system of communication and dominance hierarchy formation in crayfish that underlies intense competition for resources. In contrast, in chapter 15 Richardson explores the ecological constraints that may have allowed some crayfish to move toward a more cooperative social system. In addition, the proximate mechanisms associated with conflict or cooperation often involve the same sensory modalities such as olfactory communication (chapters 4, 6). So why have some crustaceans, particularly spiny lobsters, become social?

Spiny lobsters exhibit a wide range of ecological and behavioral adaptations related to their social interactions (Childress and Jury 2006). Although we have considerable knowledge about the ecology and population biology of commercially harvested spiny lobsters (Lipcius and Eggleston 2000), we still know very little about the range of behaviors that occur across the family Palinuridae. In this chapter, I review the current literature on spiny lobster social behaviors and attempt to reconstruct the evolutionary history of sociality using phylogenetic comparative methods. It is premature to say that we have enough information to fully describe the evolution of sociality in spiny lobsters, but we do have sufficient information to construct testable hypotheses regarding the role of particular life history and environmental characters that may influence lobster sociality. In this chapter, I review and summarize the literature on spiny lobster sociality, compare key life history and environmental characteristics most correlated with lobster sociality using a phylogenetic comparative approach, and identify central questions for future research on the evolution of spiny lobster sociality.

Ecology of Spiny Lobster Sociality

All spiny lobsters show some type of social behavior (Atema and Cobb 1980). At the minimum, males and females interact long enough to copulate and transfer sperm. However, most species of spiny lobsters also interact with conspecifics in other contexts, as well (Fig. 13.1). The most commonly observed social behavior is gregariousness, or attraction to conspecifics leading to nonrandom aggregation. Gregariousness is a widely observed phenomenon in spiny lobsters and occurs in nearly all species (*Jasus edwardsii*, MacDiarmid 1994; *J. lalandii*, Atkinson 2001; *Palinurus elephas*, Goñi et al. 2003; *Panulirus argus*, Herrnkind et al. 1975; *Panulirus cygnus*, Cobb 1981; *Panulirus homarus*, Berry 1971; *Panulirus interruptus*, Lindberg 1955; *Panulirus marginatus*, MacDonald et al. 1984; *Panulirus ornatus*, Trendall and Bell 1989). These observations have led to the general conclusion that spiny lobsters are social (Atema and Cobb 1980, Lipcius and Eggleston 2000) and that gregariousness is beneficial by reducing predation risk through group defense (Zimmer-Faust et al. 1985, Butler et al. 1999).

However, a closer examination of the pattern of gregariousness reveals that spiny lobsters vary in their degree of conspecific attraction and nonrandom aggregation. For example, the coral-obligate spotted spiny lobster, *Panulirus guttatus* (Fig. 13.1a), is observed to occupy individual holes inside large coral crevices but is rarely observed to share shelters like the Caribbean spiny lobster, *P. argus* (Fig. 13.1b) (Hunt et al. 1991,

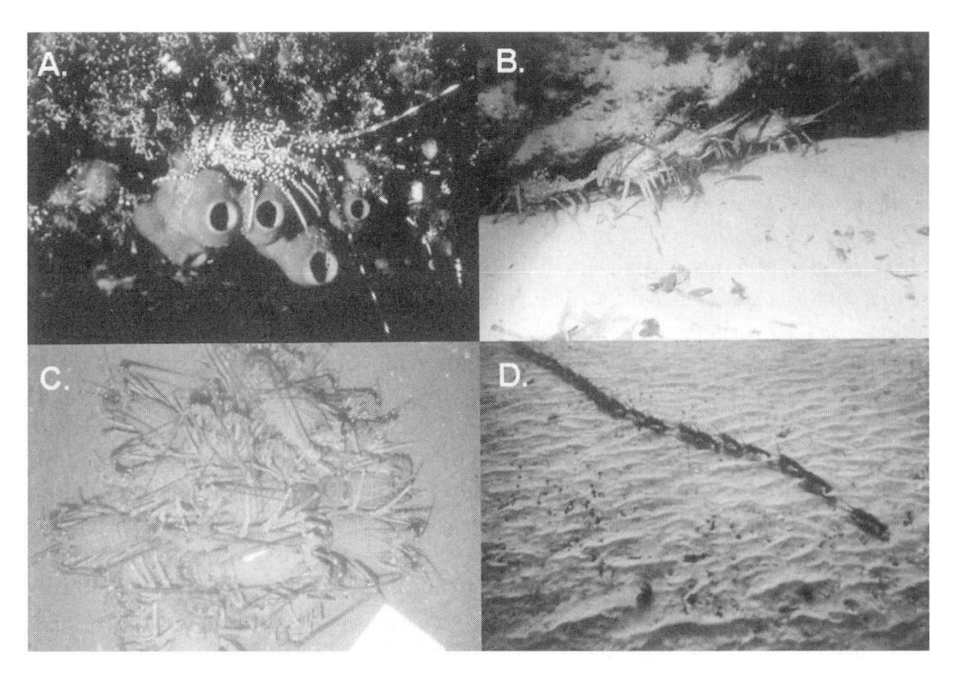

Figure 13.1 Social and asocial behaviors of spiny lobsters. (A) Foraging by the reef-obligate asocial spiny lobster *Panulirus guttatus*. Photo by Bill Herrnkind. (B) Den sharing by social spiny lobster *Panulirus argus*. Photo by Bill Herrnkind. (C) Aggregation in the open by the social spiny lobster *Jasus edwardsii*. Photo by Shane Kelly. (D) Single-file mass migration in queue formation by the social spiny lobster *Panulirus argus*. Photo by Eric Freshee.

Sharp et al. 1997). This subtle difference in the expression of den sharing behavior compels us to look closely at the characteristics of lobster aggregations and both the proximate and ultimate causation of gregariousness.

Phylogenetic History

Spiny lobsters are members of the family Palinuridae and share a number of morphological and life history characteristics (Holthuis 1991). Spiny lobsters have long, robust first antennae usually armed with many sharp spines. They also possess two forward-curved spines, or horns, above the eyes. But unlike the clawed lobsters of the family Nephropidae, spiny lobsters lack an enlarged claw on their first periopods (walking legs) and have a very different range of social interactions (Childress and Jury 2006).

Spiny lobsters can be subdivided into two major clades, the Stridentes that have a stridulatory organ at the base of their antennae, and the Silentes that lack this structure (George and Main 1967, Patek 2002). Recent analysis of mitochondrial and nuclear gene sequences support this subdivision (Ptacek et al. 2001) and suggests that furry (coral) lobsters of the family Synaxidae (genus *Palinurellus*) may be members of the Silentes clade (Patek and Oakley 2003).

Life History

Spiny lobsters share a long-lived, predatory phyllosome larva and a short-lived, nonfeeding puerulus postlarval stage (Phillips and Sastry 1980). The phyllosome larva is positively phototactic and may travel in surface currents for 6–12 months (Booth and Phillips 1994). The puerulus postlarva is an active swimmer and, through a combination of passive transport on tidal currents and active choice, seeks an appropriate settlement substrate (Butler and Herrnkind 2000).

Spiny lobsters are generalist omnivores that grow rapidly while feeding on a wide range of invertebrates and algae (Phillips et al. 1980). Growth from settlement to sexual maturity can occur in less than two years for warm water species, but may take as long as 6–10 years for large-bodied, cold-water species (Aiken 1980). Spiny lobster diets change throughout ontogeny, because larger animals usually undergo ontogenetic shifts in habitat (Lipcius and Eggleston 2000).

Spiny lobsters have a modified form of external fertilization where males deposit a sticky spermatophore on the ventral surface of the female during mating (Kittaka and MacDiarmid 1994). The female then scratches open the sperm packet when she is ready to spawn anytime from 1 to 30 days later, depending on the species. The eggs and sperm are mixed together by the female, and the mass of fertilized eggs is attached to her pleopods (abdominal swimmerets). The eggs hatch after several weeks of development and the first stage phyllosomes are usually released in areas of high current flow (MacDiarmid and Kittaka 2000).

Shelter Sharing

A number of studies have examined individual preferences for shelter characteristics. For example, Spanier and Zimmer-Faust (1988) found that shaded cover was more important than the presence of shelter walls for *P. interruptus*. Individual shelter preferences may also change with ontogeny or with increasing risk of predation. For example, *P. argus* shift from small shelters to large shelters in the presence of conspecifics and/or predatory nurse sharks (Eggleston and Lipcius 1992) or octopus (Berger and Butler 2001). Even those spiny lobsters species that share shelters often use different parts of the shelter. For example, when *P. guttatus* and *P. argus* are found in the same shelter, *P. guttatus* is usually upside down clinging to the ceiling in the darkest and most recessed crevice, while *P. argus* usually occurs in groups resting on the floor and crowding the shelter entrance (Sharp et al. 1997, Lozano-Álvarez and Briones-Fourzán 2001). These observations suggest that shelter selection is not random and may be adaptive if the choice of the right shelter increases an individual's probability of survival.

Shelters provide an important diurnal resting site out of the view of day-active predators (Fig. 13.1b). Many studies have found that crevice shelters reduce the risk of predation on juvenile lobsters (Eggleston et al. 1990, Smith and Herrnkind 1992). For some size classes of lobsters (>40 mm carapace length [CL]), the presence of conspecifics further reduces the risk of predation in comparison to sheltering alone (*J. edwardsii*, Butler et al. 1999; *P. argus*, Mintz et al. 1994). But for smaller lobsters (<40 mm CL), conspecifics do not influence the relative risk of predation (*J. edwardsii*, Butler et al. 1999; *P. argus*, Childress and Herrnkind 2001a). These observations suggest that there are potentially other benefits that could explain shelter sharing by early benthic juvenile lobsters.

To examine these alternative benefits of shelter sharing, it is important to first consider the proximate causation of spiny lobster shelter sharing. Most spiny lobsters are attracted to shelters containing the odors of conspecifics (*J. edwardsii*, Butler et al. 1999; *P. argus*, Nevitt et al. 2000; *P. interruptus*, Zimmer-Faust and Spanier 1987; *P. guttatus*, Briones-Fourzán and Lozano-Álvarez 2005). This attraction to conspecific cues may or may not lead to gregarious den sharing. For example, *P. argus* regularly shares crevice shelters with conspecifics (Berrill 1975, Herrnkind et al. 1975), but *P. guttatus* does not (Sharp et al. 1997). The gregarious behavior of *P. argus* has been exploited by the Florida Keys commercial fishery, where juvenile lobsters in traps significantly increase the catch of adults (Heatwole et al. 1988). This observation also suggests that lobsters might benefit from attraction to conspecific odors by reducing the time required to find a high-quality crevice shelter (the guide effect; Childress 1995). When captive lobsters were allowed to search for a single crevice shelter in a large mesocosm, individuals were able to find it three times faster when a conspecific was already present in the shelter than when the shelter was empty (Childress and Herrnkind 2001a).

These ontogenetic changes in the benefit of conspecific attraction and den sharing may also reveal insights into the role of habitat complexity in the evolution of sociality. Most newly settled spiny lobsters are found alone in small holes in the substratum or in attached vegetation such as algae, seagrass, surfgrass, or mangrove roots (Butler and Herrnkind 2000). In the Caribbean spiny lobster, *P. argus*, early benthic juveniles (<15 mm CL) are not attracted to the odor cues of similarly sized conspecifics (Ratchford and Eggleston 1998), nor do they influence the activity patterns of similarly sized conspecifics (Childress and Herrnkind 1996). However, they are attracted to the odors of larger conspecifics or the equivalent mass of small conspecifics

(Ratchford and Eggleston 1998). The functional onset of conspecific attraction and increased locomotory behavior is concomitant with the ontogenetic habitat shift from vegetation to crevice sheltering (~15 mm CL; Marx and Herrnkind 1985, Childress and Herrnkind 1997). Early benthic juveniles in the presence of conspecifics make this ontogenetic habitat shift at a smaller size (~13 mm CL) than do individuals reared in the absence of conspecifics (Childress and Herrnkind 2001b). Thus, the early onset of conspecific attraction in *P. argus* might be beneficial through a reduction in search time for a crevice shelter at the size of initial habitat transition.

A similar pattern of ontogenetic change in sociality is observed in *J. edwardsii*, although the size at transition is much larger. Early benthic juveniles (<20 mm CL) rarely share shelters (Booth 2001, Booth and Ayers 2005), but shelter sharing increases with body size to a peak of nearly 100% cohabitation at 50 millimeters CL (MacDiarmid 1994). Butler et al. (1999) examined the proximate and ultimate causation of this pattern by testing for attraction to conspecific odor and visual cues in a three-way shelter choice test. They found that only juveniles >40 millimeters CL significantly preferred shelters containing the odors of conspecifics. These lobsters were also the smallest size class to benefit from a reduced rate of predation when tethered in shelters with conspecifics. These results suggest that attraction to conspecific odor cues and subsequent shelter sharing may significantly reduce predation risk through either the dilution effect or cooperative group defense.

The tendency to share shelters continues to change with ontogeny, sex, season, and reproductive condition. For example, adult spiny lobsters are generally less likely to share shelters than are juveniles and subadults (*J. edwardsii*, MacDiarmid 1994; *P. interruptus*, Zimmer-Faust and Spanier 1987). One notable exception to this rule is *P. ornatus*. Juvenile *P. ornatus* are usually found solitarily in crevice shelters of deep-water (14–32 m) limestone pavement, whereas adult lobsters are usually found solitarily in rubble piles or gregariously in coral crevices on the shallow (<10 m) reefs (Trendall and Bell 1989, Dennis et al. 1997, Skewes et al. 1997). Also, during the reproductive season, the tendency for mature males to share dens decreases and the sex ratio becomes female biased in shelters containing both sexes (*P. argus*, Hunt et al. 1991; *J. edwardsii*, MacDiarmid 1994). Single large males have been observed sharing dens with multiple females in *P. argus* (Kanciruk 1980), but the length of these associations is unknown. Tagged, reproductively active *J. edwardsii* males and females were also found to change dens frequently during the reproductive season (MacDiarmid et al. 1991).

Homing and Nomadism

Spiny lobsters have a remarkable homing ability and often return to the same crevice shelter, or one very close by, each morning after an evening of foraging hundreds to thousands of meters away from their den (*J. edwardsii*, MacDiarmid et al. 1991; *P. argus*, Herrnkind et al. 1975; *P. cygnus*, Cobb 1981). The proximate causation underlying this ability is likely to involve multiple cues including visual references such as rocks or coral heads, tactile references such as wave surge (Herrnkind and McLean 1971), olfactory scent plumes (Nevitt et al. 2000), and even differences in magnetic fields (Lohmann et al. 1995, Boles and Lohmann 2003). Although *P. guttatus* can find its home shelter when displaced less than 100 meters along the reef, it was unable to return to its home if displaced 500 meters away from the reef (Lozano-Álvarez et al. 2002), suggesting that this reef-obligate species has a limited home range relative to the other palinurid lobsters. The ability to return to a central place shelter allows lobsters to exploit food resources of adjacent habitats where suitable crevice shelters may be limited (*J. edwardsii*, MacDiarmid et al. 1991; *P. argus*, Cox et al. 1997; *P. cygnus*, Jernakoff et al. 1987).

This tendency to return to the same shelter (den fidelity) varies with species, sex, size, and season. For example, Lozano-Álvarez et al. (2003) estimated den fidelity of tagged *P. argus* to be 18–32% for nearshore seagrass sites and 30–47% for offshore hardbottom sites. Juvenile *P. argus* may utilize the same den for 4–10 days before moving on to a new den during the spring and summer months, but may move to new den every night during the fall (Childress and Herrnkind 1994). This increased tendency for nomadism in the fall mirrors the behavioral patterns observed in adult *P. argus* (Kanciruk and Herrnkind 1973) and corresponds to the timing of annual mass migrations to deeper water (Kanciruk and Herrnkind 1978). In the Florida Keys off the coast of the United States, juvenile *P. argus* appear to have a net movement away from their shallow-water nursery habitat toward their adult habitat of offshore coral reef over a period of 2–3 years (Gregory and Labisky 1986, Davis and Dodrill 1989).

Some lobster species show a seasonal movement cycle, moving from inshore to offshore areas and back at particular times of the year (Herrnkind 1980). *Jasus edwardsii* adult females tend to move offshore after egg laying and prior to egg hatching—August and September (MacDiarmid 1991). Aggregations of ovigerous females are occasionally observed resting in the open in tight rosette formations (Fig. 13.1c), facing outward with antennae raised (McKoy and Leachman 1982). Such coordinated formations have also been observed when queues of migrating *P. argus* were attacked by predators (Kanciruk 1980). The ultimate cause of these female aggregations is unknown but may be adaptive in reducing the risk of predation on females prior to larval release or increasing the probability of survival and dispersal of the larvae. Interestingly, *J. edwardsii* adult males also are found in similar aggregations on open substrates, but usually in different months than the females—July and December (Kelly et al. 1999). Although male and female *J. edwardsii* make seasonal movements from inshore to offshore reefs and travel up to 12 kilometers per year, they regularly return to the same patch reef they occupied the previous year (Kelly 2001). This suggests that the majority of individuals, although capable of migration for tens to hundreds of kilometers, are residential on a scale of $<10\,km^2$ rather than nomadic (Gardner et al. 2003).

Migration and Queuing

Despite a strong homing ability, a tendency to return to the same patch reef seasonally, and an ability to return to the same den daily, spiny lobsters sometimes undertake long-distance ($>10\,km$) migrations (*J. edwardsii*, McKoy 1983; *J. verreauxi*, Booth 1984; *P. argus*, Herrnkind et al. 1973; *P. cygnus*, Phillips 1983; *P. ornatus*, Moore and MacFarlane 1984). Migrations are directed locomotory movements, usually confined to a specific period of time, over relatively long distances that usually have a predictable periodicity (Herrnkind 1980). For example, *P. cygnus* has an ontogenetic migration where preadult lobsters (age 4–5 years) move offshore northwest to their spawning grounds in November and December (Phillips 1983). Other migrations such as those by *J. verreauxi* appear to be associated with movement of ovigerous females toward deeper water and/or more favorable currents for larval release and retention (Booth 1997). However, these mass migrations associated with reproduction rarely involve the entire population. For example, adult *P. ornatus* in the Torres Strait undergo a long-distance spawning migration, while individuals along the northeast coast of Queensland do not (Bell et al. 1987), and the vast majority of gravid *J. edwardsii* release their larvae within a home range of less than $10\,km^2$ (MacDiarmid 1991, Gardner et al. 2003). Reef-obligate lobsters such as *P. guttatus* have not been observed to migrate and mark–recapture studies have found no evidence of even short distance movements among adjacent patch reefs (Robertson and Butler 2003).

Other lobster migrations occur in any season and appear to be related to tracking optimal habitat conditions (Herrnkind 1980). For example, *P. argus* adults in the Bahamas migrate from shallow water to deeper water with the arrival of the first winter storms (Kanciruk and Herrnkind 1978). The proximate cause of this migration appears to be a combination of shorter length days, a sudden drop in water temperature, and an increase in turbidity (Kanciruk and Herrnkind 1973).

Perhaps the most fascinating spiny lobster social behavior is coordinated mass migration in single-file queues (Herrnkind and Cummings 1964). This peculiar head-to-tail formation (Fig. 13.1d) has been observed in both the juvenile (Berrill 1975) and adults of *P. argus* (Herrnkind 1969) and in *P. marginatus* (MacDonald et al. 1984). Queuing behavior in *P. argus* is associated with the sudden departure of large groups of individuals from shallow reefs during the annual fall migration to deeper water. During migration these lobsters will travel both night and day and expose themselves to diurnal predators while on open substrates (Herrnkind 1985). The ultimate cause of queuing probably derives from two likely benefits, energy conservation through drag reduction (Bill and Herrnkind 1976) and efficiency in mounting cooperative group defense when attacked by predators (Herrnkind et al. 2001).

Antisocial Behaviors

Spiny lobsters can be aggressive toward conspecifics when defending food, shelter, or mates (Atema and Cobb 1980). However, since they lack large prey-processing claws commonly found on many crayfish, clawed lobsters, and mantis shrimp, their aggressive interactions are much less likely to result in injury or death (see chapters 5, 6). Individuals of several species may form dominance hierarchies that determine their positions within a shelter (*J. edwardsii*, Fielder 1965; *P. argus*, Berrill 1975; *P. cygnus*, Berrill 1976). It is unknown if spiny lobsters have the ability to recognize familiar individuals whom they have previously encountered or if pairs of familiar individuals co-occur more frequently than expected by chance associations. However, a recent laboratory experiment suggested that mature male *J. edwardsii* competing for access to females do not recognize each other, because there is no difference in the duration or intensity of aggressive interactions in bouts between familiar and unfamiliar male pairs (Raethke 2005).

Evolution of Spiny Lobster Sociality

Mapping the variation of lobster social behaviors across lobster evolutionary history is not an easy task. It requires a good approximation of the phylogenetic relationships between the species in order to have confidence in our reconstructed ancestral character states (Patek and Oakley 2003). It requires simplifying assumptions that allow character states to be equally likely to be gained or lost at a rate proportional to the length of time since divergence from a common ancestor (Pagel 1999). It further assumes that closely correlated life history or ecological characteristics can be evaluated as independent evolutionary traits by first removing the nonindependence of related taxa (Harvey and Pagel 1991). Despite this potentially dangerous pyramid of assumptions, phylogenetic comparative analyses of correlated traits can lead to new hypotheses regarding which life history and ecological factors are most likely to be correlated with the development of sociality (Martins 1996).

Phylogenetic Analysis

In order to examine the potential coevolution of spiny lobster social behaviors and life history and ecological traits, I first constructed a phylogenetic hypothesis based upon a subset of taxa with sufficient behavior and life history information available to formulate a set of these characteristics for comparative analyses. The phylogeny forming the basis of these comparative analyses was constructed from an alignment of the ribosomal small subunit mitochondrial DNA gene sequences (16S rDNA) from 16 species within the family Palinuridae (Ptacek et al. 2001, Patek and Oakley 2003). The deep-water genera *Projasus*, *Linuparus*, *Puerulus*, and *Palinustus* were excluded from the phylogenetic analysis due to a lack of information on behavior or life history traits for these taxa. The slipper lobsters *Scyllarides nodifer* and *Scyllarus arctus* were used as the outgroup taxa.

I used maximum likelihood analysis to reconstruct the phylogenetic relationships among the 16 palinurid taxa. After selecting the best fit model of evolution using MODELTEST 3.04 (Posada and Crandall 1998), maximum likelihood trees were constructed by performing 100 heuristic searches with random addition in PAUP* version 4.0b10 (Swofford 2002). The HKY model (Hasegawa et al. 1985) with a proportion of sites being invariable ($I = 0$) and incorporating rate variation ($\Gamma = 1.637$) was selected by hierarchical hypothesis testing of alternative models using Akaike information criterion in MODELTEST.

The maximum likelihood topology generated from the 16S rDNA sequence data using the subset of taxa included in Patek and Oakley (2003) resolved the same relationships among ingroup taxa as their previous analysis (Fig. 13.2). Branch lengths from this tree (the number of nucleotide substitutions per site) were used as an estimate of time since divergence in reconstructing the ancestral character states and their correlation to ecological and life history characters, using the computer program DISCRETE (Pagel 1994, 1999).

Ancestral Character State Reconstruction

In order to examine the phylogenetic pattern of spiny lobster sociality, I used the presence of social behaviors such as conspicuous den sharing (more than three

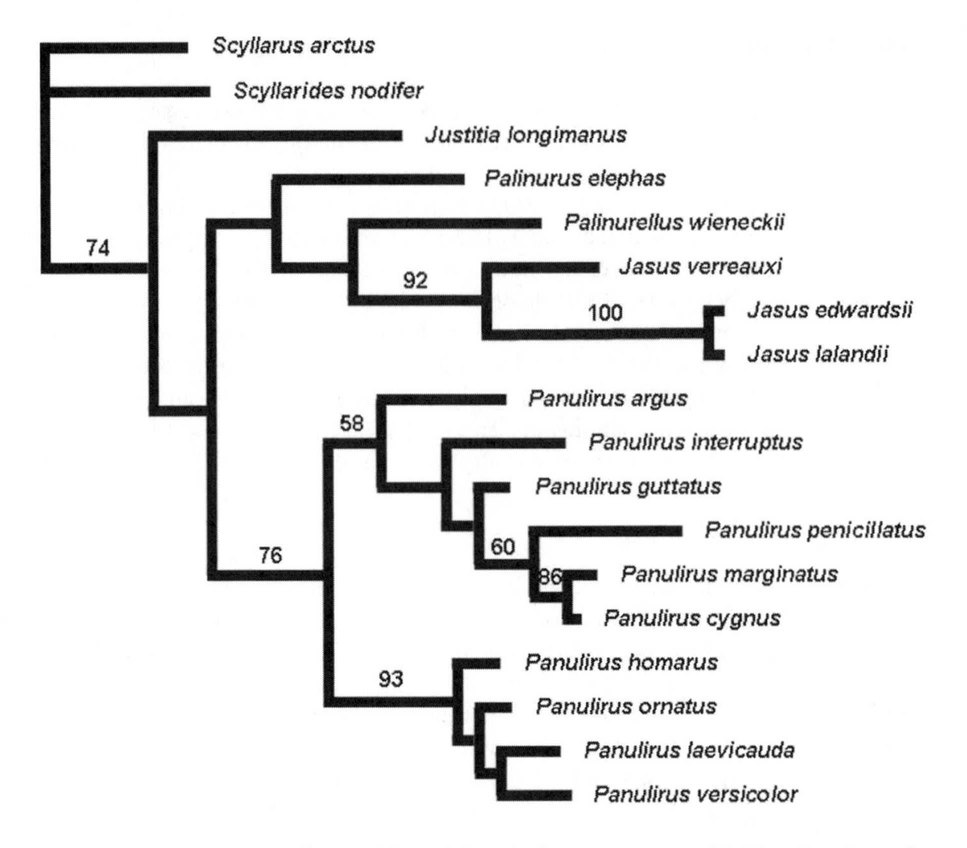

Figure 13.2 Consensus topology and branch lengths from a maximum likelihood analysis of spiny lobster 16S rDNA gene sequences. Sequence data and alignment of taxa are from Patek and Oakley (2003). Values at the nodes represent the bootstrap percentages for 500 replicates, and nodes without any values had less than 50% bootstrap support.

individuals) (G), long-range homing ability (>500 m) (H), semiperiodic nomadism (>1 km) (N), long-distance migrations (>10 km) (M), and coordinated queuing (single-file locomotion with physical contact) (Q) as characteristics of social lobsters, and the absence of such behaviors as characteristics of asocial lobsters (Fig. 13.3). For some species, the lack of evidence of these characteristics may be simply due to no observations reported in the literature to date. Therefore, it would be inappropriate to use the sum of these five characteristics as a metric of sociality. Instead, I have chosen a more conservative method that simply assigns "social" to those species with sufficient observations in at least one of the five sociality criteria. "Asocial" species are those species that have not been observed to meet any of the five sociality criteria. Although it would be considerably better to rank species according to some continuous scale of sociality from asocial to social, the lack of standard methods for estimating the degree of sociality and the constraints of inferring traits from previously published observations leave no real alternative to this crude binary designation.

Using an assumption of maximum parsimony, I calculated the ancestral character state of each branch of the phylogeny using the trace character function in the

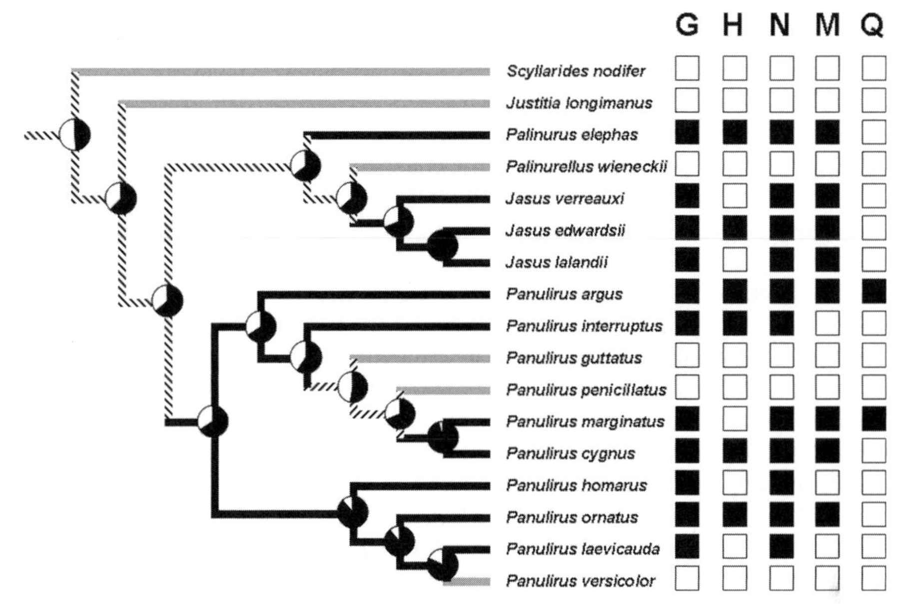

Figure 13.3 Spiny lobster behavioral traits mapped onto the phylogeny of Figure 13.2. Black boxes represent evidence of social behaviors such as gregarious den sharing (G), homing ability (H), periodic nomadism (N), long-distance migration (M), and coordinated mass migratory queuing (Q). Black bars on the branches represent social taxa, and gray bars represent asocial taxa. Hatched bars signify ancestors with equivocal sociality based on maximum parsimony character state reconstruction (MacClade). Pie charts at each node estimate the probability that the ancestor was social (black region) based on maximum likelihood character state reconstruction (DISCRETE).

program MacClade 4.02 (Maddison and Maddison 2001). Next, I estimated the probability of the ancestral character state for each node in the phylogeny using a maximum likelihood procedure incorporated in DISCRETE (Pagel 1994, 1999). This method takes into account both the character state of the terminal taxa and the genetic distance between nodes by using the maximum likelihood estimated branch lengths. This procedure attempts to ascertain if the ancestor at any given node in the tree was more or less likely to share a characteristic of the extant taxa.

The overall pattern of estimated social evolution (Fig. 13.3) suggests that the majority of spiny lobsters are social and that asocial lobsters do not make up a monophyletic group. The hypothetical ancestors of both the genera *Jasus* and *Panulirus* were most likely social (>70% probability). However, deeper than that point in the evolutionary tree, it is unclear whether the ancestor of all the palinurids was social or asocial. Likewise, it is not yet possible to conclude whether the sociality observed in *Palinurus*, *Jasus*, and *Panulirus* is from a single or multiple origins within each group. What we can determine, however, is whether social or asocial lobsters have any ecological or life history traits in common that cannot be explained by a common evolutionary ancestry.

Phylogenetic Comparative Analysis of Correlated Characters

There are dozens of potentially correlated life history and ecological characteristics that might be related to spiny lobster sociality. However, given the sparse information available for all but a handful of species, I limited myself to four characteristics that were available for all lobsters in this analysis: maximum body size, geographic range, ecological habitat type, and benthic substrate type (Holthuis 1991). For each of these continuous characters, I divided the species into five discrete bin categories (Fig. 13.4). Particular bin categories were selected for further analyses based on their level of variability between the social and asocial taxa. For example, in the comparison of sociality to benthic substrate habitat type, I chose two of the five categories most likely to be correlated, in this case mixed substrate and reef crest coral.

I used the likelihood ratio test for correlated characters in the DISCRETE software package (Pagel 1994, 1999) to examine the pattern of correlation between two binary characters using a maximum likelihood approach. It addresses the problem of nonindependence of closely related taxa by taking into consideration the relative genetic distance between taxa. Thus, a significant likelihood ratio score suggests that there is an association between sociality and a particular life history or ecological characteristic not due strictly to descent from a common ancestor.

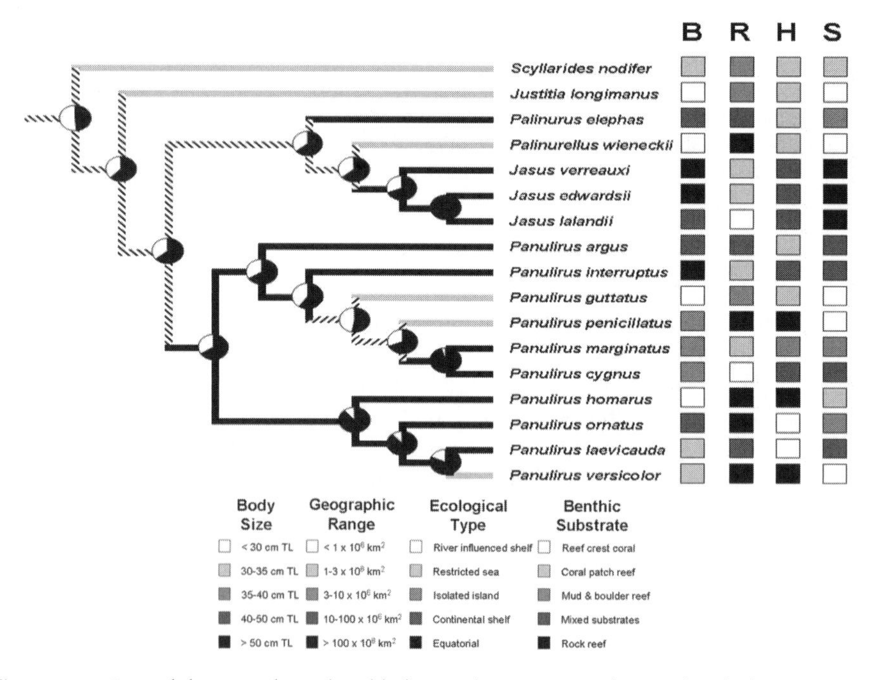

Figure 13.4 Spiny lobster ecological and behavioral traits mapped onto the phylogeny of Figure 13.2. Social taxa (black bars) and asocial taxa (gray bars) were determined as in Figure 13.3. Potentially correlated life history and environmental characters included body size in total length (B), geographic range in square kilometers (R), ecological habitat type (H), and benthic substrate habitat type (S) (see legend). Only the reef-obligate benthic substrate habitat type was significantly correlated (negatively) with lobster sociality.

Body Size

Spiny lobsters have a broad range of body sizes and indeterminate growth (Fig. 13.4). The maximum body total length (TL, in cm) is quite variable from less than 30 centimeters for *Justitia longimanus*, *Palinurellus wieneckii*, *Panulirus guttatus* and *Panulirus homarus* to more than 50 centimeters for *Jasus edwardsii*, *Jasus verreauxi*, and *Panulirus interruptus*. Neither small body size ($G = 3.337$, $P > 0.500$) nor large body size ($G = 2.984$, $P > 0.500$) was significantly correlated with lobster sociality.

Geographic Range

Some spiny lobsters are only found on a single island, while others are found in three oceans (Fig. 13.4). Some lobsters have tiny geographic ranges less than 1×10^6 km^2, such as *Jasus lalandii* and *Panulirus cygnus*, while others have extensive ranges greater than 1×10^8 km^2, such as *Panulirus homarus*, *Panulirus ornatus*, *Panulirus penicillatus*, and *Panulirus versicolor*. Neither small geographic range ($G = 2.023$, $P > 0.500$) nor large geographic range ($G = 0.852$, $P > 0.900$) was significantly correlated with lobster sociality.

Ecological Habitat Type

George (1997) developed a categorical variable of ecological habitat type to examine how habitat differences might be related to speciation in the genera *Jasus* and *Panulirus*. His five ecological habitat types were (1) river-influenced shelf, (2) restricted sea, (3) isolated island and seamount, (4) continental shelf fronting open ocean, and (5) widespread equatorial (Fig. 13.4). Neither the restricted sea habitat type ($G = 4.291$, $P > 0.100$) nor the widespread equatorial habitat type ($G = 1.791$, $P > 0.500$) was significantly correlated with lobster sociality.

Benthic Substrate Habitat Type

I used the benthic substrate habitat types listed for each lobster in Holthuis (1991) and classified them to one of five categories based on those described by George (1974): (1) rock reef, (2) mixed substrates, (3) mud and boulder reef, (4) coral patch reef, and (5) reef crest coral (Fig. 13.4). Mixed substrates were not significantly correlated with lobster sociality ($G = 3.915$, $P > 0.100$), but reef crest coral was significantly negatively correlated with lobster sociality ($G = 12.048$, $P < 0.025$).

Discussion and Conclusions

The primary conclusion of this comparative analysis of sociality is that the majority of spiny lobsters are gregarious and have homing ability and a highly variable seasonal pattern of movement. These behaviors are likely to be correlated with one another through a common proximate causation of attraction to conspecific odors (Zimmer-Faust et al. 1985). Individuals that are attracted to conspecific odors are more likely to find crevice shelters faster (Childress and Herrnkind 2001a), more likely to share shelters with conspecifics (Nevitt et al. 2000), more likely to aggregate in the open (Kelly et al. 1999), and more likely to make long-distance migrations in mass (Herrnkind 1985). These behaviors may also share a common ultimate causation by reducing an individual's risk of predation when sharing dens (Mintz et al. 1994, Butler et al. 1999), when changing habitats (Childress and Herrnkind 2001b), and when attacked during migration (Herrnkind et al. 2001).

The phylogenetic pattern of spiny lobster behavioral complexity, from simple aggregation in a den (e.g., *P. homarus*) to coordinated single-file mass migrations (e.g., *P. argus*), does not reflect a consistent pattern of increasing complexity across the phylogeny and, as in fiddler crabs (Sturmbauer et al. 1996) seems to suggest that some crown taxa (e.g., *P. versicolor*) may have lost complex social behaviors present in distant common ancestors. Furthermore, some highly social species (e.g., *J. edwardsii*) have low levels of long-distance migration (MacDiarmid 1991, Kelly 2001, Gardner et al. 2003), while others (e.g., *J. verreauxi*) have very extensive migrations (Booth 1997).

At this time, it is not possible to conclude whether this suite of correlated behaviors evolved multiple times in different lobster groups or had a single origin in an early common ancestor. Within the genus *Panulirus*, parsimony suggests that the asocial species *P. guttatus*, *P. penicillatus*, and *P. versicolor* did evolve from a social ancestor (Fig. 13.3), whereas maximum likelihood (ML) reconstructions leave some room for ambiguity. What these asocial species share with the other asocial spiny lobsters *Justitia* and *Palinurellus* is their coral-obligate lifestyle (Colin 1978, Sharp et al. 1997). Perhaps attraction to conspecific cues and subsequent social behaviors no longer confer a significant advantage for those individuals that never venture far from the safety of their coral crevice shelter. Or perhaps the cost of sociality in terms of competition for food, shelter, and mates now exceeds the benefits of protection from predators. In either case, more research is needed to critically test the prediction that life histories with ontogenetic shifts in habitat or periodic long-distance movements are more likely to have well-developed social behaviors.

Perhaps the single biggest advantage of social aggregation in spiny lobsters is the ability to exploit so many different settlement habitats (Butler and Herrnkind 2000). Having a long-lived larval stage transported by surface currents makes it difficult to specialize on just one type of benthic substrate (George 1997, 2005). Gregariousness at an early stage allows juvenile lobsters seek out conspecifics often leading to habitats with greater structural complexity and more suitable crevice shelters. By episodic and coordinated movement between patches of suitable habitats, spiny lobsters reach their reproductive habitat having already developed into the gregarious and nomadic aggregations that are commercially exploited around the world. But what about those rarely seen lobster species, such as *P. guttatus*, that settle directly on the coral reef habitat? Has the lack of an ontogenetic habitat shift led to a decreased advantage of being social? Does living deep within the intricate crevices within a coral reef remove the need to utilize conspecifics to find suitable shelter? These are the predictions that remain to be tested regarding the evolution of spiny lobster sociality.

Spiny lobsters represent a striking example of social evolution not just in the crustaceans but among all animals. Like crayfish, clawed lobsters, and snapping shrimp, chemical communication provides the critical proximate mechanism for the elaboration of their complex behavioral interactions (see chapter 6). But it is their ontogenetic shift in habitat that creates a need for reliable cues of habitat quality regarding where to go and when to move in an unfamiliar territory. Such cues have been called "public information" since these habitat quality assessments are usually based on the presence or density of conspecifics (Danchin et al. 2004). Once aggregated in crevice shelters, the potential advantages of coordinated group defense against fish predators could evolve (Butler et al. 1999) along with mechanisms to traverse long distances in coordinated groups (Herrnkind 1985), both conserving energy and maintaining group cohesion if attacked while in the open (Herrnkind et al. 2001). It seems likely that the extraordinary social behaviors of spiny lobsters are inextricably linked together with their life history characteristics and ecology.

In contrast to spiny lobsters, data on the sponge-dwelling *Synalpheus* snapping shrimps suggest that sociality has arisen multiple times in highly specialized niches where dispersal is greatly reduced and colony defense is paramount to reproductive success (Duffy et al. 2000, 2002). This dramatic difference suggests that larval dispersal more than habitat specificity plays a key role in the evolution of crustacean social systems. The social evolution of isopods (chapter 8), terrestrial crayfish (chapter 15), and snapping shrimp (chapter 18) may have been strongly influenced by limited dispersal ability and strong kin selection. Spiny lobster social aggregation appears to confer a mutual benefit of association even among those individuals that are unrelated and unfamiliar (Mesterton-Gibbons and Dugatkin 1992).

Future Directions

Herrnkind (1980) described the general behavioral pattern of spiny lobsters: "selective of their feeding and home site regions, gregarious in their residential habitat, selective of—and for long periods attached to—certain crevices, temporally organized, and spatially oriented in their movements to and from their dens." This summary is just as appropriate today as it was 25 years ago, but we have made significant progress in teasing apart both the causative mechanisms and adaptive significance of these behaviors. This literature review and the accompanying phylogenetic comparative analysis of spiny lobsters sociality raise more questions than they answer regarding the evolution of lobster behaviors. They do, however, generate a testable hypothesis regarding the relationship between benthic substrate habitat type and the probable loss or gain of sociality. This leads to some important future research questions.

Was the ancestral spiny lobster social? The analysis of ancestral character states was inconclusive on this point and could potentially be resolved with the addition of more taxa and, especially, more gene sequences or other phylogenetically informative characters to resolve the palinurid family tree. Since morphology and genes often produce different tree topologies (e.g., Patek and Oakley 2003), future studies should compare how differences in tree topology influence the conclusions of these phylogenetic comparative analyses.

Are deep-water spiny lobsters really asocial? More observations are needed to address this critical question. Many other deep-sea invertebrates show a highly aggregated distribution due to extreme habitat patchiness (Griffin and Stoddart 1995). It would not be surprising to find that deep-water lobsters are as social as shallow water lobsters.

Are coral-obligate spiny lobsters really asocial? Recent studies have revealed important social behaviors in the spotted spiny lobster, *P. guttatus*. This supposedly asocial species is sometimes attracted to conspecific odors (Briones-Fourzán and Lozano-Álvarez 2005) and shows a limited homing ability (Lozano-Álvarez et al. 2002) and low levels of aggression when sharing dens in captivity (Lozano-Álvarez and Briones-Fourzán 2001). Perhaps a better measure of sociality could be developed to estimate sociality on a continuous scale from 0 (asocial) to 1 (social). This might produce a very different pattern of correlation with the ecological and life history characteristics analyzed here.

Which came first, the coral-obligate life style or the loss of sociality? Field observations are needed to confirm the social status of the other two putative asocial species *P. versicolor* and *P. pencillatus*. The latter species has the broadest geographic range of any of the spiny lobster species and occurs across the broadest range of substrate types (George 1974). If there is a strong population level correlation between habitat type and degree of sociality in *P. penicillatus*, this would strongly support the ontogenetic habitat shift hypothesis.

Are there morphological differences between social and asocial species? This exciting question has been addressed by Bouwma (2006). The asocial *P. guttatus* has much thinner first antennae than does the same-sized *P. argus*. Since *P. argus* regularly use their antennae to fend off the attacks of fish predators at the entrance to a den, this may be a useful character to score the antipredatory strategies of other palinurids.

Does the need to change habitats during ontogeny facilitate cooperative social systems? The majority of spiny lobsters tend to settle in shallow-water habitats and migrate to deeper habitats as they grow larger (Butler and Herrnkind 2000). This constant need to relocate suitable crevice shelter in unfamiliar locations may favor response toward conspecific cues as a guide to appropriate shelter (Childress and Herrnkind 1997, 2001a). This ontogenetic habitat shift hypothesis may be similar to the evolution of sociality in other highly mobile species such as butterflies (Välimäki and Itämies 2003), song birds (Muller et al. 1997), sea birds (Podolsky 1990), and savannah-dwelling mammals (Caro et al. 2004) that use conspecific cues during migration and habitat selection.

Do spiny lobsters use public information? Spiny lobsters show a remarkable range of social behaviors, from gregarious den sharing to coordinated migratory queues. But what makes this system so interesting to the general field of sociobiology is their use of public information, such as the presence of conspecifics, to evaluate the location or quality of valuable resources (Danchin et al. 2004). Many animals form social aggregations by using conspecific cues for habitat selection (lizards, Stamps 1988; songbirds, Muller et al. 1997; spiders, Jeanson et al. 2004; fruit flies and crabs, Stamps et al. 2005). Spiny lobsters are perhaps a good system to study how the use of public information may favor social aggregation and set the stage for the evolution of extraordinary cooperative behaviors such as single-file mass migratory queues.

Like eusocial insects, social spiders, cooperatively breeding birds, and herds of mammals, social crustaceans have evolved from the interaction of many different evolutionary forces that favored close association with conspecifics (Whitehouse and Lubin 2005). It is our challenge now to find clever ways to untangle the potential costs and benefits and develop more generalized predictive models for the evolution of sociality.

Summary

Spiny lobsters show a wide range of social behaviors from sharing of crevice shelters to coordinated migratory queues. There exists considerable variation among species of spiny lobsters along a continuum from solitary asocial species to highly gregarious social species. Phylogenetic comparative analysis of social condition was unable to determine if the spiny lobster common ancestor was social or asocial, but suggests that several modern asocial taxa may have evolved from social ancestors. The only life history characteristic significantly correlated with sociality in spiny lobsters was the type of benthic substrate occupied. All the asocial spiny lobsters are coral-reef-obligate species that settle directly into their adult habitat and do not undergo an ontogenetic habitat transition. Those spiny lobsters that change habitats during their lifetime have strong attraction to conspecific cues, leading to sharing of crevice shelters, coordinated group defense behaviors, and even single-file migratory queues. Using conspecific cues as a form of public information to assess the quality of new habitats may be an important and widespread advantage to spiny lobsters and many other social species.

Acknowledgments I thank Candis Lollis for valuable assistance in finding, organizing, and cataloging the vast literature on spiny lobster behavior. My ideas on spiny lobster evolution were strongly influenced by lively discussions with my lobster colleagues Charles Acosta, Don Berringer, Rod Bertelesen, Pete Bouwma, Mark Butler, Lyn Cox, Dave Eggleston, Ray George, John Hunt, Steve Jury, Kari Lavalli, Tom Matthews, and Bill Sharp. I appreciated the helpful suggestions of Alison MacDiarmid, Sheila Patek, Darren Parsons, and Margaret Ptacek on earlier drafts of the manuscript. I am particularly grateful to Margaret Ptacek for the expert advice regarding the phylogenetic analysis, ancestral character state reconstructions, and analysis of correlated traits. Finally, I thank Bill Herrnkind for his steadfast enthusiasm, encouragement, and dedication to revealing the mysteries of these incredible crustaceans.

References

Aiken, D.E. 1980. Molting and growth. Pages 91–163 *in*: J.S. Cobb and B.F. Phillips, editors. The biology and management of lobsters, Volume 1. Academic Press, New York.

Alexander, R.D. 1974. The evolution of social behavior. Annual Review of Ecology and Systematics 5:325–383.

Allee, W.C. 1931. Animal aggregations: a study in general sociology. AMS Press, New York.

Atema, J., and J.S. Cobb. 1980. Social behavior. Pages 409–450 *in*: J.S. Cobb, and B.F. Phillips, editors. The biology and management of lobsters, Volume 2. Academic Press, New York.

Atkinson, L.J. 2001. Large and small scale movement patterns of the west coast rock lobster, *Jasus lalandii*. Unpublished MSc thesis, University of Cape Town, Cape Town, South Africa.

Avilés, L. 1997. Causes and consequences of cooperation and permanent-sociality in spiders. Pages 476–498 *in*: J.C. Choe and B.J. Crespi, editors. Social behavior in insects and arachnids. Cambridge University Press, Cambridge.

Bell, R.S., P.W. Channells, J.W. MacFarlane, R. Moore, and B.F. Phillips. 1987. Movements and breeding of the ornate rock lobster, *Panulirus ornatus*, in Torres Strait and on the north-east coast of Queensland. Australian Journal of Marine and Freshwater Research 38:197–210.

Berger, D.K., and M.J. Butler, IV. 2001. Octopuses influence den selection by juvenile Caribbean spiny lobster. Marine and Freshwater Research 52:1049–1053.

Berrill, M. 1975. Gregarious behavior of juveniles of the spiny lobster, *Panulirus argus* (Crustacea: Decapoda). Bulletin of Marine Science 25:515–522.

Berrill, M. 1976. Aggressive behaviour of post-puerulus larvae of the western rock lobster *Panulirus longipes* (Milne-Edwards). Australian Journal of Marine and Freshwater Research 27:83–88.

Berry, P.F. 1971. The biology of the spiny lobster *Panulirus homarus* (Linnaeus) off the east coast of southern Africa. Oceanographic Research Institute (Durban) Investigational Report 28:3–75.

Bill, R.G., and W.F. Herrnkind. 1976. Drag reduction by formation movement in spiny lobsters. Science 193:1146–1427.

Boles, L.C., and K.J. Lohmann. 2003. True navigation and magnetic maps in spiny lobsters. Nature 421:60–63.

Booth, J.D. 1984. Movements of the packhorse lobsters (*Jasus verreauxi*) tagged along the eastern coast of the North Island, New Zealand. New Zealand Journal of Marine and Freshwater Research 18:275–281.

Booth, J.D. 1997. Long-distance movements in *Jasus* spp. and their role in larval recruitment. Bulletin of Marine Science 61:111–128.

Booth, J.D. 2001. Habitat preferences and behaviour of newly settled *Jasus edwardsii* (Palinuridae). Marine and Freshwater Research 52:1055–1065.

Booth, J.D., and D. Ayers. 2005. Characterising shelter preferences in captive juvenile *Jasus edwardsii* (Palinuridae). New Zealand Journal of Marine and Freshwater Research 39:373–382.

Booth, J.D., and B.F. Phillips. 1994. Early life history of spiny lobster. Crustaceana 66:271–294.

Bouwma, P.E. 2006. Aspects of antipredation in *Panulirus argus* and *Panulirus guttatus*: behavior, morphology and ontogeny. PhD. Dissertation. Florida State University, Tallahassee, FL.

Briones-Fourzán, P., and E. Lozano-Alvarez. 2005. Seasonal variation in chemical response to conspecific scents in the spotted spiny lobster, *Panulirus guttatus*. New Zealand Journal of Marine and Freshwater Research 39:373–382.

Butler, M.J., IV, and W.F. Herrnkind. 2000. Puerulus and juvenile ecology. Pages 276–301 *in*: B.F. Phillips and K. Kittaka, editors. Spiny lobsters: fisheries and culture. Blackwell Science, Oxford.

Butler, M.J., IV, A.B. MacDiarmid, and J.D. Booth. 1999. The cause and consequence of ontogenetic changes in social aggregation in New Zealand spiny lobsters. Marine Ecology Progress Series 188:179–191.

Caro, T.M., C.M. Graham, C.J. Stoner, and J.K. Vargas. 2004. Adaptive significance of antipredatory behaviour in artiodactyls. Animal Behaviour 67:205–228.

Childress, M.J. 1995. The ontogeny and evolution of gregarious behavior in juvenile Caribbean spiny lobster, *Panulirus argus*. Unpublished Ph.D. thesis. Florida State University, Tallahassee.

Childress, M.J., and W.F. Herrnkind. 1994. The behavior of juvenile Caribbean spiny lobster in Florida Bay: seasonality, ontogeny, and sociality. Bulletin of Marine Science 54:819–827.

Childress, M.J., and W.F. Herrnkind. 1996. The ontogeny of social behaviour among juvenile Caribbean spiny lobsters. Animal Behaviour 51:675–687.

Childress, M.J., and W.F. Herrnkind. 1997. Den sharing by juvenile Caribbean spiny lobsters (*Panulirus argus*) in nursery habitat: cooperation or coincidence? Marine and Freshwater Research 48:751–758.

Childress, M.J., and W.F. Herrnkind. 2001a. The guide effect influence on the gregariousness of juvenile Caribbean spiny lobsters. Animal Behaviour 62:465–472.

Childress, M.J., and W.F. Herrnkind. 2001b. Influence of conspecifics on the ontogenetic habitat shift of juvenile Caribbean spiny lobsters. Marine Freshwater Research 52:1077–1084.

Childress, M.J., and S.H. Jury. 2006. Behavior. Pages 78–112 *in*: B.F. Phillips, editor. Lobsters: Biology, Management, Aquaculture and Fisheries. Blackwell Publishing, New York.

Cobb, J.S. 1981. Behaviour of the western Australian spiny lobster, *Panulirus cygnus* George, in the field and laboratory. Australian Journal of Marine and Freshwater Research 32:399–409.

Colin, P.I. 1978. Marine invertebrates and plants of the living reef. THF Publications, Neptune City, N.J.

Cox, C., J.H. Hunt, W.G. Lyons, and G.E. Davis. 1997. Nocturnal foraging of the Caribbean spiny lobster (*Panulirus argus*) on offshore reefs of Florida, USA. Marine and Freshwater Research 48:671–679.

Crespi, B.J., and J.C. Choe. 1997. Explanation and evolution of social systems. Pages 499–524 *in*: J.C. Choe and B.J. Crespi, editors. Social behavior in insects and arachnids. Cambridge University Press, Cambridge.

Danchin, E., L.-A. Giraldeau, T.J. Valone, and R.H. Wagner. 2004. Public information: from nosy neighbors to cultural evolution. Science 305:487–491.

Davis, G.E., and J.W. Dodrill. 1989. Recreational fishery and population dynamics of spiny lobsters, *Panulirus argus*, in Florida Bay, Everglades National Park, 1977–1980. Bulletin of Marine Science 44:78–88.

Dennis, D.M., T.D. Skewes, and C.R. Pitcher. 1997. Habitat use and growth of juvenile ornate rock lobsters, *Panulirus ornatus* (Fabricius, 1798), in Torres Strait, Australia. Marine and Freshwater Research 48:663–670.

Duffy, J.E., C.L. Morrison, and R. Rios. 2000. Multiple origins of eusociality among sponge-dwelling shrimps (*Synalpheus*). Evolution 54:503–516.

Duffy, J.E., C.L. Morrison, and K.S. Macdonald. 2002. Colony defense and behavioral differentiation in the eusocial shrimp *Synalpheus regalis*. Behavioral Ecology and Sociobiology 51:488–495.

Eggleston, D.B., and R.N. Lipcius. 1992. Shelter selection by spiny lobster under variable predation risk, social conditions, and shelter size. Ecology 73:992–1011.

Eggleston, D.B., R.N. Lipcius, D.L. Miller, and L. Coba-Cetina. 1990. Shelter scaling regulates survival of juvenile Caribbean spiny lobster *Panulirus argus*. Marine Ecology Progress Series 62:79–88.

Fielder, D.R. 1965. A dominance order for shelter in the spiny lobster *Jasus lalandii* (H. Milne-Edwards). Behaviour 24:236–245.

Gardner, C., S. Frusher, M. Haddon, and C. Buxton. 2003. Movements of the southern rock lobster *Jasus edwardsii* in Tasmania. Australian Bulletin of Marine Science 73:653–671.

George, R.W. 1974. Coral Reefs and rock lobster ecology in the Indo-West Pacific region. Pages 321–325 *in*: Proceedings of the Second International Coral Reef Symposium. Great Barrier Reef Committee, Brisbane, Queensland, Australia.

George, R.W. 1997. Tectonic plate movements and the evolution of *Jasus* and *Panulirus* spiny lobsters (Palinuridae). Marine and Freshwater Research 48:1121–1130.

George, R.W. 2005. Comparative morphology and evolution of the reproductive structures in spiny lobsters, *Panulirus*. New Zealand Journal of Marine and Freshwater Research 39:493–501.

George, R.W., and A.R. Main. 1967. The evolution of spiny lobsters (Palinuridae): a study of evolution in the marine environment. Evolution 21:803–820.

Goñi, R., A. Quetglas, and O. Reñones. 2003. Differential catchability of male and female European spiny lobster *Palinurus elephas* (Fabricius, 1787) in traps and trammelnets. Fisheries Research 65:295–307.

Gregory, D.R., Jr., and R.F. Labisky. 1986. Movements of the spiny lobster *Panulirus argus* in south Florida. Canadian Journal of Fisheries and Aquatic Sciences 43:2228–2234.

Griffin, D.J.G., and H.E. Stoddart. 1995. Deep-water decapod Crustacea from eastern Australia: lobsters of the families *Nephropidae, Palinuridae, Polychelidae* and *Scyllaridae*. Records of the Australian Museum 47:231–263.

Harvey, P.H., and M.D. Pagel. 1991. The comparative method in evolutionary biology. Oxford University Press, Oxford.

Hasegawa, M., H. Kishino, and T. Yano. 1985. Dating of the human-ape splitting by a molecular clock of mitochondrial DNA. Journal of Molecular Evolution 21:160–174.

Heatwole, D.W., J.H. Hunt, and F.S. Kennedy, Jr. 1988. Catch efficiencies of live lobster decoys and other attractants in the Florida spiny lobster fishery. Florida Marine Research Publication 44:1–15.

Herrnkind, W.F. 1969. Queuing behavior of spiny lobsters. Science 163:1425–1427.

Herrnkind, W.F. 1980. Spiny lobsters: patterns of movement. Pages 349–407 *in*: J.S. Cobb and B.F. Phillips, editors. The biology and management of lobsters, Volume 2. Academic Press, New York.

Herrnkind, W.F. 1985. Evolution and mechanisms of mass single-file migration in spiny lobster: synopsis. Contributions in Marine Science 27:197–211.

Herrnkind, W.F., and W.C. Cummings. 1964. Single file migration of the spiny lobster, *Panulirus argus* (Latreille). Bulletin of Marine Science of the Gulf and Caribbean 14:123–125.

Herrnkind, W.F., and R. McLean. 1971. Field studies of homing, mass emigration, and orientation in the spiny lobster, *Panulirus argus*. Annals of the New York Academy of Sciences 188:359–377.

Herrnkind, W.F., P. Kanciruk, J. Halusky, and R. McLean. 1973. Descriptive characterization of mass autumnal migrations of spiny lobster, *Panulirus argus*. Proceedings of the Gulf and Caribbean Fisheries Institute 25:79–98.

Herrnkind, W.F., J. Van DerWalker, and L. Barr. 1975. Population dynamics, ecology and behavior of spiny lobsters, *Panulirus argus*, of St. John, USVI: (IV) Habitation, patterns of movement and general behavior. Science Bulletin of the Museum of Natural History of Los Angeles County 20:31–45.

Herrnkind, W.F., M.J. Childress, and K.L. Lavalli. 2001. Cooperative defense and other benefits among exposed spiny lobsters: inferences from group size and behaviour. Marine and Freshwater Research 52:1113–1124.

Holthuis, L.B. 1991. FAO species catalogue, Volume 13. Marine lobsters of the world. An annotated and illustrated catalogue of species of interest to fisheries known to date. FAO Fisheries Synopsis No. 125. Food and Agriculture Organization of the United Nations, Rome, Italy.

Hunt, J.H., T.R. Matthews, D. Forcucci, B.S. Hedin, and R.D. Bertelsen. 1991. Management implications of trends in the population dynamics of the Caribbean spiny lobster, *Panulirus argus*, at Looe Key National Marine Sanctuary. NOAA Technical Report. U.S. Department of Commerce, Washington, D.C.

Jeanson, R., J.-L. Deneubourg, and G. Theraulaz. 1994. Discrete dragline attachment induces aggregation in spiderlings of a solitary species. Animal Behaviour 67:531–537.

Jernakoff, P., B.F. Phillips, and R.A. Maller. 1987. A quantitative study of nocturnal foraging distances of the western rock lobster *Panulirus cygnus* George. Journal of Experimental Marine Biology and Ecology 113:9–21.

Kanciruk, P. 1980. Ecology of juvenile and adult Palinuridae (spiny lobsters). Pages 59–96 *in*: J.S. Cobb and B.F. Phillips, editors. The biology and management of lobsters, Volume 2. Academic Press, New York.

Kanciruk, P., and W.F. Herrnkind. 1973. Preliminary investigations of the daily and seasonal locomotor activity rhythms of the spiny lobster, *Panulirus argus*. Marine Behavioral Physiology 1:351–359.

Kanciruk, P., and W.F. Herrnkind. 1978. Mass migration of spiny lobster, *Panulirus argus* (Crustacea: Palinuridae): behavior and environmental correlates. Bulletin of Marine Science 28:601–623.

Kelly, S. 2001. Temporal variation in the movement of the spiny lobster *Jasus edwardsii*. Marine and Freshwater Research 52:323–331.

Kelly, S., A.B. MacDiarmid, and R.C. Babcock. 1999. Characteristics of spiny lobster, *Jasus edwardsii*, aggregations in exposed reef and sandy areas. Marine and Freshwater Research 50:409–416.

Kittaka, J., and A.B. MacDiarmid. 1994. Breeding. Pages 384–401 *in*: B.F. Phillips, J.S. Cobb, and J. Kittaka, editors. Spiny lobster management. Blackwell Scientific, Oxford.

Lindberg, R.G. 1955. Growth, population dynamics, and field behavior in the spiny lobster, *Panulirus interruptus* (Randall). University of California Publications in Zoology 59:157–248.

Lipcius, R.N., and D.B. Eggleston. 2000. Ecology and fishery biology of spiny lobsters. Pages 1–41 *in*: B.F. Phillips and J. Kittaka, editors. Spiny lobsters: fisheries and culture. Blackwell Science, Oxford.

Lohmann, K.J., N.D. Pentcheff, G.A. Nevitt, G.D. Stetten, R.K. Zimmer-Faust, H.E. Jarrard, and L.C. Boles. 1995. Magnetic orientation of spiny lobsters in the ocean: experiments with undersea coil systems. Journal of Experimental Biology 198:2041–2048.

Lozano-Álvarez, E., and P. Briones-Fourzán. 2001. Den choice and occupation patterns of shelters by two sympatric lobster species, *Panulirus argus* and *Panulirus guttatus*, under experimental conditions. Marine and Freshwater Research 52:1145–1155.

Lozano-Álvarez, E., G. Carrasco-Zanini, and P. Briones-Fourzán. 2002. Homing and orientation in the spotted spiny lobster, *Panulirus guttatus* (Decapoda, Palinuridae), towards a subtidal coral reef habitat. Crustaceana 75:859–873.

Lozano-Álvarez, E., P. Briones-Fourzán, and M.E. Ramos-Aguilar. 2003. Distribution, shelter fidelity, and movements of subadult spiny lobsters (*Panulirus argus*) in areas with artificial shelters (Casitas). Journal of Shellfish Research 22:533–540.

MacDiarmid, A.B. 1991. Seasonal changes in depth distribution, sex ratio and size frequency of spiny lobster *Jasus edwarsii* on a coastal reef in northern New Zealand. Marine Ecology Progress Series 70:129–141.

MacDiarmid, A.B. 1994. Cohabitation in the spiny lobster *Jasus edwardsii* (Hutton, 1875). Crustaceana 66:341–355.

MacDiarmid, A.B., and J. Kittaka. 2000. Breeding. Pages 485–507 *in*: B.F. Phillips and J. Kittaka, editors. Spiny lobsters: fisheries and culture. Blackwell Science, Oxford.

MacDiarmid, A.B., B. Hickey, and R.A. Maller. 1991. Daily movement patterns of the spiny lobster *Jasus edwardsii* (Hutton) on a shallow reef in Northern New Zealand. Journal of Experimental Marine Biology and Ecology 147:185–205.

MacDonald, C.D., S.C. Jazwinski, and J.H. Prescott. 1984. Queuing behavior of the Hawaiian spiny lobster, *Panulirus marginatus*. Bulletin of Marine Science 35:111–114.

Maddison, D.R., and W.P. Maddison. 2001. MacClade 4: analysis of phylogeny and character evolution, version 4.02. Sinauer Associates, Sunderland, Mass.

Martins, E.P., editor. 1996. Phylogenies and the comparative method in animal behavior. Oxford University Press, Oxford.

Marx, J.M., and W.F. Herrnkind. 1985. Macroalgae (Rhodophyta: *Laurencia* spp.) as habitat for young juvenile spiny lobsters, *Panulirus argus*. Bulletin of Marine Science 36:423–431.

McKoy, J.L. 1983. Movements of rock lobsters, *Jasus edwardsii* (Decapoda: Palinuridae), tagged near Steward Island, New Zealand. New Zealand Journal of Marine and Freshwater Research 17:357–366.

McKoy, J.L., and A. Leachman. 1982. Aggregations of ovigerous female rock lobsters, *Jasus edwardsii* (Decapoda: Palinuridae). New Zealand Journal of Marine and Freshwater Research 16:141–146.

Mesterton-Gibbons, M., and L.A. Dugatkin. 1992. Cooperation among unrelated individuals: evolutionary factors. Quarterly Review of Biology 67:267–281.

Mintz, J.D., R.N. Lipcius, D.B. Eggleston, and M.S. Seebo. 1994. Survival of juvenile Caribbean spiny lobster: effects of shelter size, geographic location and conspecific abundance. Marine Ecology Progress Series 112:255–266.

Moore, R., and J.W. MacFarlane. 1984. Migrations of the ornate rock lobster, *Panulirus ornatus* (Fabricius), in Papua New Guinea. Australian Journal of Marine and Freshwater Research 35:187–212.

Muller, K.L., J.A. Stamps, V.V. Krishnan, and N.H. Willits. 1997. The effects of conspecific attraction and habitat quality on habitat selection in territorial birds (*Troglodytes aedon*). American Naturalist 150:650–661.

Nevitt, G.A., N.D. Pentcheff, K.J. Lohmann, and R.K. Zimmer. 2000. Den selection by the spiny lobster *Panulirus argus*: testing attraction to conspecific odors in the field. Marine Ecology Progress Series 203:225–231.

Pagel, M. 1994. Detecting correlated evolution on phylogenies: a general method for the comparative analysis of discrete characters. Proceedings of the Royal Society of London, Series B 255:37–45.

Pagel, M. 1999. The maximum likelihood approach to reconstructing ancestral character states of discrete characters on phylogenies. Systematic Biology 48:612–622.

Patek, S.N. 2002. Squeaking with a sliding joint: mechanics and motor control of sound production in palinurid lobsters. Journal of Experimental Biology 205:2375–2385.

Patek, S.N., and T.H. Oakley. 2003. Comparative tests of evolutionary trade-offs in a palinurid lobster acoustic system. Evolution 57:2082–2100.

Phillips, B.F. 1983. Migrations of pre-adult western rock lobsters, *Panulirus cygnus*, in Western Australia. Marine Biology 76:311–318.

Phillips, B.F., and A.N. Sastry. 1980. Larval ecology. Pages 11–57 *in*: J.S. Cobb, and B.F. Phillips, editors. The biology and management of lobsters, Volume 2. Academic Press, New York.

Phillips, B.F., J.S. Cobb, and R.W. George. 1980. General biology. Pages 1–82 *in*: J.S. Cobb and B.F. Phillips, editors. The biology and management of lobsters, Volume 1. Academic Press, New York.

Podolsky, R.H. 1990. Effectiveness of social stimuli in attracting Laysan albatross to a new potential nesting sites. Auk 107:119–125.

Posada, D., and K.A. Crandall. 1998. MODELTEST: testing the model of DNA substitution. Bioinformatics 14:817–818.

Ptacek, M.B., S.K. Sarver, M.J. Childress, and W.F. Herrnkind. 2001. Molecular phylogeny of the spiny lobster genus *Panulirus* (Decapoda: Palinuridae). Marine and Freshwater Research 52:1037–1047.

Raethke, N. 2005. Chemosensory communication in the rock lobster *Jasus edwardsii*. Unpublished Ph.D. thesis, University of Auckland, Auckland, New Zealand.

Ratchford, S.G., and D.B. Eggleston. 1998. Size- and scale-dependent chemical attraction contributes to an ontogenetic shift in sociality. Animal Behaviour 56:1027–1034.

Robertson, D.N., and M.J. Butler, IV. 2003. Growth and size at maturity in the spotted spiny lobster, *Panulirus guttatus*. Journal of Crustacean Biology 23:265–272.

Sharp, W.C., J.H. Hunt, and W.G. Lyons. 1997. Life history of the spotted spiny lobster, *Panulirus guttatus*, an obligate reef-dweller. Marine and Freshwater Research 48:687–698.

Skewes, T.D., D.M. Dennis, C.R. Pitcher, and B.G. Long. 1997. Age structure of *Panulirus ornatus* in two habitats in Torres Strait, Australia. Marine and Freshwater Research 48:745–750.

Smith, K.N., and W.F. Herrnkind. 1992. Predation on early juvenile spiny lobsters *Panulirus argus* (Latreille): influence of size and shelter. Journal of Experimental Marine Biology and Ecology 157:3–18.

Spanier, E., and R.K. Zimmer-Faust. 1988. Some physical properties of shelter that influence den preference in spiny lobsters. Journal of Experimental Marine Biology and Ecology 121:137–149.

Stamps, J.A. 1988. Conspecific attraction and aggregation in a territorial species. American Naturalist 131:329–347.

Stamps, J.A., R. McElreath, and P. Eason. 2005. Alternative models of conspecific attraction in flies and crabs. Behavioral Ecology 16:974–980.

Strassmann, J.E., and D.C. Queller. 1989. Ecological determinates of social evolution. Pages 81–101 *in*: M.D. Breed and R.E. Page, Jr., editors. The genetics of social evolution. Westview Press, Boulder, Colo.

Sturmbauer, C., J.S. Levinton, and J. Christy. 1996. Molecular phylogeny analysis of fiddler crabs: test of the hypothesis of increasing behavioral complexity in evolution. Proceedings of the National Academy of Sciences, USA 93:10855–10857.

Swofford, D.L. 2002. PAUP*. Phylogenetic analysis using parsimony (* and other methods), beta version 4.0b10. Sinauer Associates, Sunderland, Mass.

Trendall, J., and S. Bell. 1989. Variable patterns of den habitation by the ornate rock lobster, *Panulirus ornatus*, in the Torres Strait. Bulletin of Marine Science 45:564–573.

Trivers, R.L. 1985. Social evolution. Benjamin/Cummings Publications. Menlo Park, Calif.

Välimäki, P., and J. Itämies. 2003. Migration of the clouded Apollo butterfly *Parnassius mnemosyne* in a network of suitable habitats—effects of patch characteristics. Ecography 26:679–691.

Whitehouse, M.E.A., and Y. Lubin. 2005. The functions of societies and the evolution of group living: spider societies as a test case. Biological Reviews 80:347–361.

Wilson, E.O. 1975. Sociobiology. Belknap Press of Harvard University Press. Cambridge, Mass.

Zimmer-Faust, R.K., and E. Spanier. 1987. Gregariousness and sociality in spiny lobsters: implications for den habitation. Journal of Experimental Marine Biology and Ecology 105:57–71.

Zimmer-Faust, R.K., J.E. Tyre, and J.F. Case. 1985. Chemical attraction causing aggregation in the spiny lobster, *Panulirus interruptus* (Randall), and its probable ecological significance. Biological Bulletin 169:106–118.

Social Behavior of Parent–Offspring Groups in Crustaceans

Martin Thiel

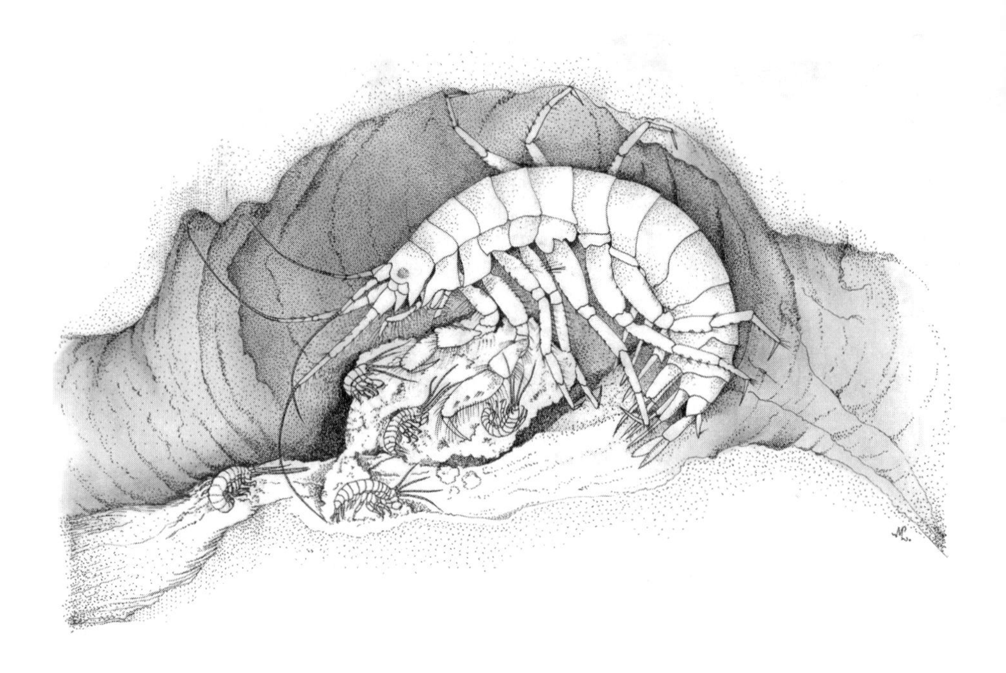

The evolutionary roads of presociality have been traveled by many kinds of arthropods besides the hymenopterans and termites. Time and again phyletic lines have pressed most of the way to eusociality, in some cases to the very threshold, and then unaccountably stopped.

—**E.O. Wilson**

14

Subsocial Behavior and Evolution of Eusociality

Parental care is considered an important evolutionary precursor to eusocial behavior (Wilson 1975). Many researchers agree that eusociality (i.e., overlapping adult generations, cooperative care for offspring, reproductive skew, and division of labor) in most species has evolved via the subsocial route (i.e., parents cohabiting with offspring) (Wilson 1971, Alexander et al. 1991). Crustaceans seem to follow this rule (see chapters 17, 18). However, while eusocial behavior has evolved multiple times in terrestrial animals (Wilson 1975, Alexander et al. 1991, Crespi and Mound 1997), it is rare in aquatic animals (Spanier et al. 1993). Study of crustaceans, which are found in marine, freshwater, and diverse terrestrial systems, may help to elucidate why eusocial behavior is more common in terrestrial than in aquatic animals.

The evolution of eusociality can be studied from two different perspectives: (1) by investigating the biology of eusocial species (e.g., Andersson 1984, Myles and Nutting 1988, Queller and Strassmann 1998), or (2) by examining the biology of subsocial species that may illuminate the biology and environmental characteristics hypothesized to select for the origin of eusociality (see, e.g., O'Connor and Shine 2003, Thorne and Traniello 2003, Field and Brace 2004). In the latter context, it can be useful to apply a decision approach (*sensu* Helms Cahan et al. 2002) and ask why these subsocial species did not evolve more advanced social behavior. Presumably, either the selective environment did not favor the evolution of eusociality, or phylogenetic constraints impeded the adaptive radiation of advanced social behaviors. In almost all metazoan taxa, some species provide parental care for their offspring. However, even though in some taxa there are many species with extended parental care (e.g., in crustaceans; Spanier et al. 1993), few if any cases of eusociality are known from these taxa (Duffy 1996, 2003). Thus, it seems clear that, in addition to extended parental care, other traits are required on the evolutionary path to eusociality.

The objective of this contribution is to synthesize our present knowledge of social behaviors in parent–offspring groups of crustaceans, using a comparative approach. Most subsocial and eusocial organisms inhabit terrestrial environments or feature particular life habits. For example, most social insects live in plant-dominated terrestrial habitats (e.g., Tallamy and Wood 1986, Crespi and Mound 1997, Schwartz et al. 2005). Similarly, advanced social evolution in subterraneous rodents is restricted to arid terrestrial environments (Bennett and Faulkes 2000). Crustaceans are a notable exception to this. Subsocial and eusocial species have been reported from diverse marine, freshwater, and terrestrial environments (Linsenmair 1987, Duffy 2003, Thiel 2003a; see also chapter 17). Here I take advantage of the diversity of crustacean habitats to compare parental care behaviors in the context of the evolution of advanced social behavior.

Extended Parental Care Behavior among Crustaceans

Parent–offspring associations have been reported from a variety of aquatic and terrestrial crustacean species (Thiel 1999a). Most crustacean species perform some form of post-zygotic parental care. Females are the principal caregivers during these early phases of parental care, and their main activities are carrying, grooming, and ventilation of the embryos (Förster and Baeza 2001, Fernández et al. 2003). In general, embryos are maintained on the mother's body during early stages, followed by release of pelagic larvae, which develop in the water column without the care of their parents. In some species, however, parental care extends beyond offspring emergence from the egg or maternal brood chamber. These family groups are most commonly found among crustaceans that inhabit confined dwellings (Thiel 2003a), in which parents and offspring may cohabit until the latter have reached subadult or even adult stages (see, e.g., chapters 16–18). However, also in some free-living species, offspring remain for relatively long time periods with parents and reach advanced juvenile stages (Aoki and Kikuchi 1991, Svavarsson and Davidsdottir 1995). During the period of extended parental care, parents may feed and groom offspring and, most important, protect them against predators, either by providing a safe dwelling (Thiel 1999b) or by actively fending off enemies. As in many other subsocial invertebrates (e.g., Park and Choe 2003), extended parental care can improve offspring growth and survival (Aoki 1997, Thiel 1999b, 2003a). Otherwise, knowledge of the behavioral interactions among family members is limited.

This contribution is based on a database of 136 crustacean species in which parent–offspring groups have been reported. These reports include species from the orders Amphipoda, Isopoda, Tanaidacea, Mysidacea, and Decapoda (including the infraorders Brachyura, Caridea, Astacoidea, Thalassinidea, and Stenopodida) (see also chapter 1, Table 1.1). Whereas previous reviews focused on the demographic characteristics of parent–offspring groups (Thiel 1999a, 2003a), the present analysis focuses on social behaviors (defending, shepherding, grooming, feeding) in these parent–offspring groups. Most such observations are anecdotal; that is, they represent brief remarks on occasional observations of interactions. Specific studies on the social behavior of crustaceans that live in parent–offspring groups are only available for few species (crayfish, Ameyaw-Akumfi 1976, Hazlett 1983, Figler et al. 1997, 2001; *Hemilepistus reaumuri*, Linsenmair 1984, 1987; *Synalpheus regalis*, Duffy 1996, Duffy et al. 2002, Tóth and Duffy 2005; *Metopaulias depressus*, Diesel 1989, 1992, Diesel and Schuh 1993). In the following, I use the term "family" to refer to a parent–offspring group that consists of one (or both) parents and their offspring from one or more clutches.

Behavioral Interactions in Crustacean Parent–Offspring Groups

Information on social interactions of family groups is available for only 40 (29%) of the 136 known crustacean species that live in parent–offspring groups (Table 14.1). These reports span several of the principal malacostracan taxa, most being from peracarids (Amphipoda, 20; Isopoda, 8; Tanaidacea, 2), and only 10 are from decapods (Astacidea, 5; Brachyura, 4; Caridea, 1). Interactions among family members have been observed or inferred for 36 crustacean species, most of which (34) refer to interactions between parents and their offspring (Table 14.1).

Female–Male Interactions

In many crustacean species, male and female parents interact only early in the reproductive cycle, that is, shortly before and after fertilization. Specific observations on

Table 14.1. Crustacean species for which social interactions among family members or between family members and non-related individuals has been reported.

	Social Interactions Reported among			
	Parent–Offspring	Both Parents	Offspring	Family–Nonkin
Amphipoda				
Caprella decipiens Mayer, 1890	+			
Caprella monoceros Mayer, 1890	+			+
Caprella scaura typica Mayer, 1890	+			+
Casco bigelowi (Blake, 1929)	+			
Corophium bonnellii (Milne Edwards, 1830)	+			
Dulichia rhabdoplastis McCloskey, 1970				+
Dyopedos monacanthus (Metzger, 1875)	+	+		+
Dyopedos porrectus (Bate, 1857)				+
Gammarus finmarchicus Dahl, 1938				+
Gammarus palustris (Bousfield, 1969)	+			
Gammarus pulex (L.)	+			
Lembos websteri Bate, 1856	+	+		
Leptocheirus pilosus Zaddach, 1844			+	
Gammarus obtusatus Dahl, 1938				+
Neohaustorius schmitzi Bousfield, 1965	+			
Paraceradocus gibber Andres, 1984	+			
Parallorchestes ochotensis (Brandt, 1851)	+			
Phronima sedentaria (Forskål, 1775)	+			
Pseudoprotella phasma (Montagu, 1804)	+			
Siphonoecetes dellavallei Stebbing, 1899	+			+
Isopoda				
Hemilepistus aphganicus Borutzkii, 1958	+			
Hemilepistus elongatus (Brandt, 1880)	+			
Hemilepistus reaumuri (Adóuin, 1826)	+	+		+
Hemilepistus rhinoceros Borutzkii, 1958	+			
Porcellio albinus Budde-Lund, 1885	+			
Porcellio Fuert-P.spec.1.	+			
Porcellio Fuert-P.spec.2.	+			
Porcellio simulator Budde-Lund, 1885	+			
Tanaidacea				
Heterotanais oerstedi (Krøyer, 1842)	+	+		+
Tanais dulongii (Adóuin, 1826)	+	+		
Astacidea				
Orconectes propinquus (Girard, 1852)	+			
Orconectes virilis (Hagen, 1870)	+	+		+
Pacifastacus trowbridgi (Stimpson, 1857)	+			+
Procambarus alleni (Faxon, 1884)	+			+
Procambarus clarkii (Girard, 1852)	+			+

(Contd.)

Table 14.1. (*Contd.*)

	Social Interactions Reported Among			
	Parent–Offspring	Both Parents	Offspring	Family–Nonkin
Brachyura				
Geosesarma notophorium Ng and Tan, 1995	+			
Metopaulias depressus Rathbun, 1896	+			+
Paranaxia serpulifera Guérin	+			
Paratelphusa (Barytelphusa) jacquemontii Rathbun	+			
Caridea				
Synalpheus regalis Duffy,1996			+	+

+, Species for which published observations are available. For other species with extended parental care (but no behavioral observations), see Thiel (1999a, 2003a).

these interactions have been made for six species. In three of these, males leave the females immediately after fertilization, apparently without any aggressive interaction between them. For example, S.B. Johnson and Attramadal (1982) noted that male tanaids *Tanais cavolinii* leave the burrow of the female shortly after copulation. Similar observations have been made for a burrow-dwelling amphipod (Shillaker and Moore 1987). Males of the amphipod *Casco bigelowi* and of the isopods *Sphaeroma terebrans*, *S. wadai*, and *Limnoria chilensis* are also thought to leave the females shortly after copulation, but no observations of behavioral interactions are available (Thiel 1998, 1999c, 2003b, Murata and Wada 2002). Very little is known about the mating systems of these species or the opportunities for males to obtain additional fertilizations with other females. At least for *S. wadai*, it has been suggested that males may have the chance to mate with a second female during the annual reproductive period (Murata and Wada 2002). Males apparently are actively ejected by the female after copulation in the tanaid *Heterotanais oerstedi* (Bückle-Ramírez 1965) and the stomatopod *Gonodactylus bredini* (Dingle and Caldwell 1972). It is not known why males do not assist with parental care in these species. Most likely, continuing male presence interferes with juvenile survival due to resource (space or food) limitation or male aggression toward offspring.

In contrast to these species with very brief mating associations, in other species the female and male cohabit for prolonged periods after copulation. Males of the freshwater crayfish *Orconectes virilis* construct burrows that are later used by females while caring for their offspring (Ameyaw-Akumfi 1976, Hazlett 1983). Following copulation, a male may continue to cohabit with the female through the winter but leaves the burrow before offspring are released by the female (Hazlett 1983). Prolonged cohabitation of the male and female was also reported for the crayfish *Procambarus clarkii* and *P. acutus*, but it is unknown whether males participate in parental care (Hazlett 1983). Finally, in the epibenthic amphipod *Dyopedos monacanthus*, females construct large mud whips on which they harbor their offspring (Thiel 1997a). Males apparently are tolerated on these mud whips, but the female always maintains a position between her offspring and the male, not permitting direct contact between them (Mattson and Cedhagen 1989). The role of the males is not clear, but it has been suggested that they might defend the mud whips against intra- and interspecific intruders (Stephan 1980, Mattson and Cedhagen 1989). Also in other species in which males have been observed together with offspring-caring females, the juveniles are often shielded by the female against direct contact with the male (e.g., in boring isopods, see below). Possibly, the male that cohabits with a caring female is not the father of her present brood but rather a male awaiting the chance to fertilize her next clutch of eggs.

Active cooperation between the male and female has been reported for only one crustacean species, the desert isopod *Hemilepistus reaumuri* (see chapter 16). But cooperation between parents is also likely for peracarids that bore into wood or algal tissues in which males and females cohabit during extended periods of parental care (*Limnoria lignorum*, Henderson 1924; *Limnoria algarum*, Menzies 1957; *Peramphithoe stypotrupetes*, Conlan and Chess 1992). Similarly, intriguing observations of extensive family groups with several offspring clutches in the mysid *Heteromysis harpax* (Vannini et al. 1993) suggest some form of collaboration between parents during extended parental care. Finally, in several species of semiterrestrial crayfish, males and females are found together with large offspring (see chapter 15). All these observations suggest biparental care, but unfortunately, at present the only species for which details of the social behavior has been uncovered and male participation in parental care has been clearly shown is *H. reaumuri* (chapter 16).

Parent–Offspring Interactions

Parental interactions with offspring in crustaceans usually involve exclusively the mother (Thiel 2003a). In general, paternity of a female's embryos by the attending male is uncertain, if not unlikely, and so it is not surprising that males often behave aggressively toward small offspring. For example, both male and nonovigerous female crayfish readily consume juveniles (Scudamore 1948, Mason 1970), and consequently, juvenile crayfish move away from adult males (Little 1975, 1976) (Fig. 14.1). Similarly, juveniles of the tube-dwelling amphipod *Lembos websteri* actively sought out contact with their mother but avoided the attending male: "Often when a hatchling *Lembos* crawled toward the male it would stop, turn around and crawl back to the female" (Shillaker and Moore 1987). Yet in other crustacean species males apparently show no aggressive behavior toward juveniles. For example, in the tube-dwelling amphipod *Siphonoecetes dellavallei*, small juveniles may attach their first tiny tubes to the tube of their mother, glued to the dwelling of a male, who behaves indifferently toward them (Richter 1978a). This author suggested that harem-guarding males are fathers of these juveniles, although convincing evidence is lacking.

Females of seven crustacean species were observed to actively manipulate their offspring by gently grasping them with the chelae and placing them either on the mother's body or on the substratum (Morgan 1987). Many authors use descriptive adverbs such as "gently," "carefully," or "cautiously" to refer to these interactions. An observation by Ng and Tan (1995) for the land crab *Geosesarma notophorum* describes the nature of these female-offspring interactions: "When the young become very

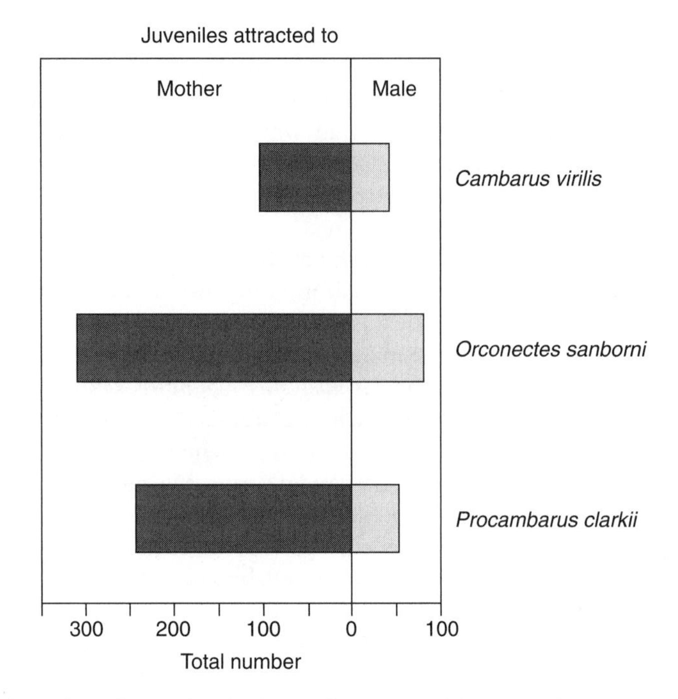

Figure 14.1 Number of juvenile (third-stage larvae) crayfish attracted to chemical stimuli of their mother or of a conspecific male in a two-choice maze for three different species. Data from Little (1975).

active, the female often removes them from her mouth and eyes with her chelae, occasionally picking them up and placing them on the substrate." Many researchers interpret active manipulation of small juveniles by the female as grooming behavior (e.g., Richter 1978b, Lim and Alexander 1986, Mattson and Cedhagen 1989, Aoki and Kikuchi 1991, Aoki 1997). Females of some crayfish will selectively eat dead offspring but do no harm to the living (Ameyaw-Akumfi 1976, Hazlett 1983). Based on these descriptions, it appears that females of these species engage in a specific and directed care behavior toward their offspring when handling them.

For 11 crustacean species, females have been reported to frequently gather their offspring, often in cases of danger (e.g., Bovbjerg 1956, Ameyaw-Akumfi 1976). An account of this shepherding behavior is provided by Lim and Alexander (1986) for *Caprella scaura typica*: "On a number of occasions a female was observed to transfer her young from the substratum to her body before crawling rapidly away from predators." These interactions are accompanied by active and rapid aggregation of juveniles around the female (Andrews 1904). Chacko and Thyagarajan (1952) reported on small offspring of the freshwater crab *Paratelphusa jacquemontii* crawling around the mother: "They make a quick retreat into the pouch at the least sign of danger, when the mother also assumes a hostile attitude by spreading her chelate legs." Thus, mothers seem to communicate with their offspring during shepherding, and both chemical and visual signals have been invoked (Hazlett 1983, Aoki and Kikuchi 1991). Signals appear to be specific, because juveniles quickly gather on or under their mother or in the family dwelling when the mother recognizes danger or is about to move to a different place (Hazlett 1983, Aoki 1997, Gherardi 2002). In the tube-dwelling tanaids *Heterotanais oerstedi* and *Tanais cavolinii*, females have been observed to grasp stray juveniles and return them to the rest of their offspring group (Bückle-Ramírez 1965, S.B. Johnson and Attramadal 1982).

Food transfer from parents to offspring has been reported (or inferred) for 19 crustacean species. As in other invertebrates in which juveniles participate in meals from parents (Park and Choe 2003), offspring of these crustaceans may feed on resources that their parents are processing or ingesting. This form of food-sharing is not uncommon among aquatic crustaceans and has been reported from several amphipods and tanaids. Juveniles may emerge halfway from the marsupium and feed on food particles held by their mothers (Coleman 1989, Kobayashi et al. 2002), but food sharing may also occur during later stages. In the detritus-feeding amphipod *Lembos websteri*, females frequently tolerate their offspring while feeding themselves (Shillaker and Moore 1987): "On two occasions more than half the young in the tube converged on the maternal gnathopods when the female caught and started to eat a large bolus of detritus." Similar observations were made for the deposit-feeding amphipod *Casco bigelowi* (see frontispiece), and for other amphipod species (e.g., Harrison 1940). Females of the pelagic hyperiid amphipod *Phronima sedentaria* also share their prey with their developing brood (Richter 1978b). In general, offspring-caring females appear to be relatively tolerant toward advances of small juveniles that snatch food particles while the female is feeding.

Active foraging for food by parents, which they then return to the offspring nest, has been reported exclusively for terrestrial crustaceans and is described in detail for the desert isopod *H. reaumuri* (chapter 16) and the tree-dwelling crab *Metopaulias depressus* (chapter 17). Parents of both species leave their dwellings, forage in the vicinity, and carry food back to their dwelling, where offspring feed on it. Such an active food provisioning may also occur in semiterrestrial crayfish since some species that provide parental care in their burrows also forage outside and retrieve food to their burrows (see chapter 15).

For three crustacean species, researchers reported that females behaved indifferently toward their offspring—they neither engage in any specific parental care behavior nor show aggression toward their offspring. Females of these species simply seem to tolerate their offspring in their dwellings or on their body. This form of parental tolerance toward small offspring may be most widespread among crustaceans that engage in extended parental care.

Parent–Offspring Conflicts and Parental Aggression

Conflicts over the extent and duration of parental care tend to become most pronounced toward the end of the care phase in animals (Trivers 1974). Although the degree to which juveniles or parents control the outcome of these conflicts is poorly understood in crustaceans, the few available reports provide some information on when and why they arise and how they are resolved. In four species, the female actively evicted her offspring from her dwelling. For three of these, it appears likely that these females actively expel juveniles in order to feed efficiently or to create space for a new offspring clutch. Bückle-Ramírez (1965) observed that females of the tanaid *Heterotanais oerstedi* evicted their offspring from their dwelling and suggested that most females will molt afterward and either continue reproductive life as a female or change sex and reproduce as a male. Mothers of some crayfish species become a menace to their offspring near the end of the extended care period (Mason 1970, Hazlett 1983). However, juveniles of some species also leave the family group without apparent intervention from the female. In *Tanais cavolinii*, recently molted juveniles first fed on diatoms on the tube wall and feces of their mother, but then ripped a small hole into the female's tube and immediately started building their own tubes on the outside (S.B. Johnson and Attramadal 1982). No aggressive eviction by the female was reported, but she quickly repaired her own tube, thereby effectively sealing off the juveniles from the maternal tube.

Prolonged cohabitation, active grooming, and food sharing by parents probably are facilitated by the suppression of parental aggression toward offspring during parental care, but the proximate mechanisms for this suppression are not well known. Parental tolerance is likely under hormonal influence since both female feeding and molting appear to be suppressed in several species where females cohabit with fully developed juveniles. In some species, in fact, females do not feed at all while caring for offspring (e.g., Ng and Tan 1995). In *Tanais cavolinii*, females stop feeding shortly before the offspring are released from the marsupium into their dwelling, that is, when extended parental care starts (S.B. Johnson and Attramadal 1982). Similar observations have been made for another tube-dwelling tanaid, *Heterotanais oerstedi*, in which the female closes the tube opening and thus cannot feed during parental care (Bückle-Ramírez 1965). Maternal feeding during extended parental care may also be restricted in burrow-dwelling isopods in which offspring occupy the terminal part of the burrows (*Sphaeroma terebrans*, Thiel 1999c; *Limnoria chilensis*, Thiel 2003b). While in some species maternal feeding appears to be completely restricted during extended parental care (see also Hazlett 1983), in others the females continue to feed but do not attack their offspring (see also above). For example, Richter (1978b) reported that female *Phronima sedentaria* do not prey on their offspring, even though they usually would feed on any prey item that has the size and shape of developing offspring. Similarly, brood-caring female crayfish *Procambarus clarkii*, which usually would cannibalize conspecifics, leave their own offspring and those of similar size unharmed (Figler et al. 1997).

While aggression by parents toward juveniles appears to be suppressed during extended parental care, aggression toward other adults may be strongly enhanced, as observed in many crustaceans that exhibit parental care (McCloskey 1970, Mason 1970, Hazlett 1983; see also chapter 17). In crayfish *Procambarus clarkii* and in the lobster *Homarus americanus*, brooding females show higher levels of aggression than nonreproductive individuals and usually win agonistic interactions (Figler et al. 1995, 1997, 2001) (Fig. 14.2). These observations provide further indication that parental care behavior is under hormonal control, even though present knowledge on this topic remains inconclusive (Figler et al. 2004).

Hazlett (1983) mentioned that crustacean mothers increasingly turn aggressive toward juveniles when contact with offspring has been experimentally inhibited. In several studies maternal behavior was maintained when mother–offspring contact was experimentally inhibited for 24 hours or less, but when separation lasted longer, mothers started to react increasingly aggressively toward offspring (Figler et al. 1997; see also

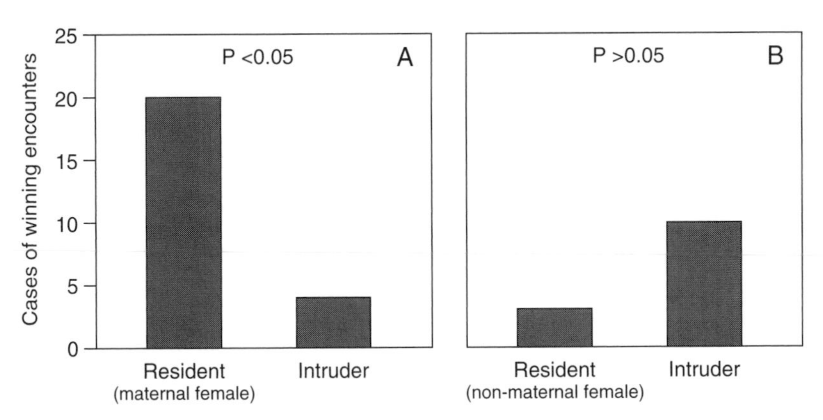

Figure 14.2 Agonistic encounters in the red swamp crayfish, *Procambarus clarkii*: number of encounters won by maternal resident females versus intruders (male or nonmaternal female) (A) and nonmaternal resident females versus nonmaternal female intruders (B). Data from Figler et al. (1995).

chapter 16). Linsenmair (1987) mentioned (unidentified) aggression-inhibiting chemicals on recently molted juvenile *H. reaumuri* that reduce aggression by reproductive adults. Figler et al. (2004), who studied maternal behavior in ovigerous lobster females, suggested that mechanical stimuli by the developing embryos (or juveniles) on the female pleopods may induce maternal behaviors. A similar mechanism had been proposed for freshwater crayfish by Little (1975, 1976). Reporting on parental care in the bromeliad-breeding crab *M. depressus*, Diesel (1989) suggested that "young seek maternal care." Similarly, Aoki and Kikuchi (1991) expressed that "juveniles insist on clinging to their mother." These observations indicate that specific (chemical or mechanical) cues produced by offspring actively induce and maintain parental care behavior in crustaceans.

Movement activity of females may also be reduced during extended parental care. In the freshwater crayfish *Pacifastacus trowbridgi*, females confine themselves to a den while hosting their offspring (Mason 1970). Hazlett (1983) reported that daily movements of female *Orconectes virilis* almost ceased completely when they were carrying eggs. Reduced foraging excursions were also observed for brood-caring females of *Parastacoides tasmanicus tasmanicus* by Hamr and Richardson (1994).

Offspring presence may not only influence the aggressive and foraging behavior but also the molting frequency and future reproduction of their parents. For *Caprella monoceros*, Aoki and Kikuchi (1991) reported that females did not molt while carrying offspring but rapidly molted after juveniles had been experimentally removed from the female's body. Interestingly, in a closely related species (*C. decipiens*) in which females also care for offspring but not on their body, molting frequency apparently is not affected by the presence of offspring (Aoki and Kikuchi 1991). The high load of epibionts on females of the isopod *Arcturus baffini*, which carry offspring on their antennae (Svavarsson and Davidsdottir 1995), may also suggest that female molting is suppressed during maternal care in this species. These observations indicate that carrying offspring on the mother's body negatively affects her future reproductive potential. Some crayfish that cohabit with offspring for long time periods (several months) only reproduce every other year (see chapter 15), but it is not known whether offspring presence influences female molting and production of a subsequent brood in these burrow-dwelling species. Reports of overlapping offspring cohorts in a parental dwelling, though, suggest that continuing female reproduction is not suppressed by the presence of her offspring (see also below). The existence of a family dwelling thus alleviates potential parent–offspring conflicts (see also Trivers 1974) by allowing parents to tolerate the presence of older offspring while starting to care for a subsequent brood (chapters 17, 18).

Interactions among Offspring

In crustaceans with extended parental care, the number of juveniles in parent–-offspring groups commonly ranges between 10 and 100 individuals. Several observations strongly suggest that juveniles within a clutch compete for resources and that some do better than others. For example, after a developmental period of three months, the juveniles in a brood of *Procambarus clarkii* ranged from 7 to 21 millimeters in body length (Ameyaw-Akumfi 1976). High variance in body sizes within offspring cohorts also occurs in the tree-dwelling crab *Metopaulias depressus* (chapter 17). In *Phronima sedentaria*, in which juveniles are attached to the maternal dwelling, those in a central position may have better access to food and grow faster than juveniles at the edge (Richter 1978b), although no aggressive interactions among cohabiting juveniles were reported (Richter 1978b).

In some species, aggressive interactions among cohabiting juveniles have been observed or inferred (Hazlett 1983). Reports on siblicide are not uncommon. For example, Morgan (1987) suggested that loss of juvenile *Paranaxia serpulifera* crablings during maternal care may be due to cannibalism among offspring. Juveniles of the crayfish *Pacifastacus trowbridgi* can also cannibalize siblings during later phases of maternal care (Mason 1970). Synchronous molting and development of the juveniles in a family group can diminish the risk of cannibalism (Richter 1978b), as has been suggested for subsocial bugs (Kudo and Nakahira 2004).

In the amphipod *Leptocheirus pinguis*, some juveniles leave the maternal burrow after synchronous molting events, and it has been inferred that this is the result of aggressive interactions (Thiel 1999d). Video observations showed that juveniles in the maternal burrow occasionally engaged in aggressive interactions (Thiel 1997b, 1997c). Competition among juvenile *L. pinguis* appeared to increase toward the end of the rearing period when resources in the parental dwelling (space and food) became increasingly limited.

Parental Care for Several Subsequent Offspring Cohorts

The occurrence of overlapping generations is considered an important driver of social evolution, because it offers the opportunity for helping by older siblings (Wilson 1971). One necessary condition for cohabitation of several generations is that parents reproduce repeatedly and that some offspring reach adulthood in the family dwelling. Of the 136 crustacean species for which extended parental care has been reported, the reproductive strategy (iteroparous vs. semelparous) has been reported for 47 species. Most of these latter species (39; 83%) are iteroparous. Cohabitation of two

Table 14.2. Crustacean species that care for several subsequent broods simultaneously, their environment, offspring dwelling, and number of cohabiting broods reported.

Species[a]	Environment[b]	Dwelling[c]	Parent	Number of Broods	References
Caprella decipiens (Am)	M	—	Mother	2	Aoki 1997, 1999
Dyopedos monacanthus (Am)	M	Mud whips	Mother	2	Thiel 1997a
Leucothoe spinicarpa (Am)	M	Ascidian	Biparental	2	Thiel 2000a
Peramphithoe stypotrupetes (Am)	M	Burrow (kelp)	Biparental	3	Conlan and Chess 1992
Leptocheirus pinguis (Am)	M	Burrow (soil)	Mother	3	Thiel 1997c
Heteromysis harpax (My)	M	Hermit shell	Biparental	3+	Vannini et al. 1993
Synalpheus regalis (Ca)	M	Sponge	(Biparental)	3+	Duffy 2003
Engaeus leptorhynchus (As)	T	Burrow (soil)	(Mother)	2+	Horwitz et al. 1984
Metopaulias depressus (Br)	T	Bromeliad pool	Mother	2+	Chapter 17

[a]Am – Amphipoda; As – Astacidea; My – Mysidacea; Ca – Caridea; Br – Brachyura. [b]M – marine; T – terrestrial; [c]Burrows may be excavated in soil/sediment, or in kelp stipes.

offspring cohorts is nevertheless uncommon in free-living crustaceans, and one of the few known cases is the caprellid amphipod *Caprella decipiens*, extensively studied by Aoki (1997, 1999). This author reported that females, the sole parent in this species, need to remate while providing care for one brood. Most reports on overlapping offspring cohorts come from crustacean species that care for offspring in a dwelling (Table 14.2). In the mysid *Heteromysis harpax*, several (up to six) cohorts may live together with their parents in the shells of their hermit crab hosts (Vannini et al. 1993). Similarly, in the amphipods *Leucothoe spinicarpa* and *Leptocheirus pinguis*, parents cohabit with several subsequent offspring cohorts, but the parental dwellings either are not or only to a limited extent expandable (Thiel et al. 1997, Thiel 1999d, 2000a). In the kelp-boring amphipod *Peramphithoe stypotrupetes*, parents that share their dwelling with several offspring cohorts apparently continuously increase the size of their dwelling (Conlan and Chess 1992). Possibly, older juveniles cooperate in maintenance and expansion of the dwelling, but nothing is known about the behavior of this species. In the crab *Metopaulias depressus* and in the desert isopod *Hemilepistus reaumuri*, large offspring have been observed to support their mothers during care for subsequent offspring cohorts mainly by aiding in defense of the family dwelling (chapters 16, 17). Similarly, in social snapping shrimp, older offspring defend the sponge inhabited by mothers and younger siblings (chapter 18).

Family Recognition and Reaction Toward Unrelated Conspecifics

One of the most important questions in the context of social evolution is whether and how individuals recognize each other (Gamboa et al. 1986). In parent–offspring groups, parents need to recognize their mates and offspring, and offspring need to recognize their parents and siblings. Recognition cues may be simple or highly complex (see, e.g., chapter 16). For example, in species with a low probability of contact with unrelated conspecifics parents may simply accept any juvenile that they encounter within their nest or that approximates the size of their own juveniles. Evidence for this comes from experimental studies and anecdotal observations (e.g., Thiel 2000b). Among crustaceans acceptance of unrelated juveniles is relatively common early in offspring development. Several researchers who removed embryos or juveniles from the marsupium of amphipod mothers noted that the juveniles would reenter the brood structure, with aid or tolerance of their mother (Croker 1968, Borowsky 1980, Kobayashi et al. 2002). In many crustaceans, unrelated juveniles are easily accepted by females if they are of similar developmental stages as their own offspring (Sheader and Chia 1970). Figler et al. (1997) replaced small offspring of maternal crayfish *Procambarus clarkii* with unrelated offspring of the same age. Females immediately adopted these unrelated juveniles, and their behavior was indistinguishable from that shown toward their own juveniles. Similar experiments have been conducted by Linsenmair in the desert isopod *H. reaumuri* (see chapter 16). While parents of this species are capable of recognizing their own offspring, there are time windows when parents would adopt unrelated offspring (see also below). Also, when unrelated offspring were placed in the family dwelling of the bromeliad crab *Metopaulias depressus*, females would immediately adopt them; a follow-up experiment showed that females would accept unrelated juveniles up to a specific size (~10 mm carapace width) but would aggressively reject any individual larger than this (see chapter 17).

In many crustacean species, juveniles appear incapable of recognizing their parents. Hazlett (1983) synthesized several studies on freshwater crayfish and remarked that juveniles would "not differentiate between their own mother and another conspecific female that is maternal." In the reports described above on females adopting unrelated offspring, researchers never reported aggressive behavior of offspring toward a foreign (but maternal) female, further supporting this suggestion. Offspring interacting with unrelated juveniles have been reported for three species: in two species, juveniles aggressively rejected an outsider (*Hemilepistus reaumuri*, Linsenmair 1987) or even killed it (*Metopaulias depressus*, chapter 17), while in the other species, juveniles behaved indifferently toward it (*Dyopedos monacanthus*, Stephan 1980). Lack of family recognition and adoption of unrelated offspring are most likely for species where either the probability of invasion by unrelated individuals is low, or the costs for family members of adopting an occasional unrelated individual are limited.

Some of the parents that adopt unrelated offspring placed directly within their dwelling, react highly aggressively toward individuals approaching a family dwelling from the outside (see above). In these species, aggressive behavior is usually heightened in the vicinity or near the entrance of their dwelling. Figler et al. (2001) observed that maternal crayfish *P. clarkii* show strong territorial behavior, aggressively defending a shelter against intruders. Also, Linsenmair (1987) noted that young desert isopods react highly aggressively toward foreign offspring in the immediate vicinity of their family burrow but showed avoidance behavior when farther away. These observations suggest that the spatial context of an encounter, whether inside or at the entrance to the nest, has important influence on how individuals behave toward one another.

In some crustaceans that inhabit dwellings, family members appear capable to distinguish between individual kin and unrelated conspecifics. The nature of these interactions has been revealed in experimental studies with desert isopods and snapping shrimp (chapters 16, 18). The researchers suggest that family affiliation is recognized on the basis of chemical "badges." In desert isopods, in particular, individuals without such a family badge are aggressively rejected by family members (chapter 16). An intriguing report by Mattson and Cedhagen (1989) on maternal behavior of the epibenthic amphipod *Dyopedos monacanthus* also suggests the existence of chemical cues: "the mother quickly returned to the young, seized a couple of them (in turn) gently between gnathopods 2, raised them slightly, and immediately released them again." This behavior was observed while the female aggressively defended her mud whip against a neighboring female (Mattson and Cedhagen 1989). In most other species, aggressive rejection of intruders has been reported without explicitly discussing the mechanisms used to discriminate against unrelated intruders.

Social Evolution in Crustaceans and Other Organisms

Evolution of Subsocial Behavior

This review reveals that crustacean parents show similar care behaviors as parents in many other arthropod or vertebrate species (Tallamy and Wood 1986, Clutton-Brock 1991). Moreover, many of these behaviors have evolved independently in different crustacean taxa (Table 14.3). For example, grooming or parental manipulation of individual juveniles has been observed in amphipods, brachyuran crabs, and crayfish. This behavior, which is also found in many subsocial insects (Tallamy and Wood 1986), is thought to have evolved in response to the risk of parasite infection in dense aggregations of organisms (Clutton-Brock 1991). Similarly, shepherding of juveniles, as reported for crustacean mothers from diverse taxa (Table 14.3), has typically been observed in situations of danger, and correspondingly, this behavior is interpreted as active protection of juveniles against enemies (Clutton-Brock 1991). Food sharing, as found in diverse crustaceans (Table 14.3), is also observed in many subsocial insects, where small juveniles are thought incapable of acquiring or processing their main food resources (e.g., Wilson 1971, Tallamy and Wood 1986, Park and Choe 2003). Most intriguing and revealing are the reports of central place foraging and active food provisioning of crustacean parents to their offspring, which is exclusively known from (semi)terrestrial crustaceans (Fig. 14.3). Active food provisioning most likely has evolved from food sharing in species that retrieve food materials into their dwellings (see also chapter 16). This form of active food provisioning is also very common in many subsocial insect species (Tallamy and Wood 1986, Rauter and Moore 2002). Lack of this behavior in all truly aquatic crustaceans probably is due to environmental constraints and the inherent difficulties of accumulating, preserving, and relocating food bounties stored in dwellings (Fig. 14.3). The lack of active food provisioning to a central place (dwelling) may be one of the most important impediments for the evolution of advanced social behavior in many aquatic crustaceans.

Dwellings are important for extended parental care because they allow parents to engage in other activities (molting, mating, feeding) while caring for offspring. When dwellings are costly to construct or to acquire and defend, often both parents

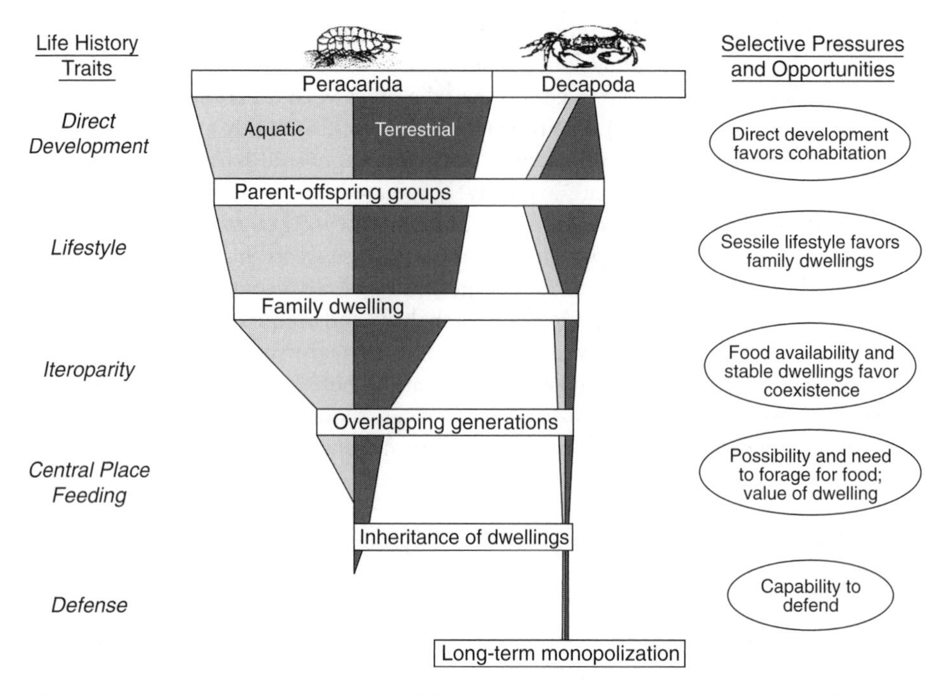

Figure 14.3 Schematic representation of life history traits and selective pressures and opportunities affecting the evolution of social behaviors via the subsocial route in peracarid and decapod crustaceans. Shaded areas indicate increasing or decreasing tendency that the respective social level is represented in the two crustacean groups.

participate in care activities. Paternal assistance is common in plant-boring arthropods that feed on their burrow substratum. In these species, males are primarily involved in burrow maintenance and defense (Nalepa 1984, Reid and Roitberg 1994, Kirkendall et al. 1997). This is also paralleled in crustaceans, where cohabiting male partners are frequently reported from species that live in wood or soil burrows, or in otherwise valuable dwellings with internal food resources.

Stable dwellings also permit extensive care periods during which offspring may reach subadult or in some cases even adult stages (Thiel 2003a; see also Duffy 2003,

Table 14.3. Crustacean taxa and the respective parental behaviors reported.

Species	Manipulation/Grooming	Shepherding	Feeding[a]
Amphipoda	+	+	+ (S)
Isopoda			+ (P)
Tanaidacea		+	+ (S)
Astacidea		+	
Caridea			
Brachyura	+	+	+ (P)

+, The respective parental behavior has been observed in this taxon. For references, see Thiel 1999a and 2003a. [a]S, food sharing; P, food provisioning.

and chapters 16, 17). This generates the opportunity for overlapping offspring generations in iteroparous species (Fig. 14.3). Dwellings in biotic microhabitats (ascidians, bivalves, sponges, bromeliads) may be very stable and long-lived, but they usually are not expandable or only to a very limited degree (Table 14.4). Nevertheless, these dwellings have a high resource value, because they provide efficient protection against predators and often predictable food resources (e.g., chapter 18), as has also been suggested for wood-dwelling termites (Myles and Nutting 1988) or gall-dwelling thrips and aphids (Crespi and Mound 1997). In several crustacean species that inhabit biotic dwellings, a few offspring of subadult and even adult stages have been observed together with smaller offspring in the parental dwelling (Thiel 2000a; Table 14.4; see also chapters 15, 17). In eusocial synalpheid shrimp, hundreds of adult offspring may cohabit in the natal sponge (chapter 18).

From Subsocial to Eusocial Behavior: A Narrow Road

Extended parental care inevitably results in social interactions among kin (and nonkin) individuals. Crustaceans are clearly capable of engaging in specific and directed communication with related (or unrelated) conspecifics. Recognition of individual conspecifics occurs not only in parent–offspring groups but also between partners of monogamous pairs (V.R. Johnson 1977, Seibt and Wickler 1979), in mating interactions (chapter 6), or between opponents during agonistic encounters (chapter 5). This suggests that the neural capabilities of crustaceans permit interactions that would facilitate the persistence of parent–offspring groups and evolution of advanced social behaviors.

The observations reviewed herein indicate that the evolutionary step from subsocial behavior (parent–offspring interactions) to advanced social behavior depends on the selective environment and, in particular, on the existence of a highly valuable and long-lived family dwelling (Alexander et al. 1991, Crespi 1994, Queller and Strassmann 1998). Even though the dwellings of snapping shrimp and bromeliad crabs are not (or only little) expandable, eusocial behavior with overlapping cohorts and sharing of reproductive tasks has evolved, most likely because the longevity of the dwellings substantially exceeds the life time of individual residents. One important behavior accompanying the evolution of eusocial behavior is the use of the family dwelling as a central place, which appears the crucial step in the evolution of kin recognition. Temporary leaving (or presence at the periphery) of the family dwelling is known for five crustacean taxa that provide extended parental care (Table 14.4; Diesel and Horst 1995). In three of these taxa (*Hemilepistus reaumuri*, *Metopaulias depressus*, *Synalpheus* spp.), recognition of family members has been demonstrated (chapters 16–18). It thus appears that the existence of a stable family dwelling in combination with central place behavior is an important condition for the evolution of eusocial behavior (Fig. 14.3). Desert isopods from the genus *Hemilepistus* may not have achieved entry into the eclectic assemblage of eusocial organisms, because they are semelparous and their burrows persist for only one reproductive season; that is, there is no chance for overlapping generations.

When dwellings are monopolized by family units, efficient mechanisms of defense are necessary to fend off enemies (Wilson 1971, Andersson 1984). Defense of offspring or a dwelling can be achieved either by shielding offspring against intruders or by actively attacking and fending off enemies (Kudo and Hasegawa 2003, Park and

Table 14.4. Crustacean taxa that care for their offspring in dwellings.

Species	Environment[a]	Offspring Generations	Nonreproductive Adults[b]	Dwelling Persistence	Expandable Dwellings	Defense of Dwelling	Central Place[c]	Family Recognition
Tanais spp.	M	1	No	Weeks[d]	(No)[d]	?	No[d]	?
Dyopedos spp.	M	2	No	Months	(Yes)[d]	Conspecifics	No[d]	?
Limnoria spp.	M	1	No	Months[d]	(Yes)[d]	Conspecifics	No[d]	?
Sphaeroma spp.	M	1	No	Months[d]	(Yes)[d]	Conspecifics	No[d]	?
Porcellio spp.	T	1	No	Months	No	?	Yes	(No)
Cambarus spp.	ST	1	No	Months	(Yes)[d]	Conspecifics	?	?
Parastacus spp.	ST	1	No	Months	(Yes)[d]	?	?	?
Casco bigelowi	M	1	No	Months	Yes, but[d]	?	No	?
Leptocheirus pinguis	M	3	No	Months	(Yes)[d]	?	No	?
Peramphithoe stypotrupetes	M	3	No	Months	Yes, but[d]	?	No	?
Sesarma jarvisi	ST	1	No	Months	No	?	(Yes)	?
Leucothoe spinicarpa	M	2	(Yes?)	~1Year	No	?	No	?
Hemilepistus spp.	T	1	No	~1 Year	(No)	Conspecifics	Yes	Yes
Engaeus spp.	ST	2+	?	Years	(Yes)	?	?	?
Parastacoides spp.	ST	(2+)	?	Years	(Yes)	?	?	?
Metopaulias depressus	ST	2+	Yes (TS)	Years	No	Conspecifics	Yes	Yes
Synalpheus spp.	M	3+	Yes (TS)	Years	No	Conspecifics/ congeners	(Yes)	Yes

Only species for which information on cohabiting offspring generations and approximate persistence of dwellings is available are considered. For references see Thiel (1999a, 2003a). [a]M, marine; T, fully terrestrial; ST, terrestrial, but dependent on sources of open water. [b]TS, task sharing. [c]Family members leave burrows occasionally. [d]Not explicitly stated by authors, but strong inference.

Choe 2003, Lin et al. 2004), as also seen in crustaceans (e.g., Table 14.4). Some of the most effective defense mechanisms have been reported for eusocial organisms, such as hymenopterans (Wilson 1975), termites (Shellman-Reeve 1997), and gall thrips (Crespi and Mound 1997). Painful venoms and stings are considered important precursors for the evolution of eusocial behavior (Hunt 1999). Members of the two crustacean groups for which eusocial species have been reported (chapters 17, 18) feature strong claws that may even serve to attack larger predators (Diesel 1989, Versluis et al. 2000). The powerful snapping claw, in combination with a well-defined and defendable opening (sponge osculum), permits snapping shrimp to block their family dwellings successfully against many intruders (Duffy 2003). The lack of efficient weapons in most crustacean species that care for their offspring in dwellings (e.g., amphipods) may have restricted the prolonged persistence of parent–offspring groups in valuable dwellings (Fig. 14.3).

Eusocial behavior has evolved several times independently within many different taxa (e.g., Wilson 1971, Stern 1994, Crespi and Mound 1997, Hunt 1999, Cameron and Mardulyn 2001). For example, eusocial behavior has evolved at least twice independently in soil-burrowing rodents that live in arid conditions with clumped food distribution (Faulkes et al. 1997). Crustaceans show a similar pattern—evolution of eusociality in crustaceans is correlated with certain traits and ecological conditions (Duffy et al. 2000). However, eusocial behavior has also evolved in crustacean species from very different environments. Superficially, there are many differences between the habitats of sponge-dwelling snapping shrimp and an arboreal bromeliad-breeding crab, but both taxa share particular key traits and environmental constraints, the most important ones being extended parental care for subsequent offspring cohorts, central place behavior, and efficient defense and monopolization of a long-lived dwelling.

Future Directions

While extended parental care is not uncommon in crustaceans, few studies have explored the factors leading to the evolution of this and more advanced social behaviors. In particular, the role of environmental factors (stress, predation, availability of dwellings, food supply) should be experimentally tested. Furthermore, the importance of social conflicts (between mates, between parents and offspring, among siblings, between subsequent cohorts) should be explored. For example, during long-lasting family association, parent–offspring conflicts may intensify. The fact that offspring removal in some species immediately leads to molting and the production of a new brood (e.g., Aoki and Kikuchi 1991) indicates that offspring presence can suppress future reproduction of parents. In such a case, functional semelparity of parents (*sensu* Tallamy and Brown 1999) could be a consequence rather than the cause of extended parental care. This suggests that the outcome of parent–offspring conflicts during extended parental care strongly influence the expression of social behaviors—this hypothesis should be exposed to experimental tests in the future. The active defense of dwellings against enemies is also considered crucial in social evolution (Andersson 1984), and whether efficient defense (or lack thereof) of family dwellings affects the evolution of social behaviors in crustaceans should be examined.

While anecdotal information on the interactions within family groups is available, next to nothing is known about the social behavior of most crustacean species that engage in extended parental care. The fact that detailed knowledge about the parental care behaviors commonly is only available for single species within a particular crustacean taxon (e.g., a genus or a family) prohibits comparative phylogenetic methods and thereby complicates the understanding of social evolution. Future studies on behavioral interactions within parent–offspring groups should therefore strive to thoroughly describe parental care behaviors and include several phylogenetically related species.

Kin recognition appears to have only evolved in species where family members frequently leave the family dwelling (e.g., during central place foraging). In order to expose this hypothesis to a rigorous test, the presence and degree of family recognition should be examined in additional species. Central place behavior (motivated by foraging activities) appears to be more prevalent in terrestrial than in aquatic crustacean species. Given the importance of foraging and food provisioning during social evolution (Andersson 1984, Thorne and Traniello 2003), it appears particularly worthwhile to focus future research attention on terrestrial and semiterrestrial crustaceans. Some of the most intriguing groups of crustaceans awaiting study of social behavior are semiterrestrial crayfish, which inhabit long-lived family burrows (see also Gherardi 2002). Many semiterrestrial crayfish are iteroparous, and their burrow systems persist for several years (chapter 15). It can thus be expected that they gather food outside their burrows, that family members recognize each other, and that older offspring inherit burrows from their parents, possibly after helping them to raise subsequent cohorts.

In summary, experimental studies are suggested to reveal the responses of parents and offspring under different environmental conditions and in different social scenarios. Furthermore, thorough natural history studies are necessary, because these provide the information on behavioral traits that is required for phylogenetic comparative methods (see also chapter 3).

Summary and Conclusions

Extended parental care is found among diverse crustacean species from aquatic and terrestrial environments and in most cases care is provided exclusively by the females. Biparental care has primarily been reported from burrow-living species, where males engage in establishment, maintenance, or defense of burrows. The aggressiveness of parents is often higher than that of nonbrooding adults. The observations reviewed herein demonstrate that family members show specific social behaviors (defense, shepherding, grooming, food sharing) while cohabiting. Advanced social behavior, where parents and offspring cooperate, have only been reported from snapping shrimp, bromeliad crabs, and desert isopods, all of which inhabit long-lived family dwellings. Available data indicate that family recognition has evolved in species where family members frequently leave and return to the family dwelling (desert isopods and semiterrestrial crayfish), and where dwellings are rare and valuable (bromeliad crabs and snapping shrimp). Prolonged cohabitation of parents and sexually maturing offspring is rare among crustaceans, most likely because resources become limiting and are difficult to replenish, and because family members are unable to maintain and defend stable dwellings. This review suggests that the presence of a stable dwelling and central place behavior around this dwelling may have been important steps during social evolution. Extended parental care, iteroparous reproduction, and possession of an effective defense mechanism seem to be crucial preadaptations for the evolution of eusocial behavior in crustaceans.

Acknowledgments I am grateful to C. Correa, J.T.A. Dick, an anonymous reviewer, and in particular, J.E. Duffy for many helpful comments on the manuscript. Financial support for studies on crustacean parental care has been obtained over the years from the University of Maine, the Smithsonian Marine Station at Fort Pierce, and FONDECYT Chile.

References

Alexander, R.D., K.M. Noonan, and B.J. Crespi. 1991. The evolution of eusociality. Pages 3–44 *in*: P.W. Sherman, J.U.M. Jarvis, and R.D. Alexander, editors. The biology of the naked mole-rat. Princeton University Press, Princeton, N.J.

Ameyaw-Akumfi, C.E. 1976. Some aspects of breeding biology of crayfish. Unpublished Ph.D. thesis, University of Michigan, Ann Arbor, Mich.

Andersson, M. 1984. The evolution of eusociality. Annual Review of Ecology and Systematics 15:165–189.

Andrews, E.A. 1904. Breeding habits of crayfish. American Naturalist 38:165–206.

Aoki, M. 1997. Comparative study of mother-young association in caprellid amphipods: is maternal care effective? Journal of Crustacean Biology 17:447–458.

Aoki, M. 1999. Morphological characteristics of young, maternal care behaviour and microhabitat use by caprellid amphipods. Journal of the Marine Biological Association of the United Kingdom 79:629–638.

Aoki, M., and T. Kikuchi. 1991. Two types of maternal care for juveniles observed in *Caprella monoceros* Mayer, 1890 and *Caprella decipiens* Mayer, 1890 (Amphipoda: Caprellidae). Hydrobiologia 223:229–237.

Bennett, N.C., and C.G. Faulkes. 2000. African mole-rats: ecology and eusociality. Cambridge University Press, New York.

Borowsky, B. 1980. Factors that affect juvenile emergence in *Gammarus palustris* (Bousfield, 1969). Journal of Experimental Marine Biology and Ecology 42:213–223.

Bovbjerg, R.V. 1956. Some factors affecting aggressive behaviour in crayfish. Physiological Zoology 29:127–136.

Bückle-Ramírez, L.F. 1965. Untersuchungen über die Biologie von *Heterotanais oerstedi* Kröyer Crustacea, Tanaidacea). Zeitschrift für Morphologie und Ökologie der Tiere 55:714–782.

Cameron, S.A., and P. Mardulyn. 2001. Multiple molecular data sets suggest independent origins of highly eusocial behavior in bees (Hymenoptera: Apinae). Systematic Biology 50:194–214.

Chacko, P.I., and S. Thyagarajan. 1952. On the development and parental care in the potamonid crab, *Paratelphusa* (*Barytelphusa*) *jacquemontii* (Rathbun). Journal of the Bombay Natural History Society 52:289–291.

Clutton-Brock, T.H. 1991. The evolution of parental care. Princeton University Press, Princeton, N.J.

Coleman, C.O. 1989. Burrowing, grooming and feeding behaviour of *Paraceradocus*, an Antarctic amphipod genus (Crustacea). Polar Biology 10:43–48.

Conlan, K.E., and J.R. Chess. 1992. Phylogeny and ecology of a kelp-boring amphipod, *Peramphithoe stypotrupetes*, new species (*Corophioidea*: Ampithoidae). Journal of Crustacean Biology 12:410–422.

Crespi, B.J. 1994. Three conditions for the evolution of eusociality: are they sufficient? Insectes Sociaux 41:395–400.

Crespi, B.J., and L.A. Mound. 1997. Ecology and evolution of social behavior among Australian gall thrips and their allies. Pages 166–180 *in*: J.C. Choe and B.J. Crespi, editors. Social behavior in insects and arachnids. Cambridge University Press, Cambridge.

Croker, R.A. 1968. Return of juveniles to the marsupium in the amphipod *Neohaustorius schmitzi* Bousfield. Crustaceana 14:215.

Diesel, R. 1989. Parental care in an unusual environment: *Metopaulias depressus* (Decapoda: Grapsidae), a crab that lives in epiphytic bromeliads. Animal Behaviour 38:561–575.

Diesel, R. 1992. Maternal care in the bromeliad crab, *Metopaulias depressus*: protection of larvae from predation by damselfly nymphs. Animal Behaviour 43:803–812.

Diesel, R., and D. Horst. 1995. Breeding in a snail shell—ecology and biology of the Jamaican montane crab *Sesarma jarvisi* (Decapoda, Grapsidae). Journal of Crustacean Biology 15:179–195.

Diesel, R., and M. Schuh. 1993. Maternal care in the bromeliad crab *Metopaulias depressus* (Decapoda): maintaining oxygen, pH and calcium levels optimal for the larvae. Behavioral Ecology and Sociobiology 32:11–15.

Dingle, H., and R.L. Caldwell. 1972. Reproductive and maternal behavior of the mantis shrimp *Gonodactylus bredini* Manning (Crustacea: Stomatopoda). Biological Bulletin 142:417–426.

Duffy, J.E. 1996. Eusociality in a coral-reef shrimp. Nature 381:512–514.

Duffy, J.E. 2003. The ecology and evolution of eusociality in sponge-dwelling shrimp. Pages 201–215 *in*: T. Kikuchi, N. Azuma, and S. Higashi, editors. Genes, behaviors and evolution of social insects. Hokkaido University Press, Sapporo, Japan.

Duffy, J.E., C.L. Morrison, and R. Ríos. 2000. Multiple origins of eusociality among sponge-dwelling shrimps (*Synalpheus*). Evolution 54:503–516.

Duffy, J.E., C.L. Morrison, and K.S. Macdonald. 2002. Colony defense and behavioural differentiation in the eusocial shrimp *Synalpheus regalis*. Behavioral Ecology and Sociobiology 51:488–495.

Faulkes, C.G., N.C. Bennett, M.W. Bruford, H.P. O'Brien, G.H. Aguilar, and J.U.M. Jarvis. 1997. Ecological constraints drive social evolution in the African mole-rats. Proceedings of the Royal Society of London, Series B 264:1619–1627.

Fernández, M., N. Ruiz-Tagle, S. Cifuentes, H.-O. Pörtner, and W. Arntz, 2003. Oxygen-dependent asynchrony of embryonic development in embryo masses of brachyuran crabs. Marine Biology 142:559–565.

Field, J., and S. Brace. 2004. Pre-social benefits of extended parental care. Nature 428:650–652.

Figler, M.H., G.S. Blank, and H.V.S. Peeke. 1997. Maternal aggression and post-hatch care in red swamp crayfish, *Procambarus clarkii* (Girard): the influences of presence of offspring, fostering and maternal molting. Marine and Freshwater Behaviour and Physiology 30:173–194.

Figler, M.H., G.S. Blank, and H.V.S. Peeke. 2001. Maternal territoriality as an offspring defense strategy in red swamp crayfish (*Procambarus clarkii*, Girard). Aggressive Behavior 27:391–403.

Figler, M.H., H.V.S. Peeke, M.J. Snyder, and E.S. Chang. 2004. Effects of egg removal on maternal aggression, biogenic amines, and stress indicators in ovigerous lobsters (*Homarus americanus*). Marine and Freshwater Behaviour and Physiology 37:43–54.

Figler, M.H., M. Twum, J.E. Finkelstein, and H.V.S. Peeke. 1995. Maternal aggression in red swamp crayfish (*Procambarus clarkii*, Girard)—the relation between reproductive status and outcome of aggressive encounters with male and female conspecifics. Behaviour 132:107–125.

Förster, C., and J.A. Baeza. 2001. Active brood care in the anomuran crab *Petrolisthes violaceus* (Decapoda: Anomura: Porcellanidae): grooming of brooded embryos by the fifth pereiopods. Journal of Crustacean Biology 21:606–615.

Gamboa, G.J., H.K. Reeve, and D.W. Pfennig. 1986. The evolution and ontogeny of nestmate recognition in social wasps. Annual Review of Entomology 31:431–454.

Gherardi, F. 2002. Behaviour. Pages 258–290 *in*: D.M. Holdich, editor. Biology of freshwater crayfish. Blackwell Scientific, Oxford.

Hamr, P., and A.M.M. Richardson. 1994. Life history of *Parastacoides tasmanicus tasmanicus* Clark, a burrowing freshwater crayfish from south-western Tasmania. Australian Journal of Marine and Freshwater Research 45:455–470.

Harrison, R.J. 1940. On the biology of the Caprellidea. Growth and moulting of *Pseudoprotella phasma* Montagu. Journal of the Marine Biological Association of the United Kingdom 24:483–493.

Hazlett, B.A. 1983. Parental behavior in decapod Crustacea. Pages 171–193 *in*: S. Rebach and D.W. Dunham, editors. Studies in adaptation—the behavior of higher Crustacea. John Wiley and Sons, New York.

Helms Cahan, S., D.T. Blumstein, L. Sundstrom, J. Liebig, and A. Griffin. 2002. Social trajectories and the evolution of social behavior. Oikos 96:206–216.

Henderson, J.T. 1924. The gribble: a study of the distribution factors and life-history of *Limnoria lignorum* at St. Andrews, N.B. Contributions to Canadian Biology 2(Part 1):309–327.

Horwitz, P.H.J., A.M.M. Richardson, and P. Cramp. 1984. Aspects of the life history of the burrowing freshwater crayfish, *Engaeus leptorhynchus* at Rattrays Marshes, north east Tasmania. Tasmanian Naturalist 82:1–5.

Hunt, J.H., 1999. Trait mapping and salience in the evolution of eusocial vespid wasps. Evolution 53:225–237.

Johnson, S.B., and Y.G. Attramadal. 1982. Reproductive behaviour and larval development of *Tanais cavolinii* (Crustacea: Tanaidacea). Marine Biology 71:11–16.

Johnson, V.R. 1977. Individual recognition in the banded shrimp *Stenopus hispidus* (Olivier). Animal Behaviour 25:418–428.

Kirkendall, L.E., D.S. Kent, and K.F. Raffa. 1997. Interactions among males, females and offspring in bark and ambrosia beetles: the significance of living in tunnels for the evolution of social behavior. Pages 181–215 *in*: J.C. Choe and B.J. Crespi, editors. Social behavior in insects and arachnids. Cambridge University Press, Cambridge.

Kobayashi, T., S. Wada, and H. Mukai. 2002. Extended maternal care observed in *Parallorchestes ochotensis* (Amphipoda, Gammaridea, Talitroidea, Hyalidae). Journal of Crustacean Biology 22:135–142.

Kudo, S., and E. Hasegawa. 2003. Diversified reproductive strategies in *Gonioctena* (Chrysomelinae) leaf beetles. Pages 727–738 *in*: P. Jolivet, J.A. Santiago-Blay, and M. Schmitt, editors. New developments in the biology of the Chrysomelidae. SPB Academic Publishing, The Hague, The Netherlands.

Kudo, S., and T. Nakahira. 2004. Effects of trophic-eggs on offspring performance and rivalry in a sub-social bug. Oikos 107:28–35.

Lim, S.T.A., and C.G. Alexander. 1986. Reproductive behaviour of the caprellid amphipod *Caprella scaura* typica Mayer, 1890. Marine Behaviour and Physiology 12:217–230.

Lin, C.-P., B.D. Danforth, and T.K. Wood. 2004. Molecular phylogenetics and evolution of maternal care in membracine treehoppers. Systematic Biology 53:400–421.

Linsenmair, K.E. 1984. Comparative studies on the social behaviour of the desert isopod *Hemilepistus reaumuri* and of a *Porcellio* species. Symposium of the Zoological Society of London 53:423–453.

Linsenmair, K.E. 1987. Kin recognition in subsocial arthropods, in particular in the desert isopod *Hemilepistus reaumuri*. Pages 121–208 *in*: D.J.C. Fletcher and C.D. Michener, editors. Kin recognition in animals. John Wiley and Sons, Chichester.

Little, E.E. 1975. Chemical communication in maternal behavior of crayfish. Nature 255:400–401.

Little, E.E. 1976. Ontogeny of maternal behavior and brood pheromone in crayfish. Journal of Comparative Physiology 112:133–142.

Mason, J.C. 1970. Maternal-offspring behavior of the crayfish, *Pacifastacus trowbridgi* (Stimpson). American Midland Naturalist 84:463–473.

Mattson, S., and T. Cedhagen. 1989. Aspects of the behaviour and ecology of *Dyopedos monacanthus* (Metzger) and *D. porrectus* Bate, with comparative notes on *Dulichia tuberculata* Boeck (Crustacea: Amphipoda: Podoceridae). Journal of Experimental Marine Biology and Ecology 127:253–272.

McCloskey, I.R. 1970. A new species of *Dulichia* (Amphipoda, Podoceridae) commensal with a sea urchin. Pacific Science 24:90–98.

Menzies, R.J. 1957. The marine borer family Limnoridae (Crustacea, Isopoda). Part I: Northern and Central America: systematics, distribution and ecology. Bulletin of the Marine Science of the Gulf and Caribbean 7:101–200.

Morgan, G.J. 1987. Brooding of juveniles and observations on dispersal of young in the spider crab *Paranaxia serpulifera* (Guerin) (Decapoda, Brachyura, Majidae) from Western Australia. Records of the Western Australian Museum 13:337–343.

Murata, Y., and K. Wada. 2002. Population and reproductive biology of an intertidal sandstone-boring isopod, *Sphaeroma wadai* Nunomura, 1994. Journal of Natural History 36:25–35.

Myles, T.G., and W.L. Nutting. 1988. Termite eusocial evolution: a re-examination of Bartz's hypothesis and assumptions. Quarterly Review of Biology 63:1–23.

Nalepa, C.A. 1984. Colony composition, protozoan transfer and some life-history characteristics of the woodroach *Cryptocercus punctulatus* Scudder. Behavioral Ecology and Sociobiology 14:273–279.

Ng, P.K.L., and C.G.S. Tan. 1995. *Geosesarma notophorum* sp. nov. (Decapoda, Brachyura, Grapsidae, Sesarminae), a terrestrial crab from Sumatra, with novel brooding behaviour. Crustaceana 68:390–395.

O'Connor, D., and R. Shine. 2003. Lizards in "nuclear families": a novel reptilian social system in *Egernia saxatilis* (Scincidae). Molecular Ecology 12:743–753.

Park, Y.C., and J.C. Choe. 2003. Effects of parental care on offspring growth in the Korean wood-feeding cockroach, *Cryptocercus kyebangensis*. Journal of Ethology 21:71–77.

Queller, D.C., and J.E. Strassmann. 1998. Kin selection and social insects: social insects provide the most surprising predictions and satisfying tests of kin selection. BioScience 48:165–175.

Rauter, C.M., and A.J. Moore. 2002. Evolutionary importance of parental care performance, food resources, and direct and indirect genetic effects in a burying beetle. Journal of Evolutionary Biology 15:407–417.

Reid, M.L., and B.D. Roitberg. 1994. Benefits of prolonged residence with mates and brood in a bark beetle (Coleoptera: Scolytidae). Oikos 70:140–148.

Richter, G. 1978a. Einige Beobachtungen zur Lebensweise des Flohkrebses *Siphonoecetes della-vallei*. Natur und Museum 108:259–266.

Richter, G. 1978b. Beobachtungen zu Entwicklung und Verhalten von *Phronima sedentaria* (Forskål), Amphipoda. Senckenbergiana Maritima 10:229–242.

Schwartz, M.P., S.M. Thierney, J. Zammit, P.M. Schwarz, and S. Fuller. 2005. Brood provisioning and colony composition of a Malagasy species of *Halterapis*: implications for social evolution in the allodapine bees (Hymenoptera: Apidae: Xylocopinae). Annals of the Entomological Society of America 98:126–133.

Scudamore, H.H. 1948. Factors influencing molting and sexual cycles in the crayfish. Biological Bulletin 95:229–237.

Seibt, U., and W. Wickler. 1979. The biological significance of the pair-bond in the shrimp *Hymenocera picta*. Zeitschrift für Tierpsychologie 50:166–179.

Sheader, M., and F.-S. Chia. 1970. Development, fecundity and brooding behaviour of the amphipod, *Marinogammarus obtusatus*. Journal of the Marine Biological Association of the United Kingdom 50:1079–1099.

Shellman-Reeve, J.S. 1997. The spectrum of eusociality in termites. Pages 52–93 *in*: J.C. Choe and B.J. Crespi, editors. Social behavior in insects and arachnids. Cambridge University Press, Cambridge.

Shillaker, R.O., and P.G. Moore. 1987. The biology of brooding in the amphipods *Lembos websteri* Bate and *Corophium bonnellii* Milne Edwards. Journal of Experimental Marine Biology and Ecology 110:113–132.

Spanier, E., J.S. Cobb, and M.J. James. 1993. Why are there no reports of eusocial marine crustaceans? Oikos 67:573–577.

Stephan, H. 1980. Lebensweise, Biologie und Ethologie eines sozial lebenden Amphipoden (*Dulichia porrecta*, *Dulichia monacantha* und *Dulichia falcata*—Crustacea, Malacostraca). Unpublished Ph.D. thesis, University of Kiel, Kiel, Germany.

Stern, D.L. 1994. A phylogenetic analysis of soldier evolution in the aphid family Hormaphididae. Proceedings of the Royal Society of London, Series B 256:203–209.

Svavarsson, J., and B. Davidsdottir. 1995. *Cibicides* spp. (Protozoa, Foraminifera) as epizoites on the Arctic antenna-brooding *Arcturus baffini* (Crustacea, Isopoda, Valvifera). Polar Biology 15:569–574.

Tallamy, D.W., and W.P. Brown. 1999. Semelparity and the evolution of maternal care in insects. Animal Behaviour 57:727–730.

Tallamy, D.W., and T.K. Wood. 1986. Convergence patterns in subsocial insects. Annual Review of Entomology 31:369–390.

Thiel, M. 1997a. Reproductive biology of an epibenthic amphipod (*Dyopedos monacanthus*) with extended parental care. Journal of the Marine Biological Association of the United Kingdom 77:1059–1072.

Thiel, M. 1997b. Extended parental care in estuarine amphipods. Unpublished Ph.D. thesis, University of Maine, Orono, Me.

Thiel, M. 1997c. Reproductive biology of a filter-feeding amphipod, *Leptocheirus pinguis*, with extended parental care. Marine Biology 130: 249–258.

Thiel, M. 1998. Reproductive biology of a deposit-feeding amphipod, *Casco bigelowi*, with extended parental care. Marine Biology 132:107–116.

Thiel, M. 1999a. Parental care behaviour in crustaceans—a comparative overview. Crustacean Issues 12:211–226.

Thiel, M. 1999b. Extended parental care in marine amphipods. II. Maternal protection of juveniles from predation. Journal of Experimental Marine Biology and Ecology 234:235–253.

Thiel, M. 1999c. Reproductive biology of a wood-boring isopod (*Sphaeroma terebrans*) with extended parental care. Marine Biology 135:321–333.

Thiel, M. 1999d. Duration of extended parental care in marine amphipods. Journal of Crustacean Biology 19:60–71.

Thiel, M. 2000a. Population and reproductive biology of two sibling amphipod species from ascidians and sponges. Marine Biology 137:661–674.

Thiel, M. 2000b. Juvenile *Sphaeroma quadridentatum* invading female-offspring groups of *Sphaeroma terebrans*. Journal of Natural History 34:737–746.

Thiel, M. 2003a. Extended parental care in crustaceans—an update. Revista Chilena de Historia Natural 76:205–218.

Thiel, M. 2003b. Reproductive biology of *Limnoria chilensis*: another boring peracarid species with extended parental care. Journal of Natural History 37:1713–1726.

Thiel, M., S. Sampson, and L. Watling. 1997. Extended parental care in two endobenthic amphipods. Journal of Natural History 31:713–725.

Thorne, B.L., and J.F.A. Traniello. 2003. Comparative social biology of basal taxa of ants and termites. Annual Review of Entomology 48:283–306.

Tóth, E., and J.E. Duffy. 2005. Coordinated group response to nest intruders in social shrimp. Biology Letters 1:49–52.

Trivers, R.L. 1974. Parent-offspring conflict. American Zoologist 14:249–264.

Vannini, M., G. Innocenti, and R.K. Ruwa. 1993. Family group structure in mysids, commensals of hermit crabs (Crustacea). Tropical Zoology 6:189–205.

Versluis, M., B. Schmitz, A. von der Heydt, and D. Lohse. 2000. How snapping shrimp snap: through cavitating bubbles. Science 289:2114–2117.

Wilson, E.O., 1971. The Insect Societies. Belknap Press, Cambridge, Mass.

Wilson, E.O. 1975. Sociobiology. The abridged edition. Belknap, Harvard Press, Cambridge, Mass.

Behavioral Ecology of Semiterrestrial Crayfish

Alastair M.M. Richardson

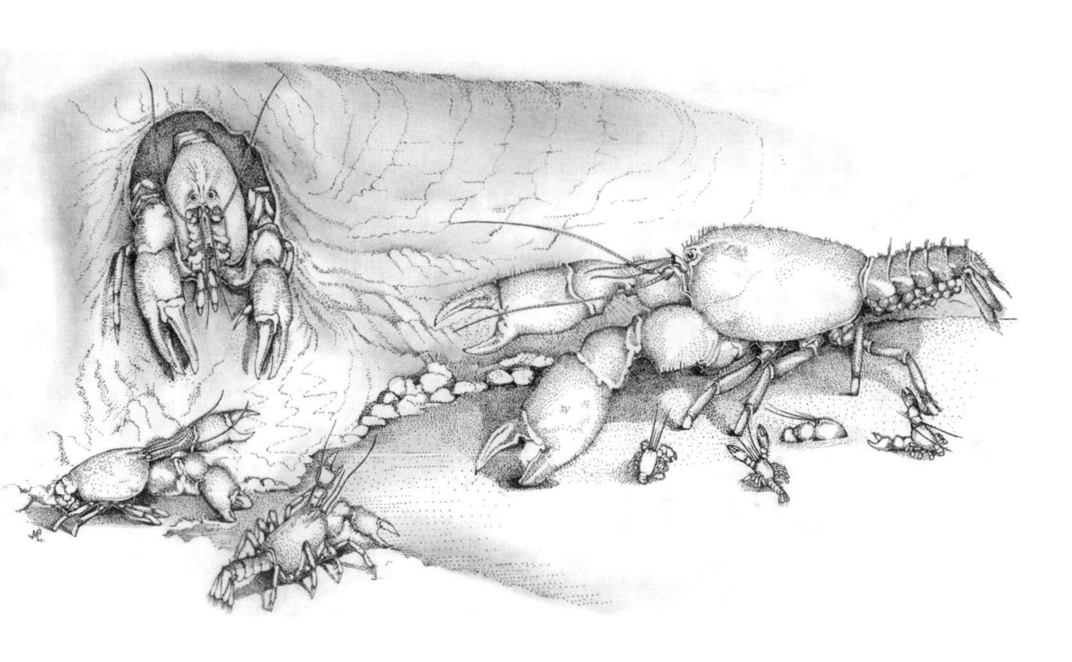

15

Freshwater crayfish occupy a range of habitats from open water in rivers and lakes, through the margins of open waters and swamps, to situations that may be kilometers from the nearest surface water. Although they remain dependent on free water, crayfish in these latter situations are almost entirely confined to a burrow and can be described as semiterrestrial rather than aquatic. This lifestyle presents a contrast to that of many of the other crustaceans discussed in this volume but provides some of the conditions under which the evolution of social behavior might be expected via the subsocial route (Wilson 1975).

Being confined to a burrow clearly imposes severe restrictions on the reproductive behavior of crayfish. Opportunities to leave the burrow are often limited by climatic conditions and the risk of predation, and this constrains foraging, mating, and dispersal. On the other hand, the burrow may be seen as a self-contained system that supplies all the animal's needs. Under these circumstances, there will be selective pressure for the inhabitants to coexist and for the generations to overlap, especially where opportunities for dispersal are infrequent, creating two of the three conditions that have been suggested to be necessary for the evolution of parental care (Wilson 1975, Clutton-Brock 1991; see also chapter 16): the juveniles face harsh physical conditions, are exposed to severe competition or predation, or rely on resources that they are unable to obtain by themselves.

This brief review summarizes the reproductive behavior and ecology of burrowing crayfish and discusses what little is known about the development of social behavior in the group, rather than providing a model for the evolution of social behavior. Because of their burrowing habit, limited information is available about the behavior of semiterrestrial crayfish; some of what follows has been extrapolated from observations of open-water species.

Taxonomy and Ecology

Systematics

Freshwater crayfish are found on all the continents except Africa and Antarctica and on a number of oceanic islands (Madagascar, New Zealand, and islands in Melanesia and the Caribbean) (Hobbs 1988). Two phylogenetically well-supported (Scholtz 2002) superfamilies of freshwater crayfish, Astacoidea and Parastacoidea, are found in the northern and southern hemispheres, respectively; Astacoidea currently includes two families, Astacidae and Cambaridae, while Parastacoidea consists of a single family, Parastacidae. The world fauna includes more than 570 species of freshwater crayfish (Fetzner 2003), with local centers of diversity in the southeast of North America and the southeast of Australia, where most of the semiterrestrial species are found. Crayfish are largely absent from the tropics, but some species inhabit New Guinea and the surrounding islands, Madagascar, and Central America.

The world fauna is quite well described at the species level, following recent studies in Australia (Morgan 1986, 1988, 1997, Horwitz 1990, Austin 1996, Austin and Knott 1996, Horwitz and Adams 2000, Hansen and Richardson 2006), but the generic framework of the North American fauna is in need of revision (Fetzner 1996; K.A. Crandall, personal communication). Phylogenetic relationships at the generic and family levels are emerging (Crandall and Fetzner 1995, Crandall et al. 1999, 2000).

In the Parastacidae, at least, these suggest that high dependence on a burrow has evolved more than once within the family (Crandall et al. 1999).

Burrowing Habits and Burrow Types

The colonization of land by the Crustacea has been modest compared with that of other arthropod groups, and among the decapods, the astacidean crayfish have established only a very limited presence on land. Like the other crustacean land colonists (isopods, amphipods, anomuran and brachyuran crabs; Bliss and Mantel 1968), the astacideans have adapted to life on land not only through morphological and physiological changes but also by developments in their behavior. The principal behavioral characteristic that has enabled crayfish to live independently of surface water has been the further development of their ancestral capacity to construct burrows. While burrowing is also seen in some of the other terrestrial crustaceans, such as the desert crab *Holthusiana transversa* (Greenaway 1984) and desert isopods (chapter 16), all semiterrestrial crayfish construct burrows that are often complex, and the vast majority are confined to their burrows for most of their lives. Perhaps the only exception is the Lamington spiny crayfish, *Euastacus sulcatus*, which lives in subtropical rain forests in Queensland, Australia. Although it constructs only simple burrows at the edges of streams, it makes extensive excursions on land through the forest to forage, returning to its home burrow (Furse and Wild 2002).

All crayfish burrow to a greater or lesser extent, whether they live permanently in open waters or on land (Berrill and Chenoweth 1982). The burrows range in structure from very simple shelters under rocks or logs in streams and lakes to extensive and complex systems on land that may ramify horizontally for several meters or descend more than 4 meters into the soil. Burrows primarily provide shelter, initially from predation (especially immediately postmolt and during brooding) in open-water species, and increasingly from harsh physical conditions in the more terrestrial species. For some species, the burrow may also supply much of the inhabitants' food, in the form of plant and animal material (Growns and Richardson 1988, Gutiérrez-Yurrita et al. 1998).

Crayfish have been classified on the basis of their burrowing capabilities, originally by Hobbs (1942). Primary burrowers are those that are restricted to their burrows for the majority of their lives; secondary burrowers wander into open water during rainy seasons; tertiary burrowers primarily live in open water and only occupy burrows during periods of drought, or sometimes in the mating season. In the North American fauna, primary burrowers are found mostly in the Cambaridae, particularly the genera *Cambarus*, *Procambarus*, *Distocambarus*, and *Fallicambarus*. In the Parastacidae, primary burrowers are found in *Engaeus*, *Engaewa*, *Parastacoides*, *Parastacus*, *Tenuibranchiurus*, and *Virilastacus*.

It is also possible to classify the types of burrow that crayfish construct on the basis of their habitat, particularly their water supply (Fig. 15.1) (Horwitz and Richardson 1986); this approach recognizes that the same species may construct different burrow types in different locations. Type 1 burrows are those found in or at the edges of open water bodies; they may have all their entrances normally under water (type 1a) or some of them normally above water (type 1b). Type 2 burrows are not associated with surface water but extend downward, sometimes several meters, to meet the water table (e.g., in the "prairie crayfish," *Procambarus hagenianus* and related species; Fitzpatrick 1975). These burrows can be found many kilometers from open water, provided there is an accessible water table. Type 3 burrows are also found away from open water, but in this case the burrow does not extend to the water table. Instead, it contains water that has drained into the burrow from the surface and is stored in chambers; these burrows are necessarily found in impervious clay soils. Type 3 burrows were thought to be confined to southeast Australia, where they are constructed by several species in the genus *Engaeus* (Fig. 15.2), but recent work in southeastern North America (S. Welch and A. Eversole, personal communication) suggests that similar burrows may be found there, constructed by species from the genus *Cambarus*. Type 3 burrows (and type 2 burrows in some situations, e.g., peatlands and heavy clay soils) are likely to be much longer lived than type 1 burrows, outlasting many generations of their inhabitants. The characteristics of these burrow types and some of the ecological correlates discussed below are summarized in Table 15.1.

In some situations, crayfish burrows may represent a valuable resource that must be defended, like the dwellings of snapping shrimps (chapter 18) or desert isopods (chapter 16). This is most likely when the conditions do not allow burrows to be readily excavated, as in type 2 and especially type 3 habitats. However, the extensive nature of most crayfish burrows, with multiple entrances, suggests that they are not designed

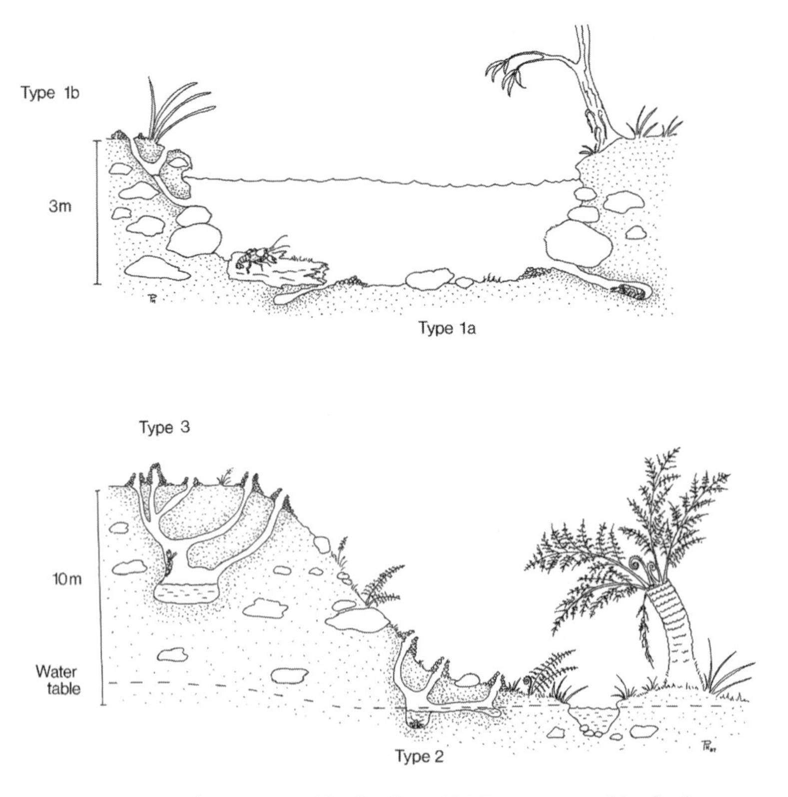

Figure 15.1 Burrow types (Horwitz and Richardson 1986) constructed by freshwater crayfish. (a) Type 1a and 1b burrows associated with open water. (b) Type 2 and 3 burrows. Drawings by Premek Hamr.

Figure 15.2 Openings of freshwater crayfish burrows. (a) *Astacopsis tricornis* at the entrance of a type 1a burrow in Lake St. Clair, Tasmania. Photo by Jon Bryan. (b) Type 2 burrow of *Parastacoides tasmanicus* in sedgeland in western Tasmania. (c) Cryptic entrance of type 2 burrow of *Engaewa reducta* in Western Australia. (d) Type 3 burrow of *Engaeus cisternarius* in temperate rainforest, northwest Tasmania. (e) Large "chimney" at entrance of type 3 burrow of *E. orramakunna* in northeast Tasmanian rainforest. Photo by Niall Doran.

to be as readily defensible as, for example, the burrows of the desert isopod *Hemilepistus reaumuri*, which have only one single opening (chapter 16).

Morphological Adaptations

Burrowing crayfish show morphological adaptations to life in burrows, and these tend to be exaggerated in the more terrestrial species. Crayfish living in low oxygen situations, which may include burrows in the beds of streams and lakes, increase their gill area by vaulting of the carapace and increasing the volume of the gill chamber, producing a tall narrow carapace with a highly reduced areola, that is, the dorsal area

Table 15.1. Burrow classification of Horwitz and Richardson (1986) (see Fig. 15.1), with ecological, morphological, and reproductive correlates.

Burrow Type	Water Source	Burrow Development	Range and Typical Genera	Burrow Occupation	Burrow Function	Typical Crayfish Morphology	Brood Size (Range)[a]	Duration of Mother–Young Association	Dispersal of Young
1a	Open water	Simple, usually short and unbranching, under rock or log; all entrances under water	All continents:[b] *Orconectes*, *Austropotamobius*, *Pacifastacus*,	Temporary	Breeding, predator refuge	Body and claws spiny; claws held horizontally, usually isomorphic	90–1,000	Weeks	Open water
1b	Open water	Simple, openings under water and on land	All continents: *Euastacus*, *Cherax*,	Temporary	Breeding, predator refuge	Body and claws spiny; claws held horizontally, usually isomorphic	36–126	Weeks	Open water
2	Water table	Simple to complex; several openings; sometimes with ramifying subsurface tunnels; depth may be > 4 m	N. and S. America, Australia: *Cambarus*, *Procambarus*, *Fallicambarus*, *Engaeus*, *Cherax*, *Parastacoides*	Permanent; sometimes surface excursions to forage and find mates	Desiccation, predator refuge; food source	Body and claws smooth; claws held at 45°, dimorphic or isomorphic; tail fan sometimes terminally spiny	22–108	<1 year	Overland in wet season (annual)?
3	Surface runoff	Complex; sometimes with ramifying subsurface tunnels; one to several water storage chambers	S.E. Australia (and S.E. U.S.?) only: *Engaeus*, *Cambarus?*	Permanent; sometimes surface excursions to forage and find mates	Desiccation, predator refuge; food source	Body and claws smooth; eyes small; claws carry sensory setae; claws held vertically, dimorphic or isomorphic; abdomen reduced	13–61	<3 years	Overland, following infrequent (>1 year) rainfall events?

[a]Australian species only.
[b] Freshwater crayfish do not occur naturally in Africa or Antarctica.

between the cervical grooves that define the upper limit of the gill chamber (Holdich and Lowery 1988). The orientation of the chelae is correlated with burrowing habit, at least within the Australian fauna (Riek 1969); open-water species hold the chelae in the horizontal plane, whereas with increasing dependence on a burrow, the chelae adopt an increasingly vertical position, which allows them to be held closely against the carapace, perhaps so that the animal can fit tightly into its tunnels. The chelae of primary burrowers are often dimorphic and bear tufts of long sensory setae; pronounced claw dimorphism implies functional differences between the claws, and while little is known about their function (Horwitz 1990), burrow defense is one possibility. Burrowing crayfish are generally less spiny and tuberculate than open-water species, though terminal spination of the tail fan may be associated with burrowing (Richardson and Swain 2002).

Particularly in the strongly terrestrial *Engaeus* species from Australia (Hobbs 1988), but also in some of the North American primary burrowers (e.g., *Fallicambarus devastator*: Hobbs and Whiteman 1991), there is a marked reduction in the development of the abdomen (Fig. 15.3). Among the 34 species of *Engaeus* treated by Horwitz (1990), there is a clear relationship between carapace/abdomen ratio and burrow type (Fig. 15.4). These short-abdomen species do not employ the typical astacuran tailflip, and carrying eggs appears to be the main function of the abdomen. The reduction in abdomen size is weakly associated with a reduction in brood size (Fig. 15.5), but there is no apparent compensation in egg size (Fig. 15.6) in *Engaeus*

Figure 15.3 *Engaeus orramakunna* Horwitz, the Mt. Arthur burrowing crayfish from northeast Tasmania, showing the setose chelae and short abdomen typical of *Engaeus* species that construct type 3 burrows. Photo by Niall Doran.

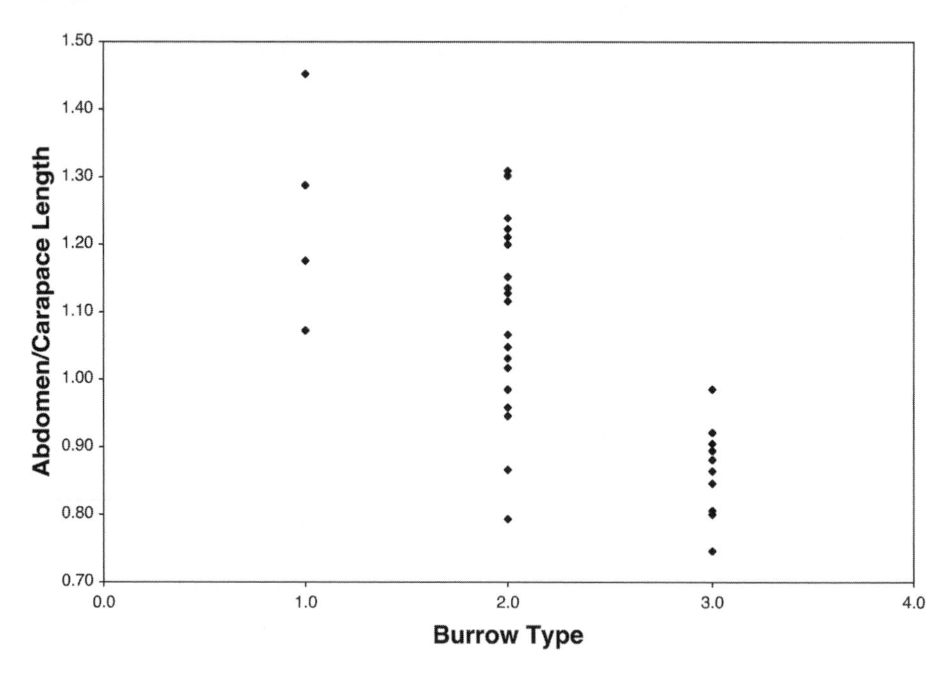

Figure 15.4 Relationship between abdomen:carapace ratio and burrow type in *Engaeus* species (data from Horwitz 1990). Many species can be found in more than one burrow type, but they have been allocated to a single type on the balance of the ecological observations recorded by Horwitz (1990).

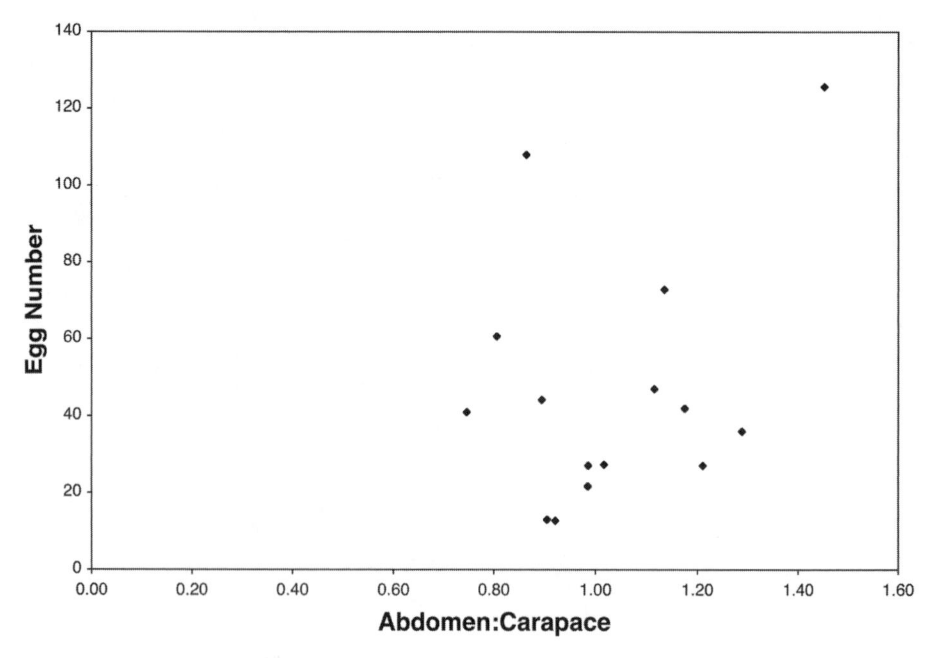

Figure 15.5 Relationship between abdomen:carapace ratio and egg number (brood size) for 15 *Engaeus* species (data from Horwitz et al. 1984, Horwitz 1990).

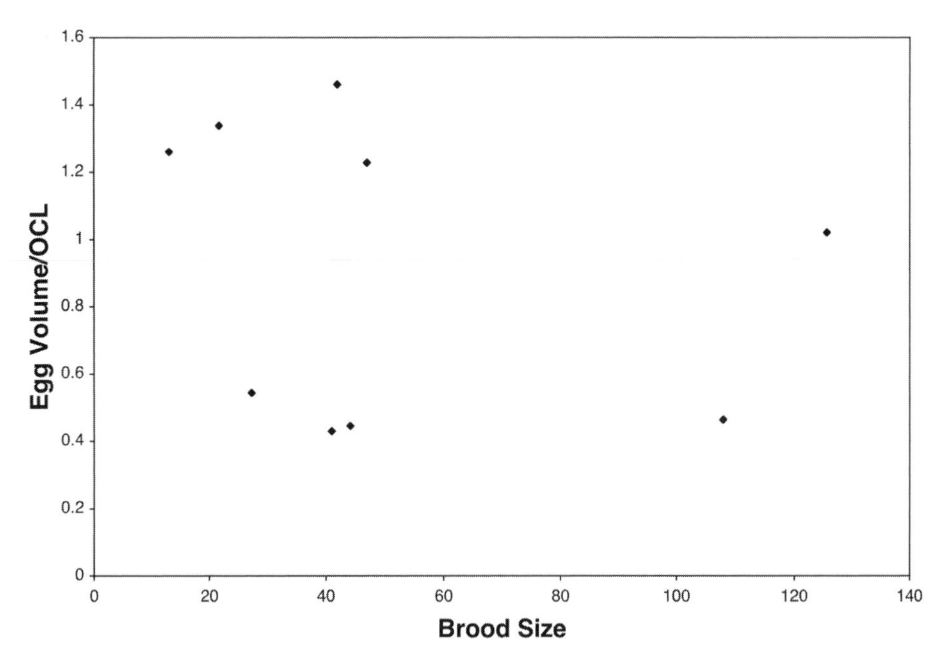

Figure 15.6 Relationship between brood size and egg volume (scaled by orbit carapace length [OCL]) in nine *Engaeus* species (data from Horwitz 1990).

species for which data are available (Horwitz 1990). In general, the fecundity of burrowing crayfish seems to be lower than that of open-water species (Lowery 1988).

Reproductive Biology

Mating

The life history of freshwater crayfish follows that of many decapods, except that there is no free-living larval stage. After hatching, juveniles remain attached to their mother's pleopods for two or three molts before becoming independent as miniature adults.

Few data are available on the life history and breeding biology of semiterrestrial crayfish compared to the wealth of information available for open-water species, especially those exploited commercially, such as *Procambarus clarkii* (Huner and Barr 1991). On reaching sexual maturity, most semiterrestrial species breed annually, but it is likely that warm climate species produce more than one brood in a year, as open-water tropical species do, while in colder climates breeding may only be biennial. Hamr and Richardson (1994) found that *Parastacoides tasmanicus*, a primary burrower, which lives in type 2 burrows in Tasmanian peatlands, mates in March (austral autumn) and carries eggs during the austral winter. The eggs hatch in late spring (November), and the young stay attached to the female's pleopods until mid-summer. They then remain in the maternal burrow for a further 12 months until the following summer (January), over which time the mother's ovaries mature. The female then mates again in the following autumn, two years after the previous mating. Population surveys showed that approximately half of the females are carrying eggs at any time in winter. Circumstantial evidence suggests a similar pattern in other Tasmanian *Parastacoides* species, though at least one species mates in spring rather than autumn. *Engaeus* species for which evidence is available (Horwitz 1990) mostly appear to mate in spring, though some species from the mainland of Australia (e.g., *E. tuberculatus*, *E. fultoni*) may mate in winter. There is insufficient evidence to assess whether any *Engaeus* species show biennial breeding.

Most North American primary burrowers appear to mate in spring, or in spring and autumn (Hobbs 1981, Hobbs and Robison 1989). Johnston and Figiel (1997) found that *Fallicambarus gordoni*, in Mississippi, mated in late spring and extruded eggs in early autumn or winter, since the bogs in which they live are dry over summer and the animals are inactive during that time. Mating season may depend on latitude; *F. fodiens* mates in summer–autumn in Ohio (Norrocky 1991) but in May in Canada (Crocker and Barr 1968, Williams et al. 1974).

Hobbs (1981) suggested that semiterrestrial crayfish live solitary lives as adults in isolated burrows, even though these burrows may be in close proximity (Hobbs and Whiteman 1991). However, Norrocky (1991) used burrow-specific traps to sample populations of *Fallicambarus fodiens*, a secondary burrower, and found extensive evidence of burrow sharing and turnover of burrow ownership; every combination of sex and reproductive stage were found together, and *F. fodiens* was also found to cohabit with *Cambarus diogenes*. No equivalent data seem to be available for other semiterrestrial species.

Although females of the open-water species *Cherax quadricarinatus*, when kept in aquaria, apparently visit males for mating when they are sexually receptive (Barki and Karplus 1999), in burrowing crayfish it is generally believed that the males travel to the females' burrows to mate (*Fallicambarus fodiens*, Hobbs and Robison 1989; *F. devastator*, Hobbs and Whiteman 1991; *F. fodiens*, Norrocky 1991). It is likely that males detect female burrows chemically, since sex pheromones have been identified in *Procambarus clarkii* (Ameyaw-Akumfi and Hazlett 1975). In regular surveys of a population of *Parastacoides tasmanicus*, Hamr and Richardson (1994) found that burrow occupancy dropped and males and females were found together in burrows for a two-week period in autumn; females carried spermatophores and eggs by the end of this period, after which males and females were no longer found together.

In *Orconectes virilis*, the male stays with the female in their burrow after mating and over winter (Ameyaw-Akumfi 1976 as cited in Gherardi 2002), while in *Procambarus clarkii* and *P. acutus* male and female may both occupy a burrow during the dry season while the female is carrying eggs, but the male's role is unclear.

Mating behavior in burrowing crayfish has rarely been described (Hamr 1991) but appears similar to that of open-water species (see descriptions in Gherardi 2002, Holdich 2002). Despite the differences in the mechanism of sperm transfer (parastacids lack the modified first pleopod of male Astacoidea), mating in parastacid crayfish is basically the same as in Astacoidea, but with the deposition of a spermatophore between the bases of the female's fourth walking legs, from which the eggs are fertilized as they are extruded. Thus, there is no apparent mechanism for sperm storage in parastacids, in contrast to the North American Cambaridae, in which the spermatophore is deposited into a receptacle, the annulus ventralis, and eggs may be extruded and fertilized weeks or months after mating (Albaugh 1973 as cited in Walker et al. 2002). Walker et al. (2002) present genetic evidence that broods of the stream-dwelling cambarid *Orconectes placidus* were often sired by two or more males, but no similar data are available for semiterrestrial burrowers.

Once they are cemented to her pleopods, the female grooms the eggs with her thoracic appendages and aerates them by slow movements of the pleopods. The eggs hatch after two to eight months or more depending on species and temperature, and the hatchlings remain attached to their mother's pleopods, using special hooks on the tips of the dactylus of the fourth and fifth pereopods (parastacids) or hooks on the tips of the dactylus and propodus of the first cheliped (Astacoidea). After the second or third intermolt stage, the juveniles start to make excursions from the female's abdomen.

Mother–Brood Interactions

Juveniles of some aggressive open-water species disperse from the mother immediately and show aggression to other juveniles from the start. However, it is likely that all burrowing crayfish spend some time associated with their mother after leaving the pleopods. Juveniles of the secondary burrower *Procambarus alleni* forage in the immediate vicinity of their mother but for a few days return to her pleopods when disturbed (Bovbjerg 1956). To facilitate the return of the juveniles, the female adopts an unusual posture, raising her body with her walking legs and extending her abdomen (Ameyaw-Akumfi 1976 as cited in Gherardi 2002). Juvenile crayfish orient toward other juveniles visually; their return to the female may also be visually controlled since they orientate toward mother-sized objects (Pieplow 1938 as cited in Gherardi 2002).

Brooding cambarid females produce a pheromone that is attractive to the juveniles (Little 1975, 1976). Juveniles will orientate toward water from recently berried females; they cannot distinguish their own mother from other brooding females, but they do distinguish between species (Pieplow 1938 as cited in Gherardi 2002; Little 1975, 1976). Production of this maternal pheromone decreases from the time when the final stage juveniles are on the female, but in *Procambarus clarkii* it can be produced over seven stages of juvenile development, lasting more than three months. Independence from the female develops with size rather than age; among a same-aged brood of *P. clarkii*, the largest animals became independent first, while middle-sized ones occasionally returned to their mother, and the smallest remained on pleopods most of the time (Ameyaw-Akumfi 1976 as cited in Gherardi 2002).

Figler et al. (2001) demonstrated that female *Procambarus clarkii* defended their offspring against invading males in an area around a shelter. This maternal aggression was observed in ovigerous females, those carrying stage 1 offspring and those tending stage 2 offspring, some of which were foraging freely.

Longer term associations between mother and juveniles of burrowing crayfish may be imposed by climatic conditions. *Procambarus clarkii* mothers and juveniles can be confined in the burrow for three months or more by dry conditions (Huner and Barr 1991). Juvenile *Fallicambarus fodiens* in Canada are forced to remain with the female over winter (Williams et al. 1974). Horwitz and Knott (1983) recorded males of the Western Australian *Cherax plebejus* cohabiting with gravid females and proposed that short-lived family units would be formed before the juveniles disperse during the winter rains.

In primary burrowers, especially those in type 3 burrows, the association between mother and offspring may be much longer. Juveniles of the biennially breeding *Parastacoides tasmanicus* remain in the maternal burrow (type 2) for at least 14 months, living in the terminal chamber (Hamr and Richardson 1994). When the female mates again, the young apparently disperse from the chamber but can still be found, albeit in reduced numbers, in small side burrows and cavities near the top of the burrow. Juveniles of *Fallicambarus gordoni* and *F. fodiens* are found in the maternal burrow (Norrocky 1991, Johnston and Figiel 1997). Horwitz et al. (1984) recorded the presence of four generations (mother plus three year classes, the oldest animals being half to two-thirds the size of the adult) in type 3 burrows of *Engaeus leptorhynchus* in northeast Tasmania, and similarly overlapping generations were found in *E. cisternarius* burrows from northwest Tasmania (Suter and Richardson 1977). This behavior is almost certainly the source of earlier observations of "communally dwelling" *Engaeus* species (Clark 1936a, 1936b, Riek 1969). Hobbs and Whiteman (1991) record the presence together of male, female, and juvenile *Fallicambarus devastator* in burrows in prairie regions of eastern Texas, perhaps as a result of drought, and multiple cohorts have been found in the burrows of *Distocambarus crockeri* in South Carolina (S. Welch, personal communication). Hobbs and Whiteman (1991) did not observe any cannibalism amongst groups of *F. devastator* and noted that a family of another primary burrower, *Procambarus* (*Hagenides*) *pygmaeus*, raised in an aquarium showed no maternal–offspring aggression, or offspring–offspring aggression after the juveniles became independent of their mother.

These observations of prolonged mother–brood associations raise the question of how the female's behavior is modified, since in most open-water species the female will consume her young after the initial few weeks of association if they are confined together. Little (1975, 1976) showed that cannibalism was inhibited in female crayfish (*Orconectes sanborni*, *O. virilis*, and *Procambarus clarkii*) carrying young and suggested that this "maternal" state was maintained by the physical presence of at least some young on the pleopods. Blinded females showed the same response, and increasing the number of young in the female's water did not maintain the maternal state; further, the maternal state started to decline when the number of young on her pleopods decreased below a certain number, all suggesting that in these species it is the physical presence of the young on the pleopods that induces the maternal state. Females can, however, distinguish dead juveniles at any stage and will eat them (Little 1976).

Since it persists long after the young have left their mother's pleopods, the prolonged female tolerance in semiterrestrial crayfish must be controlled by some other mechanism than the presence of young on the pleopods. Juvenile open-water crayfish are able to learn that males and nonbreeding females are dangerous, avoiding water conditioned by them after a first encounter (Little 1976). Kinship does not appear to affect the survival of juvenile of the open-water Australian species *Cherax quadricarinatus*, perhaps because the dispersal of the young is normally rapid in the field (Karplus et al. 1995). It remains to be seen whether the lack of aggression between the very closely related animals in type 2 and 3 burrows is chemically mediated, or by some other sensory means, or whether, as Hobbs and Whiteman (1991) suggest in *Procambarus (Hagenides) pygmaeus*, it is now genetically controlled.

Dispersal

Direct observations of dispersal by burrowing crayfish are difficult to obtain. When juveniles leave the maternal burrow is unclear, and in some species opportunities for dispersal must be very rare, because of the relatively dry surrounding habitats, reflecting the situation for eusocial mole-rats (Jarvis et al. 1994). A rainy season, when there is surface water, probably provides this opportunity, but even then the chances of establishment in a new burrow are small. In Tasmanian peatlands (an unusual habitat for burrowing crayfish), the construction of new burrows by juveniles seems very rare: over a 10-year period, no new burrow systems (juvenile or adult) of *Parastacoides tasmanicus* were observed in a 432 m^2 plot (Richardson and Swain 1991), and it is rare to see the small-diameter burrows of juvenile animals anywhere other than around the entrance of the maternal burrow. However, Growns and Richardson (1988) noted a clear correlation between the volume of the burrow systems of this species and the size of the occupant. Because the creation of new burrows seems to be so rare, and the rate of increase in the volume of adult burrows is very slow (Richardson and Swain 1991), this implies a process of burrow swapping, loosely analogous to shell exchange and eviction by hermit crabs (Hazlett 1981), and this in turn suggests that dispersing juveniles mostly take over the vacated burrows of slightly larger animals. Hobbs and Whiteman (1991) also note a loose correlation between burrow volume and size of occupant in *Fallicambarus devastator* (but in this case, there may be much more rapid burrow turnover).

In wetter habitats in clay or sandy soils, where it is easier for juveniles to establish, new burrows of juvenile animals are often seen (Hobbs 1981), and species in these habitats seem to have innate burrowing behaviors (Grow and Merchant 1980). Hobbs (1981) remarked that he had never seen lateral passages that might have been constructed by dispersing juveniles in the burrows of *Cambarus (Hagenides)* species, but Hobbs and Robison (1989) report that *Fallicambarus fodiens* juveniles are often seen dispersing over the surface following rains.

Since the chances of survival of dispersing juveniles are likely to increase with body size, there is probably strong selective pressure on juveniles in the more terrestrial burrowing species in type 2 and 3 burrows to reach as large a body size as possible before attempting to disperse. This will eventually lead to competition with their mother for food and space within the maternal burrow, which is likely to lead either to the evolution of early dispersal or cooperative foraging with their mother (chapter 14).

Comparison with Other Taxa

If social behavior has evolved in the semiterrestrial crayfish, it has done so in a situation that is more comparable with the eusocial insects (Wilson 1975) and eusocial mammals (Jarvis et al. 1994) than with the aquatic crustaceans, that is, in relatively arid terrestrial environments dominated by vascular plants (chapter 14). Among the crustaceans, the closest parallels are with the desert isopods (chapter 16) and bromeliad crabs (chapter 17). Crayfish that burrow in underwater substrata face quite different conditions than do the semiterrestrial burrowers. Atkinson and Taylor (1988) pointed out that aquatic burrowers gain protection from predators but face problems of hypoxia and hypercapnia, problems that can be relieved be leaving, or ventilating, the burrow. Terrestrial burrowers, on the other hand, while rarely affected by hypoxia, are always challenged by the problems of living in air if they leave their burrow, so their burrows are even more a refuge from desiccation than they are from predation. Consequently, there may be greater selective pressure on semiterrestrial crayfish for coexistence between mother and young, since while the young of open-water species can at least survive the physical conditions outside the burrow, those of semiterrestrial species cannot, except in rare circumstances. Not surprisingly, in most truly aquatic burrowers, mother–offspring associations are of relatively short duration. In the river-dwelling *Pacifastacus trowbridgi*, for example, the female retreats to a shelter or shallow burrow after mating; the young start to make excursions from the female in molt stage 2 and are almost immediately at risk of cannibalism by their mother (Mason 1970). Bechler (1981) observed the brooding behavior of the troglobitic crayfish *Orconectes pellucidus* and found that 17 days after hatching the young began to crawl about their mother's body; 27 days after hatching, they made excursions from the female, and by 40 days they had left completely and the female began to eat them. A remarkable exception is *Paranephrops zealandicus*, a stream-dwelling New Zealand species, in which juveniles apparently remain with their mother for 15 months (Whitmore and Huryn 1999), an association as prolonged as in many primary burrowing species.

Life in a semiterrestrial burrow entails not only an environment outside the burrow that is more hostile than in the aquatic realm, but also a higher cost of constructing a new burrow. For example, desert isopods can construct burrows only during the rainy period (chapter 16), similar to what has been reported for African mole-rats (Bennett and Faulkes 2000). These time restrictions enhance the value of a burrow, especially during those times when no new burrows can be constructed. Under these circumstances, it may be advantageous for small individuals to remain with their mothers until conditions improve (e.g., with the onset of the rainy season; see chapter 16) or until mothers give up their brooding habitat (e.g., due to death; see chapter 17).

Many aquatic crustaceans of all sizes readily construct burrows in benthic sediments, suggesting that sediment burrows may be less costly to construct than in the terrestrial environment. However, in certain environmental situations, the value of burrows may increase, and not all size classes of a species may be able to construct or maintain these burrows. This occurs where burrows are very deep or excavated in very hard substrata (e.g., wood or rock). Under these conditions, small juveniles may be better off remaining with their parents rather than leaving. When costs of producing a burrow are very high, selection may favor other options for organisms seeking refuge. Many marine crustaceans associate with larger invertebrate hosts (e.g., chapter 12). For example, sponge-dwelling shrimp in the genus *Synalpheus* live in sponges, which represent a highly valuable resource, because they offer both food and protection from predators. These sponges (and other hosts) may be scarce in the environment, however, again favoring long-lasting associations among the inhabitants of a given host (chapter 18).

In summary, the burrow or refuge represents a critical resource for (family-dwelling) semiterrestrial crayfish, as it does for many other crustaceans. Estimating the value of this burrow (cost of construction, protection from predators or adverse conditions, access to food resources) would be an important contribution to understanding the evolution of social behavior in semiterrestrial crayfish.

Future Directions

Clearly, there is much to be discovered about the behavioral ecology of semiterrestrial crayfish. For many species, the basic parameters of life history (e.g., life span, reproductive frequency and seasonality, brood sizes and survival) remain unknown, and it is probably unwise to extrapolate from a few species in such a phylogenetically and geographically diverse group.

The degree to which individuals of semiterrestrial species are confined to their burrows, particularly with regard to finding and selecting mates, and the ability of young to disperse from the parental burrow could be investigated indirectly using molecular techniques. However, this will depend on being able to catch animals and relate them to a particular burrow system without destroying the burrow, and to date, burrow-specific trapping has rarely been achieved (e.g., Norrocky 1991, Welch and Eversole 2006).

The importance of the burrow to a crayfish or family group of crayfish needs to be investigated: exactly what resources the burrow provides, whether burrow structure is adaptive, and the degree of competition for burrows in various situations. Some information may be obtained indirectly through correlation of morphological traits, particularly claw dimorphism and morphology, with burrowing habit. Laboratory experiments in artificial burrows may also be useful. It is also important to discover the extent to which the animals depend on the resources within the burrow, how much they forage outside the burrow, and, if so, whether foraging (and burrow defense) is a task performed by particular members of family groups.

Gherardi (2002) noted that *Engaeus* species are likely to be good candidates for studies of parental care. Certainly, the nature of the interactions between mother and offspring and between generations of offspring in type 3 burrows is an intriguing question in the context of social behavior. However, it is particularly difficult to make direct observations of crayfish in their burrows. While the structure of the burrow and to some extent the location of animals within the burrow can be revealed by careful excavation (particularly when the burrow is located on a slope and can be excavated from the side), once it has been excavated it is effectively destroyed. In structurally simple burrows, fiber optic devices may offer some chance of direct observation, but many type 3 burrows are too large, complex, and ramifying for such devices. Little use seems to have been made of artificial burrows, apart from physiological studies (McMahon and Hankinson 1993, McMahon and Stuart 1995), but they offer some promise if animals can be acclimatized to them.

If it becomes possible to make observations within natural or artificial burrows, the spatial relations between mother and brood can be investigated. When burrows are excavated the female is usually found close to the brood, but it is unclear whether this is the normal relationship or whether the adult animal has merely been driven to the lowest point of the burrow by the disturbance during excavation. It is clearly of interest to know whether crayfish mothers tend their brood in the complex way that bromeliad crabs do (e.g., Diesel 1992; see also chapter 17) or whether the overlapping of generations is simply a forced cohabitation as the animals await an opportunity to disperse.

At present there is no direct evidence to suggest that burrowing crayfish provide food for their brood; indeed, dietary information of any sort is scarce. It seems likely that type 3 burrows supply most of the requirements of their occupants (Suter and Richardson 1977), but Growns and Richardson (1988) noted that *Parastacoides tasmanicus* collects sections of the leaves of sedges from the surface and stores them underground. Juvenile *P. tasmanicus* have a higher proportion of animal food in their diet, but it is unknown whether this is provided to them by their mother.

Rearing broods of burrowing crayfish can be difficult, perhaps because laboratory conditions do not provide the food resources available in a burrow. But if broods can be reared, simple laboratory experiments will show the relationship between burrowing crayfish females and brood of different ages and the sensory modes that mediate coexistence between them. It would be of interest to compare the duration of recognition of the brood in females from a series of species across the range of burrow types: it should be expected that, in species from habitats where burrows are difficult to construct (e.g., drier habitats, harder substrates), mothers and offspring would remain together for longer time periods.

Summary and Conclusions

Like a number of the other crustaceans discussed in this volume, burrowing crayfish live in a situation that tends to lead to prolonged associations between mother and offspring and, in the case of primary burrowers in type 3 burrows, overlapping generations. Under these situations, social behaviors are likely to evolve, but at this stage the only ones that have been identified among burrowing crayfish are between mother and offspring, in terms of defense and grooming. Burrowing has been a preadaptation that has allowed crayfish to move out of open waters onto land, but it has also imposed severe restrictions on their movements and dispersal. The development of social behaviors may compensate for these restrictions to some extent. Although these are interesting possibilities, studying social behavior in situ in these species is always likely to be difficult because of the depth and complexity of their burrows.

Acknowledgments I am grateful to my colleague Roy Swain for access to his reprint collection and for comments on the manuscript, and to Shane Welch for making available his unpublished manuscript on North American crayfish burrows.

References

Ameyaw-Akumfi, C., and B.A. Hazlett. 1975. Sex recognition in the crayfish *Procambarus clarkii*. Science 190:1225–1226.

Atkinson, R.J.A., and A.C. Taylor. 1988. Physiological ecology of burrowing decapods. Symposia of the Zoological Society of London 59:201–226.

Austin, C. 1996. Systematics of the freshwater crayfish genus *Cherax* Erichson (Decapoda: Parastacidae) in northern and eastern Australia: electrophoretic and morphological variation. Australian Journal of Zoology 44:259–296.

Austin, C., and B. Knott. 1996. Systematics of the freshwater crayfish genus *Cherax* Erichson (Decapoda: Parastacidae) in south-western Australia: electrophoretic, morphological and habitat variation. Australian Journal of Zoology 44:223–258.

Barki, A., and I. Karplus. 1999. Mating behavior and a behavioral assay for female receptivity in the red-claw crayfish *Cherax quadricarinatus*. Journal of Crustacean Biology 19:493–497.

Bechler, D.L. 1981. Copulatory and maternal-offspring behavior in the hypogean crayfish, *Orconectes inermis inermis* Cope and *Orconectes pellucidus* (Tellkampf) (Decapoda, Astacidea). Crustaceana 40:136–143.

Bennett, N.C., and C.G. Faulkes. 2000. African mole-rats: ecology and eusociality. Cambridge University Press, New York.

Berrill, M., and B. Chenoweth. 1982. The burrowing ability of non-burrowing crayfish. American Midland Naturalist 108:199–201.

Bliss, D.E., and L.H. Mantel. 1968. Adaptations of crustaceans to land: a summary and analysis of new findings. American Zoologist 8:673–685.

Bovbjerg, R.V. 1956. Some factors affecting aggressive behaviour in crayfish. Physiological Zoology 29:127–136.

Clark, E. 1936a. The freshwater and land crayfishes of Australia. Memoirs of the National Museum of Victoria 10:5–58.

Clark, E. 1936b. Notes on the habits of land crayfishes. Victorian Naturalist 53:65–68.

Clutton-Brock, T.H. 1991. The evolution of parental care. Princeton University Press, Princeton, N.J.

Crandall, K.A., and J.W. Fetzner, Jr. 1995. The tree of life: Astacidea. Available at http://tolweb.org/tree?group = Astacidea&contgroup = Decapoda (accessed October 16, 2004).

Crandall, K.A., J.W. Fetzner, Jr., S.H. Lawler, M. Kinnersley, and C.M. Austin. 1999. Phylogenetic relationships among the Australian and New Zealand genera of freshwater crayfish (Decapoda: Parastacidae). Australian Journal of Zoology 47: 199–214.

Crandall, K.A., J.W. Fetzner, Jr., C.G. Jara, and L. Buckup. 2000. On the phylogenetic positioning of the South American freshwater crayfish genera (Decapoda: Parastacidae). Journal of Crustacean Biology 20:530–540.

Crocker, D.W., and D.W. Barr. 1968. The crayfishes of Ontario. University of Toronto Press, Toronto.

Diesel, R. 1992. Managing the offspring environment: brood care in the bromeliad crab, *Metopaulias depressus*. Behavioural Ecology and Sociobiology 30:125–134.

Fetzner, J.W., Jr. 1996. Biochemical systematics and evolution of the crayfish genus *Orconectes* (Decapoda: Cambaridae). Journal of Crustacean Biology 16:111–141.

Fetzner, J.W., Jr. 2003. Checklist of crayfish species worldwide. Available from http://crayfish.byu.edu/species_list.htm (accessed October 16, 2004).

Figler, M.H., G.S. Blank, and H.V.S. Peeke. 2001. Maternal territoriality as an offspring defense strategy in Red Swamp Crayfish (*Procambarus clarkii*, Girard). Aggressive Behavior 27:391–403.

Fitzpatrick, J.F., Jr. 1975. The taxonomy and biology of the prairie crawfishes, *Procambarus hagenianus* (Faxon) and allies. Freshwater Crayfish 2:381–389.

Furse, J.M., and C.H. Wild. 2002. Terrestrial activities of *Euastacus sulcatus*, the Lamington spiny crayfish (Decapoda: Parastacidae). Freshwater Crayfish 13:604.

Gherardi, F. 2002. Behaviour. Pages 258–290 *in*: D.M. Holdich, editor. Biology of freshwater crayfish. Blackwell Scientific, Oxford.

Greenaway, P. 1984. Survival strategies in desert crabs. Pages 145–152 *in*: H.G. Cogger and E. Cameron, editors. Arid Australia. Australian Museum, Sydney.

Grow, L., and H. Merchant. 1980. The burrow habit of the crayfish *Cambarus diogenes diogenes* (Girard). American Midland Naturalist 103:231–236.

Growns, I.O., and A.M.M. Richardson. 1988. The diet and burrowing habits of the freshwater crayfish *Parastacoides tasmanicus tasmanicus* Clark (Decapoda: Parastacidae). Australian Journal of Marine and Freshwater Research 39:525–534.

Gutiérrez-Yurrita, P.J., G. Sancho, A.Á. Bravo, Á. Baltanás, and C. Montes. 1998. Diet of the red swamp crayfish *Procambarus clarkii* in natural ecosystems of the Doñana National Park temporary fresh-water marsh (Spain). Journal of Crustacean Biology 18:120–127.

Hamr, P. 1991. Comparative reproductive biology of the Tasmanian freshwater crayfishes *Astacopsis gouldi* Clark, *Astacopsis franklinii* Gray and *Parastacoides tasmanicus* Clark (Decapoda: Parastacidae). Unpublished Ph.D. thesis, University of Tasmania, Hobart, Tasmania, Australia.

Hamr, P., and A.M.M. Richardson. 1994. The life history of *Parastacoides tasmanicus tasmanicus* Clark, a burrowing freshwater crayfish from south-west Tasmania. Australian Journal of Marine and Freshwater Research 45:455–470.

Hansen, B., and A.M.M. Richardson. 2006. A revision of the Tasmanian endemic freshwater crayfish genus *Parastacoides* (Crustacea: Decapoda: Parastacidae). Invertebrate Systematics 20:713–769.

Hazlett, B.A. 1981. The behavioral ecology of hermit crabs. Annual Review of Ecology and Systematics 12:1–22.

Hobbs, H.H., Jr. 1942. The crayfishes of Florida. University of Florida Publications in Biological Science, Series 3:1–179.

Hobbs, H.H., Jr. 1981. The crayfishes of Georgia. Smithsonian Contributions to Zoology 318:1–549.

Hobbs, H.H., Jr. 1988. Crayfish distribution, adaptive radiation and evolution. Pages 52–82 *in*: D.M. Holdich and R.S. Lowery, editors. Freshwater crayfish: biology, management and exploitation. Croom Helm, London.

Hobbs, H.H., Jr., and H.W. Robison. 1989. On the crayfish genus *Fallicambarus* (Decapoda: Cambaridae) in Arkansas, with notes on the *fodiens* complex and descriptions of two new species. Proceedings of the Biological Society of Washington 102:651–697.

Hobbs, H.H., Jr., and M. Whiteman. 1991. Notes on the burrows, behavior, and color of the crayfish *Fallicambarus* (*F.*) *devastator* (Decapoda, Cambaridae). The Southwestern Naturalist 36:127–135.

Holdich, D.M. 2002. Background and functional morphology. Pages 3–29 *in*: D.M. Holdich, editor. Biology of freshwater crayfish. Blackwell Scientific, Oxford.

Holdich, D.M., and R.S. Lowery. 1988. Functional morphology and anatomy. Pages 11–51 *in*: D.M. Holdich and R.S. Lowery, editors. Freshwater crayfish: biology, management and exploitation. Croom Helm, London.

Horwitz, P.H.J. 1990. A taxonomic revision of species in the freshwater crayfish genus *Engaeus* Erichson (Decapoda: Parastacidae). Invertebrate Taxonomy 4:427–614.

Horwitz, P.H.J., and M. Adams. 2000. The systematics, biogeography and conservation status of species in the freshwater crayfish genus *Engaewa* Riek (Decapoda: Parastacidae) from south-western Australia. Invertebrate Taxonomy 14:655–680.

Horwitz, P.H.J., and B. Knott. 1983. The burrowing habit of the koonac, *Cherax plebejus* (Decapoda: Parastacidae). Western Australian Naturalist 15:113–117.

Horwitz, P.H.J., and A.M.M. Richardson. 1986. An ecological classification of the burrows of Australian freshwater crayfish. Australian Journal of Marine and Freshwater Research 37:237–242.

Horwitz, P.H.J., A.M.M. Richardson, and P. Cramp. 1984. Aspects of the life history of the burrowing freshwater crayfish, *Engaeus leptorhynchus* at Rattrays Marshes, north east Tasmania. Tasmanian Naturalist 82:1–5.

Huner, J.V., and J.E. Barr. 1991. Red swamp crayfish: biology and exploitation. Louisiana Seagrant Program, Baton Rouge, La.

Jarvis, J.U.M., M.J. O'Riain, N.C. Bennett, and P.W. Sherman. 1994. Mammalian eusociality: a family affair. Trends in Ecology and Evolution 9:47–51.

Johnston, C.E., and C. Figiel. 1997. Microhabitat parameters and life-history characteristics of *Fallicambarus gordoni* Fitzpatrick, a crayfish associated with pitcher-plant bogs in southern Mississippi. Journal of Crustacean Biology 17:687–691.

Karplus, I., A. Barki, T. Levi, G. Hulata, and S. Harpaz. 1995. Effects of kinship on growth and survival of juvenile Australian redclaw crayfish (*Cherax quadricarinatus*). Freshwater Crayfish 10:494–505.

Little, E.E. 1975. Chemical communication in maternal behavior of crayfish. Nature 255:400–401.

Little, E.E. 1976. Ontogeny of maternal behavior and brood pheromone in crayfish. Journal of Comparative Physiology 112:133–142.

Lowery, R.S. 1988. Growth, moulting and reproduction. Pages 83–113 *in*: D.M. Holdich and R.S. Lowery, editors. Freshwater crayfish: biology, management and exploitation. Croom Helm, London.

Mason, J.C. 1970. Maternal-offspring behavior of the crayfish, *Pacifastacus trowbridgi* (Stimpson). American Midland Naturalist 84:463–473.

McMahon, B.R., and J.J. Hankinson. 1993. Respiratory adaptations of burrowing crayfish. Freshwater Crayfish 9:174–182.

McMahon, B.R., and S.A. Stuart. 1995. Simulating the crayfish burrow environment: air exposure and recovery in *Procambarus clarkii*. Freshwater Crayfish 8:451–461.

Morgan, G.J. 1986. Freshwater crayfish of the genus *Euastacus* Clark (Decapoda: Parastacidae) from Victoria. Memoirs of the Museum of Victoria 47:1–57.

Morgan, G.J. 1988. Freshwater crayfish of the genus *Euastacus* Clark (Decapoda: Parastacidae) from Queensland. Memoirs of the Museum of Victoria 49:1–49.

Morgan, G.J. 1997. Freshwater crayfish of the genus *Euastacus* Clark (Decapoda: Parastacidae) from New South Wales, with a key to all species of the genus. Records of the Australian Museum Supplement 23:1–110.

Norrocky, M.J. 1991. Observations on the ecology, reproduction and growth of the burrowing crayfish *Fallicambarus* (*Creaserinus*) *fodiens* (Decapoda, Cambaridae) in north-central Ohio. American Midland Naturalist 125:75–86.

Richardson, A.M.M., and R. Swain. 1991. Pattern and persistence in the burrows of two species of the freshwater crayfish *Parastacoides* (Decapoda: Parastacidae) in south west Tasmania. Memoirs of the Queensland Museum 31:283.

Richardson, A.M.M., and R. Swain. 2002. The sting in the tail—spination of the tail fan in freshwater crayfish. Freshwater Crayfish 13:515–524.

Riek, E.F. 1969. The Australian freshwater crayfish (Crustacea: Decapoda: Parastacidae), with descriptions of new species. Australian Journal of Zoology 17:855–918.

Scholtz, G. 2002. Phylogeny and evolution. Pages 30–52 *in*: D.M. Holdich, editor. Biology of freshwater crayfish. Blackwell Science, Oxford.

Suter, P.J., and A.M.M. Richardson. 1977. The biology of two species of *Engaeus* (Decapoda: Parastacidae) in Tasmania. III. Habitat, food, associated fauna and distribution. Australian Journal of Marine and Freshwater Research 28:95–103.

Walker, D., B.A. Porter, and J.C. Avise. 2002. Genetic parentage assessment in the crayfish *Orconectes placidus*, a high-fecundity invertebrate with extended maternal brood care. Molecular Ecology 11:2115–2122.

Welch, S.M., and A.G. Eversole. 2006. Comparison of two burrowing crayfish trapping methods. Southeastern Naturalist 5:27–30.

Whitmore, N., and A.D. Huryn. 1999. Life history and production of *Paranephrops zealandicus* in a forest stream, with comments about the sustainable harvest of a freshwater crayfish. Freshwater Biology 42:467–478.

Williams, D.D., N.E. Williams, and H.B.N. Hynes. 1974. Observations on the life history and burrow construction of the crayfish *Cambarus fodiens* (Cottle) in temporary streams in southern Ontario. Canadian Journal of Zoology 52:365–370.

Wilson, E.O. 1975. Sociobiology. The new synthesis. Belknap Press of Harvard University, Cambridge, Mass.

Sociobiology of Terrestrial Isopods

Karl Eduard Linsenmair

16

Among crustaceans, only isopods from the order Oniscoidea have evolved a great variety of entirely terrestrial species. A set of properties predisposed the marine ancestors of the Oniscoidea for the invasion of land. Among the most important were (1) the mode of locomotion, (2) direct development without planktonic larvae, and, as a key innovation, (3) the brood pouch (marsupium). The subsequent evolution of this marsupium, as well as the pseudotracheae (Hoese 1982), additional morphological structures (Warburg 1993) such as the water-conducting system (Hoese 1981), physiological mechanisms (Carefoot 1993) such as water vapor absorption (Coenen-Stass 1989, Wright and Machin 1993), and evolution of behavioral adaptations made some of these crustaceans the dominant herbivores and detritivores of vast arid areas in North Africa and Asia. The most successful representatives belong to the genus *Hemilepistus sensu stricto* (s.str.), and some also to the diverse genus *Porcellio* (Linsenmair 1984, 1985, 1987, 1989, and references therein, Röder and Linsenmair 1999; K.E. Linsenmair, unpublished observations).

In reproducing, terrestrial isopod females produce yolk-rich eggs and develop the marsupium, which in its "terrestrial type," *sensu* Hoese (1984), is not just an egg-carrying device but resembles a uterus, with regulated milieu and transfer of substances between female and brood (Hoese and Janssen 1989, Surbida and Wright 2001). The females also execute a special "parturial" molt and spend considerable time carrying their brood. These activities entail high direct costs expended for the progeny (Sutton et al. 1984, Linsenmair 1989, Röder 1995, Kight and Hashemi 2003), which curtail the female's potential future reproductive output and are thus considered maternal investments (*sensu* Trivers 1972). Since the brood pouch has only a limited capacity (e.g., Dangerfield and Telford 1995) and producing a clutch requires high energy and time allocation (Linsenmair 1989), both the maximum clutch size and the number of broods a female can raise are limited. Given these constraints, higher reproductive success may often not be attainable by increasing clutch size or number of broods. Investing still more into the already expensive progeny by prolonging brood care should often be a better choice. This hypothesis appears to explain the unusually high parental investment among terrestrial isopods.

We find different forms and intensities of brood care among the terrestrial isopods: (1) extended carrying of young in the marsupium, (2) short- to long-term maternal food provisioning, and (3) biparental care with long-lasting family cohesion, strict monogamy of the parents, and a strikingly differentiated intra- and interfamily communication system. What were the evolutionary routes, and the ecological driving forces shaping these isopods' social behavior? What are the causes for fundamental differences we find in species that live side by side to each other? Which secondary problems were created by the social behavior itself, and how were they solved? In this chapter I explore these questions.

The Studied Isopod Species and Their Geographic Distribution

Most species of the xerophilous genus *Hemilepistus* live in the central Asian steppes (Lincoln 1970). At present, data on ecology and behavior are available only for three of the nine species acknowledged by Lincoln as belonging to *Hemilepistus* s.str. and for four of the 20 species of *Hemilepistus* enumerated in the *World Catalog of Terrestrial*

Table 16.1. Some ecoethological characteristics of the investigated isopod species.

Species	Geographic Distribution	Habitat	Burrow Digging or Use of Natural Crevices	Itero-/ Semelparous	Mating System	Age at (first) Breeding[b]	Brood Care and/or Sibship Cohesion	Brood Size[d]
Hemilepistus reaumuri	North Africa, Asia Minor	Xeric, solid soil	Burrows[a] in solid Soil	Semel	Strict social and sexual monogamy	11–12 months	Up to 5 months, biparental, sibship cohesion: 10 months	88 ± 29 ($n = 79$)
H. (Desertellio) elongatus	Turkmenia, easternmost Turkey	Mesic, loose sandy and stony lava soil	Crevices under stones, some active scratching	Itero	Polygamy	50–65 days	12–14 days[c] only ♀	104 ± 26 ($n = 27$)[c]
Porcellio albinus	North Africa, Sahara, Niger	Xeric to desert, loose sandy soils, dunes	Burrows in sand	Itero	Polygamy	60–110 days	20 to >40 days[c] only ♀	23 ± 11 ($n = 22$)[c]
P. olivieri	North Africa	Mesic to xeric solid stony soil	Crevices under stones/rodent burrows, similar available retreats	Itero	Polygamy	~60 days	No brood care	50–83 ($n = 5$)[c]
P. simulator	Algeria	Mesic, solid stony soil	Crevices under stones, some active scratching	Itero	Polygamy	~75 days	30–40 days[c] only ♀	28 ± 16 ($n = 30$)[c]
Fuert.-P.1	Fuertaventura	Xeric dunes, loose sandy soil	Burrows in sand, sandy soil	Itero	Polygamy, transient mate guarding by large males	120–150 days	20–40 days[c] only ♀	34 ± 13 ($n = 23$)[c]
Fuert.-P.2	Fuertaventura	Coastal sand dunes	Burrows in sand dunes	Itero	Polygamy	~90 days	2–8 days postpartum, main provisioning before delivery, sibs together ~14 days	20–55 ($n = 4$)

[a]Burrows are always dug by the respective isopod species, although not always by the owner.
[b]Apart from *H. reaumuri*, all data refer to laboratory bred individuals (≥50 females).
[c]Data on the length of brood care and brood size, respectively, were obtained from laboratory bred individuals.
[d]Brood size is closely correlated to the female's body size (see, e.g., Linsenmair 1989). Values refer to a spectrum of size classes and are therefore highly variable.

Isopods (Schmalfuss 2003). Schmalfuss abandoned the subdivision in *Hemilepistus* s.str. and *Desertellio*, which I consider unjustified in the light of the fundamental behavioral differences (see below). The other two species from the group of *Hemilepistus* s.str., *H. rhinoceros* (Marikovsky 1969) and *H. aphganicus* (Schneider 1971; K.E. Linsenmair, unpublished observations), seem to behaviorally resemble *H. reaumuri*, the only thoroughly investigated species. *Hemilepistus reaumuri*, the main subject of this review, was predominantly studied by us in southern Tunisia and, to a lesser extent, in the Negev. *Hemilepistus* (*Desertellio*) *elongatus* was observed in the lava fields around Mount Ararat (Turkey), but most data were obtained from individuals bred in captivity. The investigated species belonging to *Porcellio* mentioned below (*P. albinus, P. olivieri*) dwell in similar climatic zones as *H. reaumuri* (the range of distribution of *P. albinus* reaching far into real desert, i.e., Saharan sand seas; K.E. Linsenmair, unpublished observations). *Porcellio simulator* dwells in stony and climatically somewhat milder areas in Algeria. The three *Porcellio* species we investigated on the Canary Island Fuertaventura, are still undescribed. The two species mentioned here are denoted as Fuert.-P.1, which is a large, highly color-polymorphic species, living in sandy areas of the peninsula Jandia not directly adjacent to the sea, and Fuert.-P.2, which is found mainly in sand dunes adjacent to the sea. This latter species seems to be very closely related to *P. albinus*. All these species were observed in the field and also kept and bred in the laboratory. See Table 16.1 for an overview of ecological and behavioral characteristics of the species treated in this chapter.

Extended Brood Care in Oniscoids

Two Differing Evolutionary Routes to Extended Brood Care

In the oniscoids, two main types of extended brood care can be distinguished, one where hatched juveniles remain and develop in the marsupium of their mother for extended time periods (extended offspring carrying time [ECT]; Linsenmair 1989) and one where females care for their progeny after releasing them from the marsupium (extended brood care after delivery [EBC]).

Very little is presently known about ECT; in particular, we do not know how widespread it is, and we lack data on all pertinent physiological or behavioral aspects (e.g., how the large young in the marsupium are nurtured). The squamiferid isopods that associate with ant or termite colonies seem to have adopted this evolutionary path. In the European ant guest *Plathyarthrus hoffmannseggi*, the female delivers a mean of four to five large offspring per brood. The dry weight of each brood amounts to ~50% that of their mother. Only two broods are produced during a three-year lifetime (Sutton et al. 1984). In the most extreme case (Taiti and Ferrara 1988) of the Southeast Asian *Exalloniscus maschwitzi*, two offspring were found within the marsupium at an early stage. One disappears, and the remaining one stays until it reaches half of its mother's length (Fig. 16.1). This species lives in colonies of *Leptogenys distinguenda* (V. Witte 2001), a ponerine ant with the life habit of an army ant. During the frequent host migrations, the isopods cling to ant pupae that are transported by workers to the new bivouac.

More information is available on the second type of extended brood care, EBC. Species exhibiting this syndrome release their progeny at the same stage as oniscoids

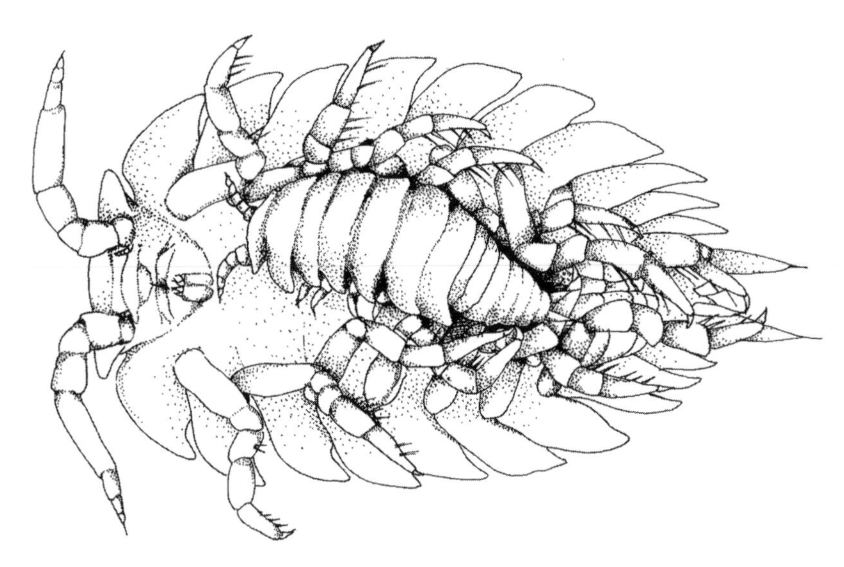

Figure 16.1 The Southeast Asian myrmecophilous *Exalloniscus maschwitzi* as an example of an isopod exhibiting "extended carrying time" (ECT). The single surviving young is not delivered before attaining about half its mother's length.

without ECT or EBC, but females then remain for days to months with their progeny (Linsenmair 1984; K.E. Linsenmair, unpublished observations). The brood-caring adults procure a favorable retreat for their offspring, provision them with food, and defend them against competitors for food and the vital shelter. In *H. reaumuri*, offspring are also defended against small enemies, cannibalistic conspecifics representing the most serious threat. Herein, I report mostly on *H. reaumuri* and other selected oniscoid species engaging in EBC.

Behavioral Aspects of Extended Maternal Brood Care in Xerophilous Oniscoids

The route to an ECT allows a nomadic life. It is also compatible with the widespread behavior of woodlice to aggregate with any conspecifics encountered in suitable retreats (Takeda 1984). EBC, on the contrary, requires site fidelity and spatial exclusivity to confine the accruing benefits to the caring individuals' progeny (see discussion of behavior of *Metopaulias depressus* in chapter 17; see also discussion on the potential of burrowing for developing social behavior in semiterrestrial crayfish in chapter 15). Digging a burrow or acquiring a natural defendable crevice that serves as an offspring nest provides solutions to this problem in several species. The digging behavior, used for constructing retreats by all mentioned *Porcellio* species (but not by *Hemilepistus* s.str., which employs a different digging behavior—see below), consists of synchronized scratching at the soil with the pereopods, alternating between the two body sides, often followed by a scooping behavior where the body serves as a blade pushing the loosened material out of the burrow. This digging behavior requires loose soil. It is best suited for excavating horizontal or only slightly vertically inclined burrows into sand dunes and other loose substrata. For solid soil, it is completely ineffective.

A reliable site specificity requires also that brood delivery be under the female's control. Hardly anything was known about the details of this process in oniscoids. According to Sutton (1980) and Schneider (1991), the mancae (i.e., the isopod "larvae" or newborn juveniles) determine when to leave the brood pouch. In EBC species, on the contrary, delivery is controlled by the female (Röder 1995; K.E. Linsenmair, unpublished observations). *Hemilepistus elongatus*, for instance, takes on a special upward bent posture and then slowly moves the hind part of its body repeatedly forward. The resulting pressure, supported by equidirectional pereopod movements, empties the brood pouch through the anterior oostegites.

Food provisioning for the progeny is a prominent behavior in all oniscoids with EBC. The brood-caring isopods, especially in *H. reaumuri*, compose a diverse diet by selection among available food items. What could have been the phylogenetic origin of food provisioning? A first step may have been a behavior we observed in different porcellionid and trachelipid species. When foraging, these isopods, independent of sex, age, and breeding status, occasionally carry pieces of food back to their retreats at the end of excursions. These food items were then immediately fed upon by the collecting individual. This occasional food gathering appears highly adaptive in arid environments, where high desiccation rates forbid long foraging excursions, and could have been the precursor for selection of the regular food provisioning of EBCs.

With food provisioning established, allowing young mancae to remain within favorable retreats during their most sensitive stage, the isopods had acquired a key adaptation (combined with digging behavior) for greatly extending their range toward more xeric habitats. More demanding ecological conditions in turn should have exerted selection pressures to improve EBC and modify other traits of social behavior. Accordingly, the intensity, duration, and spectrum of action patterns performed in EBCs differ among species, well matching their corresponding ecological conditions. For example, where climatic conditions are comparatively mild and suitable retreats for small young isopods are regularly available or easy to fabricate, the progeny might merely need an initial supply of food. We have found such a behavior in Fuert.-P.2 (K.E. Linsenmair, unpublished observations). This species lives in sand dunes adjacent to the sea where the sand is always moist 4–8 centimeters below the surface. Females carry large amounts of mainly halophytic plant detritus into their brood dens. After delivery, females stay with their offspring only for two to eight days. The young usually remain for up to about two weeks in their natal burrow before they start to disperse. Where climatic conditions are less favorable, as, for example, in Fuert.-P.1, dwelling in sand areas just 50 to a few hundred meters farther inland and higher (20 to >50 m above sea level), a more costly brood care lasting for up to more than a month (especially during dry spells) is realized. Where food within retreats moulds quickly, and inter- and intraspecific competitors for food or retreats are common, additional services beside regular food gathering, such as cleaning and continuously deepening the burrow, guarding, and actively protecting the young, have evolved (Linsenmair 1984).

EBC females, who perform dozens of foraging excursion daily, must be able to find their way back to their burrow reliably. These isopods' orientation capabilities are exemplified by *H. reaumuri* (Hoffmann 1989 and references therein), consisting of

Figure 16.2 Before starting foraging excursions, all observed *Porcellio* species (here *P. albinus*) scrape out sand from the burrow's floor and pile it up before the entrance. This sand is marked with the owner's individual-specific chemical signature. This area is much larger than the burrow entrance itself, thus greatly facilitating the relocation of its burrow by a homing isopod.

a combination of celestial and path-integrating mechanisms. These mechanisms work only with a limited acuity and are insufficient to account more than coarsely for some deviations from the isopod's intended path (e.g., slipping some distance down on a steep dune, or flight runs provoked by external disturbance), causing mistakes in direction and distance when homing. The resulting orientation errors are then compensated by a systematic search behavior. This, however, is only suited to find sooner or later each burrow within a limited area. These isopods often reach high densities with burrows built next to one other (often <10 cm apart). To distinguish one's own from a foreign burrow requires individual recognition. This is mediated by chemical signatures, which are produced in all brood caring *Porcellio* and *Hemilepistus* species. Identifying the individual-specific marks is via chemotactic receptors of the "apical organ" (Hoese 1989) on the tip of the second antenna (Seelinger 1977).

Do searching isopods need to inspect the burrow entrance for recognition? Burrow owners of the sand-dwelling *Porcellio* species, before leaving their dwelling to return again, always scrape out sand from the burrow's floor and pile it up directly before the entrance in the form shown in Figure 16.2. A touching of this area, considerably larger than the entrance to the burrow itself and therefore easier to find

Figure 16.3 *Hemilepistus reaumuri* piles up a permanent, conspicuous feces embankment around its burrow entrance, marked with the family-specific signature. It serves principally the same function, but due to its large size more effectively, as the scraped out sand shown in Figure 16.2. Species from the genus *Hemilepistus* often undertake very extended foraging excursions (reaching lengths of up to 10 m), and major homing mistakes are common.

in a search, allows the isopod to identify its den reliably (Linsenmair 1979; for the corresponding structure in *Hemilepistus*, see below and Fig. 16.3). Thus, females (or parents) of oniscoids engaging in EBC need to be able to construct a burrow and to relocate it reliably after foraging excursions to supply their progeny with sufficient food resources.

Soil Substratum, Burrows, and the Mating System in Xerophilous Oniscoids

The Soil Substratum Determining the Value of the Burrow

By comparing the different *Porcellio* species, we can draw an outline (Linsenmair 1989 and references therein) of how maternal EBC probably originated and further developed to solve increasingly demanding ecophysiological problems while adapting to more xeric habitats. But how to explain the evolution of biparental brood care and the more advanced social behavior within the genus *Hemilepistus* s.str.? *Porcellio albinus*, the species with the most EBC in the *Porcellio* lineage, often lives side by side with *H. reaumuri*. Can we find distinctive ecological factors that may have caused a disruptive selection leading to the major differences in social behavior and life history traits?

Contrary to the strictly semelparous *H. reaumuri*, all presently known species of *Porcellio* with EBC, and also *H. elongatus*, are iteroparous and polyvoltine (i.e., produce consecutively several generations without a clear annual rhythm). While maternal families in *Porcellio* species dissolve after one to about six weeks of brood care (in *P. albinus* in summer the period can be longer), with their young still small and immature, *Hemilepistus* families stay together for about 9–10 months. While *H. reaumuri* forms monogamous pairs, practicing biparental care with few sex-specific differences, males never participate in parental care in *Porcellio* species and in *H. elongatus*. In *Porcellio* species, individuals with half the adult length or even smaller (2 to ~4 months of age) may already reproduce, while *Hemilepistus* s.str. start breeding only when fully grown, that is, at an age of 10–12 months.

What may have caused these differences? According to circumstantial evidence, one factor seems to be of decisive importance: the type of soil in which the different species dwell and the differing replacement value of burrows. *Hemilepistus reaumuri* avoids any loose soil, and its digging behavior fits only solid soil. In contrast, *P. albinus* (like all other *Porcellio* spp., as well as *H. elongatus*) can dig only in loose substratum. Although all sand dwellers are poor burrowers, the mechanical properties of sand make their digging comparatively efficient. Even during the hotter part of the year a half-grown or older *P. albinus* living in a Saharan dune field will not need more than 3–6 hours to dig a burrow to a protective depth. Apart from its easier workability, sand has additional advantages: it has, on the one hand, a low water-holding capacity per volume, as its water adsorption and capillary forces are low, but it lacks, on the other hand, continuous capillaries connecting the dry surface with the deeper humid layers, reducing evaporation. Thus, in desert-like areas, a larger amount of the stored scarce rain water is available for a longer time for plant/animal use and for saturating the air in shallower layers in sandy soil, compared to more solid types of soil (Walter 1960). The fine-grained sand inhabited by *P. albinus* has, furthermore, a low thermal conductivity (H.J.L. Witte et al. 2002). Thus, a protective depth with near 100% relative humidity can be reached at shallower depths in fine sand than in solid soil.

According to the ease with which new burrows can be dug, it is not surprising that fighting for burrow possession is rare and mild in the sand-dwelling species. Foreign isopods entering the burrows of brood-caring females were quickly expelled (in >50 directly observed encounters in the field) after only very short aggressive interactions. This is in contrast to what is seen in *H. reaumuri*, where the burrow holder can wedge itself firmly against the solid walls of the burrow entrance (thus very efficiently fending off any persistent intruders), which is not possible in a burrow in loose soil.

Hemilepistus reaumuri lacks any specialized morphological devices for burrowing. Its digging behavior consists of biting and scraping off small soil particles with the mandibles and the anterior leg pairs. This digging is very inefficient and time-consuming. Even under the most favorable conditions, when soil is very moist, hours are needed to reach a length of 2 centimeters, barely sufficient to allow the burrower to withdraw from the soil surface. Only during cool weather, that is, from the end of February/beginning of March until mid to late April, can new burrows be dug in

North Africa. Only during that season can *H. reaumuri* survive extended periods outside (or inside very short) burrows. *Hemilepistus* lives in soils that become very hard when dry. During the hot season, only deep burrows (depending on soil type and soil moisture, ranging from 40 cm to about 1 m deep and up to >2 m long) offer protection. Only in their deepest part does the temperature remain below the critical maximum of 36–37°C and relative humidity remain within the required 93–100% to allow balancing water losses partly via water vapor absorption (see Coenen-Stass 1989, Wright and Machin 1993) and partly via eating humid soil from the deepest part of the burrow.

Hemilepistus s.str. are incapable of digging a new, sufficiently deep burrow during the dry season (May through September/October). During this long period, no individual will attempt to dig a new burrow, and it can survive only if it is able to withdraw into an already existing protective den. After the initial phase, the burrow is only slowly extended and deepened because family members "eat" their way down into the substratum. The eaten soil is defecated in form of small solid tilelike packages (see Fig. 16.3). These can effectively be removed from the den, contrary to fine-grained loose dry soil, and are used to pile up an embankment around the burrow. This serves the same function as the above mentioned scraped-out sand in sand-dwelling *Porcellio*. It greatly facilitates homing to the burrow: without the feces embankment, only 12% of the returning isopods find their burrow without switching to search behavior, whereas the success rate with embankment is 72% (Hoffmann 1985a, 1985b).

Given the high value of a burrow and the great effort to be expended for its construction, it is not surprising to find strong intraspecific competition, and also interspecific competition from insects, spiders, and occasionally other isopods. Before digging a burrow, *Hemilepistus* regularly tries to either find an empty one or to acquire one by force. Therefore, beside its timely acquisition, a permanent securing of the burrow is an absolute prerequisite for survival and reproduction. A single isopod would be unable to guard its burrow permanently. It cannot fast for months. To forage, it has to leave its burrow—and this always entails the danger of losing it to a competitor. This holds especially true for the early phase of reproduction when, on the one hand, females have to forage especially intensively to compensate energy losses during hibernation and acquire the needed matter and energy for producing clutches and performing parturial molts, and on the other hand, many conspecifics are also searching for burrows. Similarly, leaving the burrow unguarded while foraging or during molting is also very risky later in the year (for details, see Linsenmair 1979, Baker 2004). In *Hemilepistus*, the problem of permanent burrow guarding is very efficiently solved: the desert isopods establish monogamous cooperative pairs. After the delivery of their progeny, they form strictly closed family units. By division of labor, a permanent guarding, prolonging, and cleaning of the burrow is secured.

Burrows and the Mating Behavior of Xerophilous Oniscoids

The isopods' mode of reproduction resembles that of mammals regarding the high direct energy and time investment of females. Therefore, we expect females to be the sex on which selection should favor an intensified brood care (Trivers 1972). A male's monogamous behavior and full participation in brood care are surprising. The mate-guarding behavior of many aquatic and amphibious isopods (see reviews in Jormalainen

1998, Zimmer 2001; see also chapter 8), which could provide a starting point for permanent monogamy, has been lost during the adaptation of oniscoids to terrestrial life (for possible reasons, see chapter 8). During this evolutionary transition, the very short period for copulations during parturial molt also has been abandoned. In all closely studied terrestrial oniscoids, copulations do repeatedly take place. We registered up to more than 30 complete copulations (i.e., double-sided: isopods possess two fully developed, entirely separated, reproductive tracts) in different *Porcellio* species (Linsenmair 1989), and in *H. elongatus* up to about 60 in 24 hours (Röder 1995). Contrary to the situation in marine species (see, e.g., Zimmer 2001), sperm can be stored long enough for the fertilization of at least a second clutch in terrestrial oniscoids (see Röder 1995, Moreau et al. 2002, and references in these). While mate guarding in aquatic peracarids is well suited to secure fertilization privileges, decisive prerequisites for such a behavior are lacking in oniscoids. Males are unable to judge the female's reproductive state (in regard to the amount of stored sperm), and therefore a polygynous/promiscuous tactic in males should be favored. Under these circumstances, males cannot secure fertilization privileges at reasonable costs against the females' interest. Mate guarding therefore should have evolved anew only where it also promotes the female's fitness, which may be of particular importance in EBC species. What might have prompted the decisive evolutionary steps?

In Fuert.-P.1, all males that were not fully grown searched during activity phases for receptive females. When a male detected a female, he grasped her, and succeeding copulations often appeared rapelike. Large females regularly and successfully warded off smaller males. Some of the large males, however, behaved totally differently.

Figure 16.4 On its way back to the burrow, a food-carrying Fuert.-P.1 female, a *Porcellio* species we investigated on the Canary Island Fuertaventura, is being inspected by a large male searching for receptive females.

These males dug especially deep holes, furnished them with large quantities of food, and waited for females—also large ones—who deliberately associated with them for periods between 8 and 25 days (Linsenmair 1984; K.E. Linsenmair and E. Herlein, unpublished observations). According to field observations, females seemed to take over the deep burrow as a breeding den, which might be one benefit of their consorting. Because only males of the largest size class built these burrows, joining them might benefit females by choosing males with "good genes" (e.g., Hamilton and Zuk 1982, Andersson 1994). In contrast to *H. reaumuri*, in which male and female stay together during the entire reproductive period, these pairs of Fuert.-P.1 always separate a few days before the female's parturial molt. Females in this and the other observed *Porcellio* species lose their attractiveness to males a few days before the parturial molt. Because breeding seems to be year-round in the Canary Islands, with females being polyvoltine and not synchronized, males of Fuert.-P.1 always have chances to find a new receptive female (Fig. 16.4).

Monogamy and Biparental Care in Xerophilous Oniscoids: *Hemilepistus reaumuri*

Mating Behavior in H. reaumuri

What may have led to the abandonment of polygyny in the ancestors of the genus *Hemilepistus* s.str.? After the dissolving of families in spring, predominantly females start to dig and, later on, to lengthen burrows. Males invest more time in searching empty burrows or those with single females. After pair formation, they spend, during activity phases, most of their time guarding (Linsenmair and Linsenmair 1971, Linsenmair 1979). This could indicate that, at an earlier evolutionary stage, females always started digging, because they were more dependent on burrows than were males. With self-dug holes, females gained full control over mate choice and also a male's fertilizations (see below). Climate change to more xeric conditions may have enforced an increasing synchronization among reproducing females and the abandoning of polyvoltine breeding, thereby reducing the males' chances of maximizing fitness by successively pairing with different females.

Pair formation requires that one partner owns a burrow. The owner of a new burrow stops digging at a depth of about 1.5–3 centimeters and then just sits in its burrow. A potential pair partner, which early in the breeding phase is usually a male (e.g., Linsenmair and Linsenmair 1971, Baker 2004), who arrives at the burrow will have to strive many hours for admission and may have to fight frequently with competitors. The female thus achieves a mate selection according to the "best of *n* males" tactic (Janetos 1980; see also chapters 7, 8). After a certain time span, depending on temperature and reproductive state of the burrow owner, the potential partner gets admission (Fig. 16.5). The two then frequently change their position within the burrow and thereby learn to recognize each other individually (Linsenmair 1989 and references therein). From that time onward, partners practice division of labor with the guard fending off any con- or heterospecific party interested in taking over the burrow. After an early pair formation under the climatic conditions in early spring, a rather long period of up to 8–14 days passes during which females are not yet ready to copulate (Fig. 16.5). This period is followed by a phase of 6–12 days, depending on the prevailing

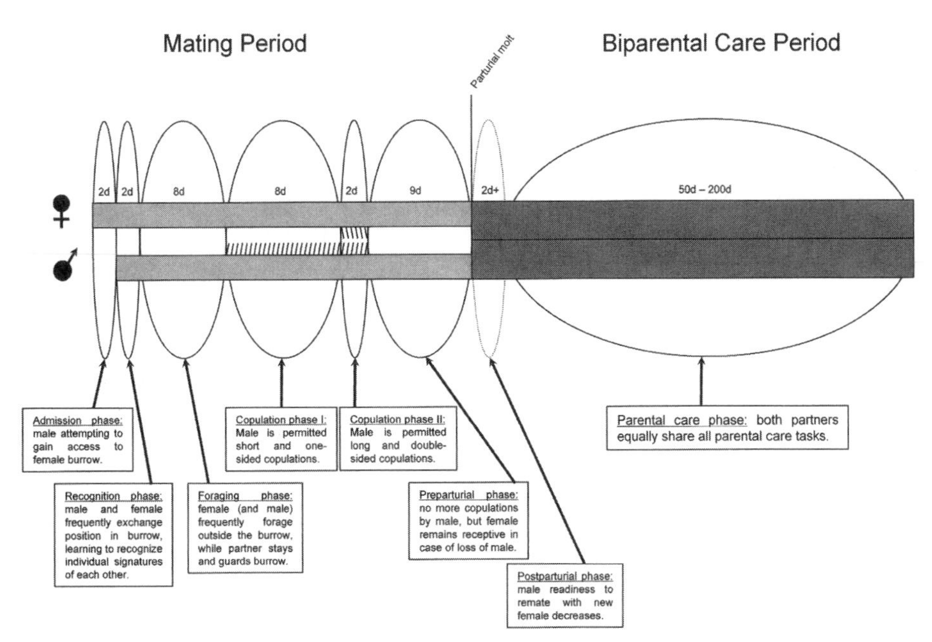

Figure 16.5 Schematic illustration of the most important phases during the breeding period of *Hemilepistus reaumuri*. At the beginning of the actual mating period, in spring, males and females roam (for up to 4 weeks) in search of an appropriate site for a burrow or an existing burrow that they may take over or share with a heterosexual partner. Most often, females start to dig burrows in which they await males. Once a mate arrives at a burrow, the admission phase is initiated. The entire mating period lasts (depending on prevailing temperatures) from about two weeks (pair formation during second half to end of April) to more than six weeks (pair formation beginning of March); parturial molt to delivery of young requires three to four weeks, with biparental care until the parents' death (in rare cases until February/March of the next year, mostly until autumn). Family (mostly sibling) dissolution occurs toward mid February to the end of March of the next year, depending on the considerably variable temperature regimes in different parts of its distribution. Approximate duration of each phase is given in the upper part of the respective ellipsoids; size of ellipsoids reflects only approximate temporal scale. The presented values of about 30 days for the time period from first contact of pair partners until parturial molt correspond to the average duration revealed in southern Tunisia, with pair formation starting in mid March.

temperatures, with many short (usually one-sided) copulations, interrupted by the female. In short copulations (lasting 30 sec to about 2–6 min), no or only very small quantities of sperm are transferred in *H. reaumuri*. After about two-thirds to three-quarters of the time span between pair formation and parturial molt has passed, females engage in long, both-sided copulations (between 8 and 35 min for a single side; for details, see Linsenmair 1979, 1984). Such long copulations have, to date, not been observed in any other oniscoid. In *H. elongatus*, the average duration is 22 seconds (Röder, and Linsenmair 1999), and in all investigated *Porcellio* species, durations range from 13 seconds to rarely 6 minutes (K.E. Linsenmair, S. Dentl, A. Golla, E. Herlein, U. Würsching, unpublished observations). Apart from *H. reaumuri*, three to four short copulations are sufficient to transfer enough sperm to fertilize the next clutch.

Copulations in *H. reaumuri* take place only directly at the burrow entrance. Never did we find any indication of a paired individual trying to copulate beyond its own burrow or at its burrow with a foreigner. Females have full command not only over when to copulate, but also how long. To interrupt, females just move a little back into the burrow, and the male cannot help being slipped off. Thus, before males of *H. reaumuri* are admitted to copulations, they have to invest much time, which cannot be used to obtain additional mates. By delaying effective copulations, females benefit from the male's presence, for example, by male guarding of the burrow during the females' frequent feeding excursions. Guarding the burrow and warding off foreign females are in the male's interest, as well, because admitting an alien female would mostly lead to the loss of the time the male has already invested in the first female. Usually, long copulations terminate the actual mating phase well before the parturial molt (see Fig. 16.5).

Loss of Mating Partners and Sexual Conflicts in H. reaumuri

During the initial phase of the breeding period, both males and females are at a rather high risk of pairing with an only short-lived partner—a considerable number of recently paired *Hemilepistus* s.str. die due to infestation with parasites, and many more, both in the field and laboratory, die for unknown reasons. Many lose their mating partner by an accident (due to predation or trampling by pasturing livestock—see also Baker and Rao 2004) and then may have problems in finding a replacement partner in time. However, a female without a male partner has no chance to survive all three of the following most critical phases: (1) the time during parturial molt, (2) the period of pregnancy and delivery of the young, and (3) the early phase of brood care, when its offspring is still absolutely dependent upon the female's EBC. A single parent is unable to continuously deepen and clean the burrow, to defend it and thereby protect its brood, and to provision its offspring with a suitable diet. To realize its single reproductive chance, a female is highly dependent on a cooperative male.

Should a female lose her mate after the termination of the regular copulation phase but before her parturial molt, she will readily accept a new partner and allow long copulations rather quickly (about 24–48 hr after re-pairing—see Fig. 2.5 in Linsenmair 1989) (Fig. 16.5). Such females can be as attractive for a mate-searching male as are virgin females. However, the success of a previously mated female in a second pairing depends on the time period between the pair formation with the new male and the female's parturial molt. If this time period is longer than three days, the male will usually remain with the female. If the parturial molt of the female comes within 3 days after exchanging a male copulating partner, the male will, with only very rare exceptions, evict the newly molted female from the burrow, leading to the female's certain death under field conditions. After the parturial molt, a female will never find a replacement male.

The nondiscrimination against such females that had already terminated their regular copulation phase would be no surprise if sperm precedence with a latest-in first-out scheme occurred in *H. reaumuri*. But as experimental results have shown (Linsenmair 1989), and as also the morphology of the spermatheca proves (Röder 1995), such a form of sperm precedence does not occur. When a second pair formation took place during the second half of the potential pairing period (i.e., the time period from the first short copulation until about 24 hr before parturial molt, spanning from <15 to >30 days depending on the temperature regime; Linsenmair 1989), the first males fertilized 84% eggs, and 30% of the second males had no fertilization success at all. In replacement males gaining access to a female only during the last fifth of the pairing period, none had any fertilization success. Why do males not avoid such females and prefer those offering better reproductive chances?

To exert preferences between mated and virgin females, males would have to be able to assess a female's reproductive status. Indeed, males of *H. elongatus* show preferences for virgin females in choice experiments (Röder 1995). However, in *H. reaumuri*, the females' sexual attractiveness apparently does not change for males from the beginning of the pairing phase until shortly before the parturial molt, and it is not altered by the preceding sexual behavior of the female. Female *H. reaumuri* apparently are practicing "concealed sperm storage" and also "concealed ovulation" (*sensu* Wickler and Seibt 1981), a concept discussed in our own species (Strassmann 1981) and also in other crustacean species, for example, *Alpheus heterochelis* (Rahmann et al. 2003) and the rock shrimp *Rhynchocinetes typus* (Diaz and Thiel 2004). Could the readiness of a female for long copulations shortly after pair formation not serve as an indicator of reduced paternity chances? Would it not pay a male to desert such a female immediately afterward—given it would still have a chance to find a female offering better reproductive prospects? A female's immediate readiness, however, indicates only that she is under pressure of time since she is near her parturial molt. A male pairing with such a female would very often not incur any reproductive loss because there are different reasons in females for being late and for being then under pressure of time (e.g., parturial molts cannot be delayed arbitrarily but are induced by, e.g., rising temperatures independent of any male influence). Even though a male's inability to classify females according to their reproductive value may in some circumstances be a disadvantage, on average he still gains by fathering a part of the brood (Linsenmair 1989). In order to achieve at least a partial paternity in those late females that have not yet fully filled their spermatheca with sperm from previous copulations, males may have to insert a certain amount of sperm within a certain period (i.e., >3 days) into the female's reproductive tract before the respective female executes its parturial molt. This minimum period, however, does not protect males against fully mistaken investments in females that had already terminated their regular copulation phase and filled their spermatheca completely (Linsenmair 1989). Males may attempt to retake some control by "counting" the days until the parturial molt, thus avoiding a misinvestment in each case, independent of the late females' previous history. However, since females are totally dependent on the cooperation of a male partner, selection should strongly favor the female maintaining her option for re-pairing after loss of her male as long as possible. Concealing her sexual status and remaining sexually attractive until shortly before the parturial molt have given females advantages in this intersexual conflict of interests.

The only chance for a male in finding a (new) mating partner is to wander in search of a female without a partner and, when successful, to strive for access to its burrow. After finding and copulating with a female (at least 3 days before her parturial molt), male *H. reaumuri* never leave her in search of an additional mating opportunity. Why not? First, the risk that all reproductive investment in the first female will be lost (since the female cannot rear the brood alone) is extremely high. Also, the chance to find an additional virgin female during the pairing period may be low. Furthermore, leaving the pair burrow would leave the male without the protective den, exposing it to a high mortality risk (Baker 2004, Baker and Rao 2004; K.E. Linsenmair, unpublished observations). And finally, since males are unable to distinguish between virgin females and those that had already been paired, the potential gains in fathering additional offspring may be very low. Although the male's risks are so high that remaining with its mate is easy to understand in an ultimate view, the question arises: How is the sudden loss of the female's sexual attractiveness dealt with on the proximate level?

Pair Maintenance in H. reaumuri *after the Parturial Molt*

How do females prevent males from deserting them after the parturial molt? In experiments where paired males were repeatedly separated immediately before their females' parturial molt and offered a new female (K.E. Linsenmair, unpubished results), all ($n = 22$) surviving males readily formed new pairs on four to five subsequent occasions. In contrast, when males remained with a female until about two to three days after her parturial molt before separation, the males behaved very differently toward sexually still attractive females. During the first five days after removal of their first female, 60% of these males ($n = 30$) consistently warded off all these females within the first two to four days. Only when separated for more than five days, 75% ($n = 23$) of the males eventually became ready to accept a new female. Apparently, males receive a (probably) chemical signal from the molting female, causing a motivational change, which originally might have been purely manipulative, serving only the female's interests. In the actual situation, it serves the male's interests usually as well, since deserting would not pay on average. Nevertheless, it remains advantageous for a female to manipulate the male, thus avoiding a conflict of interests to its disadvantage.

Family Cohesion and Communication in H. reaumuri

After delivery of the young *H. reaumuri*, both parents care for their developing young and remain with them in the family burrow for up to 200 days (Fig. 16.5). If one parent dies, the other continues to care for the brood alone. As soon as the young *Hemilepistus* are at least half-grown (after ~70–90 days), they have a chance of surviving on their own, even after a loss of both parents. Yet, contrary to the situation in their *P. albinus* neighbors, the young maintain a closed unit, bound to their natal burrow, and they neither disperse nor start reproducing. The most likely reason is that they have no chance of excavating a new protective den. With this indispensable prerequisite for reproduction lacking, the best they can do is remain together and optimize through cooperation their own and their siblings' survival prospects for the next reproductive phase. Because the burrow is size limited, with only the deepest sections offering a favorable microclimate during the hot part of the year, and since social units trading valuable goods and services should always establish exclusivity, to avoid being parasitized, it is not surprising that families of *Hemilepistus* s.str. remain closed units beyond the time when the progeny are dependent on EBC by their parents. What is more surprising is the strictness of this exclusivity, being stricter than in most eusocial insect societies.

The chemical compounds used as discriminators in the recognition system of *H. reaumuri* are nonvolatile, easily transferred during direct body contact (see details in Linsenmair 1987). Therefore, any closer contact with aliens must be avoided, because this entails the danger of immediate and strong alienation, resulting in rejection from the burrow. Any compromise to this exclusivity, even if only for a short period or the admission of a single foreign individual, may lead, at best, to the alienation and death of one or more family members that came into contact with the alien isopod because such individuals will acquire a different discriminator pattern and therefore will be treated as aliens and warded off from the protective den. At worst, such a compromise can result in breakdown of the entire family recognition system. Since available space does not allow treating this aspect more thoroughly, only a few additional facets of this highly sophisticated communication are mentioned here (for an extensive account, see Linsenmair 1987 and references therein).

Exclusivity requires, above all, a permanent, very attentive burrow guard. In order to guarantee, on the one hand, a selective admittance of every family member returning from any outside activity and to prevent, on the other hand, all foreigners from entering, a very reliable recognition system must have been developed. This is no minor task since, in dense populations, family members must be unambiguously discerned from thousands of conspecifics belonging to hundreds of families that may be encountered within the family's activity range.

The Signature System of H. reaumuri: *Properties, Intrinsic Problems, and Their Solutions*

In various contexts, we have performed more than 100,000 tests in which members from different families were reciprocally exchanged. No two families were ever found to use identical signatures. These signatures are blends of different discriminator substances. Their patterns are purely genetically determined (Linsenmair 1972, 1987). All the compounds identified to date (1994) are polar, nonvolatile or only poorly volatile substances and thus differ from the mostly nonpolar, volatile carbohydrates that are assumed (but rarely definitely proven in behavioral tests) to be responsible for the colony-specific odors of social insects (e.g., Lahav et al. 1999, Gamboa 2004). Discriminators in *Hemilepistus* s.str. are evenly spread over the entire body surface. For recognizing a foreigner, the touching of any part of the body with the apical organ (see above) is usually sufficient. The fact that discriminator patterns are genetically determined and extremely variable results in at least three main sets of problems: (1) members of a family may have different discriminator patterns, requiring a mechanism of "family cohesion"; (2) given that (1) is true, molting individuals will lose substances obtained from family members, making it necessary to reacquire these; and (3) adults are cannibalistic, requiring mechanisms to avoid infanticide vis-à-vis their newborn progeny. The solutions to these problems are briefly explained below.

Intrafamiliar Variability in Discriminator Patterns

Family members differ in their self-produced discriminator patterns, not only between parents and young but also among full siblings. The family-specific discriminator "badge" originates from transferring and mixing individual patterns by close body contact that family members maintain in their den. In *Hemilepistus* s.str., only substances produced by several members become elements of the common badge of a family (a brood may contain up to 120 individuals). Many members produce, however, "private" discriminators not showing up in the common family signature. Individual patterns, deviating from the common family badge, have to be learned. These patterns are not learned as a series of compounds that may deliberately be permutated, but rather, deviating patterns have to be learned as differing "gestalts" (Linsenmair 1987). This requires a very astonishing learning capacity of an animal possessing not more than 10,000 neurons in its central nervous system, of which 6,000, at best, may be involved in processing chemical information (Seelinger 1977, Krempien 1983).

Loss of Substances during Molting

When molting, family members lose all substances they have acquired from other family members. For some time, they possess only what they themselves are producing. This sudden loss of many of the common compounds and the appearance of a very deviating individual pattern may cause problems. Even slight aggression toward a newly molted individual is life threatening due to its extreme vulnerability. Presumably for this reason, a protection against intrafamily aggression has developed: the newly molted family member is, during the first hour after each of its bipartitioned molts, well protected by a very strong aggression-inhibiting property (transferable substances of unknown chemical nature; for more details on the very important role of learning and additional safeguards for molting individuals, see Linsenmair 1987).

Avoiding Infanticide

Because adult *Hemilepistus* are highly cannibalistic toward alien young conspecifics, how is the progeny protected against infanticide by its own parents? Parents separated from their own young immediately after delivery and reunited with them 24–70 hours later cannibalized their progeny without exception and independent of whether they met them in their own burrow or elsewhere. However, when parents (with females being very close to delivery or having already delivered their own offspring) were offered alien young, released from the brood pouch not more than 6 hours before, these showed no aggression in 85% (of $n = 40$ pairs) and in the remaining 15% only very slight, never life-threatening aggression, even in alien burrows. Thus, the young of a pair are not recognized with their delivery, nor are they accepted site specifically.

These, newborn young also possess aggression-inhibiting properties. Because these are not transferable to other young, the properties are different from those used in molting family members. This inhibition functions only in those individuals that either are already parents or are about to become parents, with very pronounced sex-specific differences (for the full account, see Linsenmair 1987). The inhibitory properties of these young do not only protect the bearers themselves, but induce a more general tolerance. Those adults that are ready to accept newborn young (but had not yet had any contact with them) will always fiercely attack and cannibalize foreign young 2 days up to several weeks of age, wherever these are met and caught. However, after spending 1–3 hours with newly delivered young, adult *H. reaumuri* behave completely benignly when confronted with young up to the age of three weeks. How to explain this abrupt change of behavior? The inhibitors obviously not only inhibit aggression but also seem to transiently inhibit the learning of the newly delivered progeny's signature. Despite the fact that the offspring is extremely often touched with the antennae—especially by the female—it takes up to 48 hours before it can be proven that parents have learned their progeny's discriminator pattern. In other situations, learning of alien signatures is achieved much more quickly. My assumption is that discriminators are distributed over the entire body, most probably also including the female's marsupium. If this holds true, then the young cannot help but be contaminated with their mother's discriminators and, in addition, with marsupial fluids partly licked off by the female. The only way to get rid of this contamination is via molting. All young regularly molt during the first 12 hours after delivery. After the young lose the maternal contamination, the parents can, and apparently do, learn the pure badge of their own offspring.

The discriminator system evolved by *Hemilepistus* allows—under natural conditions—a very reliable discernment between kin (i.e., nest mates) and aliens. It is very stable against environmental influences, and it allows extreme settling densities without causing any recognition problems. But discriminators are very easily transferred between individuals in every direct body contact. Since discriminators are used in the form of "gestalt" patterns, which are highly susceptible to any qualitative modifications (especially against additions of any new substance), families must remain absolutely exclusive if this system is to function properly.

Clearly, the long-lasting association between parents and their offspring in *Hemilepistus* s.str. required the evolution of a very efficient recognition system. In the other xerophilous oniscoids with EBC, recognition between family members is much less developed. In Fuert.-P.2, *H. elongatus*, and *P. simulator*, one's own progeny is not recognized and can be exchanged with foreign young; only the burrow/retreat is marked and clearly distinguished from those of other females. In the first two species, females leave their progeny early (2–14 days), before the latter are active outside the burrow. In none of these species did we ever observe any cannibalism. In Fuert.-P.2, the only one of the three species extensively observed in the field, we never found any alien young trying to enter a breeding den. Under such conditions, there is no need to distinguish one's own from alien offspring. Field and laboratory observations in Fuert.-P.1 and *P. albinus* point to a recognition by the parents of their own progeny, but further experiments are needed for a definite clarification.

Discussion and Comparison with Other Taxa

Space does not allow discussion of more than a few selected aspects of the sociobiology of these terrestrial isopods. For more exhaustive treatments, readers may refer to the original publications, especially the detailed treatment in Linsenmair (1987). According to theory (Trivers 1972), parental investment reducing the caretaker's residual reproductive value is mainly to be expected where juveniles (1) face physically harsh environments, (2) are exposed to intense competition and/or predation, and (3) depend on special resources they are unable to procure and secure alone (see Wilson 1975, Clutton-Brock 1991, and references in these). When more than one of these factors is effective, as demonstrated above for *H. reaumuri*, living in harsh environments, being temporally totally dependent upon digging and maintaining a costly burrow, and having to deal with strong competition and also predation, the probability of taking such an evolutionary route should be high.

Once on this way, further fitness enhancements may be achievable only by prolonging and/or intensifying care or increasing the expenditure for the single progeny at the cost of their number. Ovoviviparous species (in a broad sense, including most of the brood-caring crustaceans; Thiel 2000, 2003a; see also chapters 14, 17, 18) and viviparous species are mostly caught in this one-way route. The most advanced taxa along this route include some insects, such as tsetse flies (*Glossina* spp.), which produce a single larva per reproductive episode that is fully grown when delivered and ready for immediate pupation (Barraclough and Londt 1986/1989), and the mammals (Clutton-Brock and Godfray 1991).

In the Peracarida, the brood pouch was the decisive device predisposing this large taxon to develop both ECT and EBC. Where ecological conditions change strongly within short time periods, where food is scarce and only temporally available, a central place foraging mode of progeny provision (Ydenberg 1994) may be prohibitive. Here, the ECT developmental route should be ideal, when young born/released at an advanced stage of development and/or with a larger body mass have cost-compensating fitness advantages vis-à-vis young released at smaller sizes. Such advantages have, for instance, been proven in regard to locomotor development for viviparous scincid lizards (Shine 2003). Another advantage may be obtained through larger body size, as found in most precocial mammals (Derrickson 1992). Peracarids practicing ECT will probably have both of these and additional advantages (by, e.g., reduced enemy pressure, competitive advantages in acquiring much contested microhabitats; see Thiel 2003a and references therein). Carrying one's progeny on the body, using at the same time a protective dwelling and releasing there the progeny, provides an ideal basis for the selection of EBC (see chapters 14, 18). Simple tolerance of the progeny may thereby deliver further fitness benefits and prepare the ground for subsequent evolutionary steps (see also discussion of the contribution on brood care in the bromeliad-living crab *Metopaulias depressus* in chapter 17).

Many of the marine brood-caring crustaceans are bound to microhabitats offering abundant food and favorable, comparatively stable habitats, such as bivalves, ascidians, sponges, and kelp holdfasts (Thiel 2000, Duffy et al. 2000; see also chapters 14, 18). Really harsh abiotic ecological conditions seem to be less important drivers of social evolution in the aquatic species. Although intertidal habitats are demanding and stressful environments, I am not aware of a study that convincingly demonstrates which of the typical problems (of desiccation, osmotic stress, temperature extremes) in a crustacean is solved by social behavior. This is different in species burrowing into hard substrata. They have to deal with a difficult habitat property that can more effectively be handled by adults. This holds true, for example, for marine peracarids (Thiel 2003b and references therein), as well as terrestrial isopods (see above), insects (Seelinger and Seelinger 1983, Tallamy and Wood 1986 and references therein, Kent and Simpson 1992), and many of the burrowing vertebrates, especially the social mole-rats (Burda and Kawalika 1993, Jarvis et al. 1994, Faulkes et al. 1997). Where adults and young live for prolonged periods together in such burrows, opportunity should often arise for selecting fitness-enhancing forms of interaction. The longer parent(s) and young live together, due, for example, to increasingly harsher ecological conditions, the more advanced and differentiated should these interactions become, on average. With prolonged life together, young become more independent, changing their behavior from a helpless-altricial type to more or less full-fledged with a much broader spectrum of action patterns. As long as parent(s) and/or young are dependent on their common dwelling and thus in close contact with each other, they will gain inclusive fitness by restricting selfishness and developing mutually beneficial behavior. The more diverse and complex their behavior becomes with approaching the adult stage, the more evolutionary possibilities should arise for advanced social behavior.

Harsh environmental conditions in strongly seasonal habitats with only short periods available for securing vital resources for reproduction, as is characteristic of deserts and semideserts, should favor especially extended caring behavior (Emlen 1982). To date, we know very little about the behavioral ecology of most invertebrates living in xeric environments, and we thus have no basis for guessing how often EBC has independently evolved. However, it fits the presumption that the highest level of social complexity in a noneusocial insect has been found in the Namib-dwelling desert beetle *Parastizopus armaticeps* (Rasa 1999), closely resembling *Hemilepistus* in many facets: showing extended biparental care with division of labor, being dependent on a burrow that can be dug only within a very short interval after rain fall, with foraging for the larvae and digging to deepen the burrow being vital services for the progeny.

All organisms foraging on the outside of home burrows must be able to relocate and recognize their burrows. This is also true for parents caring for offspring in a burrow or home dwelling and feeding or collecting their forage outside. Apart from visual recognition in many of the brood-caring or eusocial hymenopterans (Tinbergen and Linde 1938), chemical signatures are the most common means of marking burrows individually, for example, in the earwig *Labidura riparia* (Radl and Linsenmair 1991) or solitary hymenopterans (Steinmann 1976). If the parents' brood care is restricted to altricial young, these seem never to be individually recognized. If brood care, however, extends into the time where young start outside activity or fledge in birds (see, e.g., Beecher et al. 1981), social parasites can only be detected, if one's own progeny can be recognized reliably. In eusocial hymenopterans, hydrocarbons have been proven to be, at least to some major extent, responsible for recognition of nests, nest mates (e.g., Lahav et al. 1999, Gamboa 2004), and siblings (avoiding incestuous mating in bee wolves, Herzner et al. 2006). Signatures differ in the extent to which they derive from nest environment and food (Liang and Silverman 2000, Wagner et al. 2000) or in which they are genetically determined (see summary in Hölldobler and Wilson 1990). The volatile nature of these signatures facilitates their acquisition, allowing occasionally alien conspecifics and guests/parasites/predators (Elgar and Allan 2004) a relatively easy integration. *Hemilepistus reaumuri* uses nonvolatile discriminators, shows a very high intrafamily variability, and possesses therefore many individual and a common family discriminator "gestalt," causing problems through frequent molts and in regard to the primary recognition of offspring. No other signature system in arthropods has to date been more thoroughly investigated on the behavioral level as that of *Hemilepistus reaumuri*.

Future Directions

Extended offspring carrying time (ECT) merits more attention to clarify under which ecological conditions this form of brood care has been selected and how widely it is distributed. With respect to extended brood care after delivery (EBC), we find among trachelipid isopods, on the one hand, *H.* (*Desertellio*) *elongatus* with its short-term, purely maternal brood care and no recognition of its progeny, and on the other hand, the highly developed social and communication behavior in *H. reaumuri*. It would be worthwhile to look for relatives with an intermediate social behavior to examine whether our assumptions on the probable evolutionary route to highly developed social behavior are supported. Most important, however, is to advance the study of the chemical communication. With the techniques now available, we hope to "break the code" and ideally identify the signatures on an individual level. This would open a very wide field of experimentation and allow us to critically check all our behavioral results and come to a unique understanding of any such system.

Summary

Isopods are the only crustacean group that has evolved many truly terrestrial species and, surprisingly, also highly successful xerophilous forms. Among these, the North African and Near Eastern *Hemilepistus reaumuri* is the most thoroughly investigated. This species shows a highly developed social behavior without which it could never thrive under the harsh conditions prevailing during many months in its distributional range. For survival and reproduction, species of *Hemilepistus* s.str. depend on burrows that are costly to produce and can be dug anew only in spring. The vital dens have to be continuously defended against intra- and interspecific competitors. This is achieved by division of labor between the sexually and socially monogamous pair partners and later with the progeny's participation. Families of *H. reaumuri* are strictly closed units. To discern family members from aliens, an extremely variable chemical recognition system is used, which requires for its reliable functioning a surprising learning capacity. The use of purely genetically determined signatures perceived via contact chemical receptors guarantees a reliable recognition of nest mates also under extremely high population densities where thousands of aliens originating from many hundred families may be encountered. Since signatures within a family are highly diverse, and the nonvolatile secretions are easily transferred by any close body contact, several intriguing problems arise, for example, in the context of the frequent molts of the growing progeny and in regard to the primary recognition of the offspring in parents, which are highly cannibalistic toward alien young. Some *Porcellio* species, distributed according to their habitat choice and their geographic distribution, along a climatic and edaphic gradient leading from less xeric and less seasonal to strongly arid and very seasonal areas, have developed similar behavioral adaptations. However, there are some remarkable deviations, presumed to be mainly due to differences in the burrow substratum they are dwelling in and on. A reconstruction of the probable evolutionary path of *H. reaumuri* social behavior was attempted in the light of our findings on these *Porcellio* species. Especially remarkable is the strict monogamous behavior of the *Hemilepistus* males. Since no intermediate mating systems between a polygamous/promiscuous and the strict monogamous mating system is known, inferences of the probable evolutionary route can be drawn only by scrutinizing the actual behavior of females and males, revealing highly interesting female tactics. Among the most remarkable ones are the females' full control over the temporal distribution of copulations, the extremely prolonged copulation times needed for successful sperm transfer, and the total concealment of the moment of their parturial molt and the therewith connected ovulation. Male *H. reaumuri* appear to retake at least partial control of pair formation by "measuring" the time between first association with a female and the female's parturial molt. Once potential sexual conflicts during the mating period are overcome, both partners engage in long-lasting biparental care. It is concluded that the narrow temporal window during which the extremely valuable family burrow can be constructed has resulted in the sophisticated social behavior found in this semelparous oniscoid isopod.

References

Andersson, M. 1994. Sexual selection. Princeton University Press, Princeton, N.J.

Baker, M. 2004. Sex biased state dependence in natal dispersal in desert isopods, *Hemilepistus reaumuri*. Journal of Insect Behavior 17:579–598.

Baker, M., and S. Rao. 2004. Incremental costs and benefits shape natal dispersal: theory and example with *Hemilepistus reaumuri*. Ecology 85:1039–1051.

Barraclough, D., and J. Londt. 1989. Order Diptera. Pages 283–321 *in*: C. Scholtz and E. Holm, editors. Insects of southern Africa. Butterworths Publishers, Durban, South Africa (originally published 1985, revised reprint 1986).

Beecher, M., I. Beecher, and S. Lumpkin. 1981. Parent-offspring recognition in bank swallows (*Riparia riparia*): development and acoustic basis. Animal Behaviour 29:86–94.

Blaschke, C. 1994. Über die chemische Ökologie der Wüstenassel *Hemilepistus reaumuri*. Unpublished Ph.D. thesis, University of Heidelberg, Heidelberg, Germany.

Burda, H.W., and M.W. Kawalika. 1993. Evolution of eusociality in the Bathyergidae. The case of the giant mole rats (*Cryptomys mechowi*). Naturwissenschaften 80:235–237.

Carefoot, T. 1993. Physiology of terrestrial isopods. Comparative Biochemistry and Physiology 106A:413–429.

Clutton-Brock, T. 1991. The evolution of parental care. Princeton University Press, Princeton, N.J.

Clutton-Brock, T., and C. Godfray. 1991. Parental investment. Pages 234–262 *in*: J. Krebs and N. Davies, editors. Behavioral ecology. Blackwell Scientific, Oxford.

Coenen-Stass, D. 1989. Transpiration, vapor absorption and cuticular permeability in woodlice (Oniscidea, Isopoda, Crustacea). Pages 253–267 *in*: G. Benga, editor. Water transport in biological membranes, Volume 2. From cells to multicellular barrier systems. CRC Press, Boca Raton, Fla.

Dangerfield, J., and S. Telford. 1995. Tactics of reproduction and reproductive allocation in four species of woodlice from southern Africa. Journal of Tropical Ecology 11:641–649.

Derrickson, E. 1992. Comparative reproductive strategies of altricial and precocial eutherian mammals. Functional Ecology 6:57–65.

Diaz, E.R., and M. Thiel. 2004. Chemical and visual communication during mate searching in rock shrimp. Biological Bulletin 206:134–143.

Duffy, J.E., C. Morrison, and R. Ríos. 2000. Multiple origin of eusociality among sponge-dwelling shrimps (*Synalpheus*). Evolution 54:503–516.

Elgar, M., and R.A. Allan. 2004. Predatory spider mimics acquire colony-specific cuticular hydrocarbons from their ant model prey. Naturwissenschaften 91:143–147.

Emlen, S.T. 1982. The evolution of helping. I. An ecological constraints model. American Naturalist 119:40–53.

Faulkes, C.G., N.C. Bennett, M.W. Bruford, H.P. O'Brien, G.H. Aguilar, and J.U.M. Jarvis. 1997. Ecological constraints drive social evolution in the African mole-rats. Proceedings of the Royal Society of London, Series B 264:1619–1627.

Gamboa, G.J. 2004. Kin recognition in eusocial wasps. Annales Zoologici Fennici 41:843–858.

Hamilton, W.D., and M. Zuk. 1982. Heritable true fitness and bright birds: a role for parasites? Science 218:384–387.

Herzner, G., T. Schmitt, F. Heckel, P. Schreier, and E. Strohm. 2006. Brothers smell similar: variation in the sex pheromone of male European beewolves and its implications for inbreeding avoidance. Biological Journal of the Linnean Society. 89:433–442.

Hoese, B. 1981. Morphologie und Funktion des Wasserleitungssystems der terrestrischen Isopoden (Crustacea, Isopoda, Oniscoidea). Zoomorphologie 98:135–167.

Hoese, B. 1982. Morphologie und Evolution der Lungen bei den terrestrischen Isopoden (Crustacea, Isopoda, Oniscoidea). Zoologische Jahrbücher Anatomie 109:487–501.

Hoese, B. 1984. The marsupium in terrestrial isopods. Pages 65–76 *in*: S. Sutton and D. Holdich, editors. The biology of terrestrial isopods. Clarendon Press, Oxford.

Hoese, B. 1989. Morphological and comparative studies on the second antennae of terrestrial isopods. Monitore Zoologico Italiano (NS) Monographia 4:127–152.

Hoese, B., and H. Janssen. 1989. Morphological and physiological studies on the marsupium in terrestrial isopods. Monitore Zoologico Italaliano (NS) Monograph 4:153–173.

Hoffmann, G. 1985a. The influence of landmarks on the systematic search behavior of the desert isopod *Hemilepistus reaumuri*. I. Role of the landmark made by the animal. Behavioral Ecology and Sociobiology 17:325–334.

Hoffmann, G. 1985b. The influence of landmarks on the systematic search behavior of the desert isopod *Hemilepistus reaumuri*. II. Problems with similar landmarks and their solution. Behavioral Ecology and Sociobiology 17:335–348.

Hoffmann, G. 1989. The orientation of terrestrial isopods. Monitore Zoologico Italiano 4:489–512.

Hölldobler, B., and E. Wilson. 1990. The ants. Harvard University Press, Cambridge, Mass.

Janetos, A. 1980. Strategies of female choice: a theoretical analysis. Behavioral Ecology and Sociobiology 7:107–112.

Jarvis, J.U.M., M.J. O'Riain, N.C. Bennett, and P.W. Sherman. 1994. Mammalian eusociality: a family affair. Trends in Ecology and Evolution 9:47–51.

Jormalainen, V. 1998. Precopulatory mate guarding in crustaceans: male competitive strategy and intersexual conflict. Quarterly Review of Biology 73:275–304.

Kent, D., and J. Simpson. 1992. Eusociality in the beetle *Austroplatypus incompertus*. Naturwissenschaften 79:86–87.

Kight, S., and A. Hashemi. 2003. Diminished food resources are associated with delayed reproduction or increased post-reproductive mortality in brood-bearing terrestrial isopods *Armadillidium vulgare* Latreille. Entomological News 114:61–68.

Krempien, W. 1983. Die antennale Chemorezeption von *Hemilepistus reaumuri* (Audouin & Savigny) (Crustacea, Isopoda, Oniscoidea). Unpublished Ph.D. thesis, University of Würzburg, Würzburg, Germany.

Lahav, S., V. Soroker, A. Hefetz, and R.K. Vander Meer. 1999. Direct behavioral evidence for hydrocarbons as ant recognition discriminators. Naturwissenschaften 86:246–249.

Liang, D., and J. Silverman. 2000. "You are what you eat": diet modifies cuticular hydrocarbons and nestmate recognition in the Argentine ant, *Linepithema humile*. Naturwissenschaften 87:412–416.

Lincoln, R. 1970. A review of the species of *Hemilepistus* s. str. Budde-Lund, 1885 (Isopoda, Porcellionidae). Bulletin of the British Museum (Natural History) Zoology 20:111–130.

Linsenmair, K.E. 1972. Die Bedeutung familienspezifischer "Abzeichen" für den Familienzusammenhalt bei der sozialen Wüstenassel *Hemilepistus reaumuri* Audouin u. Savigny (Crustacea, Isopoda, Oniscoidea). Zeitschrift für Tierpsychologie 31:131–162.

Linsenmair, K.E. 1979. Untersuchungen zur Soziobiologie der Wüstenassel *Hemilepistus reaumuri* und verwandter Isopodenarten (Isopoda, Oniscoidea): Paarbindung und Evolution der Monogamie. Verhandlungen Deutsche Zoologische Gesellschaft 1979:60–72.

Linsenmair, K.E. 1984. Comparative studies on the social behavior of the desert isopod *Hemilepistus reaumuri* and of a *Porcellio* species. Symposium Zoological Society London 53:423–453.

Linsenmair, K.E. 1985. Individual and family recognition in subsocial arthropods, in particular in the desert isopod *Hemilepistus reaumuri*. Pages 411–436 *in*: B. Hölldobler and M. Lindauer, editors. Fortschritte der Zoologie, Band 31. Experimental Behavioral Ecology and Sociobiology. Gustav Fischer Verlag, Stuttgart, New York.

Linsenmair, K.E. 1987. Kin recognition in subsocial arthropods, in particular in the desert isopod *Hemilepistus reaumuri*. Pages 121–207 *in*: D. Fletcher and C. Michener, editors. Kin recognition in animals. John Wiley and Sons, Chichester.

Linsenmair, K.E. 1989. Sex-specific reproductive patterns in some terrestrial isopods. Pages 19–47 *in*: A. Rasa, C. Vogel, and E. Voland, editors. The sociobiology of sexual and reproductive strategies. Chapman and Hall, London.

Linsenmair, K.E., and C. Linsenmair. 1971. Paarbildung und Paarzusammenhalt bei der monogamen Wüstenassel *Hemilepistus reaumuri* (Crustacea, Isopoda, Oniscoidea). Zeitschrift für Tierpsychologie 29:134–155.

Marikovsky, P. 1969. A contribution to the biology of *Hemilepistus rhinoceros* [in Russian]. Zoologitcheskii Zournal 48:677–685.

Moreau, J., S. Seguin, Y. Cubet, and T. Rigaud. 2002. Female remating and sperm competition patterns in a terrestrial crustacean. Animal Behaviour 64:569–577.

Radl, R., and K.E. Linsenmair. 1991. Maternal behavior and nest recognition in the subsocial earwig *Labidura riparia* Pallas (Dermaptera: Labiduridae). Ethology 89:287–296.

Rahmann, N., D. Dunham, and C. Govind. 2003. Social monogamy in the big-clawed snapping shrimp, *Alpheus heterochelis*. Ethology 109:457–473.

Rasa, O.A.E. 1999. Division of labour and extended parenting in a desert tenebrionid beetle. Ethology 105:37–56.

Röder, G. 1995. Untersuchungen zur Geschlechtsbestimmung und zur reproduktiven Strategie bei der türkischen Wüstenassel *Hemilepistus* (*Desertellio*) *elongatus* (Crustacea, Isopoda, Oniscoidea). Ph.D. thesis, Würzburg University. Tectum-Verlag, Marburg, Germany.

Röder, G., and K. Linsenmair. 1999. The mating system in the subsocial desert woodlouse, *Hemilepistus elongatus* (Isopoda Oniscidea): a model of an evolutionary step towards monogamy in the genus *Hemilepistus* sensu stricto? Ethology, Ecology and Evolution 11:349–369.

Schmalfuss, H. 2003. World catalog of terrestrial isopods (Isopoda: Oniscoidea). Stuttgarter Beiträge zur Naturkunde Serie A (Biologie) 654:1–341.

Schneider, P. 1971. Lebensweise und soziales Verhalten der Wüstenassel *Hemilepistus aphganicus* Borutzky 1958. Zeitschrift für Tierpsychologie 29:121–133.

Schneider, P. 1991. Asseln (Ernährung, Gangart und Fortpflanzung). Verhandlungen Deutsche Zoologische Gesellschaft 84:395.

Seelinger, G. 1977. Der Antennenendzapfen der tunesischen Wüstenassel *Hemilepistus reaumuri*, ein komplexes Sinnesorgan (Crustacea, Isopoda). Journal of Comparative Physiology 113:95–103.

Seelinger, G., and U. Seelinger. 1983. On the social organisation, alarm and fighting in the primitive cockroach *Cryptocercus punctulatus* Scudder. Zeitschrift für Tierpsychologie 61:315–333.

Shine, R. 2003. When to be born? Prolonged pregnancy or incubation enhances locomotor performance in neonatal lizards (Scincidae). Journal of Evolutionary Biology 16:823–832.

Steinmann, E. 1976. Über die Nestorientierung solitärer Hymenopteren: Individuelle Markierung der Nesteingänge. Mitteilungen der Schweizerischen Entomologischen Gesellschaft 49:253–258.

Strassmann, B. 1981. Sexual selection, paternal care, and concealed ovulation in humans. Ethology and Sociobiology 2:31–40.

Surbida, K., and J. Wright. 2001. Embryo tolerance and maternal control of the marsupial environment in *Armadillidium vulgare* (Isopda: Oniscidea). Physiological and Biochemical Zoology 74:894–906.

Sutton, S.L. 1980. Woodlice. Pergamon Press, Oxford.

Sutton, S.L., M. Hassal, R. Willows, R.C. Davis, A. Grundy, and K.D. Sunderland. 1984. Life histories of terrestrial isopods: a study of intra- and interspecific variation. Symposium Zoological Society London 53:427–434.

Taiti, S., and F. Ferrara. 1988. Revision of the genus *Exalloniscus* Stebbing, 1911 (Crustacea: Isopoda. Oniscidea). Zoological Journal of the Linnean Society 94:339–377.

Takeda, N. 1984. The aggregation phenomenon in terrestrial isopods. Symposium Zoological Society London 53:381–404.

Tallamy, D., and T. Wood. 1986. Convergence patterns in subsocial insects. Annual Review of Entomology 31:369–390.

Thiel, M. 2000. Extended parental care behavior in crustaceans—a comparative overview. Crustacean Issues 12:211–226.

Thiel, M. 2003a. Extended parental care in crustaceans—an update. Revista Chilena de Historia Natural 76:205–218.

Thiel, M. 2003b. Reproductive biology of *Limnoria chilensis*: another boring peracarid species with extended parental care. Journal of Natural History 37:1713–1726.

Tinbergen, N., and R. Linde. 1938. Über die Orientierung des Bienenwolfes (*Philanthus triangulum* Fabr.). Biologisches Zentralblatt 58:425–435.

Trivers, R. 1972. Parental investment and sexual selection. Pages 136–179 *in*: B. Campbell, editor. Sexual selection and the descent of man. Aldine Publishing Company, Chicago, Ill.

Wagner, D., M. Tissot, W. Cuevas, and D. M. Gordon. 2000. Harvester ants utilize cuticular hydrocarbons in nestmate recognition. Journal of Chemical Ecology 26:2245–2257.

Walter, H. 1960. Einführung in die Phytologie, III/I. Standortslehre. Eugen Ulmer Verlag, Stuttgart, Germany.

Warburg, M. 1993. Evolutionary biology of land isopods. Springer-Verlag, Berlin.

Wickler, W., and U. Seibt. 1981. Monogamy in crustacea and man. Zeitschrift für Tierpsychologie 57:215–234.

Wilson, E. 1975. Sociobiology. Harvard University Press, Cambridge, Mass.

Witte, H.J.L., G.J. van Gelder, and J.D. Spitler. 2002. In situ measurement of ground thermal conductivity: the Dutch perspective. ASHRAE Transactions 108:263–272.

Witte, V. 2001. Struktur und Organisation des Treiberameisenverhaltens bei südostasiatischen Ponerinen der Gattung *Leptogenys*. Unpublished Ph.D. thesis, University of Frankfurt, Frankfurt, Germany.

Wright, J., and J. Machin. 1993. Atmospheric water absorption and the water budget of terrestrial isopods (Crustacea, Isopoda, Oniscidea). Biological Bulletin 184:243–253.

Ydenberg, R. 1994. The behavioral ecology of provisioning in birds. Ecoscience 1:1–14.

Zimmer, M. 2001. Why do male terrestrial isopods (Isopoda: Oniscoidea) not guard females. Animal Behaviour 62:815–821.

The Social Breeding System of the Jamaican Bromeliad Crab *Metopaulias depressus*

Rudolf Diesel

Christoph D. Schubart

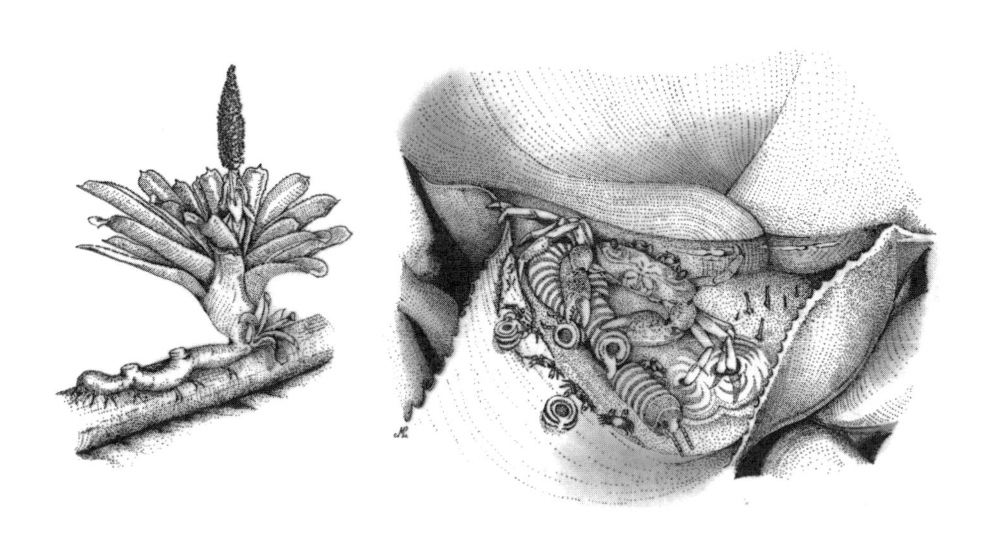

17

In most brachyuran crabs, brood care appears to be limited to the care for the eggs carried under the female's abdomen. Interactions between the mother and larvae or young are extremely rare. One major impediment for brood care to evolve in crabs is their predominant reproductive mode via planktonic larvae. The larvae of marine crab species—except a very few species with direct development (see Rabalais and Gore 1985)—are swept away from their mothers immediately after hatching. The same applies for land-living crabs that retain a marine planktonic development (e.g., Gecarcinidae).

With the colonization of freshwater and terrestrial habitats, most crab species abandoned planktonic larval development because of the unfavorable conditions in the available freshwater habitats (osmotic pressure, downstream transport, predation). Large vitellogenic eggs and a direct development evolved in various lines of freshwater and terrestrial groups independently (see Diesel et al. 2000). The miniature crabs hatching from the eggs usually remain attached to the mother for several days before they disperse. In these species, there is opportunity for brood care to evolve. However, in only a few of the presently studied species, the young and mothers remain together for a protracted period. A common trait of these insufficiently studied species is their distinct breeding habitat: they live in particular microhabitats, in which the mother and the young are safe from predators or find a supporting microclimate in otherwise adverse surroundings. Examples of crabs that use such microhabitats are *Sesarma jarvisi* breeding in the shells of large land snails (Diesel and Horst 1995), *Potamonautes raybouldi* living in water-filled tree holes (Bayliss 2002), and *Metopaulias depressus*, an obligate inhabitant of water-storing bromeliads (Diesel 1989). In these nestlike breeding habitats, the offspring remain with the mother because conditions in the nest are more favorable than elsewhere.

The bromeliad crab *Metopaulias depressus* Rathbun 1896 is one of 10 described endemic species of nonmarine crabs from the Caribbean island of Jamaica. Mothers and offspring of this species live together in the water-storing leaf axils of large bromeliads (Diesel 1989). Large bromeliads of the species *Aechmea paniculigera*, which qualify as breeding sites, live for several years and represent a stable and reliable water source, even during prolonged droughts. They provide the food resources to support large crab families, water, and shelter from predators and are expandable at the same time. Alexander et al. (1991) proposed that extended parental care and occupation of particular microhabitats or nests, which are safe, provide food, and are expandable, are common preconditions for the evolution of eusociality. Small-colony eusocial insects and cooperative breeding vertebrates share the traits of (1) overlapping adult generations or delayed dispersal from the natal group, (2) reproductive division of labor or reproductive suppression, and (3) cooperative or alloparental care of young (Michener 1969, Wilson 1971, Sherman et al. 1995, Lacey and Sherman 1997, Choe and Crespi 1997). The results of earlier studies show that *Metopaulias depressus* combines the traits of extended parental care and inhabiting and defending a well-maintained nest, which both favor the evolution of advanced social life (Alexander et al. 1991, Crespi 1994). In this contribution, we focus on the question whether the bromeliad crab has followed the evolutionary route that Alexander et al. (1991) predicted for species showing these life history features and evolved traits, characteristic of advanced social species. We specifically study (1) colony composition, dispersal, and the evidence for overlapping generations; (2) the existence and potential breeding activity of subordinate adults in a colony; (3) territory/nest defense by adult and juvenile colony members; and (4) alloparental care of broods. In addition to a summary of published information, we present previously unpublished results.

The Model Organism: *Metopaulias depressus*

General Biology

The freshwater and terrestrial crabs endemic to Jamaica (family Sesarmidae) are descendents from a marine ancestor that colonized Jamaica about 4.5 million years ago (Schubart et al. 1998). While their closest relatives are marine species with a larval development in the marine plankton, the Jamaican endemics have an abbreviated larval development in fresh water and thus live in independence from the sea. These Jamaican crabs occupy various ecological niches, such as rivers and streams, mountain rain forests, limestone caves, and bromeliad plants (Diesel et al. 2000).

The bromeliad crab is the most unusual of these endemics in terms of habitat choice. The crab is an obligate inhabitant of bromeliad plants. Its life history, morphology, and physiology are adapted to life on bromeliads. The crabs measure up to about 80 mm in width, including the legs (20 millimeters carapace width [mm CW]) and have a dark brownish red color resembling leaf litter and a depressed flat body adapting them to better fit between the narrow leaf bases of the bromeliad plants. Despite their terrestrial habits, bromeliad crabs require access to fresh water for breathing and breeding, which they encounter in the water retained in the leaf axils of the bromeliad. In general, the bromeliads provide all the essential resources in an otherwise hostile environment: water for gill respiration, molting, and reproduction, but also food and, in the narrowing leaf axils, shelter against various predators such as lizards or birds. The bromeliad crab occurs in the central and western mountainous parts of the island, for example, in Cockpit Country, where locations with patches of relatively high bromeliad densities are common. The selected bromeliads, usually *Aechmea paniculigera* and *Hohenbergia* species, grow in locations with high precipitation (Laessle 1961, Diesel 1989).

Despite providing all the crabs' needs, the quality of bromeliads as habitat for bromeliad crabs varies considerably. Plants can roughly be categorized as follows: (1) small plants with few leaves and a low capacity to store water, mostly of the genus *Hohenbergia*, which are usually uninhabited; if they harbor a crab, then usually a single, not fully grown individual; (2) medium-sized plants with several leaf axils holding water, which may contain one or a few juveniles or adults, but rarely breeding females—adult males usually live in this plant category (Diesel 1989); (3) high-quality, large, ground-borne and epiphytic bromeliads, usually the species *Aechmea paniculigera*, which measure on average 1.5 m in diameter between the tips of the leaves, have up to 22 rainwater-filled leaf axils, and can store up to 4 liters of water from rain and dew (Fig. 17.1a). These are usually occupied by breeding females, which prefer the largest bromeliads (Diesel 1989). Most of these large bromeliads grow on the ground and up to about 3 meters on the trunk of a tree. Only occasionally, females breed in the canopy 20 meters above ground. Large *A. paniculigera*, which qualify as breeding sites, live for several years and represent a stable and reliable water resource

Figure 17.1 (a) Large terrestrial bromeliad, *Aechmea paniculigera*. These plants are growing as clones from a rhizome. Rhizomes might be decades old and constantly grow to large plants at the same location. Sprouts grow rapidly to a large plant after the older clone has flowered and died. (b) Nursery axil with bromeliad crab (*Metopaulias depressus*) larvae. This is one of one to two dozen water-holding axils on a plant. The nursery axil has been cleared of leaf litter and debris by the mother crab. (c) Mother crab with last brood. The young crabs stay in the nursery axil for up to three months and are cared for by the mother. Later they disperse to other leaf axils over the plant. (d) The mother has killed and deposited a beetle in the nursery with the young crabs.

even during long periods of drought because they collect dew as well as rainwater. Most of these large *A. paniculigera* grow from strong old rhizomes, which rapidly produce a new plant (clone) about every three years.

Resource Availability

Bromeliad crabs rely entirely on the food resources on their host plant, because they search for food exclusively on the plant (Diesel 1989). Food consists of (1) plant material and insects that are caught between the long leaves of the bromeliad and funneled into the water-filled axil; (2) the organic matter and associated aquatic fauna in the leaf axil; (3) the leaf-litter—associated fauna, such as millipedes, in the lower leaf axils that have lost their capacity to hold water; and (4) smaller invertebrates that visit the plant, for example, in search for water. There is a strong positive correlation between the size of an *A. paniculigera* and the numbers of leaf axils, the amount of water stored, and the amount of available nutritional resources (Laessle 1961, Diesel 1989). As the bromeliad plant grows, the numbers of leaf axils and the associated resources increase, too. The bromeliad crabs continue to grow throughout their life, and female fecundity increases with body size (Diesel 1989). Hence, nest size and the amount of resources increase with every breeding, since the crabs usually remain on the same plant throughout their reproductive life. Therefore, the size of bromeliads is proportional to their quality and suitability for raising large broods. Since large bromeliads are scarce, there is probably strong competition among crabs for these resources.

Breeding Biology

Metopaulias depressus breeds annually; in the Cockpit Country, the breeding season starts in December and January, when the males leave their bromeliad to pay visits to the females in their vicinity for mating. Females start to produce eggs in January. The large vitellogenic eggs (20–100 eggs per brood, depending on female size) are carried under the abdomen, where they develop for about 10–12 weeks and then hatch (Diesel 1989). All the larvae are released together into one of the water-filled leaf axils, the so-called nursery axil, holding on average 240 milliliters of water (Fig. 17.1b). Female reproduction is relatively synchronous within one area, and most females release larvae within one to two weeks (Diesel 1989). The larvae develop within 13 days into small juvenile crabs (Fig. 17.1c; Diesel and Schuh 1993). For up to three months, mothers defend their offspring from predatory damselfly larvae and spiders, provide food, and improve the microclimate of the nursery axil by enhancing pH, calcium, and dissolved oxygen levels (Diesel 1989, 1992a, 1992b, 1997, Diesel and Schuh 1993).

After about three months, the young crabs start leaving the nursery axil and disperse over the other axils of their natal bromeliad. After one year, they have reached about 7–10 mm CW and belong to cohort 1 (C-1). Although they hatched on the same day, the body size within the C-1 may now vary considerably.

Metopaulias depressus attains sexual maturity at about 12 mm CW in males and 13 mm in females (Hartnoll 1964; R. Diesel, unpublished observations). Sexual maturity in females is marked by a distinct "molt of puberty" in which they develop genital openings, change the shape of the abdomen, and modify their pleopods to facilitate attachment of eggs. Likewise, males develop functional gonopods. To determine sexual maturity in the 11–14 mm CW size group, the individuals have to be removed from the bromeliad, except females that carry eggs, which remain visible under the abdomen for about 2.5 months. After the eggs have hatched, some empty egg cases remain attached to the pleopods underneath the abdomen until the next molt, allowing identification of these females as reproductives.

The maximum reproductive lifespan of a female is about three years. Therefore, she can produce about three broods during her lifetime. This coincides with the life expectancy of single shoots of the preferred bromeliad *A. paniculigera*, which exists as a grown plant for about three years, flowers, and then dies (R. Diesel, unpublished observations). Simultaneously with the inflorescence, a lateral shoot (clone) is produced at the base of the dying plant, which grows rapidly and later offers a good-quality territory at the same location within less than a year, which may then be available to a daughter of the previous breeding female.

Objectives

For *M. depressus*, bromeliads comprise a microcosm providing a safe place from predators, a favorable microclimate, and all resources the crabs need for life, but suitable large, unoccupied bromeliads are in short supply. Animals living under similar conditions of habitat limitation frequently refrain from dispersing and have evolved an advanced social organization involving cooperation among groups of relatives, many of which do not breed (Alexander et al. 1991, Queller and Strassmann 1998). Earlier work has shown that the bromeliad crab shows extended parental care and that offspring remain for a protracted period on the natal plant, forming family groups (colonies), consisting of a mother and her young from more than one brood (Diesel 1989). Based on these findings, we here investigate the presence in the bromeliad crab of several conditions and behaviors frequently found in eusocial or cooperative breeding species (see Crespi 1994), namely, overlap of generations, alloparental care, and cooperative colony defense.

Overlapping adult generations and nonbreeding adults are characteristic of many eusocial and cooperatively breeding species. Here, we describe colony composition and the dispersal of young crabs from the natal colony. Do overlapping generations occur in bromeliad crab colonies? When do young females become sexually mature, and how do they breed for the first time?

A second trait common in eusocial species, defense of the nest or colony against unfamiliar conspecific invaders, was studied in a set of experiments. Firstly, we investigated whether bromeliads are defendable resources. To a resident female, the highest risk to lose its territory comes from other females that search for a breeding site. Therefore, we studied whether females can defend their nest against other females. An important predisposition for defending the colony is that invaders were recognized as nonmembers by the colony members. These experiments included also interactions between older (C-1) juvenile crabs from other colonies versus the colony mother and versus the older juveniles of the focal colony. For resident C-1 individuals, the participation in colony defense could be their contribution (payment) for staying in the nest.

Cooperatively breeding or eusocial species frequently show alloparental care. In this context, we were interested in whether older C-1 young of the bromeliad crab help their mother in caring for the brood or in colony defense. Field experiments to address these questions in detail are difficult to perform. Hence, we concentrated on the more general question, whether the most recent brood may benefit from the presence of older C-1 young in the colony.

At the end of this chapter, we review our findings and discuss the social system of *M. depressus* with respect to our new results. Finally, we briefly review the biology of other crab species that might be possible candidates for an advanced social organization worth studying, and we point out open questions and future studies that should be addressed to better understand the social biology of the bromeliad crab.

Colony Structure and Social Behavior of *Metopaulias depressus*

Colony Composition and Size at Dispersal

Composition of bromeliad crab colonies was studied in the mountains of the Cockpit Country (Trelawny) during several field seasons, usually in spring, from 1986 to 1999. Here, we define a colony as the group of crabs found on a single bromeliad plant. All individual crabs from 181 dissected bromeliads were counted and the carapace width measured with a digital caliper (±0.05 mm). Crabs larger than 6 mm CW were sexed and the reproductive status noted. Age cohorts were distinguished by size, the latest brood being referred to as cohort 0 (C-0) and previous broods referred to as cohort 1 (C-1), which consists of individuals that must be at least one year old.

The results show that at least two successive broods coexist on a single bromeliad and form colonies consisting of presumed family groups (Fig. 17.2). Individuals of different age cohorts were distinct by their size. C-0 juveniles usually attained 2–4 mm CW in May–June, whereas C-1 individuals were larger than 7 mm CW. There exists a pronounced body-size variation within C-1 (17.1–226.3%, $n = 58$ colonies with more than five C-1 members of known size). This variation of up to 226.3% suggests that C-1 consists of crabs one and two years old, but no attempt was made to further subdivide C-1, due to a continuum in body sizes, possibly a consequence of differential growth. The largest of 181 studied colonies consisted of a very large reproductive female (20.2 mm CW), 71 C-0 offspring of around 3 mm CW, and 12 C-1 offspring of 6.2–14.6 mm. In 83.4% of the colonies, 1–25 older young (C-1) were present in addition to the most recent brood (C-0).

All colonies invariably held only one single reproductive female, ranging in size from 13.2 to 20.3 mm CW. In addition to the single reproductive colony mother, 21.5% of the colonies held one and in few cases two smaller crabs of 13.2–14.6 mm CW that were large enough to reproduce, as known from solitary females of this size with eggs (see below). Our comparisons within this smaller size class revealed that 53% ($n = 15$) of the solitary females were carrying eggs, whereas similar-sized females

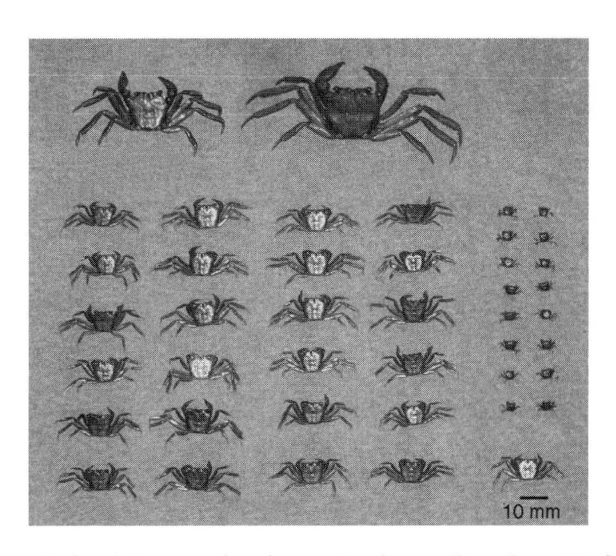

Figure 17.2 Bromeliad crab (*Metopaulias depressus*) colony collected in 1998 from a large *Aechmea paniculigera*. The colony mother (top right) is the only reproductive female, with an adult nonreproductive female, possibly her daughter (top left), the most recent brood (C-0, relatively small because of an overall high brood mortality in 1998), and the offspring of about one year of age (C-1).

in a colony never carried eggs or showed signs of reproductive activity ($n = 10$; $\chi^2 = 7.8$, df = 1; $P = 0.008$). This means that mature females did not reproduce in the presence of the dominant colony mother, thus acting as subordinate females.

Females of about 13–15 mm CW that are carrying eggs are breeding for the first time (primiparous) in their lives. An analysis revealed that only 31% ($n = 35$) of these egg-carrying young females were residing alone on a bromeliad. Most likely, they had dispersed recently from their natal colony, in an effort to start a new colony as foundresses. The other 69% of the small primiparous females were found on a bromeliad together with 1–12 smaller C-1 individuals. In this case, it seems most likely that the larger reproductive colony mother must have died and the primiparous female took over the bromeliad as an heiress. Alternatively, this female could be an invader that dispersed from another colony and now tolerates the unfamiliar juveniles in her new breeding territory. This, however, seems highly unlikely because inhabited bromeliads are widely dispersed and the risk of facing injury or death when invading a colony occupied by another female is very high. Much more probable is that a daughter of the former colony mother takes over the nest location together with her siblings.

Staying Home or Leaving

How long do young crabs stay in their natal plant? Because the exact age of a crab cannot be ascertained, we used body size as an approximation. To determine the sizes at which crabs disperse from maternal colonies, we studied the size distribution of the crabs from 329 bromeliads dissected in the period of March to June over several years. The body sizes of the crabs in colonies were compared with those found solitary on bromeliads, assuming that solitary individuals must have dispersed from their natal colony.

The size distribution of colony members shows a marked decline in the number of individuals larger than 9 mm CW (Fig. 17.3). The smallest solitary individuals

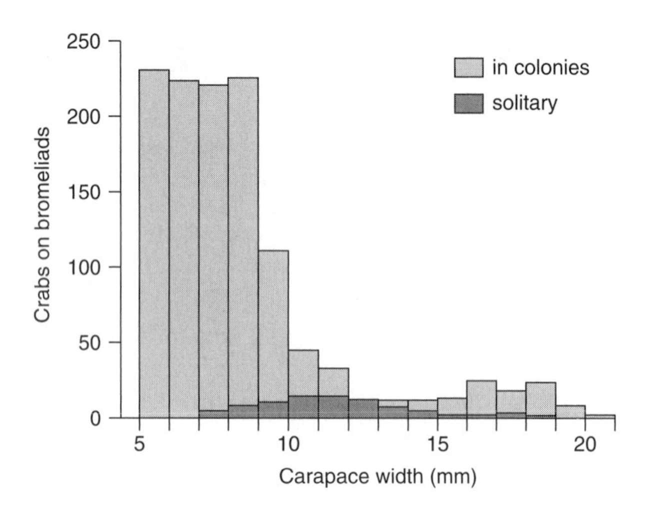

Figure 17.3 Frequency distribution of body sizes (CW) of crabs found in colonies or solitary on bromeliads. The decline in frequency of individuals larger than 9 mm CW in colonies indicates the size at which some individuals disperse. These are found solitary on bromeliads.

found were 8 mm CW; the highest abundance of solitary individuals is between 9 and 10 mm CW. There was no marked difference in this pattern between sexes. Hence, the young crabs appear to remain with their mother for at least one year, the time necessary to reach the minimum size observed in solitary juvenile crabs. Then some individuals may disperse, whereas others stay.

Colony Defense

Do resident females defend their bromeliad against colony nonmembers? Three experiments were performed to study the reaction of the resident female against unfamiliar adult males, females, and juveniles. Most experiments were conducted in the field. Only few colonies were brought for observation into the laboratory over the years, because it is difficult to transport the large plants without damage.

Resident Female Versus Unfamiliar Adult Crabs

The behavior of the resident female toward intruding large adult females was investigated in 11 colonies (one resident female was used twice against different invaders) that were transported to the laboratory in the years from 1996 to 1999. Resident females ranged from 15.9 to 19.8 mm CW, and the nonfamiliar females that were released into the colony were 15.4–19.4 mm CW in size. The invaders were released onto the outer tip of the leaf of the axil inhabited by the resident female. For distinction, invaders were marked with a black marker dot on the carapace. Following the release, the crabs were observed until one left the axil. In some cases, the axils were videotaped to document the interactions between the two females in later analyses. To study the reaction of the breeding female toward males, we released a large nonfamiliar male in each of four colonies during the nonbreeding season.

In the staged contests, the colony mothers defended their nest vigorously. Female and male invaders usually moved very slowly downward into the axil with the resident female. Within 30 minutes, all resident females invariably attacked and expelled the intruding females from the axil (sign test, $p < 0.001$, $n = 11$), even when the invader was larger than the resident ($n = 8$). For an invader, there is a high risk of becoming injured during fights. One intruding female lost walking legs, one lost both chelae, and two females fled and leapt from the plant when attacked by the resident female. Similarly, resident females attacked and expelled the intruding males ($n = 4$). The killing of invaders was observed by chance in two colonies in the field (R. Diesel, unpublished observations): in each case, an adult individual that had previously not been recorded from the colony was killed and cannibalized by the resident female.

Resident Female Versus Cohort-1 Individuals

The behavior of colony mothers toward unfamiliar and familiar C-1 individuals was studied in June 1999 in the field and in the laboratory. Fourteen individuals ranging from 7.6 to 14.2 mm CW were collected as colony nonmembers from distant bromeliads. One C-1 individual was collected from each of six focal colonies as familiar colony members (ranging from 8.9 to 12.4 mm CW). All C-1 individuals were marked, and each released into a different nonfamiliar ($n = 14$) or familiar ($n = 6$) colony.

The individuals were observed continuously for the first 30 minutes and afterward by frequent controls for up to several days.

The result of this experiment suggests that the colony mother is capable of distinguishing between colony members and unfamiliar young crabs. In the experiment in which familiar C-1 individuals were released into their natal bromeliad, they were not attacked, remained in the colony, and were seen for 12–14 days until the end of the experiments ($n = 6$). In contrast, all unfamiliar invaders larger than 10 mm CW were attacked and expelled by the mother or withdrew, mostly within 30 minutes ($n = 10$; 10.6–14.2 mm CW). Those smaller than 10 mm CW ($n = 4$; 7.6–9.6 mm CW) were not attacked and expelled. Two of these smaller nonfamiliar C-1 individuals had left the test plant four and five days after introduction and were found dead of desiccation on the laboratory floor.

Cohort-1 Defense Against Unfamiliar Versus Familiar Crabs

Do C-1 individuals participate in colony defense? Because of the secretive life of the C-1 individuals within the leaf axils, we could not directly observe their behavior. To determine their role in colony defense, we studied the fate of C-1 individuals after they were released into unfamiliar colonies with only C-1 individuals present.

In February and June 1999, C-1 individuals of 7.2–13.0 mm CW were collected from different bromeliads in the field, individually marked, and released within one hour into a foreign colony ($n = 11$ bromeliads) or their natal colony ($n = 8$ bromeliads). From these test colonies, the mother crabs had previously been removed so that the invaders encountered only juveniles and nonreproductive C-1 individuals on the plant. At the start of the experiment, it was not possible to determine the exact number of C-1 individuals in the colony, because this can only be done following dissection of the bromeliad. After seven days, the plants were dissected and the content of the leaf axils was studied.

After seven days, all familiar C-1 individuals that were introduced in their natal colony ($n_{familiar} = 8$) were found alive, whereas 73% of the unfamiliar invaders ($n_{unfamiliar} = 11$) had disappeared ($\chi^2 = 12.4$, df $= 1$; $p = 0.0004$). Body remains that were found in two colonies suggest that the colony members had killed and partly eaten the invaders.

C-1 individuals apparently do not remain in a foreign colony, irrespective of whether the colony mother was present or only the C-1 individuals. In some experimental colonies containing mothers, the colony mother reacted aggressively toward foreign C-1 individuals, but more frequently, we did not observe direct aggressive behaviors between any colony members and the C-1 invader. Our observations suggest that the C-1 individuals—as well as some of the adult females and males—that we experimentally introduced into another colony withdraw on their own. The risk of being injured or killed over the long term by the colony members appears high for unfamiliar juveniles and adults from other colonies or bromeliads.

Alloparental Care

Do the young of the most recent brood benefit from older siblings? In the field, we could not observe the behavior of C-1 toward the C-0 individuals, nor could food availability, predation pressure, or the number of C-1 individuals be manipulated without impairment to the colony. We approached this question by studying the effect of the number and size of C-1 individuals present in a colony on survival and growth of their younger siblings (C-0). Twenty-two colonies with young about two weeks old (C-0) were randomly chosen in the field in May 1997 and May 1998. We counted the C-0 offspring in the nursery axil and recorded the body size (CW) of the breeding female as a correlate of fecundity. The size of the female was also positively related to the size of the bromeliad and the number of leaf axils and thus with the availability of nutritional resources. The mother was removed from each colony to eliminate the strong effects of maternal brood care on C-0 development.

Three C-0 individuals from each colony were taken as a subsample for calculating the total dry mass of the brood (average dry weight of an individual × $N_{C\text{-}0}$) at the start of the experiment. After four weeks, the bromeliads were collected and dissected, and the C-0 and C-1 individuals counted and collected. Colonies differed in the number of C-1 crabs, which ranged from 0 to 19. Survival and dry weight increase of the C-0 cohort and the size of the C-1 was measured. The number and size of C-1 individuals present in the colony could not be established at the beginning, but after four weeks when the colonies were dissected, the size and number of C-1 were recorded and the dry mass of the brood was determined (total dry mass of surviving C-0; Sartorius R 160 P balance [Data Weighing Systems, Elk Grove, Ill.], ±0.01 mg accuracy). As a measure of the reproductive success of the colony, we used total dry weight increase of the brood. This was calculated as dry mass of the C-0 young that survived until the end of the experiment minus the estimated dry mass of the brood at start. The value of total dry weight increase is the product of average weight increase of individual young time survival.

This experiment was designed to distinguish between two possible outcomes. First, older sibs could have no or a negative effect on the orphaned brood, for example, when competing for food resources or killing C-0 individuals. In that case, a negative relationship between number of C-1 individuals and the number and growth of the C-0 would be expected. The second possible outcome would be that the older C-1 siblings enhance survival and development of the brood, which may then provide evidence for alloparental care.

The influence of older siblings on growth and survival of the brood was tested with a multiple regression model. Independent variables were the number and average size of C-1 individuals present in the colony, the initial calculated dry mass of the brood, and the size of the colony mother. The size of the mother was positively correlated with fecundity (see Diesel 1992b) and with the number of leaf axils of her plant ($p = 0.006$, $n = 97$), hence with the amount of stored water and food resources. Body size of the mother was therefore used in the model to control for fecundity and food resources. As dependent variable, we used the total dry-weight increase of the brood during the four weeks, as a combined measure of survival and growth of the young.

The results showed that the model was highly significant and explained 70% of the variation in total dry-weight increase of the brood (multiple regression, $F_{(4,17)} = 9.5$, $r^2 = 0.69$, $p = 0.0003$,). The number of C-1 individuals was the only significant variable in the equation ($t_{(17)} = 4.36$, $p = 0.0004$), whereas the size of mother ($t = 1.45$, $p = 0.16$), the size of the C-0 at start of the experiment ($t = 1.53$, $p = 0.14$), and the size of C-1 individuals ($t = 0.63$, $p = 0.54$) were not significant. That means

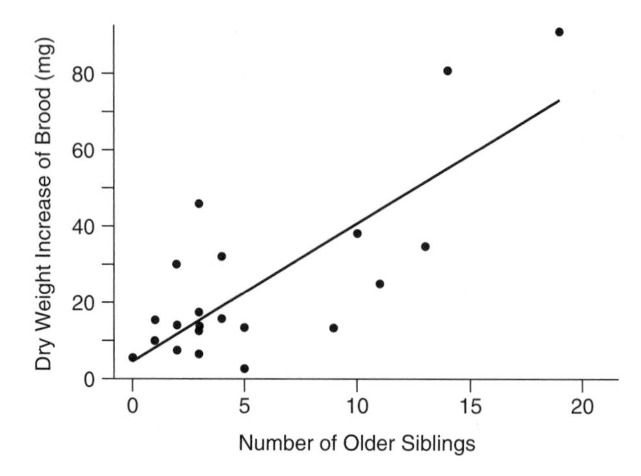

Figure 17.4 Positive correlation between number of older siblings (C-1) and estimated total dry-weight increase of the brood (C-0). Colonies were randomly chosen and revisited four weeks after removal of the mother.

that the number of older siblings present in a colony explains a very large proportion of the total weight increase of the brood (Fig. 17.4). This means that the more older siblings were present in a colony, the faster the C-0 young grew and the more survived the experiment in absence of the mother.

In this field experiment, the broods (C-0) developed well. Total dry weight of the brood increased even in the absence of their mothers. Thus, reproductive success increased with the number of C-1 siblings present in the colony. At the end of the experiment, the broods were about six weeks old. In absence of the mother, the orphaned C-0 young had left the nursery axil about one to two weeks earlier than usual and spread to other leaf axils of the plant in search of food.

Our results suggest that older siblings enhance survivorship and growth of their younger sibs. The amount of nutritional resources for the development of the brood and the probability that predators are detected and eliminated should increase with the number of defending crabs distributed over the colony. The C-1 individuals may deter and/or kill potential predators such as spiders or increase the availability of food by killing prey and allowing their younger siblings to feed on it. In the laboratory, we observed that C-1 crabs fed with a piece of millipede allowed C-0 individuals to share it. Hence, the activities of C-1 colony members result in nest defense and in exploitation of nutritional resources, which will benefit survival and growth of younger siblings.

The Social System of *Metopaulias depressus*

The results of previous research and the present studies demonstrate that the Jamaican bromeliad crab *Metopaulias depressus* clearly shows traits that are characteristic for small colony eusocial insects and cooperative breeding vertebrates (see Michener 1969, Wilson 1971, Alexander et al. 1991, Crespi 1994, Sherman et al. 1995, Lacey and Sherman 1997, Queller and Strassmann 1998, Choe and Crespi 1997): (1) *Metopaulias depressus* lives in large colonies on bromeliad plants, which provide a supporting microclimate, safety, food, and are expandable and defendable; (2) it has overlapping adult generations; and (3) there are adult but nonreproductive females in the colony together with a single reproductive female, the colony mother. Our new results furthermore suggest that (4) older offspring help in colony defense and have a significant positive impact on the development of their younger siblings.

This highly developed social system is similar to that of cooperatively breeding and eusocial species (Wilson 1971, Sherman et al. 1995) and has not been described for any other crabs. The only crustaceans with a similar social organization so far reported are marine sponge-dwelling alpheid shrimps (Duffy 1996, 2003). In the majority of marine crab species, there is no association between the mother and her offspring after eggs hatch. In many freshwater and fully terrestrial species, in contrast, egg development is direct and the young remain attached to the mother's body for several days after hatching (Diesel et al. 2000; see also chapter 14). However, no particular brood care or social behaviors had been reported so far.

The basis for the social organization in the bromeliad crabs is that young stay in their natal plant at least for one year, and some may not leave at all. As obligate bromeliad dwellers, their survival depends on regular access to water. When leaving and searching for another bromeliad, they face a high risk of desiccation in the usually dry surrounding habitat. The mortality risk decreases with size: the larger the crab, the higher the chance of surviving dispersal. Experiments with juveniles of the closely related *Sesarma jarvisi* Rathbun 1914 showed a strong positive correlation between body size and resistance to desiccation (Bäurle 1995). An additional size-dependent mortality risk is predation during dispersal (e.g., by spiders, see Diesel 1989) and a high risk of injury or death when entering an occupied bromeliad. On the other side, there is relative safety in the home plant. Hence, it should be beneficial for C-1 individuals to delay dispersal until they have reached a larger body size.

Young females need to disperse if they want to colonize a new bromeliad for breeding. However, the limited availability of noninhabited, colonizable high-quality plants appears to be a constraint on successful dispersal and colony foundation. For a young female dispersing from the natal colony, the chance to acquire a suitable bromeliad from a usually much larger territory owner is very low. Most of the dispersing females have to find vacant bromeliads, and these are extremely rare. The prospects of successful colony founding in such bromeliads are low and costly, because they are usually filled with leaf litter and because physicochemical conditions are adverse (Diesel 1992a, Diesel and Schuh 1993). It is difficult for a young, relatively small female to clean the leaf axils of leaf litter and to prepare a supporting microclimate in the nursery axil for the brood.

The Lucky Princess

Some colonies held two adult females: the large colony mother and a small subordinate female. These subordinate females were physically able to reproduce (they had passed the "molt of puberty"; see Hartnoll 1964) but never carried eggs. At the same time, other females of the same size take the risk of dispersal and become solitary breeding foundresses.

What could be the benefits in staying home? Subordinate females may gain some inclusive fitness benefits by exerting altruistic behavior and improving survival and growth of younger siblings. However, this component of inclusive fitness may be negligible compared to the gain in direct fitness from increased survival and future reproductive success (see Clutton-Brock 2002) when taking over the bromeliad location (present plant or subsequently appearing shoots) once the colony mother dies (see Emlen 1991).

Marking experiments to follow the fate of subordinate females in the field over one year were not successful, because they are difficult to locate without destroying the bromeliad. Observations on marked bromeliads with colonies, however, suggest that primiparous females took over a nest location with young crabs after the colony mother died.

Subordinate females possibly refrain from early dispersal and reproduction for the benefit of staying at home and taking over the colony, or acquiring a new, adjacent bromeliad clone. This is supported by the observations that several old rhizomes repeatedly produced large plants that served as new nest locations and were perpetually occupied by colonies for the last 14 years. Younger females accompanied by smaller conspecifics usually inhabited the new clones. We assume that most of these females are subordinates of the former colony, now in charge of her siblings. However, the exact degree of relatedness still needs to be clarified with genetic methods.

Why Defend the Home Bromeliad?

Among the possible conspecific invaders, adult females can inflict the highest costs to the resident female. Adult females seeking a suitable new breeding site because their home plant has been destroyed may evict the resident females from their home plant. In our experiment, however, colony mothers attacked and always expelled intruding adult females even if they were considerably larger. Fights among females were more vigorous than those involving females versus males. On the other hand, breeding females often tolerated adult female colony members, presumably their daughters. Males, too, were attacked and evicted from the colony. In the field, however, we occasionally observed males that spent a few days in a colony.

Colony mothers also behaved aggressively against C-1 individuals from other colonies if their size was at least 10 mm CW, but did not respond toward smaller nonfamiliar individuals. How and why do resident females distinguish between nonfamiliar larger and smaller C-1 individuals? It may reflect the situation in nature, where C-1 crabs disperse from the colony at about 9 mm CW and thus only juveniles of this size and larger may become potential invaders and be recognized as such by the resident female.

The senior author of this paper made a similar observation in an earlier experiment (R. Diesel, unpublished data), in which he exchanged the larvae between nurseries from different bromeliad crab colonies. Females did accept the unrelated offspring and continued with brood care. A similar behavior is reported from burrow-living crayfish. Figler et al. (1997) observed that females did not discriminate between their own and unrelated early offspring

By defending the plant against unrelated females, the owner eliminates the risk of investing in nonrelated young. Therefore, selection for recognition of a female's own brood is not expected as long as the offspring are small. This lack of recognition of early juveniles, however, may be a cause for evolution of alternative strategies such as brood parasitism. Preliminary molecular results based on microsatellite and mitochondrial data suggest that occasionally some individuals of the C-0 cohort in a colony may not be direct descendents of the colony mother, and thus larvae dumping by other females may occur (L. Heine, C.D. Schubart, I. Marcade, M. Klinkicht, and R. Diesel, unpublished results).

Alloparental Care

Alloparental care is one of the most prevalent cooperative tasks in social species (see Emlen 1991). Our experiments suggest that older siblings support growth and survival of their younger siblings. The behaviors of C-1 individuals responsible for this result are not known. Older siblings may increase food availability and decrease predation risk for the C-0. These activities may not be costly to the C-1 individuals but could simply be a byproduct of their normal behavior.

Social Evolution in Other Crustaceans

The documentation of the social breeding system in *Metopaulias depressus* leads to the question why similar helper systems did not evolve in the other Jamaican terrestrial crabs. One of the most important predispositions for the evolution of sociality is extended parental care. This cannot evolve in species with dispersing planktonic larvae, but species with direct development have the preconditions to evolve extended parental care (Diesel et al. 2000, Thiel 2003). This is the common reproductive mode of freshwater and terrestrial crabs throughout the world. The Jamaican endemic species of the genus *Sesarma* (closest relatives of *M. depressus*) are a rare exception in still showing an abbreviated larval development, but with the important difference of larval retention in the parental habitat. In most species for which breeding is known (*S. fossarum*, *S. windsor*, *S. ayatum*, *S. verleyi*, *S. jarvisi*, see Diesel et al. 2000; R. Diesel, unpublished observations), the larvae develop in confined, nonexpandable microhabitats such as rock pools, burrows, and snail shells. With the exception of the snail shell crab *S. jarvisi* (see below), there is so far no evidence that extended parental care occurs in these species (of the John Crow mountain crab *S. cookie*, the breeding habitat is still unknown). Only cohorts of larvae were found, and juveniles appear to disperse shortly after metamorphosis.

The only other Jamaican crab with known social behavior is the snail shell crab *Sesarma jarvisi*, which has reached the highest social level next to the bromeliad crab. This crab lives in the mountain rain forests of central Jamaica. The mother selects empty shells of large land snails for breeding (Diesel and Horst 1995, Bäurle 1995). She prepares the shell for breeding and collects about 5 milliliters of dew water in the shell. The few larvae are released into the miniature aquarium and develop rapidly into juveniles. The mother cares for the larvae and juveniles in the snail shell for months, protecting them from predators, providing food, and maintaining a supporting microclimate. Although this breeding habitat provides a suitable environment for the young and is well defendable, it lacks nutritional resources, is found in higher densities, and is used only once for raising young. After about three months, too much detritus has accumulated in the shell and the juveniles have grown to a size that space in the shell becomes limited, so the juveniles have to disperse. Hence, there is no opportunity for helping and later takeover, since the breeding habitat loses its value during the rearing of a single brood.

Breeding habitats of most other Jamaican crabs are thus either too small or ephemeral or do not contain food. In contrast, the bromeliads are large, stable, and renewable habitats that also provide food. This was probably the major key difference that promoted the social evolution of the bromeliad crab.

Little is known about the breeding behavior of other true freshwater and terrestrial crab species. However, there may be other breeding habitats favoring extended parental care. We predict that species that breed in tree holes or phytotelmata are good candidates for extended brood care. These are such species as the East African Usambara tree-hole crab, *Potamonautes raybouldi* (Potamonautidae), from Kenya and Tanzania, which rears its young in water-filled tree holes (Bayliss 2002, Cumberlidge and Vannini 2004). A similar brood care pattern may be found in the tree-hole crabs of West Africa, Madagascar, and Southeast Asia (see Cumberlidge et al. 2005), but the evolution of cooperative breeding in these cases is unlikely because of the limited size of their breeding territory and the lack of nutritional resources. Some burrowing species in particularly harsh habitats should also be candidates for extended parental care (for burrowing crayfish, see chapter 15). Considering the large number and diversity of true freshwater and terrestrial crab species in the world, their mostly concealed way of life, and our present poor knowledge about their biology, many interesting discoveries of social behavior in crabs may still await us.

With their direct development, freshwater crabs have one important predisposition to evolve extended parental care, but their breeding habitats are probably too confined or too ephemeral to evolve this pattern. In comparison, bromeliads are stable habitats with renewable resources. Comparative studies between tree-hole crabs and the bromeliad crab could help us to understand the role of the microhabitat in the evolution of extended parental care and advanced social systems.

Spanier et al. (1993) asked, "Why are there no reports of eusocial marine crustaceans?" They argued that several characteristics occasionally found in crustaceans, such as cavity nesting, overlapping generations, restricted dispersal, and some form of parental care, would favor eusociality according to Alexander et al. (1991). As shown in this study, one should expand the question to include land-living crustaceans, as well. Also, Duffy (2003) agrees that "among the Crustacea there are many taxa that meet one or more of Crespi's criteria but few that me[e]t all of them." For example, crustaceans thriving on corals can rely on a long-lived habitat providing shelter and food. However, they reproduce by planktonic larvae, probably because corals are not patchy enough to favor juvenile retention in the adult habitat.

Planktonic dispersal thus acts against the evolution of extended parental care in marine crustaceans. This conjecture is supported by the occurrence of extended parental care, or other family group formation, in direct-developing amphipods living in family groups in sponges or ascidians (see Thiel 2000) and sponge-inhabiting shrimps of the genus *Synalpheus* (see Duffy 1996). Duffy (1996, 2003; see also chapter 18) showed that some of these species show a high degree of sociality and suggested that the evolution of eusociality has been favored by the presence of four characteristics that are shared with many social insects (see also Crespi 1994): (1) direct development; resulting in kin association; (2) ecological specialization on a valuable, long-lived resource; (3) strong competition for the host resource; and (4) possession of weaponry effective in monopolizing it. Our present study on the bromeliad crab *Metopaulias depressus* shows that a similar social evolution occurred in a terrestrial environment and even with a larval development through swimming stages.

Offspring retained in a confined nestlike structure usually have limited access to food. Therefore, various species of invertebrates and vertebrates provide their eggs or embryos with nutritional resources as yolk (lecitotrophic eggs). In addition, some have evolved an active food-provisioning behavior by the parents. For example, in the desert isopod *Hemilepistus reaumuri*, the young remain in a burrow without food resources. Both parents forage for food away from the offspring nest, carry food back to the nest, and provision the offspring (Linsenmair 1972; see also chapter 16). A similar food-provisioning social behavior is well known in many social insects (Wilson 1971) and also occurs in the ambrosia beetle *Austroplatypus incompertus* (see Kent and Simpson 1992) and some passalid beetles (see Schuster and Schuster 1996).

An interesting parallel to the bromeliad crab is found in frogs. The Jamaican bromeliad frogs (*Osteopilus brunneus*) deposit their eggs in the same bromeliad species as the bromeliad crab (avoiding plants with crabs). The tadpoles develop in the leaf axils with very limited nutritional resources and are fed by their mother with unfertilized eggs (Lannoo et al. 1987). Similar provisioning of fertilized or unfertilized eggs to oophagous tadpoles also evolved in other frog species that use water-filled bromeliad axils, tree holes, or bamboo stumps to deposit their larvae (Weygoldt 1980, Brust 1993, Caldwell and Oliveira 1999, Kam et al. 2000). Food provisioning in frogs frequently involves eggs. They seem to be a more convenient and handy food item to feed the young, than carrying small prey items. This early cannibalism may not be a useful predisposition for social evolution, though.

Future Directions

Despite many years of research on the bromeliad crab, interesting questions remain to be studied in order to understand the social organization of the crab. A better understanding of the forces selecting for the social organization of the bromeliad crab should provide valuable information for understanding the evolution of higher social systems in other animal taxa.

The Mating System and Genetic Relatedness of the Colony Members

Most important for understanding the social organization of the bromeliad crab will be knowledge of the genetic relationships among the colony members. Are the individuals of the C-0 or the C-1 cohorts full or half-siblings? Is the subordinate female the daughter of the colony mother? The genetic relatedness is predicted to affect their social behavior. If, for example, older offspring have different fathers, then the young siblings are half-siblings. Is this degree of relatedness high enough for alloparental care behavior to evolve? Would a subordinate daughter help raise additional competitors with a low degree of relatedness?

Whether offspring clutches of the bromeliad crab have one or multiple fathers depends on the mating system. With the new marking, tracking, and observation techniques available today, it would be interesting to study the mating behavior of the bromeliad crab. Presently, only a few basic details on mating behavior are known (R. Diesel, unpublished observations): males usually leave their home plant during the mating season in the Cockpit Country (December and January) and visit and mate with surrounding females. They may move from bromeliad to bromeliad and return to their home plant or spend a few days on the bromeliad with the female. Visits on a bromeliad with a colony mother might be dangerous for a male, especially if the female is its size or larger. As shown in the defense experiments reported above, females do attack males and could injure or even kill a male with their strong and piercing claws (in one case we observed that a female had killed a male in the field). Two sensory cues may be important for males during mating. (1) We have observed that when the male enters a bromeliad, he sometimes employs a short series of taps with the points of his claws on the surface of the leaf (claw drumming on the bromeliad leaf). This possible courtship signal will travel over the plant and tell the female that there is a visitor and might reduce aggressive interactions between the prospective mates. (2) By sampling the axil water with the claw (see below), males could probably obtain information on the receptive state of the female.

An important determinant of paternity is the mating behavior of the female. Does she mate with different males within a mating season and over successive mating seasons? If so, how does she use the sperm of these males for fertilization? Preliminary molecular results suggest that multiple paternity is reflected within single cohorts (L. Heine, J. Heinze, C.D. Schubart, unpublished results). What is the consequent degree of relatedness among the colony members? If it is low, one has to address the question why would they still cooperate. The senior author has observed different males visiting a female, but it is not known whether multiple mating took place in these cases. There is a high chance that a female is mating repeatedly with the same male even over successive mating seasons, because males live stationary in an area and appear to have a defined home range with several females, similar to those of the marine spider crab *Inachus phalangium* (see Diesel 1986). Females store the sperm in seminal receptacles for long periods. Several females dissected six months after the mating and spawning season held large sperm reserves (R. Diesel, unpublished data). Thus, the sperm of one male may be used repeatedly. However, the influence of sperm competition is still unknown. Females may be able to manipulate the degree of genetic relatedness among the offspring by selective mating or selective sperm use.

The study of the degree of relatedness among the colony members would provide valuable information not only for the interpretation of the behavior of C-1 individuals in respect to alloparental care but also for the recognition of colony members, thereby helping to understand the evolution of the social system of the bromeliad crab.

Larva Dumping

An additional possibility for a lower degree of relatedness among the colony members would be through individuals that come from outside the colony. Preliminary studies with microsatellite and mtDNA markers showed that some colonies included small C-0 individuals that did not share alleles or haplotypes with the colony mother (L. Heine, C.D. Schubart, I. Marcade, M. Klinkicht, and R. Diesel, unpublished results).

How do these crabs come into the colony? Small juveniles of this size (about 3 mm CW) could not have invaded the colony from another bromeliad. The most likely explanation is that "larvae dumping" occurs in the bromeliad crab. Females with eggs ready to hatch may search for colonies with larvae and release some of their own larvae into the nursery while the colony mother is somewhere else on the plant. This could be an interesting alternative strategy for females that do not possess a suitable bromeliad for breeding or are yet too small and ineffective to prepare and maintain a nursery axil. Apparently, females cannot or do not discriminate between their own versus unrelated larvae. In one experiment where larvae were swapped between two colonies, both mothers continued to care for the foster larvae (R. Diesel, unpublished observations). With molecular techniques, it will be possible to establish how frequently unrelated C-0 individuals occur, and additional field observations are necessary to explain how these get into the nursery and which females are releasing them.

Recognition of Nest Mates

How do bromeliad crabs recognize an individual as familiar, unfamiliar, colony member, or possibly kin? The mechanisms underlying the recognition should be studied. We suggest that relevant cues may be water-soluble components. We frequently observed that crabs use the tips of their chelae and mouthparts to "sample" the axil water or the moist body surface of another crab. Crabs that were released on an unfamiliar bromeliad, for example, walk down the leaf to the water edge, sample the water with one claw, and move the tip of the chela to the mouthparts in a "drinking" manner. They may gain information on the reproductive state of a female and whether the bromeliad or axil is familiar or unfamiliar. The latter could explain that in the experiments on colony defense, some individuals released into an unfamiliar colony left the bromeliad on their own. Is there a "colony odor"? Do genetically related individuals produce similar odors, or excretions or do miniature ecosystems have distinct bouquets because of their different microbe assembly and thus the crabs on a bromeliad collectively achieve specific "colony odor"?

The Lucky Princess

What is the fate of an adult subordinate female? The reproductive success of dispersing primiparous females and the subordinate females that remain in their home plant should be compared. Does it pay to remain home and forfeit reproduction for the prospect of inheriting the nest location? In this system, the daughter not only may inherit directly the mother's territory but also could await a new territory that develops at the same location in a very predictable way in form of a lateral shoot. Hence, bromeliads are like territories with renewable resources. To understand this unique system, future research should include the bromeliad dynamics.

Summary

The Jamaican bromeliad crab *Metopaulias depressus* (Decapoda, Brachyura, Sesarmidae) lives exclusively in the water-storing leaf axils of bromeliad plants, which provide a supporting microclimate, safety, and food and are expandable and defendable resources.

We suggest that the species evolved extended parental care to ensure reproduction in this scattered microhabitat. Theory and comparison with other social animals predict that such limited habitat availability may predispose species for the evolution of eusociality, which is classically characterized by overlapping adult generations, reproductive division of labor (caste formation), and altruistic behavior by nonreproductives. In this study, we show that similar characteristics are found in the bromeliad crab: it lives in large colonies consisting of the colony mother and her offspring. Only the colony mother produces eggs. More than 80% of the colonies consisted of at least two annual broods, and 21.5% of the colonies held at least one additional adult crab besides the colony mother. Hence, some offspring remain in the natal colony until they become adult and thus form overlapping generations. Young adult females stayed in their natal colony as subordinate females and forfeited reproduction in the presence of the colony mother, while others of the same age cohort dispersed to other bromeliads (foundresses) and produced eggs. Colony members defended the bromeliad territory against unfamiliar invaders. The colony mother vigorously fought off unfamiliar adult females. Individuals of the C-1 cohort and the colony mother repelled unfamiliar C-1 invaders but not familiar C-1 individuals used as controls. The results of a field experiment in which the colony mother was removed showed that the C-0 broods benefit from the presence of older siblings. That is, C-1 offspring that are at least one year old or nonreproductive adults participate in colony defense and increase survival and growth of their younger siblings. Thus, the bromeliad crab has evolved traits that are characteristic of eusocial and cooperatively breeding species. It displays a high degree of sociality that is unique among crabs and represents the pinnacle of a remarkable and swift social evolution from a nonsocial marine ancestor.

Acknowledgments We thank D. Horst, W. Janetzky, and H. Weimer for help in the field; the Discovery Bay Marine Laboratory for support during the fieldwork; W. Wickler of the Max-Planck Institut für Verhaltensphysiologie for financial support; B. Knauer for preparation of the figures; and E. Duffy, H. Fricke, J. Greeff, M. Thiel, J. Schneider, P.W. Sherman, F. Vollrath and P. Wirtz, for comments on a previous draft of the manuscript. This project was supported by the Max Planck Gesellschaft and the Deutsche Forschungsgemeinschaft.

References

Alexander, R.D., K.M.D. Noonan, and B.J. Crespi. 1991. The evolution of eusociality. Pages 3–44 *in*: P.W. Sherman, J.U.M. Jarvis, and R.D. Alexander, editors. The biology of the naked mole-rat. Princeton University Press, New York.

Bäurle, G. 1995. Snail shells as islands in an inhospitable environment: the parental care of the snail shell crab *Sesarma jarvisi* in the tropical karst of Jamaica. Unpublished Diploma Thesis, University of Tübingen, Tübingen, Germany.

Bayliss, J. 2002. The East Usambara tree-hole crab (Brachyura: Potamoidea: Potamonautidae)—a striking example of crustacean adaptation in closed canopy forest, Tanzania. African Journal of Ecology 40:26–34.

Brust, D.G. 1993. Maternal brood care by *Dendrobates pumilio*: a frog that feeds its young. Journal of Herpetology 27:96–98.

Caldwell, J.P., and V.R.L. de Oliveira. 1999. Determinants of biparental care in the spotted poison frog, *Dendrobates vanzolinii* (Anura: Dendrobatidae). Copeia 1999:565–575.

Choe, J.C., and B.J. Crespi, editors. 1997. The evolution of social behavior in insects and arachnids. Cambridge University Press, Cambridge.

Clutton-Brock, T.H. 2002. Breeding together: kin selection and mutualism in cooperative vertebrates. Science 296:69–72.

Crespi, B.J. 1994. Three conditions for the evolution of eusociality: are they sufficient? Insectes Sociaux 41:395–400.

Cumberlidge, N., and M. Vannini. 2004. Ecology and taxonomy of a tree-living freshwater crab (Brachyura: Potamoidea: Potamonautidae) from Kenya and Tanzania, East Africa. Journal of Natural History 38:681–693.

Cumberlidge, N., D.B. Fenolio, M.E. Walvoord, and J. Stout. 2005. Tree-climbing crabs (Potamonautidae and Sesarmidae) from phytotelmic microhabitats in rainforest canopy in Madagascar. Journal of Crustacean Biology 25:302–308.

Diesel, R. 1986. Optimal mate searching strategy in the symbiotic spider crab *Inachus phalangium* (Decapoda). Ethology 72:311–328.

Diesel, R. 1989. Parental care in an unusual environment: *Metopaulias depressus* (Decapoda: Grapsidae), a crab that lives in epiphytic bromeliads. Animal Behaviour 38:561–575.

Diesel, R. 1992a. Managing the offspring environment: brood care in the bromeliad crab, *Metopaulias depressus*. Behavioral Ecology and Sociobiology 30:125–134.

Diesel, R. 1992b. Maternal care in the bromeliad crab, *Metopaulias depressus*: protection of larvae from predation by damselfly nymphs. Animal Behaviour 43:803–812.

Diesel, R. 1997. Maternal control of calcium concentration in the larval nursery of the bromeliad crab, *Metopaulias depressus* (Grapsidae). Proceedings of the Royal Society of London, Series B 264:1403–1406.

Diesel, R., and D. Horst. 1995. Breeding in a snail shell: ecology and biology of the Jamaican montane crab *Sesarma jarvisi* (Decapoda: Grapsidae). Journal of Crustacean Biology 15:179–195.

Diesel, R., and M. Schuh. 1993. Maternal care in the bromeliad crab *Metopaulias depressus* (Decapoda): maintaining oxygen, pH and calcium levels optimal for the larvae. Behavioral Ecology and Sociobiology 32:11–15.

Diesel, R., C.D. Schubart, and M. Schuh. 2000. A reconstruction of the invasion of land by Jamaican crabs (Grapsidae: Sesarminae). Journal of Zoology 250:141–160.

Duffy, J.E. 1996. Eusociality in a coral-reef shrimp. Nature 381:512–514.

Duffy, J.E. 2003. The ecology and evolution of eusociality in sponge-dwelling shrimp. Pages 217–252 *in*: T. Kikuchi, S. Higashi, and N. Azuma, editors. Genes, behaviors, and evolution in social insects. University of Hokkaido Press, Sapporo, Japan.

Emlen, S.T. 1991. Evolution of cooperative breeding in birds and mammals. Pages 301–377 *in*: J.R. Krebs and N.B. Davis, editors. Behavioral ecology: an evolutionary approach. Blackwell Science, Oxford.

Figler, M.H., G.S. Blank, and H.V.S. Peeke. 1997. Maternal aggression and post-hatch care in red swamp crayfish *Procambarus clarkii* (Girard): the influences of presence of offspring, cross-fostering, and maternal molting. Marine and Freshwater Behaviour and Physiology 30:173–194.

Hartnoll, R.G. 1964. The freshwater grapsid crabs of Jamaica. Proceedings of the Linnean Society of London 175:145–169.

Kam, Y.C., Y.H. Chenn, T.C. Chen, and I.R. Tsai. 2000. Maternal brood care of an arboreal breeder, *Chirixalus eiffingeri* (Anura: Rhacophoridae) from Taiwan. Behaviour 137:137–151.

Kent, D.S., and J.A. Simpson. 1992. Eusociality in the beetle *Austroplatypus incompertus* (Coleoptera: Curculionidae). Naturwissenschaften 79:86–87.

Lacey, E.A., and P.W. Sherman. 1997. Cooperative breeding in naked mole rats: implications for vertebrate and invertebrate sociality. Pages 267–301 *in*: N.G. Solomon and J.A. French, editors. Cooperative breeding in mammals. Cambridge University Press, Cambridge.

Laessle, A.M. 1961. Micro-limnological study of Jamaican bromeliads. Ecology 42:499–517.

Lannoo, M.J., D.S. Townsend, and R.J. Wassersug. 1987. Larval life in the leaves: arboreal tadpole types with special attention to the morphology, ecology, and behavior of the oophagous *Osteopilus brunneus* Hylidae larva. Fieldiana Zoology 38:1–32.

Linsenmair, K.E. 1972. Die Bedeutung familienspezifischer "Abzeichen" für den Familienzusammenhalt bei der sozialen Wüstenassel *Hemilepistus reaumuri* Audouin u. Savigny (Crustacea, Isopoda, Oniscoidea). Zeitschrift für Tierpsychologie 31:131–162.

Michener, C.D. 1969. Comparative social behavior of bees. Annual Review of Entomology 14:229–342.

Queller, D.C., and J.E. Strassmann. 1998. Kin selection and social insects: social insects provide the most surprising predictions and satisfying tests of kin selection. BioScience 48:165–175.

Rabalais, N.N., and R.H. Gore. 1985. Abbreviated development in decapods. Crustacean Issues 2:67–126.

Schubart, C.D., R. Diesel, and S.B. Hedges. 1998. Rapid evolution to terrestrial life in Jamaican crabs. Nature 393:363–365.

Schuster, J.C., and L.B. Schuster. 1996. Social behavior in passalid beetles (Coleoptera: Passalidae): cooperative brood care. Florida Entomologist 68:267–272.

Sherman, P.W., E.A. Lacey, H.K. Reeve, and L. Keller. 1995. The eusocial continuum. Behavioral Ecology 6:102–108.

Spanier, E., J.S. Cobb, and M.-J. James. 1993. Why are there no reports of eusocial marine crustaceans? Oikos 67:573–576.

Thiel, M. 2000. Population and reproductive biology of two sibling amphipod species from ascidians and sponges. Marine Biology 137:661–674.

Thiel, M. 2003. Extended parental care in crustaceans—an update. Revista Chilena de Historia Natural 76:205–218.

Weygoldt, P. 1980. Complex brood care and reproductive behavior in captive poison-arrow frogs, *Dendrobates pumilio* O. Schmidt. Behavioral Ecology and Sociobiology 7:329–332.

Wilson, E.O. 1971. The insect societies. Harvard University Press, Cambridge, Mass.

Ecology and Evolution of Eusociality in Sponge-Dwelling Shrimp

J. Emmett Duffy

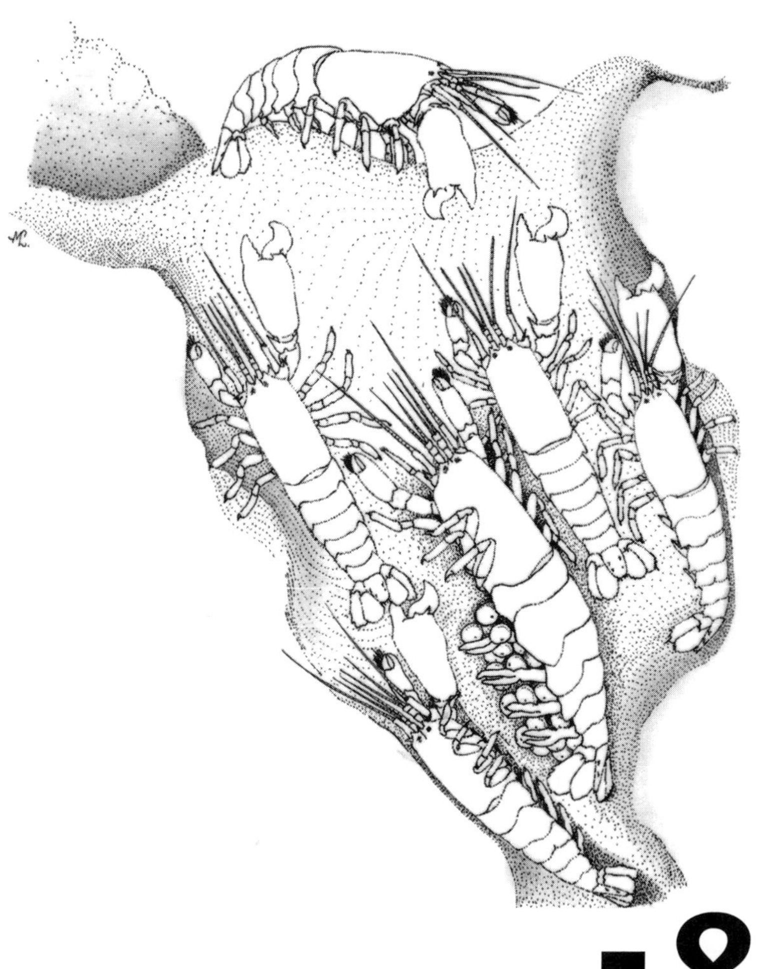

18

Eusociality refers to colonial life in which most individuals forgo reproduction to help raise and defend offspring of a lucky few of their kin. First described in ants, termites, and honeybees, eusociality is among the most striking phenomena in nature and has fostered spectacular evolutionary and ecological success in the several lineages of social insects (Wilson 1971, 1990). Eusociality also poses an enduring puzzle—explaining its evolution has vexed biologists since Darwin (1859) famously observed that sterility in social insects posed "the one special difficulty, which at first appeared to me insuperable, and actually fatal to my whole theory." In essence, the question is how, in a world of Darwinian self-interest, a species in which most individuals behave altruistically can evolve, persist, and even thrive.

Historically, eusociality was defined by three criteria: cohabitation of different adult generations, reproductive "division of labor" (i.e., reproductive skew), and cooperative care of young (Michener 1969, Wilson 1971). The traditional concept of eusociality has been criticized for, among other things, being vague, arbitrary, and biased toward arthropods; hence, various alternative definitions have been proposed subsequently that either expand or restrict the definition (Gadagkar 1994, Sherman et al. 1995, Crespi and Yanega 1995). At one end of this spectrum is the concept of the eusociality continuum (Sherman et al. 1995), which emphasizes the similarities and intergrading social systems among the various social taxa of invertebrates and vertebrates. At the other end, Crespi and Yanega (1995) emphasize the origin of irreversible sterility as a key transition that qualitatively changes the evolution of social organization, and restrict the term "eusociality" to lineages that have crossed that transition.

Efforts to solve Darwin's fundamental puzzle have ranged from individual-based fitness analyses (e.g., Bourke and Franks 1995, Crozier and Pamilo 1996) inspired by Hamilton (1964), through broad-scale comparative analyses of sociality in different animal lineages (Alexander et al. 1991, Crespi 1994, 1996, Brockmann 1997, Helms Cahan et al. 2002, Hart and Ratnieks 2004). The latter approach exploits the fact that eusociality has arisen once each in the ancestors of ants and termites, several times among the bees and wasps, with a few scattered instances among other insects (Choe and Crespi 1997), and twice in the African mole-rats (Jarvis and Bennett 1993). The most recent addition to this group is the symbiotic marine shrimp genus *Synalpheus* (Duffy 1996a), in which eusociality has arisen independently at least twice and probably several times (Duffy 1998, Duffy et al. 2000). Wilson's (1971) three criteria for eusociality have been demonstrated or inferred for at least five species of alpheids (Fig. 18.1): *Synalpheus regalis*, *S. filidigitus*, *S. rathbunae*, *S.* "*rathbunae A*," *S. chacei*, and possibly *S.* "*paraneptunus* small" (Duffy 1996a, 1998, Duffy and Macdonald 1999, Duffy et al. 2000). Although behavior of most of these species is little known, all of them consist of colonies of tens to a few hundred individuals in which one or a very few females breed at any given time. Even a restrictive definition of eusociality based on irreversible caste differentiation (Crespi and Yanega 1995) would apply to *S. filidigitus*, in which the queen typically loses the large fighting claw and develops a second, minor-form chela, rendering her morphologically unique among members of the colony and presumably dependent on them for protection (Duffy and Macdonald 1999; see chapter frontispiece). There is currently no evidence of irreversible sterility in any *Synalpheus* species, as is also true of most eusocial lower termites and many wasps, although the complete reproductive skew and large colony size of several *Synalpheus* species strongly suggests that most individuals never breed in their lifetimes.

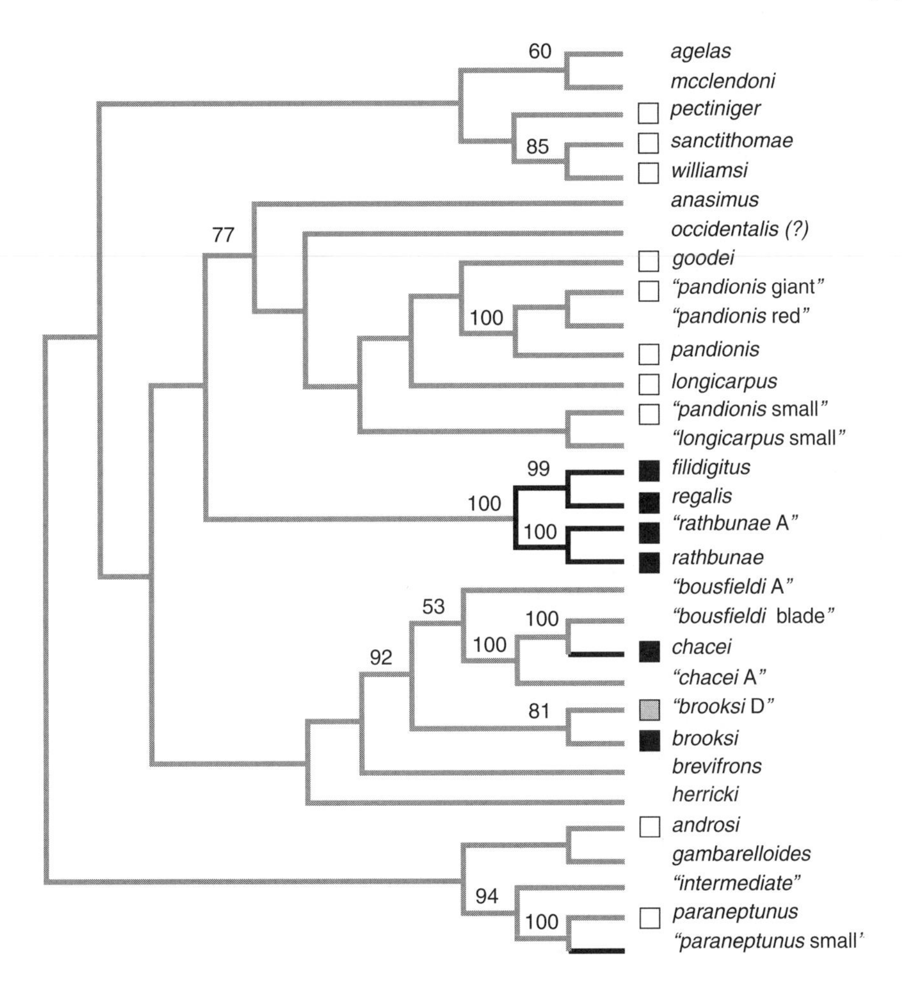

Figure 18.1 Phylogenetic hypothesis for West Atlantic *Synalpheus* species in the gambarelloides group (after Morrison et al. 2004). The tree is based on six-parameter weighted-parsimony analysis of partial mitochondrial cytochrome oxidase subunit I and 16S rRNA sequences and 45 morphological characters. Numbers above branches are bootstrap proportions ($N = 1{,}000$). Eusocial taxa are indicated as black line segments. Boxes indicate known development modes based on observations of eggs hatching in the laboratory: open boxes, swimming larvae; black boxes, crawling juveniles (direct development); gray boxes, both types of larvae have been reported. Quotation marks denote provisional names of undescribed species diagnosed in Ríos (2003).

Consistent with the eusociality continuum of Sherman et al. (1995), colony size and reproductive skew both vary more or less continuously among species of *Synalpheus* (Duffy et al. 2000). Hence I use the term "eusociality" to refer to multigenerational, cooperative colonies with strong reproductive skew and cooperative defense.

The independent origins of quite similar social organizations in disparate taxa provide a valuable sample for comparative analysis of factors promoting the evolution of altruism (Alexander et al. 1991, Crespi 1994, 1996, Hart and Ratnieks 2004). For example, exploring eusociality in a symbiotic, diploid marine arthropod with gradual metamorphosis provides insights into the generality of proposed explanations for eusociality based on haplodiploid sex determination (Hamilton 1964), parental care (Wilson 1971), and food–shelter coincidence (Alexander at al. 1991). More specifically, the multiple, relatively recent origins of eusociality in *Synalpheus* and the broad range of social systems found among its species provide a promising model system for addressing such hypotheses at a finer taxonomic resolution (Duffy et al. 2000, Duffy 2003). In this contribution, I review the phenomenon of eusociality in sponge-dwelling alpheid shrimp (*Synalpheus*), and I examine its adaptive significance and ecological and evolutionary consequences.

Diversity and Natural History

Synalpheus is a species-rich and abundant component of cryptic coral-reef faunas worldwide. In the West Atlantic, most *Synalpheus* species belong to the monophyletic "gambarelloides group" (Coutière 1909, Ríos 2003, Morrison et al. 2004; Fig. 18.1), all species of which are obligate inhabitants of sponges, living their entire lives within the host and feeding on its tissues or secretions (Ruetzler 1976, Erdman and Blake 1987, Ríos and Duffy 1999). The host thus constitutes a self-contained, highly valuable, and often long-lived (Reiswig 1973, Ayling 1983) resource, providing habitat, food, and protection from predators. Unoccupied hosts are in short supply (Duffy 1996c, 1996d, Duffy et al. 2000), so shrimp likely experience habitat saturation like that proposed to favor cooperative breeding in some social vertebrates (see Emlen 1997, Hatchwell and Komdeur 2000).

The species richness of *Synalpheus* is accompanied by considerable diversity in social organization, with species ranging from the heterosexual pair formers typical of the family Alpheidae to eusocial species living in colonies of hundreds of individuals with a single breeding queen. The current phylogenetic hypothesis suggests that eusociality has arisen three times independently within the gambarelloides species group. Four of the eusocial species (*S. regalis*, *S. filidigitus*, *S. rathbunae*, and *S. "rathbunae A"*) comprise a single clade, the ancestor of which is parsimoniously reconstructed as eusocial (Morrison et al. 2004; Fig. 18.1). For each of the two remaining eusocial taxa, *S. chacei* and *S. "paraneptunus* small," the strongly supported sister taxa are socially monogamous species (Morrison et al. 2004). Similar colony structures in Indo-Pacific species outside the gambarelloides group suggest additional origins of eusociality (Duffy 1998).

Alpheids are commonly called snapping or pistol shrimp because of their enlarged major chela or fighting claw, present in both sexes. Most alpheids are fiercely territorial, defending against any individual other than a familiar mate (Hazlett and Winn 1962, Nolan and Salmon 1970, N. Knowlton and Keller 1982, Nakashima 1987, Gherardi and Calloni 1993). Social sponge-dwelling *Synalpheus* species are a conspicuous exception in that they live in dense colonies. In these social species, snapping is used to defend and maintain the integrity of the colony against intruders (Duffy et al. 2002, Tóth and Duffy 2005).

Like decapod crustaceans generally, most species of *Synalpheus* produce planktonically dispersing larvae. Several species of *Synalpheus*, however, exhibit "direct development" in which eggs hatch into crawling juveniles with very limited dispersal potential (Dobkin 1965, 1969). These include, among others, all eusocial species (Fig. 18.1). In direct-developing social species, allozyme evidence of strong population subdivision (Duffy 1993, 1996b) suggests highly restricted dispersal. Yet dispersal clearly happens regularly as virtually all appropriate sponges are occupied in the field, at least for highly social species (Duffy 1996c, 1996d).

Reproductive Biology and Mating Systems

Mature female *Synalpheus* are easily identified by visible ovaries and morphology, whereas males show no external signs of maturity. Because sex is indistinguishable externally in individuals other than mature females, I refer here to nonovigerous individuals as either "juveniles" (small) or "adults" (large). Recent scanning electron microscopy revealed male and female gonopores in similar frequencies in social *Synalpheus* colonies across a range of body sizes, suggesting that sex is determined early in these species (E. Tóth and R.T. Bauer, personal communication), in contrast to some other alpheids that can change sex (Suzuki 1970, Gherardi and Calloni 1993). Thus, adult size classes in social *Synalpheus* appear to include both males and nonbreeding females. Allozyme studies confirm that both sexes are diploid (Duffy 1993, 1996b).

Most alpheids, including many *Synalpheus* species, exhibit social monogamy (Correa and Thiel 2003), in which adults live in heterosexual pairs but are intolerant of other individuals in a territory. Since female mating receptivity is limited to a short period after the molt (Rahman et al. 2003) and sperm storage is absent (R.E. Knowlton 1971, N. Knowlton 1980, Nakashima 1987), the female must mate each time she ovulates. Experiments suggest that social monogamy in pair-forming alpheids is maintained both by benefits to males of mate guarding (Mathews 2003, Rahman et al. 2003) and by cooperation between mates in defending the common burrow (Mathews 2002).

Synalpheus is evidently unique among alpheids in that many species depart from this social monogamy. In addition to the ancestral pair formers, the genus includes communal species in which several to hundreds of pairs cohabit in a host with multiple females breeding (e.g., *S. longicarpus*, Erdman and Blake 1987; *S. brooksi*, Duffy 1992, 1996b) as well as fully eusocial species with colonies containing hundreds of individuals and only a single female breeding (Duffy 1996a, 1998, Duffy and Macdonald 1999). It is striking that these communal aggregations, with their strong reproductive skew, occur exclusively in sponges of moderate to large size (relative to the size of their shrimp inhabitants). In chapter 12, Baeza and Thiel (see also Thiel 2000) have suggested that the size of the host resource and constraints on movement among hosts strongly influence the expected mating system, and the unusual mating and social systems of many sponge-dwelling *Synalpheus* seem consistent with this. Social monogamy appears to have been an important precursor to the evolution of eusociality in sponge-dwelling shrimp for at least two reasons:. First, in conjunction with direct development, it fosters formation of close kin groups, and more specifically, long-term parent-offspring associations. Such families are otherwise rare among decapod crustaceans (but see chapter 17), although they appear common, if temporary, among peracarids (see chapters 14, 16). The second implication of social monogamy is that it provides opportunities for long-term cooperation among individuals, for example, in maintaining and defending the shared shelter.

The Nature of Eusociality in *Synalpheus*

Colony Organization

Of the several eusocial shrimp species, *Synalpheus regalis* has the largest colonies and strongest reproductive skew. Colony size (i.e., number of individuals within a sponge) ranges up to 350 individuals and colonies invariably contain only a single breeding female, the queen (Duffy 1996a). Relatedness among colony members calculated from allozyme data averaged 0.50 (Duffy 1996a), and several colonies showed strong heterozygote excesses (Duffy 2003). These data imply that *S. regalis* colonies consist largely of full-sib offspring of a single breeding pair and that outbreeding is the norm, since heterozygote excesses are expected where colony members are offspring of a single breeding pair homozygous for different alleles at the locus. Inbreeding, in contrast, produces heterozygote deficiencies, which were never found (Duffy 2003). Long-term reproductive monopoly by a single queen is also consistent with the strong correlation between queen body size and colony size in *S. regalis* (Duffy 1996a), which suggests coordinated growth of the breeding female and her colony. Thus, eusociality in *Synalpheus* has clearly arisen via the growth of nondispersing families, that

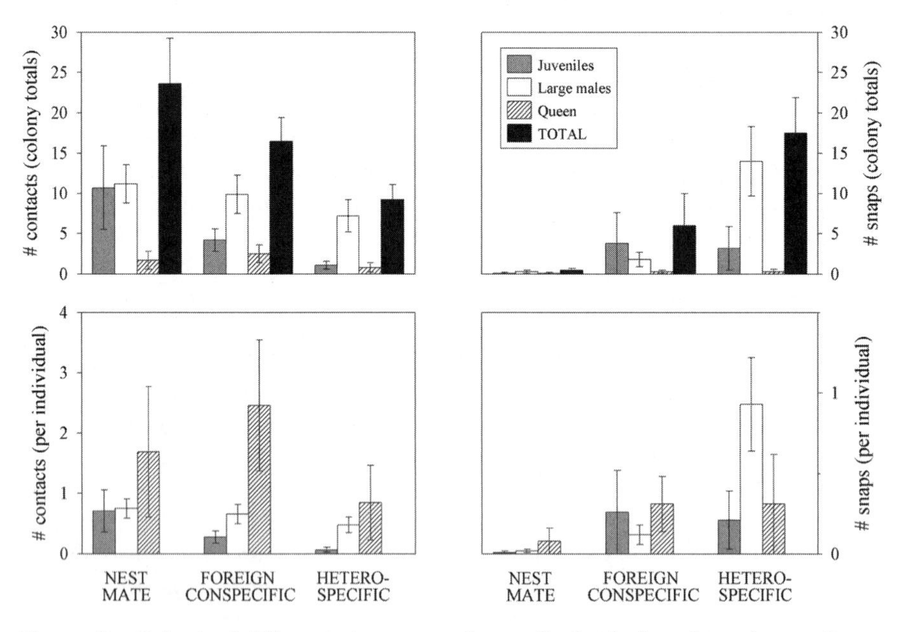

Figure 18.2 Behavioral differentiation among classes of individuals within colonies of *Synalpheus regalis*. Bars show mean numbers of contacts (left) and snaps (right) by resident small, large, and queen shrimp to each of three types of introduced intruders ($N = 13$ independent experimental colonies). Top panels show colony totals; bottom panels show average numbers per individual resident. Resampling tests demonstrated that both foreign conspecific and heterospecific intruders elicited significantly fewer contacts and more snaps than did nestmate conspecifics (pooled across all individual classes), and that large males produced significantly more snaps per individual against heterospecific intruders than did other classes of individuals (after Duffy et al. 2002).

is, through the subsocial route (*sensu* Wilson 1971), as appears true of most eusocial animals (Alexander et al. 1991).

Experiments with captive colonies show that *S. regalis* discriminates between nestmates and other conspecifics entering the sponge, with even more pronounced aggression toward heterospecific intruders (Duffy et al. 2002). When faced with a nonnestmate conspecific, resident shrimp contacted it less and snapped more frequently than they did when faced with a nestmate (Fig. 18.2). Such discrimination presumably helps maintain integrity of the kin group, as it does in the remarkably sensitive kin discrimination system of social desert isopods (chapter 16).

Caste and Division of Labor

The most fundamental feature of eusociality is "reproductive division of labor" between the queen and nonbreeding workers (Wilson 1971). In colonies of *S. regalis*, as in many eusocial insects, reproduction is restricted to the single queen and an apparently small number of male mates (Duffy 1996a). The queen also differs from other adults in being considerably less active and less aggressive (Duffy et al. 2002; J.E. Duffy, unpublished observations). In *S. filidigitus*, this reduced aggressiveness is manifested in a physical polymorphism in which mature queens lack a snapping claw, instead bearing two minor-form chelae (Duffy and Macdonald 1999). This polymorphism is the clearest example in eusocial shrimp of discrete morphological caste differentiation such as that found in social insects (Wilson 1971). Females with two minor chelae have also been reported in the eusocial *S. rathbunae* (Chace 1972) and in *S. crosnieri* (Banner and Banner 1983). This situation strongly implies that queens in these social species do not aggressively dominate other individuals and may indeed be protected by them.

Among nonbreeders in *S. regalis* colonies, the clearest evidence of division of labor involves colony defense. In captive colonies, small individuals were sedentary and often congregated in groups to feed, whereas large individuals moved more frequently around the sponge, were more aggressive, and spent more time near the sponge periphery where intruders would be first contacted (Duffy et al. 2002, Duffy 2003; Fig. 18.2). These size-based differences presumably reflect the ontogenetic development of aggressive behavior typical of most animals. Even within the large size class, however, a small proportion of individuals are responsible for most aggressive attacks against intruders, and these frequent attackers have proportionally larger major chelae than the others (E. Tóth and J.E. Duffy, unpublished observations). Thus, observations and experiments suggest that a group of behaviorally specialized and morphologically distinct large individuals in *S. regalis* shoulder the burden of colony defense, leaving small juveniles free to feed and grow and the queen free to feed and reproduce.

The Adaptive Significance of Sociality in Sponge-Dwelling Shrimp

Evolution of Shrimp Sociality: An Individual-Level Perspective

Hamilton (1964) revolutionized understanding of social behavior by putting interactions among individuals into an evolutionary context with the deceptively simple equation: $rB - C > 0$. Hamilton's rule states that natural selection will favor a gene for altruistic behavior if the benefit (*B*) of the behavior to the recipient, weighted by relatedness (*r* = proportion of genes shared) between donor and recipient, is greater than the cost (C) to the donor. Hamilton's rule emphasizes that social systems arising from such interactions depend both on genetic structure of groups and on the ecological costs and benefits associated with membership. In practice, the origin of eusocial or cooperatively breeding groups can be analyzed as the outcome of a series of individual decisions (Emlen 1994, Brockmann 1997, Helms Cahan et al. 2002): (1) whether to disperse or stay on the natal territory; (2) given the decision to stay, whether to breed; and (3) given the decision not to breed, whether to help others in the group. The term "decision" is used loosely because these outcomes may reflect either individual choices or actions forced by other individuals. I consider each of these three decisions in turn.

Whether to Disperse

The relative fitness benefits of dispersal versus remaining in the natal territory depend on ecological constraints on independent breeding and on any benefits that accrue from remaining in the natal group (Koenig et al. 1992, Emlen 1994). A primary ecological constraint on independent breeding in many social vertebrates is habitat saturation, that is, shortage of available territories (Emlen 1984, 1994, Koenig et al. 1992, Hatchwell and Komdeur 2000). Similarly, in several social insects, evidence suggests that low success of independent breeding has selected for delayed dispersal (Lin and Michener 1972, Strassmann and Queller 1989, Brockmann 1997).

In decapod crustaceans, the option to disperse is constrained in part by development mode. Species with swimming larvae seemingly have little choice but to disperse (although some terrestrial decapods have ingeniously circumvented this constraint; see chapter 17), precluding formation of kin groups. Thus, cooperative behavior and strong reproductive skew are not known from those *Synalpheus* species with swimming larvae (Fig. 18.1), even though some such species occur in large groups (e.g., *S. longicarpus*; Duffy 1992, Duffy et al. 2000). In contrast, species with direct development, including all social *Synalpheus* species, may disperse as juveniles or stay in the natal sponge.

In social sponge-dwelling shrimp, colony cohort structure (e.g., Duffy and Macdonald 1999) and relatedness estimates (Duffy 1996a) indicate that most juveniles remain for extended periods, probably permanently, in the natal sponge. Moreover, field evidence indirectly supports the hypothesis that suitable habitat is saturated in that nearly all host sponges are occupied by shrimp (Duffy 1992, 1996d). Thus, independent breeding opportunities appear limited by shortage of territories (hosts). Forgoing dispersal may also confer direct benefits in that juvenile shrimp likely benefit from the larger size and more effective defensive capabilities of resident adults. Thus, individual shrimp remaining in the natal sponge as juveniles may often have higher inclusive fitness than those that disperse.

Whether to Breed

Once an individual has chosen to remain in its natal group, it faces the decision whether to attempt breeding. In eusocial colonies, by definition, most individuals do not breed. Understanding what controls this inequity is the central problem of eusociality and is addressed formally by reproductive skew models (Keller and Reeve 1994, Johnstone 2000, Reeve and Keller 2001). The main factors hypothesized to influence reproductive skew include the severity of ecological constraints on independent breeding, genetic relatedness among individuals, and the relative power of individuals to control reproduction by others. Although genetic and behavioral data are currently insufficient to test reproductive skew models rigorously in *Synalpheus*, some tentative conclusions are possible. Transactional concession models (Reeve et al. 1998) predict that skew will be pronounced in colonies of close relatives living under strong ecological constraints, which is consistent with data from social shrimp. In small colonies of social animals such as some wasps (Michener and Brothers 1974, Jeanne 1980, Fletcher and Ross 1985), cooperatively breeding vertebrates (Emlen 1997), and the eusocial naked mole-rat (Reeve and Sherman 1991), this reproductive skew is maintained by behavioral dominance. But in eusocial *Synalpheus*, observations of captive colonies found no evidence of aggression or behavioral dominance by the queen (Duffy et al. 2002; J.E. Duffy, unpublished observations). And in *S. filidigitus*, the queen's lack of a fighting claw further implies that she cannot maintain her sole breeding status through aggression.

In large social colonies such as those of *S. regalis* and *S. filidigitus*, reproductive monopoly might be explained by the "worker policing" (Ratnieks 1988) or "majority-rules" model (Reeve and Jeanne 2003). This model predicts that a single individual—the "virtual dominant"—can come to monopolize a colony's reproduction even without being behaviorally dominant, simply by being the individual to which other colony members have the greatest average genetic relatedness (Reeve and Jeanne 2003). This situation is most likely in mother–offspring associations. In such colonies, nonbreeding workers are more closely related to the current queen's offspring (their siblings; $r = 0.50$ in diploids) than they would be to offspring of a sibling worker (their nieces/nephews; $r = 0.25$), so selection will favor subordinates ("workers") preventing one another from breeding in favor of reproduction by their mother, the queen. The majority-rules model is supported by progeny sex ratios in some social Hymenoptera that are consistent with control by workers rather than the queen (Ratnieks 1988, Queller and Strassmann 1998). Worker policing is also an attractive, albeit yet untested, hypothesis for reproductive skew in social *Synalpheus*, in which colonies appear to consist mainly of the resident queen's offspring. Finally, the large colony size and complete reproductive skew (single breeding female) typical of social *Synalpheus* suggest that many colony members never reproduce. Explaining such lifetime sterility is the most difficult problem of social biology—and indeed, of evolution generally, as recognized by Darwin. A recent model (Jeon and Choe 2003) finds that evolution of sterile castes occurs only under very restrictive conditions, namely, saturated (large), asymmetrical relatedness (parent–offspring) groups with complete reproductive skew. These conditions describe *Synalpheus regalis* colonies well and are consistent with Jeon and Choe's (2003) model.

A final potential explanation for reproductive skew in social shrimp colonies involves incest avoidance. Where social colonies consist of close relatives, such as full-sib offspring of outbred parents, avoidance of incest often can explain why offspring of the breeding pair do not attempt to breed (Emlen 1995). This hypothesis is supported, in part, by an elegant series of experiments in the eusocial Damaraland mole-rat, which lives in colonies consisting of a breeding pair and one or more cohorts of its adult offspring (Bennett et al. 1996, Cooney and Bennett 2000, Greeff and Bennett 2000). Nevertheless, recent genetic analyses show that Damaraland mole-rat colonies often contain unrelated, nonbreeding individuals, suggesting that reproductive skew cannot be maintained solely by incest avoidance (Burland et al. 2004). Maintenance of reproductive skew by incest avoidance is also consistent with some data from eusocial termites (Shellman-Reeve 2001). Incest avoidance is an attractive possibility for the strong reproductive skew in *S. regalis,* in which colonies consist mostly of full-sibs descended from a single breeding pair (Duffy 1996a, 2003). In such situations, unrelated potential mates would appear scarce. On the other hand, the heterozygote excesses common in colonies of *S. regalis* (Duffy 2003) clearly indicate that the queen breeds with an unrelated mate, which in turn implies sufficient dispersal to avoid inbreeding. If dispersal indeed brings unrelated individuals to a sponge frequently, some mechanism must prevent most of them from breeding since *S. regalis* colonies maintain strong reproductive skew. The frequency of immigration to colonies, conditions under which it occurs, and mechanisms suppressing immigrant (and resident) breeding are important frontiers for future research.

Whether to Help

Given that an individual has chosen to stay in its natal group without breeding, the final decision in the "social trajectory" (Helms Cahan et al. 2002) is whether to help, that is, participate in activities that entail individual investment and that benefit other group members. In social insects and vertebrates, helping most commonly involves provisioning young and defending the nest or territory. Currently, we have no evidence of provisioning of young in social shrimp. Even very small juvenile shrimp appear capable of feeding on their own. Thus, foraging outside the sponge is unnecessary, and our admittedly limited observations of captive colonies have found no evidence that nonbreeders care directly for the queen or her young. Instead, the principal work of social *Synalpheus* appears to be defense from competitors, and perhaps predators. In this sense, social shrimp are similar to "fortress defender" social insects (Queller and Strassmann 1998), such as gall-forming aphids and thrips that live inside their food source and have independent young (Choe and Crespi 1997, Aoki 2003). Insofar as social aggregations provide a safe place for offspring to live, colony defense might be interpreted as a form of parental care.

The most direct evidence that nonbreeding individuals help in shrimp colonies is that large individuals regularly engage in dangerous defense of the colony, despite genetic evidence (Duffy 1996a) that most of them do not breed. Experiments show how this cooperative defense enhances effectiveness of social colonies at repelling intruders, and thus of holding and dominating the host resource (Tóth and Duffy 2005). Conspecific intruders attempting to enter a sponge elicited vigorous snaps from resident *S. rathbunae* inside. Frequency of snaps by residents increased sharply after intruders were introduced (Fig. 18.3a), but sometimes these individual snaps failed to repel intruders, which caused many residents suddenly to begin snapping in unison, producing a distinctive crackling noise lasting up to tens of seconds. Up to 60% of the visible colony members were involved in these "coordinated snapping" events, which were observed exclusively after introducing intruders (Fig. 18.3b) and usually

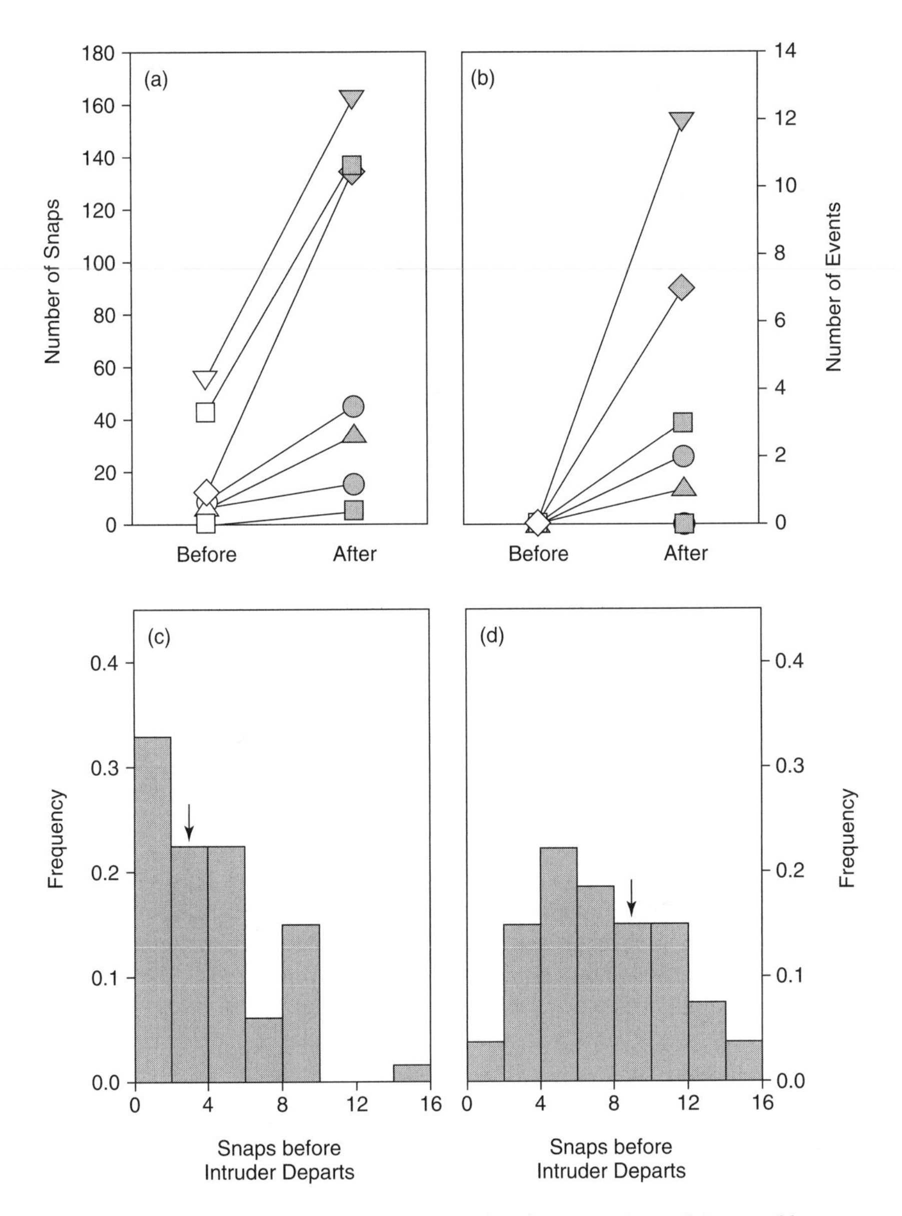

Figure 18.3 Defensive responses to nest intruders by eusocial *Synalpheus rathbunae* in intact sponges in the laboratory. (a and b) Numbers of single snaps (a; $N = 8$) and coordinated snap events (b; $N = 8$) for 30 minutes before (open symbols) and 30 minutes after (solid symbols) introduction of conspecific intruders in each of eight colonies. (c and d) Frequency distributions of numbers of snaps required to repel an intruder when the intrusion failed (c; $N = 70$) versus succeeded (d; $N = 27$) in eliciting a coordinated snapping event. Arrows above histograms show the mean across colonies of the median value per colony. Coordinated snapping was an escalated response that occurred after single snaps were unsuccessful at repelling the intruder. From Tóth and Duffy (2005, used by permission).

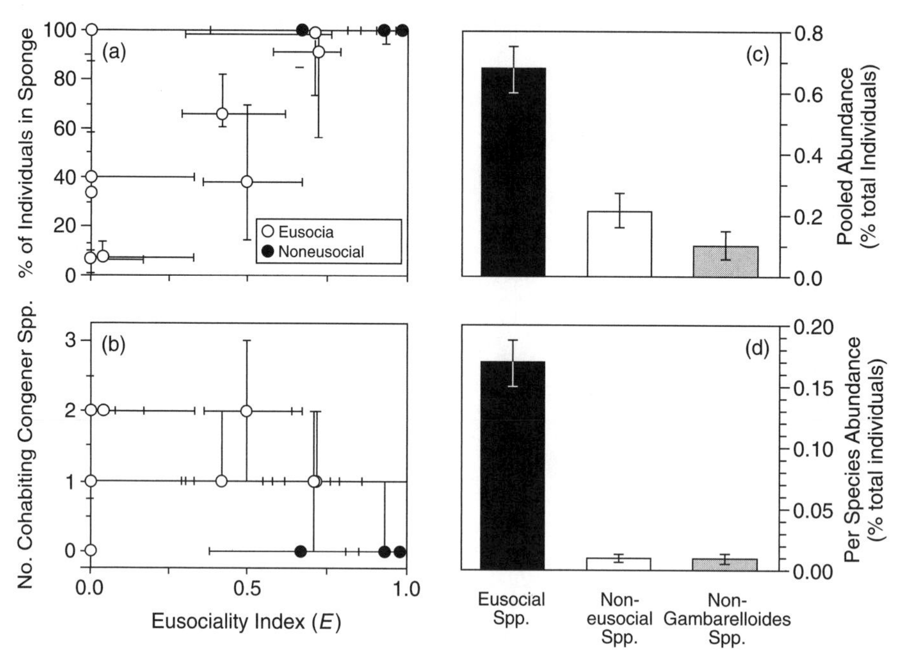

Figure 18.4 Ecological consequences of eusociality in sponge-dwelling shrimp. (a and b) Dominance of the host sponge is positively related to degree of social organization (eusociality index; Keller and Perrin 1995) among species of *Synalpheus*; phylogenetically controlled comparisons (not shown) are significant for percentage of individuals in sponge (a) and marginally nonsignificant for number of cohabiting congener species (b) (after Duffy et al. 2000). Symbols show median ± 95% confidence interval. (c and d) Relative abundance of eusocial versus noneusocial *Synalpheus* species in samples of sponge-infested coral rubble from shallow patch reefs in Belize (after Macdonald et al. 2006). (c) Mean (± standard error) proportions of sampled shrimp that belonged to eusocial versus noneusocial gambarelloides-group species and to non-gambarelloides species ($N = 13$ samples). (d) Mean (± standard error) proportional abundance per species for eusocial, noneusocial, and non-gambarelloides species.

occurred only after intruders ignored repeated snaps by single defenders (Fig. 18.3c,d). Coordinated snapping effectively repelled intruders when repeated snaps by single defenders did not (Tóth and Duffy 2005), supporting the conclusion that nonbreeding helpers enhance defensive capability of eusocial shrimp colonies. Finally, patterns of field abundance are also consistent with superior defense by eusocial colonies. Phylogenetically controlled comparisons among *Synalpheus* species show that, whereas noneusocial species typically share their host sponge with congeners, eusocial species tend to dominate numerically the sponges they occupy; indeed, eusocial species are typically found in dense, monospecific groups in their hosts (Duffy et al. 2000; Fig. 18.4).

These data support the hypothesis that defense by nonbreeders enhances fitness at the colony level. But what are the benefits of helping to the helper? Defensive behavior in eusocial shrimp colonies has both direct benefits, in protecting oneself against intruders, and indirect fitness benefits in protecting close relatives. Nonbreeders in *S. regalis* colonies are defending primarily their parents and siblings

(Duffy 1996a, 2003), with whom they share the same genetic relatedness as to their own offspring. Moreover, the defense may have low costs in terms of average offspring production: field surveys of *S. regalis* colonies showed that total colony egg production (i.e., clutch size of the resident queen) was strongly and linearly related to colony size, whereas average per capita egg production (i.e., queen clutch size divided by number of colony members) declined little as colony size grew to 350 individuals (Duffy et al. 2002).

In summary, data from eusocial shrimp offer considerable circumstantial support for the hypothesis that ecological constraints are a fundamental driver of cooperative breeding: (1) host sponges appear to be a limiting resource for sponge-dwelling shrimp species; (2) offspring of eusocial colonies generally do not disperse, thus forming families of close genetic relatives; (3) these families develop strong reproductive skew, probably as a result of both incest avoidance and the collective interest of most colony members in maintaining their parents as sole reproductives; and (4) nonbreeding individuals engage in group defense that enhances the colony's productivity and dominance of the host sponge resource. By doing so, nonbreeders gain both direct benefits of enhanced protection and inclusive fitness benefits of enhancing survival and productivity of nondescendent kin.

Evolution of Shrimp Sociality: A Comparative Perspective

The preceding discussion examined the adaptive significance of shrimp eusociality from the perspective of a nonbreeding individual. But why has eusociality evolved in *Synalpheus* and not in other crustacean taxa? Here I ask whether sociality in shrimp might be explained by two models for the evolution of eusociality/cooperative breeding, one based primarily on insects (Crespi 1994) and one on birds (Hatchwell and Komdeur 2000).

The Fortress Defense Hypothesis

Crespi (1994; see also Alexander et al. 1991) proposed that three conditions are sufficient, though not necessary, to explain most known cases of eusociality in that all taxa that met the criteria in his judgment are eusocial. (1) Coincidence of food and shelter in an enclosed habitat creates a highly valuable, often long-lived resource that promotes accumulation of kin and frees juveniles from foraging to accomplish other tasks. (2) Strong selection for defense arises from the high value of food–habitat resources, which are vulnerable to attack by competitors and/or predators and puts a high premium on cooperative defense. (3) Ability to defend the resource creates opportunities for nonbreeding individuals to specialize in protecting kin, thus gaining large indirect fitness by defending the colony. Queller and Strassmann (1998) termed eusocial colonies of this type "fortress defenders."

Eusociality in sponge-dwelling shrimp (Duffy 1996a) independently supports Crespi's (1994) hypothesis. First, the near absence of unoccupied sponges in the field (Duffy 1996d, Duffy et al. 2000) and the aggressive defense of host resources (Duffy et al. 2002, Tóth and Duffy 2005) indicate that habitat is saturated. Food and habitat are clearly coincident for sponge-dwelling shrimp generally, and some sponges reach sizes that can support thousands of shrimp (Pearse 1932) and live for decades

(Reiswig 1973, Ayling 1983). Genetic and morphometric differentiation among shrimp demes in individual sponges (Duffy 1996b, 1996c) suggests long-term occupation by particular genetic lineages of shrimp with little exchange among sponges. Thus, the sponge resource is both valuable and long-lived, fosters formation of kin groups, and provides a shrimp family with the potential for resource inheritance (Myles 1988). Second, selection for defense is supported indirectly by the evidence that suitable hosts are in short supply and by the ubiquity of aggression and territoriality among alpheids. Together, these factors suggest that pressure from enemies is likely strong and persistent for sponge-dwelling alpheids. Third, sponge-dwelling shrimp are clearly well endowed for defense by the powerful snapping claw, used in aggressive contests, and in social species by cooperative defense of the host sponge (Tóth and Duffy 2005). Andersson (1984) and Starr (1985) similarly emphasized that the transition from group living to eusociality has occurred only in taxa with effective weapons, notably, the sting of aculeate Hymenoptera. Thus, eusocial *Synalpheus* appear to meet each of Crespi's (1994) three criteria. The puzzle is that *all* sponge-dwelling *Synalpheus* appear to meet these criteria. Yet these species range from pair forming (the majority of species), through communal, to eusocial. It thus appears that additional criteria are necessary to explain eusociality in shrimp. The clearest of these involves life history.

Turnover of Breeding Opportunities

Life history has fundamental consequences for social organization and evolution. For example, parental care is an important prerequisite for eusociality in insects and cooperative breeding in vertebrates (Alexander 1974, Andersson 1984, Alexander et al. 1991), as well as crustaceans (chapter 14). Long periods of offspring dependence (Queller 1989) and delayed age of reproduction (Gadagkar 1991) appear important to social evolution in insects. In birds, recent analyses have used phylogenetically controlled comparisons to assess the importance of life history and ecology in evolution of sociality (Arnold and Owens 1998, 1999). Cooperative breeding was concentrated in particular families characterized by low fecundity and high adult survival, and within these families, ecological factors—specifically, sedentariness and a warm, stable environment—significantly predicted the incidence of cooperative breeding among species. These results were interpreted as showing that life history features predisposed lineages toward sociality, after which ecological factors influenced which species within lineages evolved cooperative breeding. This interaction of life history and ecology limits the turnover rate of breeding opportunities, which appears to be the best predictor of the distribution of cooperative breeding among species of birds (Hatchwell and Komdeur 2000).

It seems clear that a similar interaction of life history and ecology influences evolution of sociality in shrimp. In many marine invertebrate groups, direct development is associated with lower fecundity and higher juvenile survival (predictors of cooperative breeding in birds), compared with planktonic dispersal (Strathmann 1985). Among crustaceans, family and social groups occur almost exclusively in taxa with direct development (chapter 14). Indeed, the most striking exception proves the rule: terrestrial crabs of Jamaica raise and care for their planktonic larvae in carefully prepared and managed pools of rainwater in leaf axils (reviewed in chapter 17), that is, in restricted spaces that prevent dispersal. Thus, life history traits that restrict dispersal are an important prerequisite for family living in crustaceans.

In the gambarelloides species group of *Synalpheus*, unlike in birds, life history is less conserved than the fundamental ecological characteristic of life in sponges. That is, all species in the group are obligate sponge dwellers, sharing the sedentary lifestyle in warm, stable environments that fosters cooperative breeding in birds (Arnold and Owens 1998). Life history, in contrast, varies among *Synalpheus* species, which include both direct developers and species with planktonically dispersing larvae. Thus, the role of life history in promoting eusociality should be especially clear in *Synalpheus* because dispersal potential varies among species while general ecology is fairly constant. As expected, eusociality occurs only in *Synalpheus* species with direct development, supporting low dispersal as a prerequisite for eusociality in this group (Fig. 18.1).

Finally, high adult survival correlates with cooperative breeding in birds (Arnold and Owens 1998) and seems likely in sponge-dwelling *Synalpheus*, as well. The interior of sponges is well insulated from enemies and abundant with food, compared with the rock crevices in which most other alpheids live. The very low fecundity of some *Synalpheus* species (e.g., *S. filidigitus*; median, 4.5 eggs per clutch; range, 1–29; Duffy and Macdonald 1999) also indirectly suggests high survival, which is necessary to balance low fecundity in a population at equilibrium.

Comparative Analysis of Shrimp Social Evolution: Synthesis

Many crustacean taxa meet one or more of Crespi's criteria, but few meet all of them. For example, several crustaceans associate with valuable hosts that provide both shelter and food, and some of these are well equipped with powerful claws to defend their host resource (Patton 1974, Bruce 1976, 1998). But, to my knowledge, all of these groups produce planktonically dispersing larvae, precluding formation of kin groups. Other promising candidates for crustacean eusociality include the leucothoid or anamixid amphipods, which have direct development and live within sponges or ascidians. Some species are found in apparent family groups (Thomas 1997, Thiel 1999, 2000; see also chapter 14), and males bear hypertrophied gnathopods (claws), presumably sexually selected, that might be employed in cooperative defense. To date, however, I know of no reports of multigenerational colonies with strong reproductive skew (eusociality) in amphipods. Thus, for marine crustaceans, I would argue that Crespi's three criteria for eusociality must be supplemented by an additional one: life history that allows restricted dispersal and long-term kin interaction. The association between direct development and kin grouping in crustaceans generally, and among species of *Synalpheus* specifically, is consistent with this hypothesis.

In summary, eusociality in sponge-dwelling shrimps appears to have been favored by a combination of four characteristics shared with several social insect lineages but possibly unique in the Crustacea: (1) direct development resulting in limited dispersal and kin association (2) ecological specialization on a valuable, long-lived resource; (3) strong enemy pressure associated with occupation of the host resource; and (4) possession of weaponry (the snapping claw) effective in defending it. Together, these factors promote long-term occupation of specific nest sites by multigenerational family groups, conditions that result in low turnover of breeding opportunities and persistence of dynastic lineages headed by one or a few breeders of each sex, with nonbreeding adults defending the colony from intruders. The first filter in the process of social evolution among sponge-dwelling shrimp appears to have been evolution of direct development, in that eusociality has evolved in none of the clades characterized by pelagic swimming larvae but has evolved independently in at least two or three clades containing direct-developing species (Fig. 18.1). Explaining the variation in sociality among direct-developing species of *Synalpheus* is more challenging because all such species have weapons and coincident food and shelter. It remains for future research to determine whether variance among species in availability of host resources and/or the strength of enemy pressure can explain which species or populations evolve eusociality.

Ecological and Evolutionary Consequences of Eusociality in Shrimp

Ecological Dominance

Eusocial insects are prominent members of most terrestrial ecosystems, dominating not only the insect fauna but also animal biomass generally (Wilson 1990). This dominance is due largely to the superior ability of organized cooperative colonies to obtain and process resources and to defend themselves against both enemies and harsh environmental conditions. Among sponge-dwelling marine animals, a parallel trend is evident. Using phylogenetically independent contrasts, Duffy et al. (2000) showed that eusocial species had higher relative abundance within sponges than did less social congeners. Eusocial species also use a larger number of host species, on average, than do noneusocial species (Macdonald et al. 2006). In quantitative samples of sponge-encrusted reef rubble from Belize, the average eusocial shrimp species was 17 times more abundant than the average noneusocial species (Fig. 18.4d), and eusocial species made up 68% of the shrimp collected (Fig. 18.4c; Macdonald et al. 2006). Thus, eusocial shrimp species appear to be more productive and more effective competitors for the limited sponge resource than are their noneusocial congeners.

Importantly, this advantage of eusociality is realized only in the highly specialized niche within living sponges, whereas free-living alpheids that inhabit rock crevices or burrow in sediments are universally monogamous pair formers. Again, this pattern supports the argument that coincidence of food and shelter selects for the fortress-defense type of eusociality (Crespi 1994). Thus, while eusociality appears to confer advantages in the specialized symbiotic niche within sponges, it appears not to have been selected in more open marine habitats. Consequently, the ecological impacts of eusocial shrimp are much more limited than those of widely foraging eusocial insects (Queller and Strassmann 1998).

Morphological Evolution

Among some eusocial insects, the specialization of worker groups on different tasks is accompanied by morphological differentiation into discrete castes. In certain large-colony ants and termites, this caste specialization has led to spectacular morphological differentiation among groups within a colony (Wilson 1971, Hölldobler and Wilson 1990). Although such dramatically different castes are not found in shrimp (or in most social insects), eusocial shrimp species do show evidence of morphological evolution compared with less social relatives (E. Tóth and J.E. Duffy, unpublished observations). Allometry of social species differed consistently from that of pair-forming species in each of two lineages in which eusociality has arisen independently. First, queens in social colonies had smaller fighting claws than did females in pair-forming species. Second, allometry of fighting claw size among other colony members was steeper in social than in pair-forming species. Finally, social species showed a change in allometry with increasing colony size; large but not small colonies exhibited a biphasic allometry of fighting claw and finger size (i.e., slope of the relation changes abruptly between smaller and larger individuals), indicating a distinctive group of large individuals possessing relatively larger weapons than other colony members (E. Tóth and J.E. Duffy, unpublished observations). These patterns are similar to those seen in some eusocial insects and naked mole-rats and emphasize that social life has had similar evolutionary consequences in disparate lineages.

Future Directions

Many challenges remain to understanding eusociality in snapping shrimp, but some solutions should soon be within reach. For example, reproductive skew models make testable predictions about how reproductive skew and cooperation arise as a function of genetic relatedness, disparities in fighting ability among individuals, and ecological constraints on independent breeding (e.g., Keller and Reeve 1994, Reeve et al. 1998, Reeve and Jeanne 2003). Testing these predictions will require high-resolution genetic markers to measure relatedness among interacting individuals and methods for observing and manipulating shrimp interactions under quasi-natural conditions. Progress is being made: shrimp have been observed for periods up to a few weeks in captivity (Duffy et al. 2002, Tóth and Duffy 2005), and microsatellite markers are currently under development.

Some specific questions for future research include the following: Do habitat saturation and other ecological constraints foster reproductive skew and cooperation in sponge-dwelling shrimp? Similarly, do the rarity and size of host sponges influence mating systems as predicted (chapter 12)? Now that a phylogeny is available for the gambarelloides species group (Morrison et al. 2004), these hypotheses should be testable via a comparative approach. What is needed are quantitative field data on distribution and demography of host sponge populations. Similarly, more complete data on development mode will allow a rigorous comparative test of the hypothesis that restricted dispersal is the first "filter" determining which species can form social kin groups (Fig. 18.1).

Does a "majority-rules" model of social evolution explain the extreme reproductive skew typical of eusocial shrimp? Given the evidence that queens are incapable of enforcing their dominance by aggression (Duffy and Macdonald 1999), the majority-rules prediction should be testable simply by measuring whether colony members are indeed more closely related to the queen than to anyone else. Experimental manipulations of queen presence, potential chemical signals, and relatedness in captive (or wild) colonies would also yield valuable insights into mechanisms maintaining reproductive skew.

Summary

Sponge-dwelling *Synalpheus* shrimp comprise a clade of approximately 30 species that range from socially monogamous pairs to eusocial colonies of hundreds of individuals. Eusocial colonies have evolved at least three times independently within *Synalpheus* and contain multiple cohabiting generations, with one or a few breeders of each sex and nonbreeders that defend the colony from intruders. Similarly to social insects on land, eusociality in sponge-dwelling shrimp has resulted in numerical ecological dominance within their niche and in consistent changes in allometry suggestive of morphological caste evolution. Comparison of sponge-dwelling shrimp with other animal taxa reveals several shared characteristics of life history and ecology suggested to promote cooperative breeding and eusociality in insects and vertebrates. First, social shrimp share with cooperatively breeding birds a constellation of factors that lead to low turnover of breeding opportunities, including low fecundity, high adult survival, sedentariness, and life in a warm, stable climate. Second, social shrimp share three conditions hypothesized to explain the distribution of "fortress defense" eusociality in insects: coincidence of food and shelter, strong enemy pressure, and ability to defend the food–shelter resource. Combining these two approaches, I hypothesize that coincidence of four conditions favored evolution of eusociality in certain *Synalpheus* lineages: (1) direct development resulting in limited dispersal and kin association; (2) specialization on a valuable, self-contained, and long-lived resource; (3) strong competition for the host resource; and (4) possession of a weapon (the snapping claw) effective in monopolizing it. These factors allow multigenerational occupation of nest sites, resulting in low turnover of breeding opportunities and selecting for cooperative defense by philopatric offspring that lack breeding opportunities. Coincidence of these characteristics is rare within Crustacea and may explain why *Synalpheus* contains the only known eusocial marine animals. Nevertheless, explaining why certain direct-developing sponge dwellers within the genus are social and other species are not remains a challenge.

Acknowledgments I am grateful to the National Science Foundation (DEB-9201566, DEB-9815785, IBN-0131931) and the Smithsonian Institution's Caribbean Coral Reef Ecosystem (CCRE) program for long-term support of my research on sponge-dwelling shrimps; Klaus Ruetzler, Mike Carpenter, the late Brian Kensley, and the staff of the Pelican Beach resort for facilitating work in Belize; and Tripp Macdonald, Cheryl Morrison, Rubén Ríos, and Eva Tóth, whose collaboration and camaraderie have been critical to this work. I thank Jae Choe, Martin Thiel, and an anonymous reviewer for comments that improved the manuscript. This is contribution #2800 from the Virginia Institute of Marine Science and contribution #789 from the CCRE.

References

Alexander, R.D. 1974. The evolution of social behavior. Annual Review of Ecology and Systematics 5:325–383.

Alexander, R.D., K.M. Noonan, and B.J. Crespi. 1991. The evolution of eusociality. Pages 3–44 *in*: P.W. Sherman, J.U.M. Jarvis, and R.D. Alexander, editors. The biology of the naked mole-rat. Princeton University Press, Princeton, N.J.

Andersson, M. 1984. The evolution of eusociality. Annual Review of Ecology and Systematics 15:165–189.

Aoki, S. 2003. Soldiers, altruistic dispersal and its consequences for aphid societies. Pages 201–215 *in*: T. Kikuchi, N. Azuma, and S. Higashi, editors. Genes, behaviors and evolution of social insects. Hokkaido University Press, Sapporo, Japan.

Arnold, K.E., and I.P.F. Owens, 1998. Cooperative breeding in birds: a comparative test of the life history hypothesis. Proceedings of the Royal Society of London, Series B 265:739–745.

Arnold, K.E., and I.P.F. Owens, 1999. Cooperative breeding in birds: the role of ecology. Behavioral Ecology 10:465–471.

Ayling, A.L. 1983. Growth and regeneration rates in thinly encrusting Demospongiae from temperate waters. Biological Bulletin 165:243–352.

Banner, A.H., and D.M. Banner. 1983. Annotated checklist of the alpheid shrimp from the western Indian Ocean. Travaux Documents l'ORSTOM 158:1–164.

Bennett, N.C., C.G. Faulkes, and A.J. Molteno. 1996. Reproductive suppression in subordinate non-breeding female Damaraland mole-rats: two components to a lifetime of socially induced infertility. Proceedings of the Royal Society of London, Series B 263:1599–1603.

Bourke, A.F.G., and N.R. Franks. 1995. Social evolution in ants. Princeton University Press, Princeton, N.J.

Brockmann, H.J. 1997. Cooperative breeding in wasps and vertebrates: the role of ecological constraints. Pages 347–371 *in*: J.C. Choe and B.J. Crespi, editors. Social behavior in insects and arachnids. Cambridge University Press, Cambridge.

Bruce, A.J. 1976. Shrimps and prawns of coral reefs, with special reference to commensalism. Pages 37–94 *in*: O.A. Jones and R. Endean, editors. Biology and geology of coral reefs, Volume 3. Biology 2. Academic Press, New York.

Bruce, A.J. 1998. New keys for the identification of Indo-West Pacific coral associated Pontoniine shrimps, with observations on their ecology (Crustacea: Decapoda: Palaemonidae). Ophelia 49:29–46.

Burland, T.M., N.C. Bennett, J.U.M. Jarvis, and C.G. Faulkes. 2004. Colony structure and parentage in wild colonies of cooperatively breeding Damaraland mole-rats suggest incest avoidance alone may not maintain reproductive skew. Molecular Ecology 13:2371–2379.

Chace, F.A. 1972. The shrimps of the Smithsonian-Bredin Caribbean Expeditions with a summary of the West Indian shallow-water species (Crustacea: Decapoda: Natantia). Smithsonian Contributions to Zoology 98:1–179.

Choe, J.C., and B.J. Crespi, editors. 1997. Social behavior in insects and arachnids. Cambridge University Press, Cambridge.

Cooney, R., and N.C. Bennett. 2000. Inbreeding avoidance and reproductive skew in a cooperative mammal. Proceedings of the Royal Society of London, Series B 267:801–806.

Correa, C., and M. Thiel. 2003. Mating systems in caridean shrimp (Decapoda: Caridea) and their evolutionary consequences for sexual dimorphism and reproductive biology. Revista Chilena de Historia Natural 76:187–203.

Coutière, H. 1909. The American species of snapping shrimps of the genus *Synalpheus*. Proceedings of the United States National Museum 36:1–93.

Crespi, B.J. 1994. Three conditions for the evolution of eusociality: are they sufficient? Insectes Sociaux 41:395–400.

Crespi, B.J. 1996. Comparative analysis of the origins and losses of eusociality: causal mosaics and historical uniqueness. Pages 253–287 *in*: E.P. Martins, editor. Phylogenies and the comparative method in animal behavior. Oxford University Press, New York.

Crespi, B.J., and D. Yanega. 1995. The definition of eusociality. Behavioral Ecology 6:109–115.

Crozier, R.H., and P. Pamilo. 1996. Evolution of social insect colonies. Sex allocation and kin selection. Oxford University Press, Oxford.

Darwin, C. 1859. On the origin of species by means of natural selection. John Murray, London.

Dobkin, S.R. 1965. The first post-embryonic stage of *Synalpheus brooksi* Coutière. Bulletin of Marine Science 15:450–462.

Dobkin, S.R. 1969. Abbreviated larval development in caridean shrimps and its significance in the artificial culture of these animals. FAO Fisheries Reports 57:935–946.

Duffy, J.E. 1992. Host use patterns and demography in a guild of tropical sponge-dwelling shrimps. Marine Ecology Progress Series 90:127–138.

Duffy, J.E. 1993. Genetic population structure in two tropical sponge-dwelling shrimps that differ in dispersal potential. Marine Biology 116:459–470.

Duffy, J.E. 1996a. Eusociality in a coral-reef shrimp. Nature 381:512–514.

Duffy, J.E. 1996b. Resource-associated population subdivision in a symbiotic coral-reef shrimp. Evolution 50:360–373.

Duffy, J.E. 1996c. Specialization, species boundaries, and the radiation of sponge-dwelling alpheid shrimp. Biological Journal of the Linnean Society 58:307–324.

Duffy, J.E. 1996d. *Synalpheus regalis*, new species, a sponge-dwelling shrimp from the Belize Barrier Reef, with comments on host specificity in *Synalpheus*. Journal of Crustacean Biology 16:564–573.

Duffy, J.E. 1998. On the frequency of eusociality in snapping shrimps (Decapoda: Alpheidae), with description of a second eusocial species. Bulletin of Marine Science 63:387–400.

Duffy, J.E. 2003. The ecology and evolution of eusociality in sponge-dwelling shrimp. Pages 201–215 *in*: T. Kikuchi, N. Azuma, and S. Higashi, editors. Genes, behaviors and evolution of social insects. Hokkaido University Press, Sapporo, Japan.

Duffy, J.E., and K.S. Macdonald. 1999. Colony structure of the social snapping shrimp, *Synalpheus filidigitus*, in Belize. Journal of Crustacean Biology 19:283–292.

Duffy, J.E., C.L. Morrison, and R. Rios. 2000. Multiple origins of eusociality among sponge-dwelling shrimps (*Synalpheus*). Evolution 54:503–516.

Duffy, J.E., C.L. Morrison, and K.S. Macdonald. 2002. Colony defense and behavioral differentiation in the eusocial shrimp *Synalpheus regalis*. Behavioral Ecology and Sociobiology 51:488–495.

Emlen, S.T. 1984. Cooperative breeding in birds and mammals. Pages 305–339 *in*: J.R. Krebs and N.B. Davies, editors. Behavioral ecology: an evolutionary approach. Blackwell, Oxford.

Emlen, S.T. 1994. Benefits, constraints and the evolution of the family. Trends in Ecology and Evolution 9:282–285.

Emlen, S.T. 1995. An evolutionary theory of the family. Proceedings of the National Academy of Sciences, USA 92:8092–8099.

Emlen, S.T. 1997. Predicting family dynamics in social vertebrates. Pages 228–253 *in*: J.R. Krebs and N.B. Davies, editors. Behavioural ecology: an evolutionary approach, fourth edition. Blackwell Science, Oxford.

Erdman, R.B., and N.J. Blake. 1987. Population dynamics of the sponge-dwelling alpheid *Synalpheus longicarpus*, with observations on *S. brooksi* and *S. pectiniger*, in shallow-water assemblages of the eastern Gulf of Mexico. Journal of Crustacean Biology 7:328–337.

Fletcher, D.J.C., and K.G. Ross. 1985. Regulation of reproduction in eusocial Hymenoptera. Annual Review of Entomology 30:319–343.

Gadagkar, R. 1991. Demographic predisposition to the evolution of eusociality: a hierarchy of models. Proceedings of the National Academy of Sciences, USA 88:10993–10997.

Gadagkar, R. 1994. Why the definition of eusociality is not helpful to understand its evolution and what should we do about it. Oikos 70:485–488.

Gherardi, F., and C. Calloni. 1993. Protandrous hermaphroditism in the tropical shrimp *Athanas indicus* (Decapoda: Caridea), a symbiont of sea urchins. Journal of Crustacean Biology 13:675–689.

Greeff, J.M., and N.C. Bennett. 2000. Causes and consequences of incest avoidance in the cooperatively breeding mole-rat, *Cryptomys darlingi* (Bathyergidae). Ecology Letters 3:318–328.

Hamilton, W.D. 1964. The genetical evolution of social behavior I, II. Journal of Theoretical Biology 7:1–52.

Hart, A.G., and F.L.W. Ratnieks. 2004. Crossing the taxonomic divide: conflict and its resolution in societies of reproductively totipotent individuals. Journal of Evolutionary Biology 18:383–395.

Hatchwell, B.J., and J. Komdeur. 2000. Ecological constraints, life history traits and the evolution of cooperative breeding. Animal Behavior 59:1079–1086.

Hazlett, B.A., and H.E. Winn. 1962. Sound production and associated behavior of Bermuda crustaceans (*Panulirus*, *Gonodactylus*, *Alpheus*, and *Synalpheus*). Crustaceana 4:25–38.

Helms Cahan, S., D.T. Blumstein, L. Sundström, J. Liebig, and A. Griffin. 2002. Social trajectories and the evolution of social behavior. Oikos 96:206–216.

Hölldobler, B., and E.O. Wilson., 1990. The ants. Harvard University Press, Cambridge, Mass.

Jarvis, J.U.M., and N.C. Bennett. 1993. Eusociality has evolved independently in two genera of bathyergid mole-rats—but occurs in no other subterranean mammal. Behavioral Ecology and Sociobiology 33:353–360.

Jeanne, R.L. 1980. Evolution of social behavior in the Vespidae. Annual Review of Entomology 25:371–396.

Jeon, J., and J.C. Choe. 2003. Reproductive skew and the origin of sterile castes. American Naturalist 161:206–224.

Johnstone, R.A. 2000. Models of reproductive skew: a review and synthesis. Ethology 106:5–26.

Keller, L., and N. Perrin. 1995. Quantifying the level of eusociality. Proceedings of the Royal Society of London, Series B 260:311–315.

Keller, L., and H.K. Reeve. 1994. Partitioning of reproduction in animal societies. Trends in Ecology and Evolution 9:98–102.

Knowlton, N. 1980. Sexual selection and dimorphism in two demes of a symbiotic, pair-bonding snapping shrimp. Evolution 34:161–173.

Knowlton, N., and B.D. Keller. 1982. Symmetric fights as a measure of escalation potential in a symbiotic, territorial snapping shrimp. Behavioral Ecology and Sociobiology 10: 289–292.

Knowlton, R.E. 1971. Effects of environmental factors on the larval development of *Alpheus heterochelis* Say and *Palaemonetes vulgaris* (Say) (Crustacea Decapoda Caraea), with ecological notes on larval and adult Alpheidae and Palaemonidae. Unpublished Ph.D. thesis. University of North Carolina–Chapel Hill, Chapel Hill, N.C.

Koenig, W.D., F.A. Pitelka, W.J. Carmen, R.L. Mumme, and M.T. Stanback. 1992. The evolution of delayed dispersal in cooperative breeders. Quarterly Review of Biology 67:111–150.

Lin, N., and C.D. Michener. 1972. Evolution of sociality in insects. Quarterly Review of Biology 47:31–159.

Macdonald, K.S., III, R. Ríos, and J.E. Duffy. 2006. Biodiversity, host specificity, and dominance by eusocial species among sponge-dwelling alpheid shrimp on the Belize Barrier Reef. Diversity and Distributions 12:165–178.

Mathews, L. 2002. Territorial cooperation and social monogamy: factors affecting intersexual interactions in pair-living snapping shrimp. Animal Behavior 63:767–777.

Mathews, L. 2003. Tests of the mate-guarding hypothesis for social monogamy: male snapping shrimp prefer to associate with high-value females. Behavioral Ecology 14:63–67.

Michener, C.D. 1969. Comparative social behavior of bees. Annual Review of Entomology 14:299–342.

Michener, C.D., and D.J. Brothers. 1974. Were workers of eusocial Hymenoptera initially altruistic or oppressed? Proceedings of the National Academy of Sciences, USA 71:671–674.

Morrison, C.L., R. Ríos, and J.E. Duffy. 2004. Phylogenetic evidence for an ancient rapid radiation of Caribbean sponge-dwelling snapping shrimps (*Synalpheus*). Molecular Phylogenetics and Evolution 30:563–581.

Myles, T.G. 1988. Resource inheritance in social evolution from termites to man. Pages 379–423 *in*: C. Slobodchikoff, editor. The ecology of social behavior. Academic Press, San Diego, Calif.

Nakashima, Y. 1987. Reproductive strategies in a partially protandrous shrimp, *Athanas kominatoensis* (Decapoda: Alpheidae): sex change as the best of a bad situation for subordinates. Journal of Ethology 2:145–159.

Nolan, B.A., and M. Salmon 1970. The behavior and ecology of snapping shrimp (Crustacea: *Alpheus heterochelis* and *Alpheus normanii*). Forma et Functio 2:289–335.

Patton, W.K. 1974. Community structure among the animals inhabiting the coral *Pocillopora damicornis* at Heron Island, Australia. Pages 219–243 *in*: W.B. Vernberg, editor. Symbiosis in the sea. University of South Carolina Press, Columbia, S.C.

Pearse, A.S. 1932. Inhabitants of certain sponges at Dry Tortugas. Papers of the Tortugas Laboratory of the Carnegie Instiutute of Washington 27:117–124.

Queller, D.C. 1989. The evolution of eusociality: reproductive head starts of workers. Proceedings of the National Academy of Sciences, USA 86:3224–3226.

Queller, D.C., and J.E. Strassmann. 1998. Kin selection and social insects: social insects provide the most surprising predictions and satisfying tests of kin selection. BioScience 48:165–175.

Rahman, N., D.W. Dunham, and C.K. Govind. 2003. Social monogamy in the big-clawed snapping shrimp, *Alpheus heterochelis*. Ethology 109:457–473.

Ratnieks, F.L.W. 1988. Reproductive harmony via mutual policing by workers in eusocial Hymenoptera. American Naturalist 132:217–236.

Reeve, H.K., and R.L. Jeanne. 2003. From individual control to majority rule: extending transactional models of reproductive skew in animal societies. Proceedings of the Royal Society of London, Series B 270:1041–1045.

Reeve, H.K., and L. Keller. 2001. Tests of reproductive-skew models in social insects. Annual Review of Entomology 46:347–385.

Reeve, H.K., and P.W. Sherman. 1991. Intracolonial aggression and nepotism by the breeding female naked mole-rat. Pages 337–357 *in*: P.W. Sherman, J.U.M. Jarvis, and R.D. Alexander, editors. The biology of the naked mole-rat. Princeton University Press, Princeton, N.J.

Reeve, H.K., S.T. Emlen, and L. Keller. 1998. Reproductive sharing in animal societies: reproductive incentives or incomplete control by dominant breeders? Behavioral Ecology 9:267–278.

Reiswig, H.M. 1973. Population dynamics of three Jamaican Demospongiae. Bulletin of Marine Science 23:191–226.

Ríos, R. 2003. *Synalpheus* shrimp from Carrie Bow Cay, Belize: systematics, phylogenetics, and biological observations (Crustacea: Decapoda: Alpheidae). Unpublished Ph.D. thesis, College of William and Mary, Williamsburg, Va.

Ríos, R., and J.E. Duffy. 1999. *Synalpheus williamsi* (Decapoda: Alpheidae), a new sponge-dwelling shrimp from the Caribbean. Proceedings of the Biological Society of Washington 112:541–552.

Ruetzler, K. 1976. Ecology of Tunisian commercial sponges. Tethys 7:249–264.

Shellman-Reeve, J.S. 2001. Genetic relatedness and partner preference in a monogamous, wood-dwelling termite. Animal Behaviour 61:869–876.

Sherman, P.W., E.A. Lacey, H.K. Reeve, and L. Keller. 1995. The eusociality continuum. Behavioral Ecology 6:102–108.

Starr, C.K. 1985. Enabling mechanisms in the origin of eusociality in the Hymenoptera—the sting's the thing. Annals of the Entomological Society of America 78:836–840.

Strassmann, J.E., and D.C. Queller, 1989. Ecological determinants of social evolution. Pages 81–101 *in*: M.D. Breed and R.E. Page, editors. The genetics of social evolution. Westview Press, Boulder, Colo.

Strathmann, R.R. 1985. Feeding and nonfeeding larval development and life-history evolution in marine invertebrates. Annual Review of Ecology and Systematics 16:339–361.

Suzuki, H. 1970. Taxonomic review of four alpheid shrimp belonging to the genus *Athanas*, with reference to their sexual phenomena. Science Reports of the Yokohama National University, Section II. Biological and Geological Sciences 17:1–38.

Thiel, M. 1999. Host-use and population demographics of the ascidian-dwelling amphipod *Leucothoe spinicarpa*: indication for extended parental care and advanced social behaviour. Journal of Natural History 33:193–206.

Thiel, M. 2000. Population and reproductive biology of two sibling amphipod species from ascidians and sponges. Marine Biology 137:661–674.

Thomas, J.D. 1997. Systematics, ecology and phylogeny of the Anamixidae (Crustacea: Amphipoda). Records of the Australian Museum 49:35–98.

Tóth, E., and J.E. Duffy. 2005. Coordinated group response to nest intruders in social shrimp. Biology Letters 1:49–52.

Wilson, E.O. 1971. The insect societies. Belknap Press of Harvard University, Cambridge, Mass.

Wilson, E.O. 1990. Success and dominance in ecosystems: The case of the social insects. Ecology Institute Oldendorf/Luhe, Germany.

PART V

Synthesis

Anthropogenic Stressors and Their Effects on the Behavior of Aquatic Crustaceans

Thijs Christiaan van Son

Martin Thiel

19

Animal behaviors are affected either directly or indirectly by several anthropogenic activities threatening aquatic ecosystems. There are four major groups of anthropogenic stressors:

1. Pollution caused by industrial discharges, climate change, acid rain, agriculture, and aquaculture (for some behavioral consequences for crustaceans, see Tierney and Atema 1986, Hebel et al. 1997, Jones and Reynolds 1997, Waddy et al. 2002, Butterworth et al. 2004, Clotfelter et al. 2004, Zala and Penn 2004, Gilbert et al. 2005)
2. Species introductions resulting from shipping (ballast water), aquaculture, aquarium trade, and bait trade (see Gherardi and Daniels 2004, Griffiths et al. 2004, Schlaepfer et al. 2005)
3. Stock manipulation caused by fishing, stock enhancement, and aquaculture (see Hankin et al. 1997, Bannister and Addison 1998, Jamieson 2000, Jennings et al. 2001a, Jivoff 2003)
4. Habitat destruction due to shoreline construction, bottom trawling, and river impoundment (see Light et al. 1995, Bowles et al. 2000)

The use of measures of animal behavior as indicators may provide sensitive information about a population's condition. For example, toxicologists are currently using behavioral measures (e.g., activity, feeding, and antipredator behavior) as bioindicators (Clotfelter et al. 2004), and such measures are predicted to become increasingly important in ecotoxicology (Clotfelter et al. 2004, Zala and Penn 2004). In a similar way, observations of behaviors or important population parameters could be used to estimate the state of populations exposed to exploitation or other anthropogenic stressors. Crustaceans represent one group of aquatic organisms that has received relatively little attention on how anthropogenic stressors affect their behaviors, yet they are important members in all aquatic ecosystems.

For purposes of this review, we divide the behaviors of organisms into four vital functions: reproduction, nutrition and growth, habitat selection, and self-defense. These vital functions contain a number of related behaviors (see Table 19.1) that may be affected either directly (e.g., contaminants interfering with sexual communication, food searching, or refuge use) or indirectly (through demographic or other changes, e.g., biased sex ratio due to stock manipulations) by anthropogenic stressors. More specifically, we predict that chemical and sexual communication could be disrupted by various contaminants, that pollution could negatively affect an organism's ability to locate food, and that mating interactions should be affected by fishing activities, especially when these are selective for size and sex. Finally, refuge use is likely to be affected by habitat destruction and possibly by introduced species when these are occupying refuges that are normally used by native species.

Here, we explore these predictions using crustaceans as models. In order to apply behavioral measures as bioindicators for conservation decisions, it is necessary to understand how these behaviors are affected by anthropogenic stressors. First, we need to identify which behaviors are affected by which stressors. Second, we need to understand how anthropogenic stressors affect behavior to decide on what level (i.e., ecosystem, population, or individual) these effects occur. Thus, we reviewed the literature to identify the most common anthropogenic stressors affecting crustacean behavior. This review is not meant to be a complete overview, but rather a first approach toward a better understanding of how different stressors drive the observed changes in behavior. We focus on aquatic crustaceans from both marine and freshwater environments.

Table 19.1. Social behaviors affected by anthropogenic activities.

Vital Functions and Their Related Behaviors	Category (Subcategory) Observations	Total Observations	%
Reproduction		42	40.8
Mating	31		
Competition for mates	(22)		
Mate location and selection	(6)		
Coitus or spawning	(2)		
Mate guarding	(1)		
Fecundity and hatching success	6		
Sex change	4		
Hybridization	1		
Nutrition and growth		29	28.2
Foraging	12		
Food location and selection	(11)		
Competition for food	(1)		
Dominance competition	11		
Molting and limb regeneration	5		
Aggregation	1		
Habitat selection		22	21.4
Refuge use/sheltering	11		
Locomotion	4		
Settlement	4		
Migration	2		
All	1		
Self-defense		9	8.7
Avoidance/escape	6		
Antipredator behavior	3		
General		1	1.0
Total		103	100.0

Anthropogenic Stressors and Their Effects on Crustacean Behavior

As mentioned above, we classified anthropogenic stressors in four groups: pollution, species introduction, stock manipulation, and habitat destruction. A wide range of publications was examined for information on behavior (with emphasis on social behavior), important population parameters, and life history characteristics that might be affected by these stressors. Population parameters and life history characteristics are included in this review because changes in population density, size/age structure, and sex ratio can result in changes in both individual behavior and population processes. The effect of a stressor on a certain behavior was assigned an indication level: "observed" when authors documented an effect of a specific stressor on behavior and/or parameters, "likely" when authors suggested that a certain stressor may have an effect, and "possible" when not mentioned by the authors but, based on available knowledge, we infer the possibility of a stressor affecting behavior.

Table 19.2. Differences in level of indication for effect of anthropogenic activities on social behavior in crustaceans.

Anthropogenic Impact	No. observations	Indication (%)		
		Observed	Likely	Possible
Pollution	43	62.8	18.6	18.6
Species introduction	16	81.3	6.3	12.5
Stock manipulation	29	6.9	48.3	44.8
Habitat destruction	15	6.7	46.7	46.7
Total	103			

A total of 139 observations (103 on vital functions and 36 on important population and life history parameters) were included in this review. The most affected vital functions were reproduction (42 of 103 observations), followed by nutrition and growth (29 of 103) and habitat selection (22 of 103) (see Table 19.1). Reproduction is not necessarily the most affected vital function but may simply have received the most attention by researchers. With regard to the level of indication, the number of studies was quite evenly distributed among "observed" (43 out of 103 observations), "likely" (30), and "possible" (30). There were, however, proportional differences related to the anthropogenic stressors, particularly for the observed effects. In the first group of activities, pollution and species introduction, more than 60% of the studies reported observed effects on the behavior of crustaceans. In contrast, for fishing and habitat destruction, fewer than 10% of the studies reported observed effects, and evidence is more circumstantial (either likely or possible; see Table 19.2).

Pollution and Contamination

Environmental contamination represents a threat to aquatic species worldwide (Fingerman et al. 1998). In estuaries especially, water quality is often compromised by heavy influx of nutrients, organic matter, and chemical contaminants (Kennish 2002). This contamination has the potential to adversely affect chemical communication, which is an important aspect of social behavior of crustaceans (see chapter 6). Similarly, in freshwater ecosystems, acidification due to acid precipitation has caused serious declines in species at all trophic levels (Tierney and Atema 1986). Although we have relatively little information on how aquatic and especially marine benthic invertebrates are affected by sewage effluent and industrial discharges (Lye et al. 2005), some recent studies reveal how contaminants can directly affect animal behavior (see Table 19.3).

Approximately one-fourth (12 out of 43 observations) of the reports that found an effect of pollution on behavior in crustaceans, indicated that reproduction was affected. Four of these 12 addressed the effect of endocrine-disrupting chemicals (EDCs). Endocrine-disrupting chemicals are ubiquitous in the environment, and their effects on animal behavior have received increasing attention (Fingerman et al. 1998,

Table 19.3. Species and behaviors affected by Pollution and Contamination.

Species Affected	Vital Function	Main Behavior	Specific Behavior	Indication[a]	Comment	
Brachyura						
Callinectes sapidus	Self-defense	Avoidance		obs	[Eutrophication] Avoidance of hypoxic water	Bell et al. 2003a
	Nutrition and growth	Foraging	Food location and selection	obs	[Eutrophication] Feeding declined when exposed to hypoxic	Bell et al. 2003b
		Molting and limb regeneration		obs	[Hydrocarbons] Molting and limb regeneration inhibited	Cantelmo et al. 1981
	Habitat selection	Settlement		pos	[Organic] Chemcial cues during settlement disrupted	Welch et al. 1997
Carcinus maenas	Reproduction	Mating	Coitus or spawning	like	[EDCs] Demasculinisation	Ford et al. 2004a
			Mate location and selection	obs	[EDCs] Sexual communication	Lye et al. 2005
Chionoecetes opilio	Reproduction	Mating	Mate location and selection	pos	[Organic] Could affect chemical communication	Bouchard et al. 1996
Pachygrapsus crassipes	Nutrition and growth	Foraging	Food location and selection	obs	[Hydrocarbons] Feeding response inhibited	Hebel et al. 1997
	Reproduction	Mating	Mate location and selection	obs	[Hydrocarbons] Sexual communication affected	Hebel et al. 1997
Rhithropanopeus harrisii	Nutrition and growth	Molting and limb regeneration		obs	[Pesticides] Abnormally shaped regenerations of chelipeds	Clare et al. 1992
Uca pugilator	Nutrition and growth	Molting and limb regeneration		obs	[Organic] Molting and limb regeneration inhibited	Weis et al. 1992
Caridea						
Crangon crangon	Self-defense	Avoidance		obs	[Inorganic] Fast swimming followed by paralysis	Vismann 1996
Hippolyte inermis	Habitat Selection	Locomotion		obs	[Metal] Swimming velocity affected	Untersteiner et al. 2005
Palaemonetes pugio	Nutrition and growth	Molting and limb regeneration		obs	[Organic] Limb regeneration inhibited	Rao et al. 1978 as cited in Fingermen et al. 1998

(*Contd.*)

Table 19.3. (*Contd.*)

Species Affected	Vital Function	Main Behavior	Specific Behavior	Indication[a]	Comment	
ASTACIDEA						
	Nutrition and growth	Foraging	Food location and selection	pos	[Organic] Ability to locate food could be affected	Breithaupt et al. 1999
		Molting and limb regeneration		obs	[Hydrocarbons] Females molt prematurely and abort brood	Watson et al. 2004
Nephrops norvegicus	Self-defense	Avoidance		obs	[Inorganic] Rapid tell flipping, escape behavior	Butterworth et al. 2004
	Nutrition and growth	Foraging	Food location and selection	obs	[Eutrophication] Hypoxia affected feeding activity	Baden et al. 1990, Hagerman and Baden 1988
	Habitat selection	Refuge use		obs	[Eutrophication] Hypoxia elicits emergence from burrows	Baden et al. 1990, Eriksson and Baden 1997
Orconectes rusticus	Self-defense	Antipredator behavior		obs	[Pesticide] Walked toward the alarm signal	Wolf and Moore 2002
	Nutrition and growth	Foraging	Food location and selection	obs	[Pesticide] Unable to locate a food source	Wolf and Moore 2002
Orconectes virilis	Nutrition and growth	Foraging	Food location and selection	like	[Acidification] Ability to detect food cues reduced	Tierney and Aterna 1986
			Food location and selection	pos	[Organic] Ability to locate food may be affected	Tomba et al. 2001
Procambarus acutus	Nutrition and growth	Foraging	Food location and selection	like	[Acidification] Ability to detect food cues reduced	Tierney and Aterna 1986
Procambarus clarkii	Nutrition and growth	Dominance competition		pos	[Organic] Recognition of dominance hierarchies may be affected	Zulandt Schneider et al. 1999
	Reproduction	Fecundity and hatching success		obs	[Metals] Fecundity and hatching success reduced	Naqvi et al. 1993 as cited in Fingerman et al. 1998
AMPHIPODA						
Caprella danilevskii	Reproduction	Mating	Competition for mates	pos	[Organic] Induce female-biased sex ratio	Ohij et al. 2002
Corophium volutator	Reproduction	Mating	Competition for mates	like	[Metal] Juveniles unable to reach maturity	Conradi and Depledge 1998

Echinogammarus marinus	Reproduction	Mating	Competition for mates	like	[EDCs] Intersexuality	Ford et al. 2004a
		Fecundity and hatching success		like	[EDCs] Fecundity and fertility reduced	Ford et al. 2004b
Gammarus duebeni celticus	Nutrition and growth	Dominance competition		obs	[Inorganic] Interspecific interactions affected	MacNeil et al. 2004
Hyalella azteca	Reproduction	Mating	Mate guarding	obs	[Pesticide] Precopulatory mate-guarding disrupted	Blockwell et al. 1998
Monoporeia affinis	Self-defense	Avoidance		like	[Eutrophication] Severe hypoxia increased swimming activity	Johansson 1997
	Reproduction	Mating	Competition for mates	pos	[Hydrocarbons] Delayed sexual development	Sundelin et al. 2000
Pontoporeia femorata	Self-defense	Avoidance		like	[Eutrophication] Severe hypoxia increased swimming activity	Johansson 1997
Isopoda						
Saduria entomon	Habitat selection	Refuge use		obs	[Eutrophication] Severe hypoxia causes emergence onto surface	Johansson 1997
Trachelipus rathkei	Reproduction	Mating	Competition for mates	pos	[Pesticides] Sex ratio apparently affected	Paoletti and Canterino 2002
Cladocera						
Daphnia magna	Habitat selection	Locomotion		obs	[Metal] Swimming activity reduced	Untersteiner et al. 2003
		Locomotion		obs	[Metal] Positively phototactic behavior reduced	Michels et al. 2000
Mysidacea						
Neomysis integer	Self-defense	Avoidance		obs	[Eutrophication+metal] Reduced tolerance to hypoxia when combined	Roast et al. 2002
	Habitat selection	Locomotion		obs	[Pesticide] Hyperactivity, but swimming speed reduced	Roast et al. 2000

[a] Indication level: obs, observed; like, likely; pos, possible (see text for further explanation).

Depledge and Billinghurst 1999, Hutchinson 2002, Clotfelter et al. 2004, Verslycke et al. 2004, Zala and Penn 2004). Endocrine-disrupting chemicals may act both as agonists (i.e., any molecule that improves the activity of a different molecule, e.g., a hormone) and antagonists (i.e., any molecule that blocks the ability of a given chemical to bind to its receptor, preventing a biological response), inducing either underproduction or overproduction of hormones (Clotfelter et al. 2004, Zala and Penn 2004). As a consequence, EDCs can adversely affect a wide range of behaviors (Zala and Penn 2004). For example, male shore crabs (*Carcinus maenas*) exhibited reduced responses to female sex pheromone when exposed to EDCs (Lye et al. 2005). Furthermore, the demasculinization, feminized abdomens, and reduced claw depth observed in male *C. maenas* by Ford et al. (2004a) were suggested to be related to exposure to EDCs. The same is probably the case for the amphipod *Echinogammarus marinus* experiencing intersexuality (Ford et al. 2004a). Ohij et al. (2004) reported a female-biased sex ratio in a caprellid amphipod (*Caprella danilevskii*) following exposure to tributyltin during the embryonic stage and suggested that tributyltin may disturb sex determination in this species. Organic compounds (Bouchard et al. 1996, Hebel et al. 1997, Ohij et al. 2002), heavy metals (Naqvi et al. 1993 as cited in Fingerman et al. 1998), and pesticides (Paoletti and Cantarino 2002) can also negatively affect important aspects of reproduction such as sexual communication, fecundity, and sex ratio (see Table 19.3).

Foraging of crustaceans may also be directly affected by contaminants (Table 19.3). Bell et al. (2003b) reported that blue crabs (*Callinectes sapidus*) showed a decline in feeding rate when exposed to hypoxia, which is caused by heavy influx of nutrients to estuaries, and may also be related to ocean climate variability (see Gilbert et al. 2005). Similar observations have been reported in the Norway lobster, *Nephrops norvegicus* (Hagerman and Baden 1988, Baden et al. 1990). Exposure to crude oil extracts inhibited the feeding response in the crab *Pachygrapsus crassipes* (Hebel et al. 1997). In fresh water exposed to acid rain, Tierney and Atema (1986) found that the crayfish *Orconectes virilis* and *Procambarus acutus* showed a reduced ability to detect and/or respond to food cues due to the acidification.

Several crustaceans are able to detect adverse environmental conditions caused by eutrophication and subsequently exhibit avoidance or escape behavior. The blue crab (*C. sapidus*) can avoid hypoxic conditions by migrating to locations with more favorable oxygen concentrations (Bell et al. 2003a). Norway lobsters leave their burrows when the oxygen concentration becomes too low, which makes them more susceptible to fishing by trawls (Baden et al. 1990). Juveniles of this species react in a similar manner when exposed to severe hypoxia, which increases their exposure to benthic predators (Eriksson and Baden 1997). In the shrimp *Crangon crangon*, exposure to high sulfide concentrations appears to elicit not only avoidance behavior but also panic reactions (Vismann 1996).

In several crustaceans, heavy metals and organic compounds can inhibit molting and limb regeneration (Cantelmo et al. 1981, Clare et al. 1992, Weis et al. 1992, Rao et al. 1978 as cited in Fingerman et al. 1998; see also Table 19.3), which may have severe consequences for social interactions. The inhibition of molting directly affects growth and could, if the whole population is exposed to these contaminants, affect the population's size structure. Failing to regenerate lost limbs would likely have adverse effects on feeding, social status, and mating success and therefore on overall fitness.

Introduction of Nonindigenous Crustacean Species

Nonindigenous species (NIS) pose severe economic and ecological threats (Lodge et al. 2000). Many invasive species have profoundly affected the abundance and diversity of native biota. For example, in San Francisco Bay, exotic species dominate many habitats (Cohen and Carlton 1998). The damage caused by introduced species has often been suggested to result from escape from natural enemies such as predators and parasites, without which NIS often attain both high population density and large body size in their new location (Torchin et al. 2001, 2003; but see Parker et al. 2006). This may give exotic species a competitive advantage, which may affect social interactions with native species both directly and indirectly.

Introduction of NIS can generate interspecific competition and aggressive interactions with native species. Nonindigenous species may interact with native crustaceans directly through predation, cannibalism, or aggression, indirectly due to differences in life history characteristics, or through a combination (Table 19.4), potentially allowing NIS to outcompete and replace native crustaceans. For example, the introduced shore crab *C. maenas*, well known as a voracious predator (Lafferty and Kuris 1996), preys directly on juvenile *Cancer* crabs (Lafferty and Kuris 1996) and juvenile *Hemigrapsus oregonensis* (Grosholz and Ruiz 1996) on the west coast of North America. Similarly, in direct contests between similar-sized males, the resident freshwater crayfish *Orconectes virilis* is outcompeted by the introduced *O. rusticus* (Hill and Lodge 1999). This is also observed between crayfish in Italy, where the introduced *Procambarus clarkii* outcompetes the native *Austropotamobius pallipes* in contests between similar-sized males (Dardi et al. 1996 as cited in Barbaresi and Gherardi 2000). In competition for refuge under predation risk, the invading *O. rusticus* and a previous invader *O. propinquus* excluded the resident *O. virilis* from available shelter. As a consequence, the latter experienced increased mortality (Garvey et al. 1994). The final outcome of such contests between crayfish is determined by a wide range of intrinsic as well as extrinsic factors (see chapter 5).

Another direct interaction that poses a serious threat to biodiversity is hybridization. Perry et al. (2001, 2002) reported that hybridization between closely related crayfishes, the introduced *O. rusticus* and the resident *O. propinquus* (a previous invader), appears to be an important factor driving the extirpation of the latter. *Orconectes rusticus* is competitively superior to *O. propinquus* and apparently outcompetes resident males for females of both species. Unfit hybrid progeny would lead to decreased reproductive output of *O. propinquus* females. Furthermore, if adults of mixed ancestry are competitively intermediate, they will also outcompete resident *O. propinquus* males, further increasing the rate of extirpation (Perry et al. 2002).

With regard to indirect interspecific competition, several studies in crayfishes have demonstrated or suggested that differences in life history characteristics facilitate the replacement of native species by exotic species. One example involves introduction of *Pacifastacus leniusculus* to the natural habitat of *P. fortis* in California. While *P. leniusculus* reach maturity within one to three years and produce 100–400 eggs, *P. fortis* needs four years to reach maturity and produce only 10–70 eggs (Light et al. 1995). Similar differences in life history traits are also likely to have played a role in the replacement of the native penaeid prawn *Penaeus kerathurus* in the Mediterranean by exotic Lessepsian penaeid prawns (Galil 2000).

Finally, species-specific physicochemical tolerances may also play an important role in interactions between native and invasive species. For example, the invasive

Table 19.4. Species and behaviors affected by Species Introduction.

Species	Vital Function	Main Behavior	Indication[a]	Comment	References
Brachyura					
Cancer spp.	Self-defense	Antipredator behavior	obs	C. *maenas* prey on young *Cancer* crabs	Lafferty and Kuris 1996
Hemigrapsus oregonensis	Self-defense	Antipredator behavior	obs	C. *maenas* prey on *H. oregonensis*	Grosholz and Ruiz 1996
Caridea					
Penaeus kerathurus	Nutrition and growth	Dominance competition	obs	Lessepsian penaoid prawns have out competed *P. kerathurus*	Galil 2000
Astacidea					
Native spp.	Nutrition and growth	Dominance competition	obs	*Orconectes rusticus* competitively exclude native crayfish species	Holdich 2000
Austropotamobius pallipes	General health		pos	Introduction of *Pacifastacus leniusculus* w/ fungal disease	Holdich 2000
	Nutrition and growth	Dominance competition	pos	*Procambarus clarkii* competitively out compete similar-sized *A. pallipes*	Dardi et al. 1996 cited in Barbaresi and Gherardi 2000
Orconectes propinquus	Reproduction	Hybridization	obs	Introduction of closely related species result in hybridization	Perry et al. 2002
Orconectes virilis	Nutrition and growth	Dominance competition	obs	O. *rusticus* outcompetes O. *virilis*	Taylor and Redmer 1996, Hill and Lodge 1999
	Habitat selection	Refuge use	obs	O. *rusticus* excludes O. *virilis* from shelters	Taylor and Redmer 1996, Hill and Lodge 1999
		Refuge use	obs	O. *rusticus* and O. *propinquus* exclude O. *virilis* from shelters	Garvey et al. 1994
Pacifastacus fortis	Nutrition and growth	Dominance competition	like	Threatened by the introduction of *P. leniusculus* and O. *virilis*	Light et al. 1995
Procambarus acutus acutus	Habitat selection	Refuge use	obs	*P. clarkii* excludes *P. a. acutus* from shelter	Gherardi and Daniels 2004
Amphipoda					
Gammarus duebeni celticus	Nutrition and growth	Dominance competition	obs	*G. pulex* outcompetes *G. d. celticus* under certain environmental conditions	MacNeil et al. 2004
Cirripedia					
Balanus albicostatus	Nutrition and growth	Dominance competition	obs	*B. glandula* breeds at smaller size and settles earlier than *B. albicostatus*	Kado 2003

[a]Indication level: obs, observed: like, likely; pos, possible (see text for further explanation).

amphipod *Gammarus tigrinus*, despite suffering from intraguild predation by the native species *G. duebeni celticus*, outcompetes the native under certain physicochemical conditions (MacNeil et al. 2003).

Consequences of Stock Manipulation (with Focus on Fishing)

Crustacean fisheries generally act in two ways: indiscriminate harvesting with trawls (mostly demersal decapods, e.g., shrimps), and species- and size-selective extraction using traps (many large decapods, including some shrimps). Nonselective trawl fisheries can remove a large proportion of a stock, especially if it is highly aggregated (Pérez and Defeo 2003). Trap fisheries may selectively affect particular parts of the population, with preference for the largest individuals and, in most cases, with a strong bias toward males (Fig. 19.1). The consequences of such size- and sex-selective fisheries are discussed below.

Our review found that fishing primarily affects reproductive behavior of crustaceans (27 out of 29 observations, 93.1%; see Table 19.5). For example, removing

Figure 19.1 Lobsterman sorting the haul and measuring *Homarus americanus* lobsters; undersized lobsters and ovigerous females will be thrown overboard, while legal-sized lobsters are kept. Note the double gauge (lower right): oversized lobsters (lower left) will also be returned to the sea, ensuring that both large females and males remain in the population. Photographs by Carl Wilson, Maine Department of Marine Resources.

Table 19.5. Species and behaviors affected by stock manipulation.

Species Affected	Vital Function	Main Behavior	Specific Behavior	Indication[a]	Comment	
Brachyura						
Callinectes sapidus	Reproduction	Mating	Competition for mates	like	Increased participation by smaller males	Jivoff 2003
			Competition for mates	like	Spem limitation/depletion, sperm economy behavior	Kendall and Wolcott 1999, Kendall et. al. 2001, 2002
			Competition for mates	pos	Spem limitation/depletion, sperm economy behavior	Jivoff 1997
Cancer magister	Reproduction	Mating	Competition for mates	pos	Spem limitation/depletion, sperm economy behavior	Zhang et. al. 2004
Chionoecetes bairdi	Reproduction	Mating	Competition for mates	pos	Increased participation by smaller males	Paul 1992
			Competition for mates	pos	Spem limitation, small males unable to mate with large females	Donaldson and Adams 1989
Chionoecetes opilio	Reproduction	Mating	Competition for mates	like	Increased participation by smaller males	Sainte-Marie and Lovrich 1994, Sainte-Marie et al. 1997
			Mate location and selection	like	Possible change of seasonal timing of successful mating	Jamieson 2000
			Competition for mates	like	Sperm limitation, sperm economy behavior	Rondeau and Sainte-Marie 2001
Hapalogaster dentata	Reproduction	Fecundity and hatching success		like	Suggest minimum legal size be determined by the functional maturity	Goshima et al. 2000
Anomura						
Paralithodes camtschaticus	Reproduction	Mating	Competition for mates	pos	Increased participation by smaller males	Paul 1992
			Competition for mates	pos	Spem limitation/depletion, sperm economy behavior	Paul and Paul 1990, 1997

CARIDEA						
Crangon crangon	Reproduction	Sex change		pos	Intense exploitation may affect age timing of sex reversal	Boddeke et al. 1991
Pandalus borealis	Reproduction	Sex change		like	Intense exploitation may affect age timing of sex reversal	Jamieson 2000
Pandalus jordani	Reproduction	Sex change		like	Intense exploitation may affect age timing of sex reversal	Jamieson 2000
Pandalus platyceros	Reproduction	Sex change		like	Intense exploitation may affect age timing of sex reversal	Schlining and Spratt 2000
ASTACIDEA						
Homarus americanus	Reproduction	Mating	Copulation and spawning	pos	Sperm limitation, small males unable to mate with large females	Gosselin et al. 2003
	Nutrition and growth	Dominance competition		obs	Large individuals chases smaller ones away from traps	Jury et al. 2001
PALINURA						
Jasus edwardsii	Nutrition and growth	Foraging	Competition for food	obs	Large individuals more "catchable" than smaller ones	Frusher and Hoenig 2001
	Reproduction	Mating	Competition for mates	like	Spem limitation/depletion, sperm economy behavior	MacDiarmid and Butler 1999
Jasus lalandii	Reproduction	Mating	Competition for mates	pos	Spem limitation/depletion, sperm economy behavior	Newman and Pollock 1971, 1974
Palinurus elephas	Reproduction	Fecundity and hatching success		pos	Overall fecundity of population decreased	Secci et al. 2000
Panulirus argus	Reproduction	Mating	Competition for mates	like	Spem limitation/depletion, sperm economy behavior	MacDiarmid and Butler 1999
		Fecundity and hatching success		pos	Sperm limitation/depletion, economy behavior: very few egg-bearing females	Guzman and Tewfik 2004
Panulirus guttatus	Reproduction	Fecundity and hatching success		pos	Sperm limitation/depletion, economy behavior: very few egg-bearing females	Guzman and Tewfik 2004

[a]Indication level: obs, observed; like, likely; pos, possible (see text for further explanation).

only or mostly large males may instigate a cascade of undesired consequences, leading to reduced size and density of potential male mates as well as a female-biased sex ratio. This change in demography can produce behavioral changes: male sexual competition is relaxed, smaller and surviving larger males are exposed to more reproductive opportunities (Sainte-Marie and Lovrich 1994, Hankin et al. 1997, Jamieson 2000, Jivoff 2003), and female sexual competition increases. Thus, females may experience sperm limitation because of mating with smaller males (Diesel 1991, Waddy and Aiken 1991, Jivoff 1997, A.J. Paul and Paul 1997, MacDiarmid and Butler 1999, Kendall et al. 2001, Rondeau and Sainte-Marie 2001, Hinojosa and Thiel 2003) or with surviving males that either are sperm depleted (Sainte-Marie and Lovrich 1994, Kendall and Wolcott 1999, Kendall et al. 2001, 2002) or exhibit sperm economy behavior (MacDiarmid and Butler 1999, Rondeau and Sainte-Marie 2001). Eleven of 26 studies (42.3%) indicated that the sex ratio became female biased due to removal of large males. Furthermore, 12 of these 26 studies (46.2%) also reported that the size structure of the entire population was affected by preferential removal of large males (Table 19.6).

Lack of large males can result in sperm-limited populations. A field study of a commercially exploited population of blue crabs in Chesapeake Bay revealed that most males (50–90%) had extremely low vas deferens weights relative to laboratory-held males with fully recovered sperm reservoirs (Kendall et al. 2001). Female blue crabs that were fertilized by males that had recently mated twice had only one-third the brood hatching success of females fertilized by recently unmated males (Hines et al. 2003). The consequences of suboptimal rates in sperm transfer are that not all the eggs of a female may be fertilized. Such suppression of fertilization rates have been reported from natural populations of several crustacean species, including spiny lobsters (MacDiarmid and Butler 1999), snow crabs (Rondeau and Sainte-Marie 2001), and blue crabs (Hines et al. 2003, Jivoff 2003), and clearly have important implications for fishery production.

This sperm-limitation effect may be further aggravated because females of many species prefer sperm from large dominant males. For example, in the rock shrimp *Rhynchocinetes typus*, females mate preferentially with the large, dominant morph and are thought to selectively remove sperm from the small, subordinate morph (Thiel and Hinojosa 2003). If most large males were extracted by fishing activities, females would be forced to mate with small males, and possibly larger numbers of males, in order to satisfy their sperm requirements. Indeed, Gosselin et al. (2003) examined sperm receptacle loads in female American lobster (*Homarus americanus*) and found that wild-mated females had received sperm from smaller males than did laboratory-mated females. In an additional study, Gosselin et al. (2005) reported that the rate of multiple paternity for *H. americanus* was highest at the most exploited of three study sites from the Canadian coast. This latter study also underlines how molecular tools can elucidate how fishing may result in undesired selection among natural populations (see chapter 3). In summary, instead of optimizing the population's overall reproductive performance, the implementation of large-male extractions may induce a vicious circle that has an adverse effect on individual fitness as well as on important population parameters and life-history traits.

Current fishing practices may not only have short-term effects on important population parameters. Recent studies have demonstrated that natural selection can be extremely fast when affected by fishing (Zimmer 2003, Birkeland and Dayton 2005). For some brachyuran crabs (Kendall and Wolcott 1999, Abbe 2002, Lipcius and Stockhausen 2002, Jivoff 2003, Zhang et al. 2004) and caridean shrimp (Jensen 1965,

Table 19.6. Species and their population parameters and life history traits affected by Stock Manipulation.

Species Affected	Population Parameters[a]	Indication[c]	Life History[b]	Indication[c]	Comment	References
BRACHYURA						
Callinectes sapidus	a,b	like	s	like	Large males removed	Jivoff 2003
	b	obs	—	—	Large males removed	Abbe 2002
	b,c	obs	s	—	Large males removed	Lipcius and Stockhausen 2002
	a,b	pos	s	like	Large males removed	Kendall and Wolcott 1999
Cancer magister	a	like	—	—	Large males removed	Hankin et al. 1997
	a,b	pos	s	pos	Large males removed	Zhang et al. 2004
Chionoecetes opilio	a	like	e,s	pos	Large males removed	Rondeau and Sainte-Marie 2001
	a,b	like	e,s	like	Large males removed	Sainte-Marie et al. 1995
	a	like	—	—	Large males removed	Sainte-Marie and Lovrich 1994
ANOMURA						
Paralithodes camtschaticus	a	like	—	—	Decreased reproductive output	Paul and Paul 1990
CARIDEA						
Pandalus borealis	—	—	s	pos	Decrease in mean size	Jensen 1965, 1967 as cited in Bergström 2000
Pandalus platyceros	—	—	s	pos	Size selective fishery	Schlining and Spratt 2000
PALINURA						
Jasus edwardsii	b	obs	—	—	Large individuals removed	Frusher and Hoenig 2001
	a,b	like	—	—	Large individuals removed	MacDiarmid and Butler 1999
Jasus lalandii	b	like	—	—	Large individuals removed	Newman and Pollock 1971, 1974
Palinurus elephas	b	obs	—	—	Large individuals removed	Secci et al. 2000
Panulirus argus	a,b,c	like	—	—	Large individuals removed	Guzman and Tewfik 2004
Panulirus guttatus	a,b,c	like	—	—	Large individuals removed	Guzman and Tewfik 2004

[a] a, sex ratio; b, size structure; c, population density.

[b] Life history traits: e, earlier adulthood s, smaller at adulthood.

[c] Indication level: obs, observed; like, likely; pos, possible (see text for further explanation).

1967 as cited in Bergström 2000; Schlining and Spratt 2000), data suggest that individuals reach adulthood at smaller sizes compared to the time before populations were affected by intense fishing pressure (Table 19.6). Similarly, snow crabs (*Chionoecetes opilio*) are becoming sexually mature at a younger age than before intensive fishing started (Sainte-Marie et al. 1995, Rondeau and Sainte-Marie 2001). Thus, the application of minimum size limits, which induces selection for smaller individuals (i.e., unwanted phenotypes), appears to produce undesired population and possibly ecosystem consequences through rapid evolutionary change (Jamieson 2000, Grift et al. 2003). Becoming sexually mature while small gives young individuals more chances to reproduce before being killed, and hence, the overall reproductive success of these phenotypes in a population increases (Jamieson 2000, Zimmer 2003). As a consequence of this evolutionary change, the biomass and the total reproductive potential of the population as a whole may be reduced since older individuals often produce more viable larvae than do younger individuals (for discussion, see Birkeland and Dayton 2005).

Fishing pressure, especially when selective for size and sex, mostly affects mating interactions adversely and alters selective pressures on alternative life histories (Vincent and Sadovy 1998). Thus, fishing induces changes in behavior (or promotes expression or success of extreme behaviors) through alterations on the population level (i.e., population parameters, life history traits; see Tables 19.5, 19.6). However, bottom trawling is also likely to cause indirect changes in behavior at the ecosystem level, through habitat destruction (Watling and Norse 1998, Jennings et al. 2001b, Kaiser et al. 2002), and direct changes in behavior at the individual level by eliciting strong avoidance behavior as has been observed in fishes (Ryer 2004).

Habitat Destruction

Habitat destruction has far-reaching ecological consequences and is accelerating in estuaries (Kennish 2002), coral reefs (Richmond 1993), and many other habitats (Walker and Kendrick 1998) due to the continuing increase of human populations on the coast. In freshwater habitats, river impoundments (through dams and canalization) restrict the completion of life cycles of crayfishes and river shrimp (Light et al. 1995, Bowles et al. 2000). Dams pose formidable barriers to up- and downstream migration necessary for completion of the amphidromous life cycle of the river shrimp *Macrobrachium* species (Bowles et al. 2000). For example, larval river shrimp can survive only a few days in fresh water and must reach estuaries quickly (Bowles et al. 2000). Probably, water impoundment and introduction of exotic species are the main factors explaining why 36% of the native crayfish species from North America are extinct or endangered (Light et al. 1995).

Destruction or alteration of important habitat may impede crustaceans' ability to find proper refuges. For example, Steger (1987) found that any manipulation of rubble directly affected the population density of stomatopods (*Gonodactylus* spp.), probably because of increased risk of predation when available refuges were scarce. Alterations of refuge-structuring habitat may affect the crayfish *O. propinquus* in a similar way (Stein and Magnuson 1976).

It is well known that habitat loss, degradation, and fragmentation can have severe impacts on abundance and species richness, but little appears to be known about how these anthropogenic stressors affect animal behavior. In our review, only one study actually reported such behavioral effects (Table 19.7), which probably reflects the

Table 19.7. Species and behaviors affected by habitat destruction.

Species Affected	Vital Function	Main Behavior	Specific Behavior	Indication[a]	Comment	References
BRACHYURA						
Callinectes sapidus	Nutrition and growth	Foraging	Food location and selection	like	Removal/destruction of salt marshes may impede foraging/habitat use	Micheli and Peterson 1999, Jivoff and Able 2003
	Habitat selection	Settlement		like	Reduction in seagrass cover is likely to reduce juvenile settlement success	Stockhausen and Lipcius 2003
Chionoecetes bairdi	Nutrition and growth	Aggregation		pos	Habitat destruction could affect important mass molting sites	Stone 1999
CARIDEA						
Macrobrachium carcinus	Habitat selection	Migration		like	Dams impede upstream migration necessary to complete amphidromous life cycle	Bowles et al. 2000
Macrobrachium ohione	Habitat selection	Migration		like	Dams impede upstream migration necessary to complete amphidromous life cycle	Bowles et al. 2000
PALINURA						
Panulirus japonicus	Habitat selection	Settlement		pos	Habitat destruction may increase predation risk if shelter becomes limited	Yoshimura and Yamakawa 1988
ASTACIDEA						
Homarus americanus	Habitat selection	Settlement		pos	Habitat destruction may reduce suitable shelter, leading to increased predation risk	Wahle and Steneck 1991
Orconectes propinquus	Habitat selection	Refuge use		pos	Habitat destruction may reduce suitable shelter, leading to increased predation risk	Stein and Magnuson 1976
Pacifastacus fortis	Habitat selection	All		obs	Heavy decline in abundance do to water impoundment and diversion	Light et al. 1995

(*Contd.*)

Table 19.7. (*Contd.*)

Species Affected	Vital Function	Main Behavior	Specific Behavior	Indication[a]	Comment	References
ANOSTRACA						
Branchinecta sp	Reproduction	Mating	Mate location and selection	like	Habitat degradation and climate change may affect formation of vemal pools	Pyke 2005
Linderiella occidentalis	Reproduction	Mating	Mate location and selection	like	Habitat degradation and climate change may affect formation of vemal pools	Pyke 2005
STOMATOPODA						
Gonodactylus austrinus	Habitat selection	Refuge use		pos	Alterations/destruction of rubble (habitat) may cause increased predation risk	Steger 1987
Gonodactylus bredini	Habitat selection	Refuge use		pos	Alterations/destruction of rubble (habitat) may cause increased predation risk	Steger 1987
Gonodactylus oerstedii	Habitat selection	Refuge use		pos	Alterations/destruction of rubble (habitat) may cause increased predation risk	Steger 1987

[a] Indication level: obs, observed; like, likely; pos, possible (see text for further explanation).

fact that any form of habitat destruction mostly affects animal behavior in an indirect manner. However, habitat destruction has the potential to affect behavior at both the population and individual levels, especially if important habitats for recruitment, sheltering, mating, and foraging are degraded.

How Anthropogenic Stressors Affect Crustacean Behavior

Our review suggests that anthropogenic stressors affect the behavior of crustaceans at three different levels: at the ecosystem level (e.g., habitat, physicochemical parameters, species dominance, species composition), at the population level (e.g., population parameters, life-history traits), and at the individual level (parasites, hormones, health). We found clear differences in the ways in which the four main anthropogenic activities affect behaviors (Fig. 19.2). Pollution and species introduction effects occur primarily on the individual level, and they should thus be expected to cause direct changes, modifications, or impediments to particular behaviors. In contrast, fisheries and habitat destruction primarily act on the population and ecosystem level and should thus be expected to affect crustacean behaviors in a more indirect manner. Interestingly, this supports the division of the anthropogenic stressors into two main groups based on the level of indication (Table 19.2). The direct stressors (pollution and species introduction) produced many observed effects, which is understandable given that they affect behaviors at the individual level. The indirect stressors (fishing and habitat destruction) produced more circumstantial evidence of affecting behavior (Table 19.2), which also

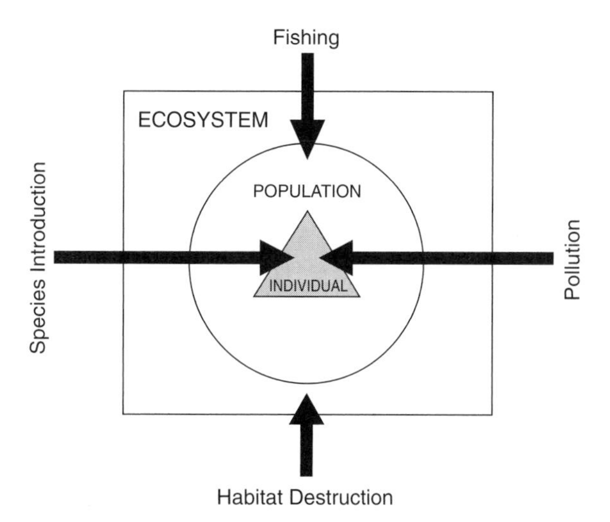

Figure 19.2 Conceptual figure indicating at which level, either ecosystem (square), population (circle), or individual (triangle), the different anthropogenic stressors affect behavior. Arrows point into the level they affect. Pollution and species introduction affect behavior directly at the individual level (e.g., contaminants affecting foraging or mating and introduced species interacting directly with native species), while fishing and habitat destruction affect behavior indirectly (e.g., through changes in population parameters and degradation of habitats) at higher levels (population and ecosystem, respectively).

seems reasonable since they rarely affect the behavior of individuals directly, but rather produce changes at a higher level (i.e., population or ecosystem).

While there appear to be distinct differences in the way that anthropogenic stressors affect the behavior of crustaceans, some stressors may have similar effects on the same behavior or population parameter even though mechanisms may differ substantially. For example, both fishing and pollution can produce female-biased sex ratios, with potential consequences for selection toward small male size and suppression of reproductive potential. At present, pollution causing female-biased sex ratios has been reported only for noncommercial crustacean species, while fishing causing female-biased sex ratio naturally is known only from commercially exploited species. It should not be excluded, though, that both pollution and stock manipulation may affect the same species simultaneously, leading to an aggravation of adverse effects.

In conjunction with other anthropogenic stressors, fishing may lead to greater changes in population size. For example, certain harvest methods, in addition to causing adverse effects on the population level, lead to devastating habitat alterations (Watling and Norse 1998, Kaiser et al. 2002). This could probably impede both successful mate finding and survival. Moreover, in areas where severe hypoxia occurs frequently, catch per unit effort (CPUE) may be directly related to an organism's behavioral responses to low oxygen concentrations. Baden et al. (1990) reported that severe hypoxia induced Norway lobsters to emerge from their burrows, which resulted in a temporary increase in CPUE. Fisheries and pollution may be beneficial to some crustaceans, as well. Guilds of crustacean scavengers may benefit from increased food resources from animals killed by pollution (e.g., Britton and Morton 1994). Synergistic effects between multiple anthropogenic stressors have also been reported for other crustaceans. For example, the native amphipod *Gammarus duebeni celticus* is not affected by the invasive amphipod *G. pulex* under natural conditions. However, with decreasing water quality, the exotic amphipod preys upon the native and ultimately replaces it (MacNeil et al. 2004). It remains to be tested whether similar interactive effects between multiple stressors can affect crustacean behavior in a synergistic way in other species, as well. Given the above examples, we consider it likely that many crustacean species are subjected to more than one of the above-mentioned stressors (see also Freire et al. 2002). This may not only complicate the identification of the causes but also expose the species to higher risks if anthropogenic stressors continue to increase their impact.

Future Directions

This literature review revealed that a wide range of behaviors of aquatic crustaceans is affected by the four principal anthropogenic stressors, but other behavioral aspects have received no or little attention so far. For example, pollution should adversely affect important behaviors related to recruitment. An example is provided by larvae of the blue crab (*Callinectes sapidus*), which during settlement are attracted to odors from seagrass beds and repelled from odors emitted by potential predators (Welch et al. 1997). Odor disruption by contaminants may inhibit larvae in finding adequate sites or locations for settlement. In a similar way, habitat destruction may degrade or change important sites for settlement of juveniles (Yoshimura and Yamakawa 1988, Wahle and Steneck 1991, Stockhausen and Lipcius 2003), sheltering (Stein and

Magnuson 1976, Steger 1987), aggregation (Stone 1999), and foraging (Micheli and Peterson 1999). Juveniles with specific habitat preferences may experience demographic bottlenecks if important sites for settlement are altered by anthropogenic stressors (Stockhausen and Lipcius 2003). Simulations by Stockhausen and Lipcius (2003) indicated that a 70% loss of seagrass habitat caused a 40–45% reduction in blue crab recruitment. Juveniles of the blue crab, as well as of *Homarus americanus* (Wahle and Steneck 1991) and *Panulirus japonicus* (Yoshimura and Yamakawa 1988), are likely to suffer from increased predation risk if shelter-providing habitat is destroyed. At present, we know little about how contamination and habitat destruction can lead to failures of recruitment. For example, reduction of available refuges may lead to increased competition for these and may cause increased levels of aggressive interactions and potentially competitive exclusion of some individuals.

The removal of large males due to harvesting appears to decrease both age and size at maturity (Jamieson 2000, Zimmer 2003), and the resulting change in sex ratios could modify the whole mating system of a species. Forsgren et al. (2004) observed that changes in the operational sex ratio (OSR) in the two-spotted goby (*Gobiusculus flavescens*) during the breeding season resulted in a reversal of sex roles. Early in the breeding season, the OSR was male biased, and fierce male–male competition and male courtship behavior were observed. This pattern, however, changed later in the breeding season as the OSR became female biased, which resulted in intensive female–female competition and females courting males. As revealed herein, EDCs and fishing can induce dramatic shifts in sex ratios. New insights into how fisheries- and contamination-induced alterations of OSR in crustaceans affect sex roles and sexual selection would prove valuable and have consequences for future management plans of fisheries.

Although more information on the effects of fishing on crustacean behavior is needed in order to better grasp its consequences, present knowledge of how fishing affects crustacean mating behavior and success appears to be more complete than in other important fisheries resources such as fishes. Rowe and Hutchings (2003) emphasized that in fishes there is a pressing need to increase the knowledge of mate competition, mate choice, and other mating components that may be negatively affected by intense fishing. In crustaceans, there is ample circumstantial and experimental evidence that mate competition relaxes when large males are selectively removed increasing the participation of smaller males (e.g., Sainte-Marie et al. 1995, Jivoff 2003; see also Table 19.5). Given the available information on crustacean mating interactions and how they are affected by intense fishing, crustaceans (especially crabs and lobsters) may prove good model species for other commercial taxa.

Several anthropogenic stressors affect crustacean behavior indirectly, making it difficult to identify how they are caused. How can indirect effects on behavior be detected at an earlier stage (i.e., before a population has reached a point of no return)? At the ecosystem level, biomarkers (which indicate or measure a biological process, e.g., levels of a specific protein in blood or genetic mutations; see Fingerman et al. 1998, Allen and Moore 2004, Marsden and Rainbow 2004, Verslycke et al. 2004) are used in order to determine in which state a particular system finds itself. Among the most efficient tools in identifying how human activities affect crustacean behaviors are behavioral studies themselves (see, e.g., Jury et al. 2001). Most crustacean species are well suited for observational studies, because (1) their limited home ranges and mobility permit field studies in their natural habitats and (2) their intermediate size and easy maintenance in laboratory environments facilitate experimental studies. We therefore consider crustaceans to be ideal model organisms that can be used to study how anthropogenic stressors affect the behavior, population dynamics, and diversity of aquatic (and terrestrial) organisms.

We have focused on the impact of human activities on the behavior of aquatic crustaceans, but it is very likely that terrestrial crustaceans are affected in similar ways. Species introduction and habitat destruction may severely affect terrestrial crustaceans on islands and in rain forests (e.g., Sodhi et al. 2004). Contaminants and selective extraction can have similar effects on the behavior of terrestrial crustaceans (e.g., Chauvet and Kadiri-Jan 1999) as revealed here for aquatic crustaceans. Future studies addressing the issues treated in this review can reveal important aspects that may serve as valuable tools for the conservation of species and hence should be implemented in management of marine protected areas and coastal zones and their counterparts in the terrestrial environment.

Summary

Four groups of anthropogenic stressors affect the behavior of crustaceans in aquatic ecosystems: pollution, introduction of NIS, stock manipulation, and habitat destruction. We recognized several important behaviors that have been affected by these stressors. The most affected vital functions were, in hierarchical order, reproduction, nutrition and growth, and habitat selection. Based on the type of evidence provided in the published studies, two principal groups of stressors are distinguished: (1) pollution and species introduction, for which most of the effects on crustacean behavior were observed directly, and (2) fishing and habitat destruction, for which most of the evidence was circumstantial. In many cases, pollution and species introduction affect individuals directly, whereas fishing and habitat destruction affect crustacean behaviors indirectly, primarily by changes in important population (e.g., sex ratio and population density) or ecosystem (e.g., habitat, physicochemical, species dominance, species composition) parameters. Interestingly, some studies showed that effects became evident only when a species was synergistically affected by two anthropogenic stressors (e.g., pollution and interference by introduced species). Based on these results, we suggest that crustaceans and their behavioral responses may serve as useful indicators to identify the impact of anthropogenic stressors on aquatic (and terrestrial) ecosystems. The information yielded by this review and that of future studies may prove useful in designing efficient conservation plans.

Acknowledgments We are grateful to the Faculty of Marine Science of Universidad Catolica del Norte for providing office space during the research stay of T.C. van Son. This chapter benefited substantially from extensive comments made by B. Sainte-Marie and J.E. Duffy.

References

Abbe, G.R. 2002. Decline in size of male blue crabs (*Callinectes sapidus*) from 1968 to 2000 near Calvert Cliffs, Maryland. Estuaries 25:105–114.

Allen, J.I., and M.N. Moore. 2004. Environmental prognostics: is the current use of biomarkers appropriate for environmental risk evaluation? Marine Environmental Research 58:227–232.

Baden, S.P., L. Pihl, and R. Rosenberg. 1990. Effects of oxygen depletion on the ecology, blood physiology and fishery of the Norway lobster *Nephrops norvegicus*. Marine Ecology Progress Series 67:141–155.

Bannister, R.C.A., and J.T. Addison. 1998. Enhancing lobster stocks: a review of recent European methods, results, and future prospects. Bulletin of Marine Science 62:369–387.

Barbaresi, S., and F. Gherardi. 2000. The invasion of the alien crayfish *Procambarus clarkii* in Europe, with particular reference to Italy. Biological Invasions 2:259–264.

Bell, G.W., D.B. Eggleston, and T.G. Wolcott. 2003a. Behavioral responses of free-ranging blue crabs to episodic hypoxia. I. Movement. Marine Ecology Progress Series 259:215–225.

Bell, G.W., D.B. Eggleston, and T.G. Wolcott. 2003b. Behavioral responses of free-ranging blue crabs to episodic hypoxia. II. Feeding. Marine Ecology Progress Series 259:227–235.

Bergström, B.I. 2000. The biology of *Pandalus*. Advances in Marine Biology 38:55–245.

Birkeland, C., and P.K. Dayton. 2005. The importance in fishery management of leaving the big ones. Trends in Ecology and Evolution 20:356–358.

Blockwell, S.J., S.J. Maund, and D. Pascoe. 1998. The acute toxicity of lindane to *Hyalella azteca* and the development of a sublethal bioassay based on precopulatory guarding behavior. Archives of Environmental Contamination and Toxicology 35:432–440.

Boddeke, R., J.R. Bosschieter, and P.C. Goudswaard. 1991. Sex change, mating and sperm transfer in *Crangon crangon* (L.). Pages 164–182 *in*: R.T. Bauer and J.W. Martin, editors. Crustacean sexual biology. Columbia University Press, New York.

Bouchard, S., B. Sainte-Marie, and J.N. McNeil. 1996. Indirect evidence indicates female semiochemicals release male precopulatory behavior in the snow crab, *Chionoecetes opilio* (Brachyura: Majidae). Chemoecology 7:39–44.

Bowles, D.E., K. Aziz, and C.L. Knight. 2000. *Macrobrachium* (Decapoda: Caridea: Palaemonidae) in the contiguous United States: a review of the species and an assessment of threats to their survival. Journal of Crustacean Biology 20:158–171.

Breithaupt, T., D.P. Lindstrom, and J. Atema. 1999. Urine release in freely moving catheterised lobsters (*Homarus americanus*) with reference to feeding and social activities. Journal of Experimental Biology 202:837–844.

Britton, J.C., and B. Morton. 1994. Marine carrion and scavengers. Oceanography and Marine Biology: An Annual Review 32:369–434.

Butterworth, K.G., M.K. Grieshaber, and A.C. Taylor. 2004. Behavioral and physiological responses of the Norway lobster, *Nephrops norvegicus* (Crustacea: Decapoda), to sulphide exposure. Marine Biology 144:1087–1095.

Cantelmo, A.C., R.J. Lazell, and L.H. Mantel. 1981. The effects of benzene on molting and limb regeneration in juvenile *Callinectes sapidus* Rathbun. Marine Biology Letters 2:333–343.

Chauvet, C., and T. Kadiri-Jan. 1999. Assessment of an unexploited population of coconut crabs, *Birgus latro* (Linne, 1767) on Taiaro atoll (Tuamotu archipelago, French Polynesia). Coral Reefs 18:297–299.

Clare, A.S., J.D. Costlow, H.M. Bedair, and G. Lumb. 1992. Assessment of crab limb regeneration as an assay for developmental toxicity. Canadian Journal of Fisheries and Aquatic Sciences 49:1268–1273.

Clotfelter, E.D., A.M. Bell, and K.R. Levering. 2004. The role of animal behavior in the study of endocrine-disrupting chemicals. Animal Behaviour 68:665–676.

Cohen, A.N., and J.T. Carlton. 1998. Accelerating invasion rate in a highly invaded estuary. Science 279:555–558.

Conradi, M., and M.H. Depledge. 1998. Population responses of the marine amphipod *Corophium volutator* (Pallas, 1766) to copper. Aquatic Toxicology 44:31–45.

Depledge, M.H., and Z. Billinghurst. 1999. Ecological significance of endocrine disruption in marine invertebrates. Marine Pollution Bulletin 39:32–38.

Diesel, R. 1991. Sperm competition and the evolution of mating behavior in Brachyura, with special reference to spider crabs (Decapoda, Majidae). Pages 145–163 *in*: R.T. Bauer and J.W. Martin, editors. Crustacean sexual biology. Columbia University Press, New York.

Donaldson, W.E., and A.E. Adams. 1989. Ethogram of behavior with emphasis on mating for the tanner crab *Chionoecetes bairdi* Rathbun. Journal of Crustacean Biology 9:37–53.

Eriksson, S.P., and S.P. Baden. 1997. Behavior and tolerance to hypoxia in juvenile Norway lobster (*Nephrops norvegicus*) of different ages. Marine Biology 128:49–54.

Fingerman, M., N.C. Jackson, and R. Nagabhushanam. 1998. Hormonally-regulated functions in crustaceans as biomarkers of environmental pollution. Comparative Biochemistry and Physiology 120C:343–350.

Ford, A.T., T.F. Fernandes, S.A. Rider, P.A. Read, C.D. Robinson, and I.M. Davies. 2004a. Endocrine disruption in a marine amphipod? Field observations of intersexuality and de-masculinisation. Marine Environmental Research 58:169–173.

Ford, A.T., T.F. Fernandes, P.A. Read, C.D. Robinson, and I.M. Davies. 2004b. The costs of intersexuality: a crustacean perspective. Marine Biology 145:951–957.

Forsgren, E., T. Amundsen, Å.A. Borg, and J. Bjelvenmark. 2004. Unusually dynamic sex roles in a fish. Nature 429:551–554.

Freire, J., C. Bernárdez, A. Corgos, L. Fernández, E. González-Gurriarán, M.P. Sampedro, and P. Verísimo. 2002. Management strategies for sustainable invertebrate fisheries in coastal ecosystems of Galicia (NW Spain). Aquatic Ecology 36:41–50.

Frusher, S.D., and J.M. Hoenig. 2001. Impact of lobster size on selectivity of traps for southern rock lobster (*Jasus edwardsii*). Canadian Journal of Fisheries and Aquatic Sciences 58:2482–2489.

Galil, B.S. 2000. Lessepsian immigration: human impact on Levantine biogeography. Crustacean Issues 12:47–54.

Garvey, J.E., R.A. Stein, and H.M. Thomas. 1994. Assessing how fish predation and interspecific prey competition influence a crayfish assemblage. Ecology 75:532–547.

Gherardi, F., and W.H. Daniels. 2004. Agonism and shelter competition between invasive and indigenous crayfish species. Canadian Journal of Zoology 82:1923–1932.

Gilbert, D., B. Sundby, C. Gobeil, A. Mucci, and G.H. Tremblay. 2005. A seventy-two-year record of diminishing deep-water oxygen in the St. Lawrence estuary: the Northwest Atlantic connection. Limnology and Oceanography 50:1654–1666.

Goshima, S., M. Kanazawa, K. Yoshino, and S. Wada. 2000. Maturity in male stone crab *Hapalogaster dentata* (Anomura: Lithodidae) and its application for fishery management. Journal of Crustacean Biology 20:641–646.

Gosselin, T., B. Sainte-Marie, and L. Bernatchez. 2003. Patterns of sexual cohabitation and female ejaculate storage in the American lobster (*Homarus americanus*). Behavioral Ecology and Sociobiology 55:151–160.

Gosselin, T., B. Sainte-Marie, and L. Bernatchez. 2005. Geographic variation of multiple paternity in the American lobster, *Homarus americanus*. Molecular Ecology 14:1517–1525.

Griffiths, S.W., P. Collen, and J.D. Armstrong. 2004. Competition for shelter among over-wintering signal crayfish and juvenile Atlantic salmon. Journal of Fish Biology 65:436–447.

Grift, R.E., A.D. Rijnsdorp, S. Barot, M. Heino, and U. Dieckmann. 2003. Fisheries induced trends in reaction norms for maturation in North Sea plaice. Marine Ecology Progress Series 257:247–257.

Grosholz, E.D., and G.M. Ruiz. 1996. Predicting the impact of introduced marine species: lessons from the multiple invasions of the European green crab *Carcinus maenas*. Biological Conservation 78:59–66.

Guzman, H.M., and A. Tewfik. 2004. Population characteristics and co-occurrence of three exploited decapods (*Panulirus argus, P. guttatus* and *Mithrax spinosissimus*) in Bocas del Toro, Panama. Journal of Shellfish Research 23:575–580.

Hagerman, L., and S.P. Baden. 1988. *Nephrops norvegicus*: field study of effects of oxygen deficiency on haemocyanin concentration. Journal of Experimental Marine Biology and Ecology 116:135–142.

Hankin, D.G., T.H. Butler, P.W. Wild, and Q.L. Xue. 1997. Does intense fishing on males impair mating success of female Dungeness crabs? Canadian Journal of Fisheries and Aquatic Sciences 54:655–669.

Hebel, D.K., M.B. Jones, and M.H. Depledge. 1997. Responses of crustaceans to contaminant exposure: a holistic approach. Estuarine, Coastal and Shelf Science 44:177–184.

Hill, A.M., and D.M. Lodge. 1999. Replacement of resident crayfishes by an exotic crayfish: the roles of competition and predation. Ecological Applications 9:678–690.

Hines, A.H., P. Jivoff, P. Bushmann, J. van Montfrans, S. Reed, D.L. Wolcott, and T.G. Wolcott. 2003. Evidence for sperm limitation in the blue crab, *Callinectes sapidus*. Bulletin of Marine Science 72:287–310.

Hinojosa, I., and M. Thiel. 2003. Somatic and gametic resources in male rock shrimp, *Rhynchocinetes typus*—effect of mating potential and ontogenetic male stage. Animal Behaviour 66:449–458.

Holdich, D.M. 2000. The introduction of alien crayfish species into Britain for commercial exploitation—an own goal? Crustacean Issues 12:85–97.

Hutchinson, T.H. 2002. Reproductive and developmental effects of endocrine disrupters in invertebrates: in vitro and in vivo approaches. Toxicology Letters 131:75–81.

Jamieson, G.S. 2000. Selective effects of fishing on the population dynamics of crustaceans. Crustacean Issues 12:627–641.

Jennings, S., M.J. Kaiser, and J.D. Reynolds. 2001a. Marine Fisheries Ecology. Blackwell, Oxford.

Jennings, S., J.K. Pinnegar, N.V.C. Polunin, and K.J. Warr. 2001b. Impacts of trawling disturbance on the trophic structure of benthic invertebrate communities. Marine Ecology Progress Series 213:127–142.

Jivoff, P. 1997. Sexual competition among male blue crab, *Callinectes sapidus*. Biological Bulletin 193:368–380.

Jivoff, P. 2003. A review of male mating success in the blue crab, *Callinectes sapidus*, in reference to the potential for fisheries-induced sperm limitation. Bulletin of Marine Science 72:273–286.

Jivoff, P., and K.W. Able. 2003. Blue crab, *Callinectes sapidus*, response to the invasive common reed, *Phragmites australis*: abundance, size, sex ratio, and molting frequency. Estuaries 26:587–595.

Johansson, B. 1997. Behavioral response to gradually declining oxygen concentration by Baltic Sea macrobenthic crustaceans. Marine Biology 129:71–78.

Jones, J.C., and J.D. Reynolds. 1997. Effects of pollution on reproductive behavior of fishes. Reviews in Fish Biology and Fisheries 7:463–491.

Jury, S.H., H. Howell, D.F. O'Grady, and W.H. Watson. 2001. Lobster trap video: in situ video surveillance of the behavior of *Homarus americanus* in and around traps. Marine and Freshwater Research 52:1125–1132.

Kado, R. 2003. Invasion of Japanese shores by the NE Pacific barnacle *Balanus glandula* and its ecological and biogeographical impact. Marine Ecology Progress Series 249:199–206.

Kaiser, M.J., J.S. Collie, S.J. Hall, S. Jennings, and I.R. Poiner. 2002. Modifications of marine habitats by trawling activities: prognosis and solutions. Fish and Fisheries 3:114–136.

Kendall, M.S., and T.G. Wolcott. 1999. The influence of male mating history on male-male competition and female choice in mating associations in the blue crab, *Callinectes sapidus* (Rathbun). Journal of Experimental Marine Biology and Ecology 239:23–32.

Kendall, M.S., D.L. Wolcott, T.G. Wolcott, and A.H. Hines. 2001. Reproductive potential of individual male blue crabs, *Callinectes sapidus*, in a fished population: depletion and recovery of sperm numbers and seminal fluid. Canadian Journal of Fisheries and Aquatic Sciences 58:1168–1177.

Kendall, M.S., D.L. Wolcott, T.G. Wolcott, and A.H. Hines. 2002. Influence of male size and mating history on sperm content of ejaculates of the blue crab *Callinectes sapidus*. Marine Ecology Progress Series 230:235–240.

Kennish, M.J. 2002. Environmental threats and environmental future of estuaries. Environmental Conservation 29:78–107.

Lafferty, K.D., and A.M. Kuris. 1996. Biological control of marine pests. Ecology 77:1989–2000.

Light, T., D.C. Erman, C. Myrick, and J. Clarke. 1995. Decline of the Shasta crayfish (*Pacifastacus fortis* Faxon) of northeastern California. Conservation Biology 9:1567–1577.

Lipcius, R.N., and W.T. Stockhausen. 2002. Concurrent decline of the spawning stock, recruitment, larval abundance, and size of the blue crab *Callinectes sapidus* in Chesapeake Bay. Marine Ecology Progress Series 226:45–61.

Lodge, D.M., C.A. Taylor, D.M. Holdich, and J. Skurdal. 2000. Reducing impacts of exotic crayfish introductions: new policies needed. Fisheries 25:21–23.

Lye, C.M., M.G. Bentley, A.S. Clare, and E.M. Sefton. 2005. Endocrine disruption in the shore crab *Carcinus maenas*—a biomarker for benthic marine invertebrates. Marine Ecology Progress Series 288:221–232.

MacDiarmid, A.B., and M.J. Butler IV. 1999. Sperm economy and limitation in spiny lobsters. Behavioral Ecology and Sociobiology 46:14–24.

MacNeil, C., E. Bigsby, J.T.A. Dick, M.J. Hatcher, and A.M. Dunn. 2003. Differential physico-chemical and intraguild predation among native and invasive amphipods; a field study. Archiv für Hydrobiologie 156:165–179.

MacNeil, C., J. Prenter, M. Briffa, N.J. Fielding, J.T.A. Dick, G.E. Riddell, M.J. Hatcher, and A.M. Dunn. 2004. The replacement of a native freshwater amphipod by an invader: roles for environmental degradation and intraguild predation. Canadian Journal of Fisheries and Aquatic Sciences 61:1627–1635.

Marsden, I.D., and P.S. Rainbow. 2004. Does the accumulation of trace metals in crustaceans affect their ecology—the amphipod example? Journal of Experimental Marine Biology and Ecology 300:373–408.

Micheli, F., and C.H. Peterson. 1999. Estuarine vegetated habitats as corridors for predator movements. Conservation Biology 13:869–881.

Michels, E., S. Semsari, C. Bin, and L. De Meester. 2000. Effect of sublethal doses of cadmium on the phototactic behavior of *Daphnia magna*. Ecotoxicology and Environmental Safety 47:261–265.

Newman, G.G., and D.E. Pollock. 1971. Biology and migration of rock lobster *Jasus lalandii* and their effect on availability at Elands Bay, South Africa. South African Division of Sea Fisheries, Investigational Report 94. Republic of South Africa, Department of Industries, Cape Town, South Africa.

Newman, G.G., and D.E. Pollock. 1974. Biological cycles, maturity and availability of rock lobster *Jasus lalandii* on two South African fishing grounds. South African Division of Sea Fisheries, Investigational Report 107. Republic of South Africa, Department of Industries, Cape Town, South Africa.

Ohij, M., T. Arai, and N. Miyazaki. 2002. Effects of tributyltin exposure in the embryonic stage on sex ratio and survival rate in the caprellid amphipod *Caprella danilevskii*. Marine Ecology Progress Series 235:171–176.

Ohij, M., T. Arai, and N. Miyazaki. 2004. Effects of tributyltin on the survival in the caprellid amphipod *Caprella danilevskii*. Journal of Marine Biological Association of the United Kingdom 84:345–349.

Paoletti, M.G., and C.M. Cantarino. 2002. Sex ratio alterations in terrestrial woodlice populations (Isopoda: Oniscidea) from agroecosystems subjected to different agricultural practices in Italy. Applied Soil Ecology 19:113–120.

Parker, J.D., D.E. Burkepile, and M.E. Hay. 2006. Opposing effects of native and exotic herbivores on plant invasions. Science 311:1459–1461.

Paul, A.J. 1992. A review of size at maturity in male tanner (*Chionoecetes bairdi*) and king (*Paralithodes camtschaticus*) crabs and the methods used to determine maturity. American Zoologist 32:534–540.

Paul, A.J., and J.M. Paul. 1997. Breeding success of large male red king crab *Paralithodes camtschaticus* with multiparous mates. Journal of Shellfish Research 16:379–381.

Paul, J.M., and A.J. Paul. 1990. Breeding success of sublegal size male red king crab *Paralithodes camtschatica* (Decapoda, Lithodidae). Journal of Shellfish Research 9:29–32.

Pérez, E.P., and O. Defeo. 2003. Time-space variation in the catchability coefficient as a function of catch per unit of effort in *Heterocarpus reedi* (Decapoda, Pandalidae) in north-central Chile. Interciencia 28:178–182.

Perry, W.L., J.L. Feder, and D.M. Lodge. 2001. Implications of hybridization between introduced and resident *Orconectes* crayfishes. Conservation Biology 15:1656–1666.

Perry, W.L., D.M. Lodge, and J.L. Feder. 2002. Importance of hybridization between indigenous and nonindigenous freshwater species: an overlooked threat to North American biodiversity. Systematic Biology 51:255–275.

Pyke, C.R. 2005. Interactions between habitat loss and climate change: implications for fairy shrimp in the Central Valley ecoregion of California, USA. Climatic Change 68:199–218.

Richmond, R.H. 1993. Coral reefs—present problems and future concerns resulting from anthropogenic disturbances. American Zoologist 33:524–536.

Roast, S.D., J. Widdows, and M.B. Jones. 2000. Disruption of swimming in the hyperbenthic mysid *Neomysis integer* (Peracarida: Mysidacea) by the organophosphate pesticide chlorpyrifos. Aquatic Toxicology 47:227–241.

Roast, S.D., J. Widdows, and M.B. Jones. 2002. Distribution and swimming behavior of *Neomysis integer* (Peracarida: Mysidacea) in response to gradients of dissolved oxygen following exposure to cadmium at environmental concentrations. Marine Ecology Progress Series 237:185–194.

Rondeau, A., and B. Sainte-Marie. 2001. Variable mate-guarding time and sperm allocation by male snow crabs (*Chionoecetes opilio*) in response to sexual competition, and impact on the mating success of females. Biological Bulletin 201:204–217.

Rowe, S., and J.A. Hutchings. 2003. Mating systems and the conservation of commercially exploited marine fish. Trends in Ecology and Evolution 18:567–572.

Ryer, C.H. 2004. Laboratory evidence for behavioral impairment of fish escaping trawls: a review. ICES Journal of Marine Science 61:1157–1164.

Sainte-Marie, B., and G.A. Lovrich. 1994. Delivery and storage of sperm at first mating of female *Chionoecetes opilio* (Brachyura: Majidae) in relation to size and morphometric maturity of male parent. Journal of Crustacean Biology 14:508–521.

Sainte-Marie, B., S. Raymond, and J.C. Brêthes. 1995. Growth and maturation of the benthic stages of male snow crab, *Chionoecetes opilio* (Brachyura: Majidae). Canadian Journal of Fisheries and Aquatic Sciences 52:903–924.

Sainte-Marie, B., J.M. Sévigny, and Y. Gauthier. 1997. Laboratory behavior of adolescent and adult males of the snow crab (*Chionoecetes opilio*) (Brachyura: Majidae) mated noncompetitively and competitively with primiparous females. Canadian Journal of Fisheries and Aquatic Sciences 54:239–248.

Schlaepfer, M.A., P.W. Sherman, B. Blossey, and M.C. Runge. 2005. Introduced species as evolutionary traps. Ecology Letters 8:241–246.

Schlining, K.L., and J.D. Spratt. 2000. Assessment of Carmel Bay spot prawn, *Pandalus platyceros*, resource and trap fishery adjacent to an ecological reserve in central California. Crustacean Issues 12:751–762.

Secci, E., D. Cuccu, M.C. Follesa, and A. Cau. 2000. Fishery tagging of *Palinurus elephas* in Sardinian seas. Crustacean Issues 12:665–672.

Sodhi, N.S., L.P. Koh, B.W. Brook, and P.K.L. Ng. 2004. Southeast Asian biodiversity: an impending disaster. Trends in Ecology and Evolution 19:654–660.

Steger, R. 1987. Effects of refuges and recruitment on gonodactylid stomatopods, a guild of mobile prey. Ecology 68:1520–1533.

Stein, R.A., and J.J. Magnuson. 1976. Behavioral response of crayfish to a fish predator. Ecology 57:751–761.

Stockhausen, W.T., and R.N. Lipcius. 2003. Simulated effects of seagrass loss and restoration on settlement and recruitment of blue crab postlarvae and juveniles in the York River, Chesapeake Bay. Bulletin of Marine Science 72:409–422.

Stone, R.P. 1999. Mass molting of tanner crabs *Chionoecetes bairdi* in a Southeast Alaska estuary. Alaska Fishery Research Bulletin 6:19–28.

Sundelin, B., C. Ryk, and G. Malmberg. 2000. Effects on the sexual maturation of the sediment-living amphipod *Monoporeia affinis*. Environmental Toxicology 15:518–526.

Taylor, C.A., and M. Redmer. 1996. Dispersal of the crayfish *Orconectes rusticus* in Illinois, with notes on species displacement and habitat preference. Journal of Crustacean Biology 16:547–551.

Thiel, M., and I.A. Hinojosa. 2003. Mating behavior of female rock shrimp *Rhynchocinetes typus* (Decapoda: Caridea)—indication for convenience polyandry and cryptic female choice. Behavioral Ecology and Sociobiology 55:113–121.

Tierney, A.J., and J. Atema. 1986. Effects of acidification on the behavioral response of crayfishes (*Orconectes virilis* and *Procambarus acutus*) to chemical stimuli. Aquatic Toxicology 9:1–11.

Tomba, A.M., T.A. Keller, and P.A. Moore. 2001. Foraging in complex odor landscapes: chemical orientation strategies during stimulation by conflicting chemical cues. Journal of the North American Benthological Society 20:211–222.

Torchin, M.E., K.D. Lafferty, and A.M. Kuris. 2001. Release from parasites as natural enemies: increased performance of a globally introduced marine crab. Biological Invasions 3:333–345.

Torchin, M.E., K.D. Lafferty, A.P. Dobson, V.J. McKenzie, and A.M. Kuris. 2003. Introduced species and their missing parasites. Nature 421:628–630.

Untersteiner, H., J. Kahapka, and H. Kaiser. 2003. Behavioral response of the cladoceran *Daphnia magna* Straus to sublethal copper stress—validation by image analysis. Aquatic Toxicology 65:435–442.

Untersteiner, H., G. Gretschel, T. Puchner, S. Napetschnig, and H. Kaiser. 2005. Monitoring behavioral responses to the heavy metal cadmium in the marine shrimp *Hippolyte inermis* Leach (Crustacea: Decapoda) with video imaging. Zoological Studies 44:71–80.

Verslycke, T.A., N. Fockedey, C.L. McKenney, S.D. Roast, M.B. Jones, J. Mees, and C.R. Janssen. 2004. Mysid crustaceans as potential test organisms for the evaluation of environmental endocrine disruption: a review. Environmental Toxicology and Chemistry 23:1219–1234.

Vincent, A., and Y. Sadovy. 1998. Reproductive ecology in the conservation and management of fishes. Pages 209–245 *in*: T. Caro, editor. Behavioral ecology and conservation biology. Oxford University Press, New York.

Vismann, B. 1996. Sulfide species and total sulfide toxicity in the shrimp *Crangon crangon*. Journal of Experimental Marine Biology and Ecology 204:141–154.

Waddy, S.L., and D.E. Aiken. 1991. Mating and insemination in the American lobster, *Homarus americanus*. Pages 126–144 *in*: R.T. Bauer and J.W. Martin, editors. Crustacean sexual biology. Columbia University Press, New York.

Waddy, S.L., L.E. Burridge, M.N. Hamilton, S.M. Mercer, D.E. Aiken, and K. Haya. 2002. Emamectin benzoate induces molting in American lobster, *Homarus americanus*. Canadian Journal of Fisheries and Aquatic Sciences 59: 1096–1099.

Wahle, R.A., and R.S. Steneck. 1991. Recruitment habitats and nursery grounds of the American lobster *Homarus americanus*: a demographic bottleneck? Marine Ecology Progress Series 69:231–243.

Walker, D.I., and G.A. Kendrick. 1998. Threats to macroalgal diversity: marine habitat destruction and fragmentation, pollution and introduced species. Botanica Marina 41:105–112.

Watling, L., and E.A. Norse. 1998. Disturbance of the seabed by mobile fishing gear: a comparison to forest clearcutting. Conservation Biology 12:1180–1197.

Watson, G.M., O.K. Andersen, M.H. Depledge, and T.S. Galloway. 2004. Detecting a field gradient of PAH exposure in decapod crustacea using a novel urinary biomarker. Marine Environmental Research 58:257–261.

Weis, J.S., A. Cristini, and K. Rao. 1992. Effects of pollutants on molting and regeneration in Crustacea. American Zoologist 32:495–500.

Welch, J.M., D. Rittschof, T.M. Bullock, and R.B. Forward, Jr. 1997. Effects of chemical cues on settlement behavior of blue crab *Callinectes sapidus* postlarvae. Marine Ecology Progress Series 154:143–153.

Wolf, M.C., and P.A. Moore. 2002. Effects of the herbicide metolachlor on the perception of chemical stimuli by *Orconectes rusticus*. Journal of the North American Benthologial Society 21:457–467.

Yoshimura, T., and H. Yamakawa. 1988. Microhabitat and behavior of settled pueruli and juveniles of the Japanese spiny lobster *Panulirus japonicus* at Kominato, Japan. Journal of Crustacean Biology 8:524–531.

Zala, S.M., and D.J. Penn. 2004. Abnormal behaviors induced by chemical pollution: a review of the evidence and new challenges. Animal Behaviour 68:649–664.

Zhang, Z., W. Hajas, A. Phillips, and J.A. Boutillier. 2004. Use of length-based models to estimate biological parameters and conduct yield analyses for male Dungeness crab (*Cancer magister*). Canadian Journal of Fisheries and Aquatic Sciences 61:2126–2134.

Zimmer, C. 2003. Rapid evolution can foil even the best-laid plans. Science 300:895.

Zulandt Schneider, R.A., R.W.S. Schneider, and P.A. Moore. 1999. Recognition of dominance status by chemoreception in the red swamp crayfish, *Procambarus clarkii*. Journal of Chemical Ecology 25:781–794.

Comparative Evolutionary Ecology of Social and Sexual Systems

Water-Breathing Insects Come of Age

Bernard J. Crespi

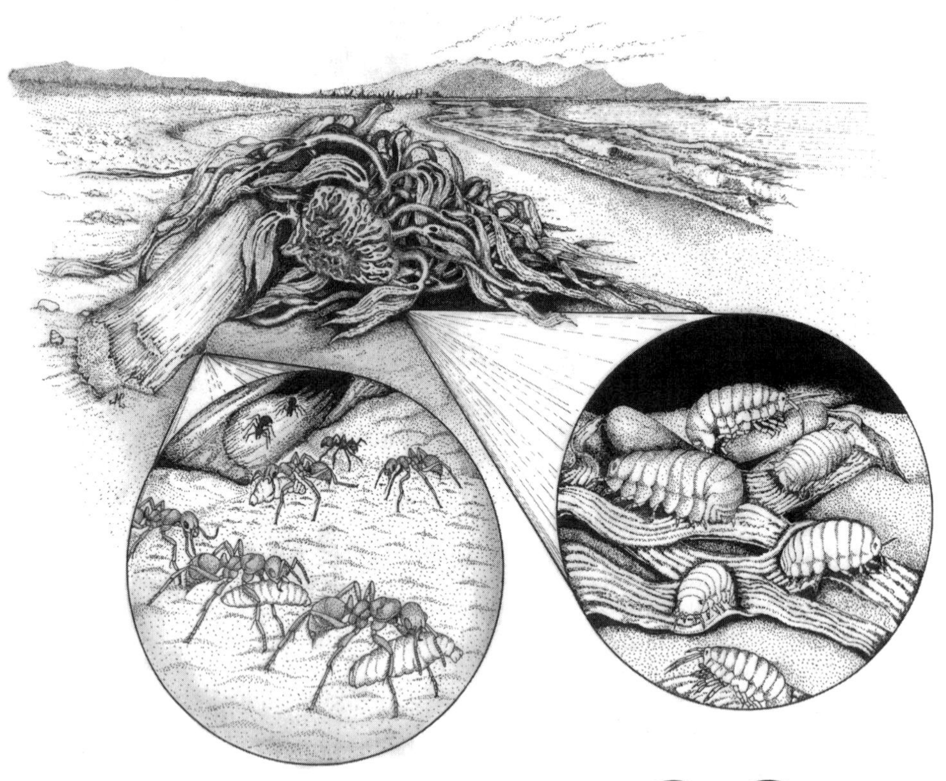

20

The evolutionary ecology of behavioral interactions is characterized by two great divides. The first divide is conceptual, in that some researchers focus on social systems and some focus on sexual selection and mating systems, and analyses of the two have seldom been combined (West-Eberhard 1983). Queller (1994) described how both types of system are characterized by competition between mutually dependent parties for limiting resources and generate complex mixtures of confluent and divergent interests. The evolution of both social and mating systems is also driven by variation in ecology (Emlen and Oring 1977, Crespi and Choe 1997, Queller and Strassmann 1998, Shuster and Wade 2003), and their evolutionary trajectories may involve complex webs of positive and negative feedback (Andersson 1994, 2004, Crespi 2004). However, this conceptual divide has seldom been crossed, except in some cases of biparental or alloparental care where cooperation and sex necessarily coincide to some degree.

The second divide in evolutionary ecology is taxonomic (Hart and Ratnieks 2005). The ever-burgeoning literature for every taxonomic group compels specialization, and indeed, in-depth analyses of particular species or clades may provide especially useful insights into microevolutionary and macroevolutionary processes (Grant 1999, Crespi et al. 2004). The main cost of specialization is limited perception of broad-scale, convergent patterns, patterns that unite the field conceptually and often lead to novel hypotheses (e.g., Hamilton 1967, 1978).

This book on crustaceans straddles both divides and brings into clear view the growing need to integrate research areas and taxa. My goal in this chapter is to draw together the concepts driving the study of sexual and social biology, and to use these "water-breathing insects" as a touchstone to reach across taxonomic groups, seeking convergences between crustaceans and other animals. Indeed, the diversity of studies represented in this volume may be considered a microcosm of evolutionary ecology as a whole, providing an ideal situation for surveying its landscape in current form.

Convergence and Divergence in Routes to Social and Sexual Inference

Analysis of the evolution per se of social and sexual behavior requires either phylogenies or studies that combine studies of selection with microevolutionary response. Phylogenetic studies have been conducted informally since Darwin (1871) began evaluating taxa for phenotypic similarities indicative of similar selective forces, and formally since Felsenstein (1985) devised the first statistically based comparative method. Informal methods are limited by their intrinsically heuristic nature, and statistical methods are limited by the difficulties of jointly analyzing far-flung taxa, such as dolphins and chimps (Connor et al. 1998), and by the loss of information in the mists of deep history for ancient clades. There are no solutions to this dilemma, only an optimality criterion of novel insight gained from any analysis.

Comparative methods are based on two processes: selection-driven convergence to common phenotypes from ones that differ, and divergence from one lineage to multiple descendant lineages that show covarying phenotypic change indicative of selective forces. For robust studies of convergence, the greater the phylogenetic distance between forms the better, because this span maximizes the odds that all traits other than the covarying ones of interest are randomized. Such randomization reduces the odds that unobserved third variables underlie inferred associations (Ridley 1989). By contrast, analysis of divergence requires species-level or population-level studies with blanket taxonomic coverage—a so-called "model clades" approach (Crespi et al. 2004)—that ensures detection of all relevant microevolutionary changes that grade into differences between populations or species—as well as avoiding secondary effects of changes that may obscure selective origins (Wilson and Hölldobler 2005). Thus, for such divergence studies phylogenetically closer is better, which also leads to more robust tests via independent contrasts or other statistical methods (Pagel 1994, 1999, Crespi 1996, Martins 2000).

One important upshot of the convergence approach to analyzing social and sexual behavior is that each major taxonomic group may contribute similar analytic "weight" to any given study, regardless of its species diversity. For example, all Hymenoptera share a constellation of traits, such as haplodiploidy, that are likely to influence the evolution of social and sexual systems (Alexander et al. 1991), but analyzing the comparative role of such traits necessitates investigating other haplodiploid taxa, because each represents only one evolutionary experiment (Doughty 1996) with such genetic systems. A crucial implication of this line of argument is that each more or less major taxon in the idiosyncratic social bestiary, from microbes to spiders, insect orders, fish, birds, and mammals, may be equally useful for addressing particular questions, such as the origin of cooperative breeding from maternal care. Thus, intuition aside, mole-rats (Sherman et al. 1991, Jarvis et al. 1994), *Austroplatypus* beetles (Kent and Simpson 1992, Kirkendall et al. 1997), spiders (Whitehouse and Lubin 2005), social lizards (Chapple 2003), or snapping shrimp (chapter 18) may each be equally conceptually useful as all termites or all ants in some analyses of social evolution. With this viewpoint in mind, the Crustacea occupy a key position, as sister group to Hexapoda (Regier et al. 2004), with comparable phenotypic, ecological, social, and sexual diversity that this book brings to light.

The evolution of social and sexual behavior involves complex interplay of ecological selective factors with phenotypic and genotypic traits. In this chapter, I first describe a general conceptual framework for understanding the evolution of cooperation. Next, I discuss the phenotypic and genetic traits of crustaceans and other animals that I consider most salient to social and sexual evolution, and discuss how they affect evolutionary trajectories in this framework. Third, I evaluate the role of ecological and life-history factors in selecting upon phenotypic variation, leading to changes in social and mating systems. Finally, I integrate phenotype with ecology and describe hypotheses on how the determinants and components of social and sexual systems evolve together.

Social and Sexual Cooperation and Conflict

Sexual and social interactions each involves a mixture of cooperation, to fertilize eggs or to rear babies, and conflict, to control fertilization or to maximize inclusive fitness via avenues that differ from those of other group members. Social cooperation can be favored by three mechanisms: (1) kinship, (2) reciprocity, and (3) byproduct benefits (Sachs et al. 2004). Kinship-based cooperation can involve altruistic or mutualistic behavior, and cooperation is normally enforced or facilitated by forms of kin recognition, policing, or colony-level benefits. Reciprocity may involve relatives or nonrelatives, and cooperation is driven by the repeated, cumulatively beneficial nature of interactions, with cheater suppression or avoidance mediated by obligate cohabitation, partner recognition, or group-level benefits. Byproduct benefits involve behavior that is directly favored by selection for all parties, with no net gains possible from cheating (Connor 1995a, 1995b).

Under the kinship and reciprocity models, conflicts of interest are virtually ubiquitous, and such conflicts are ongoing or resolved via persuasion (providing benefits to cooperators), coercion (imposing costs on noncooperators), or force (taking control of behavior away) (W.D. Brown et al. 1997, Crespi and Ragsdale 2000, Frank 2003). Which of these three routes are followed depends in turn upon colony-level costs and benefits of the alternatives (e.g., Korb and Heinze 2004) and imbalances in power or information (Crespi and Ragsdale 2000, Beekman et al. 2003, Beekman and Ratnieks 2003).

Sexual interactions can also be interpreted in this framework. Males and females may in some cases exhibit confluence of interest, with fast, efficient mating and fertilization being optimal for both parties. However, in most species, males and females should exhibit conflicts of interest over mating, fertilization, premating or postmating associations, and any associated parental investment such as resource-rich spermatophores or nuptial gifts (West-Eberhard 1983, Alexander et al. 1997, W.D. Brown et al. 1997, Fincke 1997, Chapman et al. 2003, Arnqvist and Rowe 2005). As for sociality, conflict resolution may be persuasive, coercive, or forceful, with information, power, and mutual costs or benefits as key determinants of outcome. But incest aside, sexual cooperation must be based on short-term reciprocity (e.g., parceling; Connor 1995b) or mutualistic benefits, because a set of sexual interactions (from mate detection to fertilization; W.D. Brown et al. 1997) is seldom repeated for a given pair.

The challenge for students of sociality and sex is to explain interspecific and intraspecific variation in these systems as some function of ecology and life history, in the context of evolutionary histories, expressed as current phenotypes and genotypes that delimit behavioral trajectories (e.g., Crespi and Choe 1997, Cockburn 1998). For understanding the contributions of ecology and life history to behavior, tradeoffs are key. Tradeoffs between such fundamental traits as reproduction versus survival, mating effort versus parental effort, and helping versus reproduction thus differ among taxa at all levels and structure evolutionary trajectories. Ultimately, such tradeoffs reside in the genome and are expressed as reaction norms, but for now, phenotypes are our necessary units of analysis.

The Primacy of Phenotype

Analyzing the evolution of social and sexual systems has usually focused on ecology (as the agent of selective pressures) and genes (as the units of inheritance, the key to genetic relatedness, and the progenitors of phenotypic variation). To the extent that phenotypic traits are not just the outcome of evolution but also place limits on evolutionary trajectories, they also play a causal role in how behavior evolves (e.g., Wcislo 1989).

Ever since their universal common ancestor, animal lineages have differed in phenotypic traits that influence the evolution of their social and sexual systems. Traits most salient to social evolution include modes of foraging, feeding, defense, locomotion, and sensation; aspects of life history such as how offspring are produced also both direct and respond to selective pressures on social behavior. By contrast, traits most relevant to the evolution of mating systems include structures used in mate finding, courtship, sperm transfer, and fertilization (W.D. Brown et al. 1997).

Many common insect traits, such as high potential fecundity, air breathing, flight, diverse adaptations to defense, internal fertilization, and sperm storage, have undoubtedly contributed to how their social and sexual behavior can evolve, and similar arguments can be made for other taxa. Among crustaceans, traits that I hypothesize as crucial include the following:

1. The presence of gills or breathing water via the skin, and the existence of the aquatic larval stage, which makes crustaceans the equivalent of ferns and mosses among plants, or amphibians among tetrapods. As "water-breathing insects," the habitats and basic necessary resources essential to reproduction of crustaceans are fundamentally different from those of fully terrestrial creatures. Such habitats include, of course, the seas, where symbiotic relationships flourish, and localized aquatic habitats, such as wet burrows or small pools.
2. The absence of airborne flight, which limits searching for mates and other key resources, such as genetically related cooperators, and reduces ability to transport and concentrate food (e.g., Clark and Dukas 1994) or disperse long distances in a directed way—indeed, all eusocial insects exhibit flight at least at some life-cycle stage.
3. The common presence of claws, which provides opportunities for effective altruistic or mutualistic defense, and also armor, which may reduce the strength of selection for defense by making some crustaceans less profitable or accessible as prey (e.g., crabs, lobsters).
4. Hemimetabolous development in some crustaceans, such that juveniles above a certain size are like miniature adults, which makes helping possible by juveniles (e.g., chapter 17), as in termites, aphids, and some other taxa (Alexander et al. 1991), and molting, which increases vulnerability to predation during short periods and sometimes restricts mating opportunities (e.g., chapter 7).
5. All-at-once brood production, such that offspring develop in discrete cohorts, which may reduce the duration of mother–offspring overlap and lessen the odds of older offspring helping younger ones.
6. A habit in some lineages, such as snapping shrimp and some crabs, of feeding on food of relatively low nutritional quality, such as detritus or tiny animals, which reduces any possible benefits from paternal care that involves direct feeding (Hunt and Nalepa 1994) and may contribute to "capital" breeding via all-at-once brood production.

Assessing the impact of these traits on social and mating-system evolution requires comparison of sister lineages divergent for the presence or form of the trait, and comparisons of convergent taxa, in the context of the ecological selective pressures unique to each species and clade. Such comparisons are especially useful for traits with clear functional design, such as weaponry; indeed, effective weapons appear crucial both as a predaptation to complex social evolution and as a factor allowing social groups to enlarge, and sometimes flourish, in the context of between-species interactions. Such weaponry includes bacteriocins in microbes (Crespi 2001, Gardner et al. 2004); mouthparts, forelegs, and stings in insects (e.g., Kukuk et al. 1989); venom in spiders and scorpions (e.g., Binford 2001); teeth held in strong jaws for mole-rats and social carnivores (Sherman et al. 1991); and hunting tools in humans. Crustaceans demonstrate nicely how weapons play a key role in the evolution of "factory-fortress" eusociality, which may be the only form of complex sociality in this group given the constraints on their locomotion, form of brood production, and near restriction to wet or aquatic habitats.

The Ecological Nexus of Sex and Social Cooperation

The distributions of resources for breeding in space and time are considered central drivers of both sexual and social system evolution (Emlen and Oring 1977, J.L. Brown 1978). However, the conceptual divide between mating and helping has apparently thus far precluded the joint analysis of ecological factors in both contexts. By the Emlen and Oring (1977) scheme, extended by W.D. Brown et al. (1997), Shuster and Wade (2003), and others, the distribution of resources in space and time largely determines the distribution of females, which in turn drives optimal male mating tactics; other factors, such as tradeoffs between mating and parental effort, and sexual conflict (Arnqvist and Rowe 2005), mediate the ultimate outcome. The role of ecology in social evolution has been dominated by the ecological constraints model, which asserts that cooperation evolves in habitats that are especially valuable (J.L. Brown 1978, Emlen 1982, Andersson 1984), such that independent breeding via dispersal is relatively difficult. Staying at home (and perhaps helping) may also be facilitated by an evolutionary history of high adult survivorship (Arnold and Owens 1998, 1999) and the benefits of philopatry, most notably, a safe haven and possible inheritance (Ragsdale 1999, Hatchwell and Komdeur 2000, Pen and Weissing 2000, Kokko and Ekman 2002, Solomon 2003). These considerations linking ecology with behavioral interactions imply that social systems and mating systems should coevolve closely (Fig. 20.1), and indeed, for most social vertebrates, they are intricately mixed and centered around resources of especially high value (J.L. Brown 1978, Alexander et al. 1991, Faulkes and Bennett 2001, Koenig and Dickinson 2004).

The core ecological feature of social and sexual systems is what I call the "basic necessary resource": a habitat, be it burrow, territory, nest, gall, social group members, or other requirement for successful breeding. Aspects of this resource that are crucial for its role as an arena of selection include whether or not it is created, how hard it is to acquire and defend, whether it can be replaced if lost, whether food is available in the habitat, and the longevity of the habitat in relation to generation time (Alexander et al. 1991, Crespi 1994). For any given species, these aspects of the basic necessary resource, plus ecological interactions involving competitors, predators, parasites, and mutualists, strongly affect male mating opportunities (Fig. 20.1). Mating and social systems interact in the context of the male promiscuity–care tradeoff; as this tradeoff evolves, both systems are affected. Moreover, mating system evolution influences social evolution via incest avoidance, genetic relatedness effects, and mate guarding as a predisposition to male care, and social evolution may change the distribution of receptive females, as well as the operational sex ratio and sexual selection intensity (Fig. 20.1). Empirical analysis of such predicted effects requires microevolutionary, "microphylogenetic" studies of joint changes in sociality and mating systems, as described above.

The ecological interface of social and sexual systems is structured by four main axes. First, tradeoffs between mating effort and parental effort, mediated by the scope

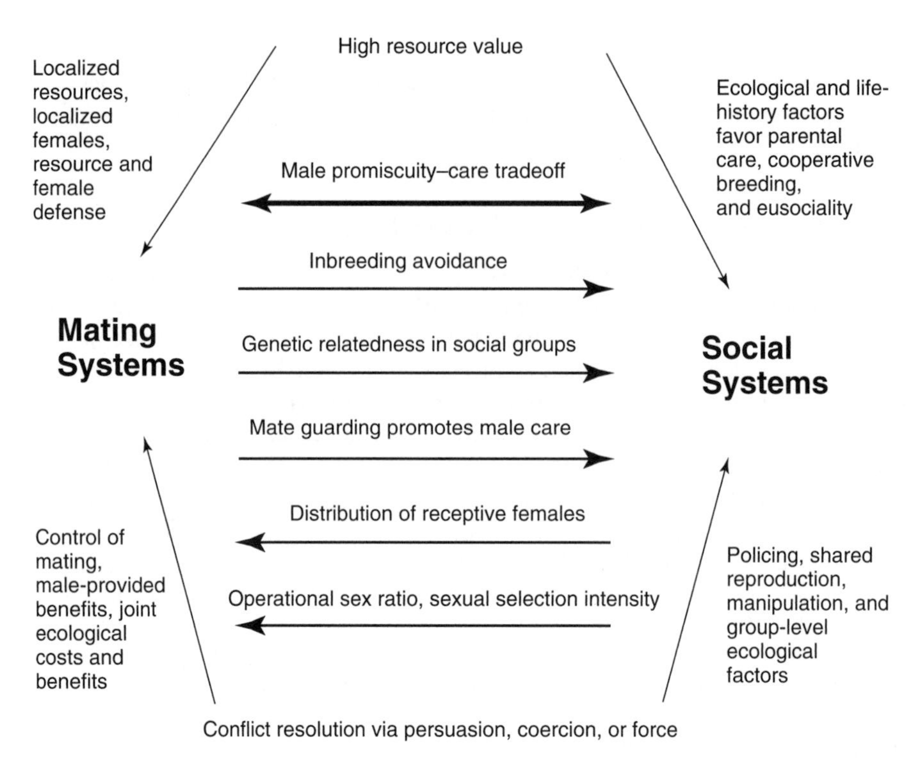

Figure 20.1 Coevolution of social and mating systems. This figure illustrates two main points. First, the evolutionary interactions between aspects of social and sexual systems are shown by the arrows at the center of the figure. These effects may be bidirectional (as with the promiscuity–care tradeoff) or unidirectional, such that changes in either mating or sexual systems can influence the evolutionary trajectory of the other. Second, both social and sexual systems are ultimately determined by (1) ecological and life-history factors, with a central role for high resource value, and (2) the nature of cooperation, conflicts and conflict resolution, which may involve provision of benefits to others ("persuasion"), imposition of costs on others ("coercion"), or force imposed on others.

for promiscuity and the benefits of helping, influence the presence and form of parental care, maternal, paternal, or biparental (e.g., Clutton-Brock 1991, Black 1996) (Fig. 20.1). Second, female foraging effort may trade off strongly with parental effort, as in some insects where females are mobile, predacious, and food limited; this tradeoff may select for exclusive paternal care, with males gaining both paternity and protection of young via guarding of eggs (Tallamy 2000, Owens 2002). Third, the tradeoff between staying, and perhaps also helping, in one's natal habitat versus dispersing to breed independently drives the initial evolution of cooperative breeding from paternal care (Fig. 20.2). This tradeoff applies to both sexes, and for males a higher intensity of sexual selection may favor staying and helping (Alexander et al. 1991). Finally, for helpers there is a tradeoff between helping and engaging in personal reproduction, within ones' social group (Fig. 20.2). This tradeoff may be energetic and physiological or social, with "cheaters," who lay eggs or engage in surreptitious matings, suffering imposed costs from dominants, including eviction at worst (Johnstone and Cant 1999).

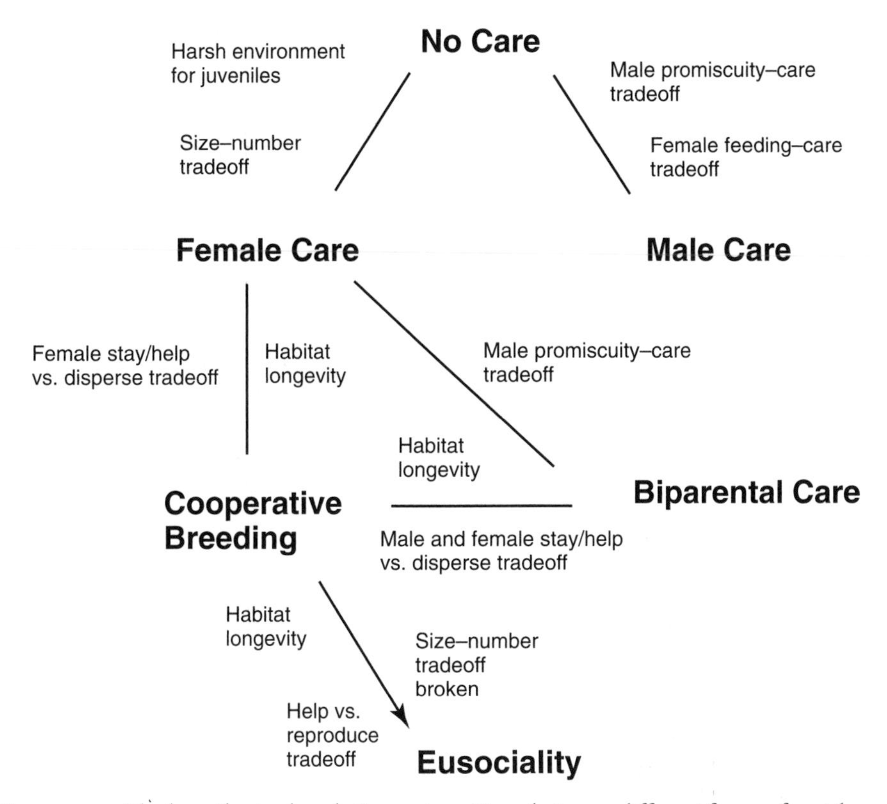

Figure 20.2 The hypothesized evolutionary transitions between different forms of social system (the lines connecting the social systems), and the most important selective factors and tradeoffs involved in each specific transition (the phrases adjacent to the lines). Only the transition from cooperative breeding to eusociality is presumed to be unidirectional; other transitions may occur in either direction. These evolutionary transitions between social systems are mediated by ecological factors and tradeoffs involving life-history and mating-system characteristics. The form of such tradeoffs is expected to differ among species, mainly as a function of their food source and the regimes of predation and competition to which they are subjected.

Each of these tradeoffs may differ in strength between species, in some cases being strong, as for the latter two tradeoffs in naked mole-rats; weak, as for helper reproduction in some factory-fortress species such as thrips (Chapman et al. 2002); or essentially nonexistent, as for some social mammals such as the Damaraland mole-rat where within-group matings must involve incest and so are precluded (Jarvis and Bennett 1993, Jarvis et al. 1994). The strength of tradeoffs for any given species depends strongly on ecological factors, including the nature of the basic necessary resource, life-history, and other phenotypic traits, as well as aspects of social power asymmetries and kinship, as exemplified in skew models (Johnstone 2000).

The ecology of crustacean social and sexual systems is dominated by several themes. As described in chapters 14 and 16, maternal care is common in this group, especially in the relatively "harsh" terrestrial environment of some isopods with desiccation risk, food provisioning, and a valuable, relatively safe microhabitat as major selective contexts. In some situations, such as that of the isopod *Hemilepistus*, such maternal care has apparently evolved into biparental care, as in many insects, many birds, and some mammals (Reynolds et al. 2002). This transition apparently involves reductions in benefits from promiscuity to males, as well as increased benefits from joint care such as division of labor in defense of especially valuable reproductive resources such as burrows (Fig. 20.2), although to what relative degree mating-system, social, and life-history factors (e.g., semelparity) are involved remains unclear. Finally, either maternal care or biparental care has evolved into cooperative breeding and eusociality in at least two clades of the Crustacea, snapping shrimp (chapter 18), bromeliad crabs (chapter 17), and possibly burrowing crayfish (chapter 15). Biparental care may provide an evolutionarily smooth route to cooperative breeding because broods of potential helpers will very likely be full sibs, and the ecological situation already involves benefits from having two helper-parents present, suggesting that additional help may be even more advantageous. Moreover, in this case both sexes have already been shown to be ecologically or sexually constrained enough (from greater fecundity with less care, or greater promiscuity) to engage in extensive parental care, such that offspring of both sexes serve as potential helpers even after they reach adulthood or breeding age. The transition from exclusive maternal care to cooperative breeding may be relatively less likely, based on these arguments, so it may require stronger effects from a basic necessary resource that constrains breeding opportunities of female dispersers.

Crustaceans provide some of the best evidence for the factory fortress model for the evolution of eusociality. Thus, social snapping shrimp (chapter 18) and bromeliad crabs (chapter 17) meet the criteria described in Crespi (1994), a claustral habitat that combines food and shelter, strong selection for defense, and ability to defend via weaponry. "Primary burrower" crayfish may also approach this situation (chapter 15), although these species must presumably forage outside of the burrow—though perhaps relatively safely. Consideration of the life histories of these three crustacean taxa also suggests that helping by juveniles, exclusively or in conjunction with helping by adults, often coincides with the factory fortress mode of sociality. Thus, in snapping shrimp and bromeliad crabs, as well as some termites, aphids with soldiers, naked mole-rats, *Austroplatypus* beetles with apparent helping by larvae (Kent and Simpson 1992, Kirkendall et al. 1997), and some social spiders (Aviles 1997, Whitehouse and Lubin 2005), juveniles engage in some degree of helping, presumably due to a combination of hemimetaboly or its equivalent in noninsects, the relatively safe nature of the habitat, and inclusive fitness gains from helping relatively soon after birth (Queller 1996). This pattern reaches its apex in some termites where all helpers senesce as juveniles (Thorne 1997, Thorne et al. 2002), and indeed, such divergence in life span between helpers and reproductives suggests that the self-reinforcing nature of senescence has helped to drive the evolution of complex sociality (Alexander et al. 1991). Helping by juveniles in cooperative vertebrates may be more or less limited to humans, which also enjoy relatively safe habitats and benefit from extensive alloparental care (Alexander 1989). Has such helping favored in part the greatly extended juvenile period of humans (because children help to raise siblings), and is it also reflected at the other end of the life cycle, with the evolution of sterility, and helping, via menopause (Foster and Ratnieks 2005)?

The absence in crustaceans of "life-insurance" forms of sociality (Queller and Strassmann 1998) or large complex societies as found in ants, wasps, bees, and termites that forage outside the nest may be due to their lack of flight, the fact that their foods are less energy dense than meat or pollen, the "low fecundity" life history of fully terrestrial species, or the lack of easy expansibility of the habitat. These hypotheses are, however, notably difficult to test, except in the context of sister-taxon comparisons of Hymenoptera and Isoptera lineages with and without complex sociality (e.g., Hunt 1999).

Joint consideration of the factory-fortress and life-insurance models and models developed for cooperative breeders suggests that one of the key variables in the evolution of helping is the degree to which helping influences mortality rates (Fig. 20.3). Thus, increased helping under cooperative breeding, as exemplified best by birds, appears to normally coevolve with decreased adult mortality. Such reduced mortality leads to increased habitat saturation, increased constraints on independent breeding, higher relatedness (Taylor and Irwin 2000, Le Galliard et al. 2005), increased chances of inheritance, and delayed senescence, all of which should further favor staying and helping and drive populations toward the "slow" end of the life-history spectrum (e.g., Härdling et al. 2003, Oli 2004, Russell et al. 2004). Indeed, under this view, ecological constraint may be a consequence and not a simple cause of cooperative breeding.

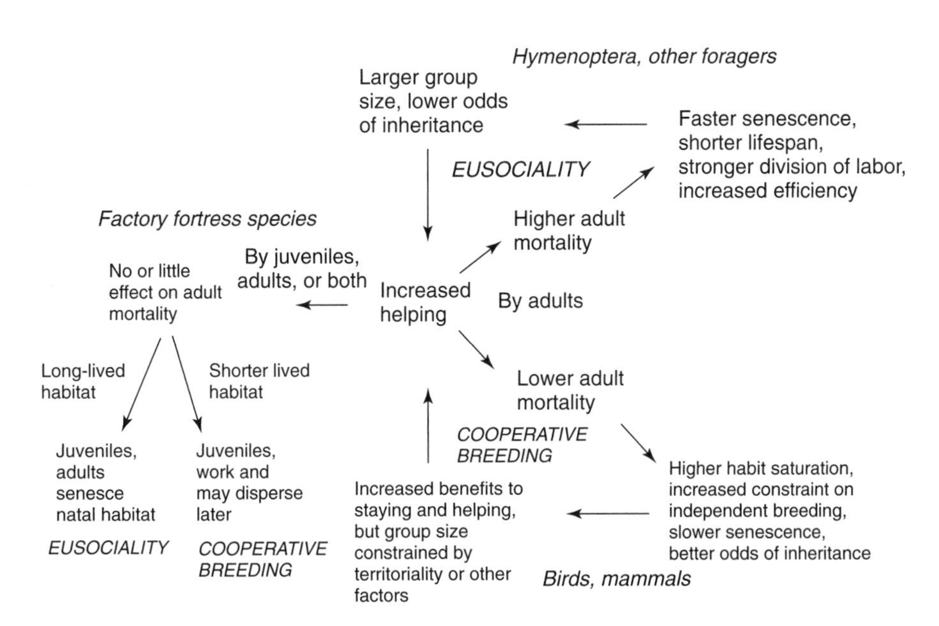

Figure 20.3 Cooperative breeders and eusocial species appear to comprise three main groups with regard to two factors: (1) their habitat (combined food and shelter "factory fortresses" vs. foraging outside the nest) and (2) whether helping involves increased versus decreased adult mortality of helpers. In Hymenoptera and other foragers, higher adult mortality and increasing group sizes lead to eusociality via a positive feedback loop. In birds and mammals, lower adult mortality follows from staying and helping and leads to habitat constraints, group sizes constrained by territoriality, and cooperative breeding. In factory fortress species, mortality rates are low in the enclosed natal habitat, and the evolution of cooperative breeding versus eusociality may be largely a function of habitat duration relative to life span.

By contrast, increased helping in many invertebrates, such as Hymenoptera that forage outside the nest, is expected to raise mortality rates of helpers, leading to accelerated senescence and shorter life span, but also enhanced efficiency via division of labor, larger groups, lower chances of inheritance, and further helping via a positive feedback cycle (Alexander et al. 1991, Keller 1997, Bourke 1999); this route thus leads directly from cooperative breeding to eusociality (Jeon and Choe 2003). Factory-fortress lineages such as snapping shrimp and bromeliad crabs may follow a life-history route whereby juveniles enjoy low mortality and gain inclusive fitness via helping full sibs; groups can increase in size until limited by density dependence in their special food–shelter habitat, after which altruistic dispersal is increasingly favored (Crespi and Taylor 1990). Under this model, individuals may remain totipotent if habitat duration is not unduly long compared to generation time, or eusociality may evolve if nepotistic effort or personal reproduction leads to senescence and death at home.

Testing these ideas requires a broad perspective from life-history theory in social evolution, models that incorporate coevolutionary dynamics, and focused study of invertebrates with small colonies and vertebrates with large ones, especially in clades with some species exhibiting totipotency (Crespi and Yanega 1995, Hart and Ratnieks 2005) but others crossing the threshold to eusociality.

Vagaries of Social and Sexual Conflict

Ecological factors exert largely deterministic effects on the evolution of social and sexual systems. By contrast, some forms of behavioral interaction that involve conflict are expected to generate antagonistic coevolution (Rice 1996, Rice and Holland 1997, Rice and Chippindale 2001) whose resolution, or ongoing nature, appears relatively difficult to predict (Arnqvist and Rowe 2005). The magnitudes of conflicts over mating and fertilization and conflicts over postzygotic investment in offspring depend on the degree of divergence of the optima for males versus females, or parents versus offspring, or intrasexual interactions (Chapman et al. 2003). For example, male–female confluence of interest can arise when harsh ecological conditions compel having two helpers raise young (Clutton-Brock 1991) or when male mating opportunities via mate searching are few and females are predatory or otherwise constrained in food acquisition, leading to exclusive paternal care (Tallamy 2000). By contrast, conflicts may persist and dominate behavioral interactions where males gain by promiscuity, females suffer costs, and neither party can fully control the mating sequence from contact until fertilization (W.D. Brown et al. 1997).

In general, the resolution of conflicts will depend on the set of strategies available (e.g., one party being able to forcibly achieve its optimum), the strength of selection on different parties, and the forms and extent of genetic variation that underlie the interactions. A key factor in conflict resolution is whether or not one party can adjust the behavior of the other via the provision of benefits, in contrast to the imposition of costs. For example, males may persuade females to mate via the offering of nuptial resources, or coerce them to mate by making them more vulnerable to predation via precopulatory guarding, and females in social groups may allow cooperators to have higher personal reproduction, or punish and evict noncooperators (e.g., Cant et al. 2001). The success of such varying strategies will depend on ecological factors as well as phenotypic and genetic asymmetries, and assessing how ecological factors interact with aspects of conflict and its resolution is a crucial challenge for future analyses of social and mating-system evolution. Moreover, whereas ecological selective pressures always generate adaptation, at least in the context of tradeoffs, antagonistic coevolution driven by behavioral conflict will often result in phenotypic traits that are maladaptive in males, females, or juveniles; moreover, such phenotypes may be difficult to recognize as maladaptive without a clear understanding of mechanisms and microevolutionary trajectories (Crespi 2000, Nesse 2005).

In crustaceans, studies of social evolution have mainly focused on cooperative behavior, which is still being characterized. Future studies of factory-fortress–inhabiting crustaceans may usefully investigate conflicts among juveniles, between mother and offspring over parental investment, and female–female conflict, especially over new breeding opportunities. Analyses of sexual interactions have, by contrast, already demonstrated complex interplay between cooperation, conflict, and ecology, especially in species with biparental care or elaborate systems of mate searching, courtship, and female choice. Thus, in fiddler crabs, predation regimes and local population density can tip the balance between different mating systems even within species, by adjusting the costs and benefits of alternative male and female strategies (chapter 10); this group nicely exemplifies how the divergence approach to analyzing evolutionary changes works best at the lowest levels. At the most divergent taxonomic levels, *Hemilepistus* isopods appear convergent with humans in exhibiting extensive biparental care, few but huge offspring, habitation of highly valuable resources, high survivorship, concealment of ovulation, control of mating largely by females, a complex system of interindividual recognition, and strong intraspecific competition (chapter 16). These traits apparently form a coadapted set, grounded in life history, at the "slow" extreme of the slow–fast continuum as has been described in mammals (Oli 2004). Indeed, the combined mating/social system involving biparental care with female control of mating and a highly valuable basic resource appears to be convergent across diverse taxa, perhaps driven by feedbacks between sexual and social system evolution.

Transitions and Feedbacks in the Evolution of Social and Sexual Systems

How do social and sexual systems coevolve? At the broadest scale, each is influenced by phenotypic and ecological factors and also drives the evolution of the other. First, consider selection on social cooperation, perhaps due to ecological change, from no parental care, to maternal care, to cooperative breeding, to eusociality. How might mating systems change in response? The origin of maternal care may localize receptive females, making them more defensible by males. Depending on the nature of any male promiscuity–care tradeoff, postcopulatory mate guarding may provide a preadaptation to biparental care (e.g., Mathews 2003), or even paternal care, depending on the female foraging–care tradeoff. Exclusive paternal care may preclude the evolution of cooperative breeding, but as noted above, biparental care may, in principle, facilitate it. How the transition from parental care to alloparental care affects mating systems must depend on which sex helps. Exclusive female helping essentially removes young females from the pool of receptive mates, which may increase the intensity of sexual selection (Andersson 1994). Exclusive male helping, as found in many birds, may have the opposite effect, unless it is driven in the first place by sharply limited mating options for young males (Alexander et al. 1991).

Finally, the transition from cooperative breeding to eusociality, as seen in some arthropods and naked mole-rats, further limits male mating opportunities in time and space. This transition has led convergently to swarming or lekking in many bees, ants, and termites with flight and foraging outside the nest, but in factory-fortress species the situation may be quite different due to their claustral habitat. Here, mating and breeding opportunities may arise within the fortress if a founder or foundress dies, leading to intense struggles given the high reproductive skew that characterizes such societies. Alternatively, inbreeding may normally occur among helpers, as in some thrips and spiders (Aviles 1997, Chapman et al. 2000, 2002), or selection against incest may effectively enforce cooperation and a lack of supersedure, as in Damaraland mole-rats (Jarvis and Bennett 1993, Jarvis et al. 1994). Such inbreeding effects may also structure the behavior and life histories of social shrimp, crabs, and crayfish.

Now consider selection on mating systems and how it may affect social evolution (Figs. 20.1, 20.2). Many mating system transitions will engender changes in the degree of multiple paternity in broods, which should affect relatedness and the likelihood that helping evolves, and the degree to which broodmates cooperate. Increases in the intensity of sexual selection may make male parental care less likely, at least by the most successfully promiscuous males, but less dominant males may be selected to care more for a female's offspring if they have some confidence of parenthood. Alternatively, stronger sexual selection on both sexes could lead to a situation where males fight for breeding resources and females compete for males, or the sexes fight jointly for resources, leading to biparental care. Increased control of mating and fertilization by females may also, in some cases, lead to increased paternal care, if females can essentially trade fertilizations for help; this may be less likely with all-at-once brood production, as in crustaceans.

It should be clear from this exercise that sexual and social systems must evolve together via mutual feedbacks, both negative feedbacks with constraining effects and positive feedbacks leading to runaway evolution and rapid shifts (Andersson 1994, 2004, Crespi 2004). The evolution of sociality may involve four main forms of positive feedback loops: (1) between the degree of helping and senescence (Alexander et al. 1991); (2) among colony size, worker reproductive potential, and queen–worker dimorphism (Bourke 1999) (Fig. 20.3), (3) among decreased adult mortality, increased habitat saturation, and increased helping (Fig. 20.3), and (4) among increased helping, increased socially generated resource, increased selection for resource usurpation via cheating or other means, and increased cooperation involving resource defense—a social arms race such as may have led to the evolution of the human psyche (Alexander 1989). The evolution of mating systems involves at least four such processes: (1) Fisherian positive feedback loops in the evolution of mate choice (Kokko et al. 2002, Mead and Arnold 2004), (2) the self-reinforced evolution of leks from other mating systems (Sutherland 1996, Bro-Jorgensen 2003), (3) coevolution between parental investment and sexual selection (Andersson 1994, 2004), and (4) runaway processes driven by other genetically covariant aspects of male and female mating phenotypes, such as promiscuity (West-Eberhard 1983, Shuster and Wade

2003). One of the main coevolutionary loops connecting social and sexual systems involves changes in the strength of sexual selection affecting the mating system and parental investment levels, which affects male and female potential reproductive rates, which feeds back to the opportunity for sexual selection via the life history (Andersson 1994). Thus far, the processes involved in this loop have been analyzed primarily in the context of the tradeoffs between parental effort and mating effort for species with biparental care, with a focus on vertebrates. Extending such analyses to species with cooperative breeding and eusociality, and to invertebrates, should lead to recognition of novel convergences, such as those between *Hemilepistus* and some vertebrates.

The degree to which feedback loops operate is an empirical matter that can only be resolved via analyses of fine-scale divergence among populations and closely related species, with a focus on the key tradeoffs involved in social and sexual behavior, and the central roles of ecology and phenotype. This is not an easy task, but it is essential to our understanding of how sex and cooperation actually evolve—theoretical models, studies of convergence, and "model species" (Dugatkin 2001) can take us only so far.

Conclusions

The two great divides of evolutionary ecology, conceptual and taxonomic, must be crossed for the study of mating systems and social systems to reach maturity. Conceptual unification should, I think, hinge on recognition that both sex and sociality involve conflict between mutually dependent parties over limiting resources, and yield complex mixtures of cooperation and conflict over different prezygotic and postzygotic processes. For any lineage, the balance of cooperation and conflict depends on how different life-history tradeoffs are expressed and how well persuasion, coercion, or force may be used by the interacting parties, as a function of their phenotypes, the strength of selection, and the ecological benefits and costs of alternative resolutions to conflict (W.D. Brown et al. 1997, Arnqvist and Rowe 2005). These considerations imply that further conceptual progress requires synthesis of two approaches: (1) the optimization analyses of behavioral ecology and (2) analyses of the evolutionary rules regarding how conflicts are resolved or ongoing, usually in ways nonoptimal to one or both parties (Beekman et al. 2003, Beekman and Ratnieks 2003). In general, I would expect that ecological "harshness" and resource limitation select for increased within-group cooperation in the evolution of sociality and for increased between-sex cooperation in the evolution of mating systems. Such cooperation is, however, coupled with enhanced between-group conflict in sociality and within-sex conflict in mating systems. Moreover, successful within-sex or between-sex cooperation creates new resources, be they "basic necessary" ones or a social group itself, which can generate new forms of intense conflict within and between species.

Taxonomic and phylogenetic divides create opportunities for recognizing convergences. Groups such as the Crustacea are highly diverse yet understudied compared to vertebrates and insects, and as such, they provide especially high returns on research investment. In particular, the presence of cooperative, claw-defended factory fortresses provides stunning comparisons with other animals, the presence of molting yields opportunities to dissect sexual conflict in precopulatory guarding and postcopulatory behavior, and their status as "water-breathing insects" means that a complex mosaic of shared and divergent traits drive the social and sexual evolution of Insecta, Crustacea, and other taxa and remain to be elucidated. This book is a crucial first step.

Acknowledgments I am grateful to A. Bourke, E. Duffy, and two anonymous reviewers for helpful comments, and I thank the Natural Sciences and Engineering Research Council of Canada for financial support.

References

Alexander, R. 1989. Evolution of the human psyche. Pages 455–513 *in*: P. Mellars and C. Stringer, editors. The human revolution: behavioral and biological perspectives on the origins of modern humans. Edinburgh University Press, Edinburgh.

Alexander, R.D., K. Noonan, and B.J. Crespi. 1991. The evolution of eusociality. Pages 3–44 *in*: P.W. Sherman, J.U.M. Jarvis, and R.D. Alexander, editors. The biology of the naked mole-rat. Princeton University Press, Princeton, N.J.

Alexander, R.D., D.C. Marshall, and J.R. Cooley. 1997. Evolutionary perspectives on insect mating. Pages 4–41 *in*: J.C. Choe and B.J. Crespi, editors. The evolution of mating systems in insects and arachnids. Cambridge University Press, Cambridge.

Andersson, M. 1984. The evolution of eusociality. Annual Review of Ecology and Systematics 15:165–189.

Andersson, M. 1994. Sexual selection. Princeton University Press, Princeton, N.J.

Andersson, M. 2004. Social polyandry, parental investment, sexual selection, and the evolution of reduced female gamete size. Evolution 58:24–34.

Arnold, K.E., and I.P.F. Owens. 1998. Cooperative breeding in birds: a comparative test of the life history hypothesis. Proceedings of the Royal Society of London, Series B 265:739–745.

Arnold, K.E., and I.P.F. Owens. 1999. Cooperative breeding in birds: the role of ecology. Behavioral Ecology 10:465–471.

Arnqvist, G., and L. Rowe. 2005. Sexual conflict. Princeton University Press, Princeton, N.J.

Aviles, L. 1997. Causes and consequences of cooperation and permanent-sociality in spiders. Pages 476–498 *in*: J.C. Choe and B.J. Crespi, editors. The evolution of social behavior in insects and arachnids. Cambridge University Press, Cambridge.

Beekman, M., and F.L.W. Ratnieks. 2003. Power over reproduction in social Hymenoptera. Philosophical Transactions of the Royal Society, Series B 358:1741–1753.

Beekman, M., J. Komdeur, and F.L.W. Ratnieks. 2003. Reproductive conflicts in social animals—who has power? Trends in Ecology and Evolution 18:277–282.

Binford, G.J. 2001. Differences in venom composition between orb-weaving and wandering Hawaiian *Tetragnatha* (Araneae). Biological Journal of the Linnean Society 74:581–595.

Black, J.M., editor. 1996. Partnerships in birds. The study of monogamy. Oxford University Press, Oxford.

Bourke, A.F.G. 1999. Colony size, social complexity and reproductive conflict in social insects. Journal of Evolutionary Biology 12:245–257.

Bro-Jorgensen, J. 2003. The significance of hotpots to lekking topi antelopes (*Damaliscus lunatus*). Behavioral Ecology and Sociobiology 53:324–331.

Brown, J.L. 1978. Avian communal breeding systems. Annual Review Ecology and Systematics 9:123–155.

Brown, W.D., B.J. Crespi, and J.C. Choe. 1997. Sexual conflict and the evolution of mating systems. Pages 352–377 *in*: J.C. Choe and B.J. Crespi, editors. The evolution of mating systems in insects and arachnids. Cambridge University Press, Cambridge.

Cant, M.A., E. Otali, and F. Mwanguhya. 2001. Eviction and dispersal in co-operatively breeding banded mongooses (*Mungos mungo*). Journal of Zoology 254:155–162.

Chapman, T., B.J. Crespi, M. Schwarz, and B. Kranz. 2000. High relatedness and inbreeding at the origin of eusociality in Australian gall thrips. Proceedings of the National Academy of Sciences, USA 97:1648–1650.

Chapman, T.W., B.D. Kranz, K.L. Bejah, D.C. Morris, M.P. Schwarz, and B.J. Crespi. 2002. The evolution of soldier reproduction in social thrips. Behavioral Ecology 13:519–525.

Chapman, T., G. Arnqvist, J. Bangham, and L. Rowe. 2003. Sexual conflict. Trends in Ecology and Evolution 18:41–47.

Chapple, D.G. 2003. Ecology, life-history, and behavior in the Australian scincid genus *Egernia*, with comments on the evolution of complex sociality in lizards. Herpetological Monographs 17:145–180.

Clark, C.W., and R. Dukas. 1994. Foraging under predation hazard: an advantage of sociality. American Naturalist 144:542–548.

Clutton-Brock, T. H. 1991. The evolution of parental care. Princeton University Press, Princeton, N.J.

Cockburn, A. 1998. Evolution of helping behavior in cooperatively breeding birds. Annual Review of Ecology and Systematics 29:141–177.

Connor, R.C. 1995a. The benefits of mutualism: a conceptual framework. Biological Reviews 70:427–457.

Connor, R.C. 1995b. Altruism among non-relatives: alternatives to the Prisoner's Dilemma. Trends in Ecology and Evolution 10:84–86.

Connor, R.C., J. Mann, P.L. Tyack, and H. Whitehead. 1998. Social evolution in toothed whales. Trends in Ecology and Evolution 13:228–232.

Crespi, B.J. 1994. Three conditions for the evolution of eusociality: are they sufficient? Insectes Sociaux 41:395–400.

Crespi, B.J. 1996. Comparative analysis of the origins and losses of eusociality: causal mosaics and historical uniqueness. Pages 253–287 *in*: E. Martins, editor. Phylogenies and the comparative method in animal behavior. Oxford University Press, Oxford.

Crespi, B.J. 2000. The evolution of maladaptation. Heredity 84:623–629.

Crespi, B.J. 2001. The evolution of social behavior in microorganisms. Trends in Ecology and Evolution 16:178–183.

Crespi, B.J. 2004. Vicious circles: positive feedback in major evolutionary and ecological transitions. Trends in Ecology and Evolution 19:627–633.

Crespi, B.J., and J.C. Choe. 1997. Evolution and explanation of social systems. Pages 499–524 *in*: J.C. Choe and B.J. Crespi, editors. The evolution of social behavior in insects and arachnids. Cambridge University Press, Cambridge.

Crespi, B.J., and J.E. Ragsdale. 2000. The evolution of sociality via manipulation: why it is better to be feared than loved. Proceedings of the Royal Society of London, Series B 267:821–828.

Crespi, B.J., and P.D. Taylor. 1990. Dispersal rates under variable patch density. American Naturalist 135:48–62.

Crespi, B.J., and D. Yanega. 1995. The definition of eusociality. Behavioral Ecology 6:109–115.

Crespi, B.J., D.C. Morris, and L.A. Mound. 2004. The evolution of ecological and behavioral diversity in phytophagous insects: Australian Acacia thrips as model organisms. Australian Biological Resources Study/Australian National. Insect Collection, Canberra, Australia.

Darwin, C. 1871. The descent of man and selection in relation to sex. John Murray, London.

Doughty, P. 1996. Statistical analysis of natural experiments in evolutionary biology: comments on the recent criticisms of the use of comparative methods to study adaptation. American Naturalist 148:943–956.

Dugatkin, L.A. 2001. Model systems in behavioral ecology: integrating conceptual, theoretical, and empirical approaches. Princeton University Press, Princeton, N.J.

Emlen, S.T. 1982. The evolution of helping. I. An ecological constraints model. American Naturalist 119:29–39.

Emlen, S.T., and L.W. Oring. 1977. Ecology, sexual selection, and the evolution of mating systems. Science 197:215–223.

Faulkes, C.G., and N.C. Bennett. 2001. Family values: group dynamics and social control of reproduction in African mole-rats. Trends in Ecology and Evolution 16:184–190.

Felsenstein, J. 1985. Phylogenies and the comparative method. American Naturalist 125:1–15.

Fincke, O.M. 1997. Conflict resolution in the Odonata: implications for understanding female mating patterns and female choice. Biological Journal of the Linnean Society 60:201–220.

Foster, K.R., and F.L.W. Ratnieks. 2005. A new eusocial vertebrate? Trends in Ecology and Evolution 20:363–364.

Frank, S.A. 2003. Repression of competition and the evolution of cooperation. Evolution 57:693–705.

Gardner, A., S.A. West, and A. Buckling. 2004. Bacteriocins, spite and virulence. Proceedings of the Royal Society of London, Series B 271:1529–1535.

Grant, P.R. 1999. Ecology and evolution of Darwin's finches. Princeton University Press, Princeton, N.J.

Hamilton, W.D. 1967. Extraordinary sex ratios. Science 156:477–488.

Hamilton, W.D. 1978. Evolution and diversity under bark. Pages 154–175 *in*: L.A. Mound and N. Waloff, editors. Diversity of insect faunas. Blackwell, Oxford.

Härdling, R., H. Kokko, and K.E. Arnold. 2003. Dynamics of the caring family. American Naturalist 161:395–412.

Hart, A.G., and F.L.W. Ratnieks. 2005. Crossing the taxonomic divide: conflict and its resolution in societies of reproductively totipotent individuals. Journal of Evolutionary Biology 18:383–395.

Hatchwell, B.J., and J. Komdeur. 2000. Ecological constraints, life history traits and the evolution of cooperative breeding. Animal Behaviour 59:1079–1086.

Hunt, J.H. 1999. Trait mapping and salience in the evolution of eusocial vespid wasps. Evolution 53:225–237.

Hunt, J.H., and C.A. Nalepa, editors. 1994. Nourishment and evolution in insect societies. Westview Press, Boulder, Colo.

Jarvis, J.U.M., and N.C. Bennett. 1993. Eusociality has evolved independently in two species of bathyergid mole-rats—but occurs in no other subterranean mammal. Behavioral Ecology and Sociobiology 33:253–260.

Jarvis, J.U.M., M.J. O'Riain, N.C. Bennett, and P.W. Sherman. 1994. Mammalian eusociality: a family affair. Trends in Ecology and Evolution 9:47–51.

Jeon, J., and J.C. Choe. 2003. Reproductive skew and the origin of sterile castes. American Naturalist 36:206–224.

Johnstone, R.A. 2000. Models of reproductive skew: a review and synthesis. Ethology 106:5–26.

Johnstone, R.A., and M.A. Cant. 1999. Reproductive skew and the threat of eviction: a new perspective. Proceedings of the Royal Society of London, Series B 266:275–279.

Keller, L. 1997. Extraordinary lifespans in ants: a test of evolutionary theories of ageing. Nature 389:958–960.

Kent, D.S., and J.A. Simpson. 1992. Eusociality in the beetle *Austroplatypus incompertus* (Coleoptera: Curculionidae). Naturwissenschaften 79:86–87.

Kirkendall, L.R., D.S. Kent, and K.F. Raffa. 1997. Interactions among males, females and offspring in bark and ambrosia beetles: the significance of living in tunnels for the evolution of social behavior. Pages 181–215 *in*: J.C. Choe and B.J. Crespi, editors. The evolution of social behavior in insects and arachnids. Cambridge University Press, Cambridge.

Koenig, W.D., and J.L. Dickinson. 2004. Evolution and ecology of cooperative breeding in birds. Cambridge University Press, Cambridge.

Kokko, H., and J. Ekman. 2002. Delayed dispersal as a route to breeding: territorial inheritance, safe havens, and ecological constraints. American Naturalist 160:468–484.

Kokko, H., R. Brooks, J.M. McNamara, and A.I. Houston. 2002. The sexual selection continuum. Proceedings of the Royal Society of London, Series B 269:1331–1340.

Korb, J., and J. Heinze. 2004. Multilevel selection and social evolution of insect societies. Naturwissenschaften 91:291–304.

Kukuk, P.F., G.C. Eickwort, M. Raveret-Richter, B. Alexander, R. Gibson, R.A. Morse, and F.L.W. Ratnieks. 1989. Importance of the sting in the evolution of sociality in the Hymenoptera. Annual Entomological Society of America 82:1–5.

Le Galliard, J.F., R. Ferriere, and U. Dieckmann. 2005. Adaptive evolution of social traits: origin, trajectories, and correlations of altruism and mobility. American Naturalist 165:206–224.

Martins, E.P. 2000. Adaptation and the comparative method. Trends in Ecology and Evolution 15:295–299.

Mathews, L. 2003. Tests of the mate-guarding hypothesis for social monogamy: male snapping shrimp prefer to associate with high-value females. Behavioral Ecology 14:63–67.

Mead, L.S., and S.J. Arnold. 2004. Quantitative genetic models of sexual selection. Trends in Ecology and Evolution 19:264–271.

Nesse, R.M. 2005. Maladaptation and natural selection. Quarterly Review of Biology 80:62–70.

Oli, M.K. 2004. The fast-slow continuum and mammalian life-history patterns: an empirical evaluation. Basic and Applied Ecology 5:449–463.

Owens, I.P.F. 2002. Male-only care and classical polyandry in birds: phylogeny, ecology and sex-differences in remating opportunities. Philosophical Transactions of the Royal Society, Series B 357:283–293.

Pagel, M. 1994. Detecting correlated evolution on phylogenies: a general method for the comparative analysis of discrete characters. Proceedings of the Royal Society of London, Series B 255:37–45.

Pagel, M. 1999. Inferring the historical patterns of biological evolution. Nature 401:877–884.

Pen, I., and F.J. Weissing. 2000. Towards a unified theory of cooperative breeding: the role of ecology and life history re-examined. Proceedings of the Royal Society of London, Series B 267:2411–2418.

Queller, D.C. 1994. Male-female conflict and parent-offspring conflict. American Naturalist 144:S84–S99.

Queller, D.C. 1996. The origin and maintenance of eusociality: the advantage of extended parental care. Pages 218–234 *in*: S. Turillazzi and M.J. West-Eberhard, editors. Natural history and evolution of paper wasps. Oxford University Press, Oxford.

Queller, D.C., and J.E. Strassmann. 1998. Kin selection and social insects. BioScience 48:165–175.

Ragsdale, J.E. 1999. Reproductive skew theory extended: the effect of resource inheritance on social organization. Evolutionary Ecology Research 1:859–874.

Regier, J.C., J.W. Shultz, and R.E. Kambic. 2004. Phylogeny of basal hexapod lineages and estimates of divergence times. Annals of the Entomological Society of America 97:411–419.

Reynolds, J.D., N.B. Goodwin, and R.P. Freckleton. 2002. Evolutionary transitions in parental care and live bearing in vertebrates. Philosophical Transactions of the Royal Society, Series B 357:269–281.

Rice, W.R. 1996. Sexually antagonistic male adaptation triggered by experimental arrest of female evolution. Nature 381:232–234.

Rice, W.R., and A.K. Chippindale. 2001. Intersexual ontogenetic conflict. Journal of Evolutionary Biology 14:685–693.

Rice, W.R., and B. Holland. 1997. The enemies within: intergenomic conflict, interlocus contest evolution (ICE), and the intraspecific Red Queen. Behavioral Ecology and Sociobiology 41:1–10.

Ridley, M. 1989. Why not to use species in comparative tests. Journal of Theoretical Biology 136:361–364.

Russell, E.M., Y. Yom-Tov, and E. Geffen. 2004. Extended parental care and delayed dispersal: northern, tropical, and southern passerines compared. Behavioral Ecology 15:831–838.

Sachs, J.L., U.G. Mueller, T.P. Wilcox, and J.J. Bull. 2004. The evolution of cooperation. Quarterly Review of Biology 79:135–160.

Sherman, P.W., J.U.M. Jarvis, and R.D. Alexander, editors. 1991. The biology of the naked mole-rat. Princeton University Press, Princeton, N.J.

Shuster, S.M., and M.J. Wade. 2003. Mating systems and strategies. Princeton University Press, Princeton, N.J.

Solomon, N.G. 2003. A reexamination of factors influencing philopatry in rodents. Journal of Mammalogy 84:1182–1197.

Sutherland, W.J. 1996. From individual behaviour to population ecology. Oxford University Press, Oxford.

Tallamy, D.W. 2000. Sexual selection and the evolution of exclusive paternal care in arthropods. Animal Behaviour 60:559–567.

Taylor, P.D., and A.J. Irwin. 2000. Overlapping generations can promote altruistic behaviour. Evolution 54:1135–1141.

Thorne, B.L. 1997. Evolution of eusociality in termites. Annual Review of Ecology and Systematics 28:27–54.

Thorne, B.L., N.L. Breisch, and M.I. Haverty. 2002. Longevity of kings and queens and first time of production of fertile progeny in dampwood termite (Isoptera; Termopsidae; *Zootermopsis*) colonies with different reproductive structures. Journal of Animal Ecology 71:1030–1041.

Wcislo, W.T. 1989. Behavioral environments and evolutionary change. Annual Review of Ecology and Systematics 20:137–169.

West-Eberhard, M.J. 1983. Sexual selection, social competition, and speciation. Quarterly Review of Biology 58:155–183.

Whitehouse, M.E.A., and Y. Lubin. 2005. The functions of societies and the evolution of group living: spider societies as a test case. Biological Reviews 80:347–361.

Wilson, E.O., and B. Hölldobler. 2005. Eusociality: origin and consequences. Proceedings of the National Academy of Sciences, USA 102:13367–13371.

Sexual and Social Behavior of Crustacea

A Way Forward

J. Emmett Duffy

Martin Thiel

21

The contributions in this book provide a sampling of the wide—and, we suspect, underappreciated—diversity of social and sexual systems among crustaceans, and they sketch the outlines of the range of environmental pressures and lifestyles that have molded that diversity. The chapters illustrate well both the value of Crustacea as models for addressing a range of general questions in evolutionary ecology, and the fascinating natural history and behavior of the group in its own right. Recurring throughout the book are tantalizing hints of novel social and sexual systems in crustaceans whose natural history and ecology remain otherwise little known. In several such cases, close relatives display quite different social or mating systems, the reasons for which are similarly poorly understood (e.g., chapters 11, 16, 18). Thus, these chapters raise at least as many questions as they answer.

A dominant theme that emerges is the pressing need for detailed, classical natural history research. Such research is necessary not only for advancing our basic understanding of crustacean sexual and social systems but also for using these animals effectively to test general theory in evolutionary ecology (e.g., chapters 2, 12). As Crespi notes in chapter 20, the historical neglect by evolutionary ecologists of crustaceans, compared with insects and vertebrates, means that crustaceans should provide high returns on research investment. Much progress could be made with creative new approaches to observing or inferring behavior in the field. Because of crustaceans' small size and cryptic habitats, most previous studies of their social behavior have been conducted in the laboratory, which can provide only imperfect mimics of the complex natural environment. In particular, simulation of natural burrows or living habitats in the lab entails substantial logistical challenges. Clues to crustacean habits in nature may be fostered by recent technological advances such as endoscopic tools (see, e.g., Richter et al. 2001) for peeking into their secret homes and biotelemetry via mounted electronic sensors or cameras to observe behavior of larger, mobile species in the field (Passaglia et al. 1997, Cooke et al. 2004). High-resolution molecular genetic markers may similarly be used to infer mating systems of animals that cannot be observed directly (e.g., Duffy 1996, Bilodeau et al. 2005, Gosselin et al. 2005).

In addition to a renewed focus on detailed natural history research, we see three promising general themes for advancing crustacean evolutionary ecology that are common to many of the disparate problems addressed in this volume: (1) closer linkage of general theory to empirical research, (2) better integration of modern molecular approaches, and (3) more phylogenetically explicit, rigorous comparative testing of evolutionary hypotheses. Combining these approaches offers much promise in addressing the more specific frontiers for understanding crustacean evolutionary ecology. In this concluding chapter, we summarize what we judge to be some of the most important open questions.

Relatedness, Conflict, and Cooperation

Family and Environment

The role of genetic relatedness in mediating the tension between conflict and cooperation among individuals has been a central theme of animal behavior since Hamilton's (1964) pioneering, and deceptively simple, formulation of the concept of inclusive fitness, later known as kin selection. Hamilton recognized explicitly that the evolution of behavioral interactions depends both on genetic relatedness among individuals and on the ecological factors that define the costs and benefits of their interactions. This interplay of genetic relatedness and ecological constraints provides a bridge between social and sexual systems (Queller 1994; see chapter 20).

In understanding the paradox of eusociality, in particular, kin selection has provided a key explanation and has stimulated four decades of highly productive research. That very success has led some authors recently to suggest that kin selection explanations have often been applied simplistically and to reemphasize the complex interplay of competition, cooperation, and sexual selection processes that mediate the evolution of sociality (Alonso and Schuck-Paim 2002, Griffin and West 2002, Wilson and Hölldobler 2005). Although kin selection theory has incorporated both genetic relatedness and ecology since its inception (Hamilton 1964, Foster et al. 2006), important questions remain about the relative importance of direct fitness benefits versus indirect, kin-selected benefits in molding social behavior and, in particular, in mediating the evolution of group living. Thus, understanding both patterns of genetic relatedness and ecology is critical to understanding the evolutionary origins and maintenance of eusocial colonies. Several crustaceans discussed in this book would be ideal candidates for application of high-resolution markers to explore this interaction of genetic and ecological influences in molding evolution of social groups (see chapter 3). What is the extent of nest parasitism and multiple paternity in bromeliad crabs (chapter 17), and how do they influence levels of cooperation and group productivity? How are breeder number, colony genetic structure, and cooperative behavior related in phylogenetically controlled comparisons among species of sponge-dwelling shrimp (chapter 18)? In most social crustaceans studied to date, sociality appears to have arisen from parent–offspring groups with a high degree of kinship (chapter 14). Although close genetic relatedness in such families fosters group cohesion and cooperation, it also raises the risk of inbreeding. How do social colonies regulate the occasional incorporation of unrelated individuals into colonies? And how does their incorporation change the strength and nature of sexual selection within colonies? Social crustaceans offer promising, and perhaps unique, opportunities for understanding the interplay of ecology, kin selection, and sexual selection in the evolution of social systems.

Status and Individual Recognition

Behavioral strategies based on kinship presuppose some ability to distinguish kin from nonkin. But kin recognition is only one facet of the broader problem of distinguishing among conspecific individuals and the potential threats or benefits they represent. Studies of a diverse array of crustaceans suggest that recognition of dominance or aggressive status of conspecifics is widespread, because individuals can recognize, evaluate status, and remember their opponents during agonistic interactions. Color, size, and concentrations of waterborne chemicals are used to obtain information about other individuals, often without the need to establish direct and potentially risky contact (e.g., Caldwell 1979, Gherardi and Daniels 2003, Bergman and Moore 2005; see also chapter 6).

Whether crustaceans are also capable of more finely tuned recognition of conspecific individuals, outside simple dominance or aggressive hierarchies, is less understood. Although individual recognition is often assumed to occur only in selected species (see, e.g., chapters 16, 18), it may be more common, particularly in species where cohabiting or neighboring individuals interact repeatedly. For example, in the monogamous shrimp *Hymenocera picta*, mates recognize one another via water-borne chemical cues (Wickler and Seibt 1972), and similar results have been reported for the shrimp *Stenopus hispidus* (Johnson 1977). Fiddler crabs also can recognize individual neighbors based on visual cues (Detto et al. 2006). Thus, there is good evidence that recognition of conspecific individuals is important in conflict resolution among crustaceans, and hints that individual recognition may also be important in facilitating cooperation. During social interactions, individuals obtain important information about potential opponents or partners, which influences their behavioral reaction toward these. Consequently, understanding the nature and sophistication of individual recognition is potentially important to interpreting social and mating systems. The Crustacea provide many species that are well suited to examine these phenomena through observation and experimentation.

Enemy Pressure and the Evolution of Social and Sexual Systems

Competition and Habitat Saturation

As discussed above, ecology is the other critical component of Hamilton's equation for understanding social interactions. Competition for limited habitat or nesting resources is thought to be a primary driver of social evolution in many cooperatively breeding birds (Selander 1964, Brown 1969, Emlen 1982, Arnold and Owens 1999, Hatchwell and Komdeur 2000) and some social insects (Lin and Michener 1972, Strassmann and Queller 1989, Brockmann 1997, Queller and Strassmann 1998, Wilson and Hölldobler 2005). Indeed, Wilson and Hölldobler (2005) argued that "the key adaptation that led to eusociality is defense against enemies, specifically predators, parasites, and competitors." Similar selection pressures resulting from the difficulty of establishing new nests—or acquiring the "basic necessary resource" for sociality (Alexander et al. 1991)—have been invoked to explain sociality in sponge-dwelling shrimp (chapter 18), bromeliad crabs (chapter 17), desert isopods (chapter 16), and burrowing crayfish (chapter 15). This growing list suggests that general theories of environmental constraints and animal social evolution developed initially for vertebrates and insects may apply more broadly. Yet, in most of the crustacean cases studied to date, the influence of environmental constraints on social systems is largely conjecture based on indirect evidence. Experiments that manipulate the availability of limiting habitat resources, or removal or addition of breeders, have been conducted with several cooperatively breeding birds (Walters et al. 1992, Komdeur 1996, Komdeur et al. 1995, Russell and Hatchwell 2001) and fishes (e.g., Bergmüller et al. 2005) and should be similarly illuminating in certain crustaceans. The common correlation in insects and vertebrates between sociality and use of resources that are scarce, discrete, or difficult to exploit also provides clues to where new cases of sociality might be expected in crustaceans (Spanier et al. 1993, Duffy 2003). Symbiotic marine crustaceans in particular should provide fertile ground for such searches. In particular, phylogenetically controlled comparative studies of closely related crustaceans differing in social organization would be rewarding (e.g., Arnold and Owens 1999,

Duffy et al. 2000). We look forward to the day when crustaceans from a broad enough range of environments have been studied to rigorously test hypotheses on the environmental controls on social organization (e.g., Emlen 1982, Alexander et al. 1991; see also chapter 12).

Predation

Predation is a pervasive threat, and consequently a strong selective pressure, in the lives of most organisms. Predation has been suggested as an important selective factor favoring cooperative breeding and sociality in birds (Hoogland and Sherman 1976), mammals (Alexander 1974, Clutton-Brock et al. 1999), cichlid fishes (Heg et al. 2004), and insects (Lin and Michener 1972, Strassmann and Queller 1989, Wilson and Hölldobler 2005). The central role of predation in social and sexual evolution of crustaceans is emphasized in several of the chapters in this book (chapters 7, 10, 12, 13). Predation appears commonly to influence the balance between sexual selection and viability selection. Its evolutionary consequences have been illuminated through comparative analyses of conspecific populations, particularly in freshwater crustaceans distributed across water bodies that differ strongly in predation pressure (e.g., Strong 1972, deMeester 1996; see also chapter 7). The ability of freshwater *Daphnia* to form "resting eggs" that accumulate in sediments has been exploited ingeniously to trace the microevolutionary history of behavioral and physiological responses of these important zooplankters to changing predation regimes (e.g., Cousyn et al. 2001, Kerfoot and Weider 2004). It would be fascinating to use this approach to ask whether mating systems have also been influenced by the changes in predation pressure. Evolutionary consequences of predation on crustaceans have also been approached more indirectly through phylogenetically controlled comparisons among species (e.g., Harrison and Crespi 1999). As the phylogenetic relationships among species become better documented, such comparative approaches should prove fruitful in discerning evolutionary impacts on mating and social systems in a wide range of crustacean taxa.

Parasitism and Its Consequences

Social groups are prime targets for parasites because they concentrate host individuals and often other vital resources such as food and shelter (Schmid-Hempel 1998). Defense against parasites and disease is thought to be a primary selection pressure maintaining, and perhaps originally giving rise to, sex (Lively 1996) because sexual recombination breaks up gene combinations in each generation and thus presents parasites with a "moving target," interfering with adaptation to the host. Persistent kin groups such as social colonies lose some of this advantage of sex because they present a temporally stable, genetically homogeneous resource that is attractive to parasites and pathogens. Defense against these enemies appears to provide an important selective pressure for multiple mating in social insects, and genetic data, experiments, and phylogenetically controlled comparisons all suggest that polyandry can reduce the impact of parasites in several social insects (e.g., Brown and Schmid-Hempel 2003, Hughes and Boomsma 2004, Tarpy and Seeley 2006). Polyandry may have similar fitness benefits and consequences for social and sexual life in crustaceans. There is also evidence from birds that parasitism and disease can be important engines of sexual selection, favoring ornaments that honestly advertise an individual's vigor and freedom from disease (Saino et al. 1997, Horak et al. 2001, Faivre et al. 2003).

Crustaceans host many parasites, some of which (e.g., bopyrid isopods, rhizocephalid barnacles) are themselves crustaceans. Bopyrids, in particular, are conspicuous large external parasites, which allows easy recognition of infected host individuals. Parasite-induced changes in crustacean behavior and physiology are well documented (Moore 1984, Bakker et al. 1997, Cezilly et al. 2000) and may indirectly affect mating interactions and thus the strength of sexual selection in crustaceans. Castrating parasites shut down sexual function in female crabs, distorting the operational sex ratio more strongly toward males and increasing the potential for mate competition among the latter (Brockerhoff 2004). Conversely, infestation with *Wolbachia* causes feminization in many crustaceans (Juchault et al. 1993, Bouchon et al. 1998), thereby distorting the operational sex ratio in the direction of females.

A reduction in proportion of males also may result from trematode parasitism of male amphipods, which increases their activity at the sediment surface and thus their predation risk (McCurdy et al. 2000). Such indirect effects of parasites on sexual selection, mediated via changing operational sex ratios, have not yet been studied in crustaceans. Several questions would repay future study in crustacean host–parasite systems. How does parasitism affect an individual's mating prospects, and the population's operational sex ratio? How, in turn, do parasite-induced changes in reproductive success influence phenotypic evolution? Has parasitism fostered or foiled the evolution of sociality in crustaceans? Is parasitism more frequent in social than in nonsocial crustaceans? Among species, or among populations within species, of social crustacean hosts, does frequency of parasitism correlate with genetic relatedness or with mate numbers? Does the risk of parasite infection favor monogamy or positive interactions among social partners (e.g., grooming)?

Finally, it is worth noting that crustacean parasites themselves may provide fascinating subjects for studying evolution of mating systems. Parasitic bopyrid isopods, for example, represent an extreme case of sexual dimorphism, with highly modified females that amount to little more than a large egg sac, and minute dwarf males. The small male, which itself lives as a parasite on the female, may produce sperm just sufficient to fertilize all the females' eggs, approaching a case of optimal adjustment of gamete numbers between the male and female partners. This intimate association between the female and male also reflects an extreme form of lifelong monogamy, which may result from the very low population densities of the parasites and the premium on finding and retaining a mate (Andersson 1994).

Crustacean Mating Systems and Sexual Selection

Consequences of Sperm Competition

Sexual selection, and sexual conflict, may also be exacerbated when both sexes mate multiply, as occurs in many crustacean species (chapters 2, 12). What are the consequences for fertilization success of different sperm donors in such cases? How has sperm competition influenced evolution of male and female behavior and morphology? How do patterns differ for species with sperm storage and internal fertilization? It has been suggested that females of some of these crustacean species are capable of choosing among sperm donations from different males (Jensen et al. 1996), but the exact mechanisms remain elusive.

Males have different mechanisms to counteract the risk of sperm competition, by (1) monitoring the reproductive status of female partners and fending off competitors (Wirtz and Diesel 1983, Murai et al. 2002), (2) reducing the number of male competitors (e.g., by mate guarding; see chapter 7), (3) sealing off the female's reproductive tract or sperm packages, or (4) diluting sperm masses from competitors with large sperm donations (Rondeau and Sainte-Marie 2001). Whether crustacean males actively remove sperm from previous males is not known at present. However, the sperm transfer apparatus of many decapod crustaceans (the modified male pleopods) contains numerous bundles of setae that give this appendix a brushlike appearance. It is very similar in appearance to that of male insects known to remove sperm of competitors from the female reproductive tract (Eberhard 1996) and may serve a similar purpose.

Female Control

As is true of animals generally, studies on mating behavior of crustaceans historically have focused on male behaviors and assumed male control of mating interactions. But recent studies in a variety of animal taxa have underscored the role of female behaviors in controlling mating interactions (Eberhard 1996). One of the most important such mechanisms of female control involves timing. In some crustaceans females appear capable of controlling the moment of the reproductive molt, physiological receptivity, ovulation, and finally fertilization (e.g., Cowan and Atema 1990, Brockerhoff and McLay 2005). Concealment of reproductive status also is employed by females in some crustaceans, allowing them to gain control over the mating process (see also chapter 16). While female control of mating interactions clearly occurs in several crustaceans, little is known of the mating behavior of social species. For example, do queens of social crustaceans mate multiply? If so, what are the consequences for relatedness of the offspring, interactions among colony members, male competition for mating, and the evolution of social behaviors generally?

Preference, Coercion, and Intersexual Conflict

Sexual selection has traditionally been understood to result from intrasexual competition (usually among males) for matings and intersexual choice (usually by females) among suitors (Darwin 1871). But interactions between the sexes can also generate antagonistic coevolution when genes at different loci enhance fitness in one sex but reduce it in the other (reviewed by Parker 2006). Under such circumstances, evolution at one locus selects for counterevolution at the other locus in a runaway chain reaction analogous to the Red Queen model of arms-race coevolution among different species (van Valen 1973). Rice and Holland (1997) argued that much of the evolution of mating and courtship signals, ornaments, and other sexual differences are driven by such antagonistic intersexual conflict evolution (ICE) and, indeed, suggested that ICE is "the dominant mode of evolution for genes controlling social behavior." They argued that many traits previously interpreted as evolving in response to female preference may actually reflect evolution in response to female resistance to mating, and that genes for sexual and social interactions, and the phenotypes they produce, should evolve much more rapidly than other genes not involved in ICE. Distinguishing preference from coercion in mating and other social behaviors is not simple, however (Kokko 2005), as shown by detailed studies of amphipod and isopod crustaceans in which female resistance to mating is well documented (e.g., chapters 7, 8). These mate-guarding crustaceans offer promising opportunities for further quantifying direct and indirect costs and benefits of mating, thus disentangling the several interacting influences on mating system evolution.

Macroevolutionary Consequences

In addition to its effects on individual morphology and behavior, sexual selection also has important macroevolutionary consequences because it tends to drive rapid coevolution between the sexes, often involving antagonistic coevolution, which results in rapid evolution of reproductive isolation among populations. In montane jumping spiders, for example, allopatric populations show strong phenotypic differentiation in male courtship-related traits and in courtship behavior (Masta and Maddison 2002). Interpopulation differentiation in these traits, in female responsiveness, and in post-zygotic viability were stronger than divergence in neutral genetic markers, confirming that evolution of sexually selected characters has been much faster than that of neutral control, and suggesting that sexual selection enhanced divergence rate. Sexual conflict may similarly enhance speciation rate. Multiple mating creates strong competition among sperm of different males to maximize fertilization, often with negative consequences for female fitness. Counterselection by the female to maximize her own fitness can set up an arms race of sexually antagonistic coevolution of reproductive traits. In contrast, when females mate singly, the interests of the male and female are basically the same. Thus, polyandry should speed up evolution of reproductive isolation and speciation. Confirming this hypothesis, Arnqvist et al. (2000) found that species richness averaged four times higher in polyandrous clades of insects than in their monandrous sister clades. Such comparisons have not yet been attempted for crustaceans, but their wide range of mating systems suggests that sexual selection and sexually antagonistic coevolution might be important drivers of speciation and diversification in crustaceans, as well.

Evolutionary Ecology and Conservation

Social and sexual phenomena are central to population dynamics. Harvesting and habitat destruction clearly have many direct impacts on animal populations but may also have indirect impacts by disrupting social or sexual interactions. Many animals that historically lived at high population density do not breed effectively at low population density either because of difficulty in mate finding or because of disrupted social interactions. This "Allee effect" creates the phenomenon known in fisheries as depensation—a decrease in per capita population growth rate at low density. Sex-specific harvesting may also change mating interactions with important consequences. Several crustacean fisheries have targeted males on the reasonable assumption that female fecundity limits population growth and that "sperm is cheap." But strong male-specific harvesting pressure has led to sperm limitation in several exploited crustaceans (chapter 9), which may disrupt normal social and mating systems. For example, in heavily exploited Canadian populations of American lobster, microsatellite data indicate higher levels of multiple paternity than in less exploited populations, suggesting that sperm limitation fosters female promiscuity (Gosselin et al. 2005). Understanding how social interactions influence mating systems is recognized as critically important in managing finfish stocks (e.g., Alonzo and Mangel 2004, Heppel et al. 2006) and is likely to be important to effective conservation and management of exploited crustaceans, as well.

Concluding Thoughts

The Crustacea represents one of the major branches in the tree of animal life, displaying diversity in form and lifestyle that rival those of the vertebrates and insects. But perhaps because of the primarily aquatic habits of crustaceans, they have received much less attention in evolutionary ecology than mostly terrestrial taxa. We hope that the chapters in this book make clear the richness of adaptations of crustaceans to social and sexual life, and their still largely untapped potential to test fundamental theory in behavioral ecology and evolution. There are rich opportunities awaiting the student willing to pursue them, both in clarifying the obscure social and sexual biology of individual crustacean species and in exploiting the Crustacea in broad comparative approaches to testing evolutionary theory. Our hope is that vigorous pursuit of these opportunities will soon render this book out of date.

Acknowledgments We thank all the contributors to this book for their unflagging support and for stimulating our ideas. We are grateful to B. Crespi for his comments on previous drafts of this chapter.

References

Alexander, R.D. 1974. The evolution of social behavior. Annual Review of Ecology and Systematics 5:325–338.

Alexander, R.D., K.M. Noonan, and B.J. Crespi. 1991. The evolution of eusociality. Pages 3–44 *in*: P.W. Sherman, J.U.M. Jarvis, and R.D. Alexander, editors. The biology of the naked mole-rat. Princeton University Press, Princeton, N.J.

Alonso, W.J., and C. Schuck-Paim. 2002. Sex-ratio conflicts, kin selection, and the evolution of altruism. Proceedings of the National Academy of Sciences, USA 99:6843–6847.

Alonzo, S.H., and M. Mangel. 2004. The effects of size-selective fisheries on the stock dynamics of and sperm limitation in sex-changing fish. Fishery Bulletin 102:1–13.

Andersson, M. 1994. Sexual selection. Princeton University Press, Princeton, N.J.

Arnold, K.E., and I.P.F. Owens. 1999. Cooperative breeding in birds: the role of ecology. Behavioral Ecology 10:465–471.

Arnqvist, G., M. Edvardsson, U. Friberg, and T. Nilsson. 2000. Sexual conflict promotes speciation in insects. Proceedings of the National Academy of Sciences, USA 97:10460–10464.

Bakker, T.C.M., D. Mazzi, and S. Zala. 1997. Parasite-induced changes in behavior and color make *Gammarus pulex* more prone to fish predation. Ecology 78:1098–1104.

Bergman, D.A., and P.A. Moore. 2005. Prolonged exposure to social odours alters subsequent social interactions in crayfish (*Orconectes rusticus*). Animal Behaviour 70:311–318.

Bergmüller, R., D. Heg, and M. Taborsky. 2005. Helpers in a cooperatively breeding cichlid stay and pay or disperse and breed, depending on ecological constraints. Proceedings of the Royal Society of London, Series B 272:325–331.

Bilodeau, A.L., D.L. Felder, and J.E. Neigel. 2005. Multiple paternity in the thalassinidean ghost shrimp, *Callichirus islagrande* (Crustacea: Decapoda: Callianassidae). Marine Biology 146:381–385.

Bouchon, D., T. Rigaud, and P. Juchault. 1998. Evidence for widespread *Wolbachia* infection in isopod crustaceans: molecular identification and host feminization. Proceedings of the Royal Society of London, Series B 265:1081–1090.

Brockerhoff, A.M. 2004. Occurrence of the internal parasite *Portunion* sp. (Isopoda: Entoniscidae) and its effect on reproduction in intertidal crabs (Decapoda: Grapsidae) from New Zealand. Journal of Parasitology 90:1338–1344.

Brockerhoff, A.M., and C.L. McLay. 2005. Factors influencing the onset and duration of receptivity of female purple rock crabs, *Hemigrapsus sexdentatus* (H. Milne Edwards, 1837) (Brachyura: Grapsidae). Journal of Experimental Marine Biology and Ecology 314:123–135.

Brockmann, H.J. 1997. Cooperative breeding in wasps and vertebrates: the role of ecological constraints. Pages 347–371 *in*: J.C. Choe and B.J. Crespi, editors. Social behavior in insects and arachnids. Cambridge University Press, Cambridge.

Brown, J.L. 1969. Territorial behavior and population regulation in birds, a review and re-evaluation. Wilson Bulletin 81:293–329.

Brown, M.J.F., and P. Schmid-Hempel. 2003. The evolution of female multiple mating in social hymenoptera. Evolution 57:2067–2081.

Caldwell, R.L. 1979. Cavity occupation and defensive behaviour in the stomatopod *Gonodactylus festai*: evidence for chemically mediated individual recognition. Animal Behaviour 27:194–201.

Cezilly, F., A. Gregoire, and A. Bertin. 2000. Conflict between co-occurring manipulative parasites? An experimental study of the joint influence of two acanthocephalan parasites on the behaviour of *Gammarus pulex*. Parasitology 120:625–630.

Clutton-Brock, T.H., D. Gaynor, G.M. McIlrath, A.D.C. Maccoll, R. Kansky, P. Chadwick, M. Manser, J.D. Skinner, and P.N.M. Brotherton. 1999. Predation, group size and mortality in a cooperative mongoose, *Suricata suricatta*. Journal of Animal Ecology 68:672–683.

Cooke, S.J., S.G. Hinch, M. Wikelski, R.D. Andrews, L.J. Kuchel, T.G. Wolcott, and P.J. Butler. 2004. Biotelemetry: a mechanistic approach to ecology. Trends in Ecology and Evolution 19:334–343.

Cousyn, C., L. deMeester, J.K. Colbourne, L. Brendonck, D. Verschuren, and F. Volckaert. 2001. Rapid, local adaptation of zooplankton behavior to changes in predation pressure in the absence of neutral genetic changes. Proceedings of the National Academy of Sciences, USA 98:6256–6260.

Cowan, D., and J. Atema. 1990. Moult staggering and serial monogamy in American lobsters, *Homarus americanus*. Animal Behaviour 39:1199–1206.

Darwin, C. 1871. The descent of man and selection in relation to sex. John Murray, London.

deMeester, L. 1996. Evolutionary potential and local genetic differentiation in a phenotypically plastic trait of a cyclical parthenogen, *Daphnia magna*. Evolution 50:1293–1298.

Detto, T., P.R.Y. Backwell, J.M. Hemmi, and J. Zeil. 2006. Visually mediated species and neighbour recognition in fiddler crabs (*Uca mjoebergi* and *Uca capricornis*). Proceedings of the Royal Society of London, Series B. 273:1661–1666.

Duffy, J.E. 1996. Eusociality in a coral-reef shrimp. Nature 381:512–514.

Duffy, J.E. 2003. The ecology and evolution of eusociality in sponge-dwelling shrimp. Pages 201–215 *in*: T. Kikuchi, N. Azuma, and S. Higashi, editors. Genes, behaviors and evolution of social insects. Hokkaido University Press, Sapporo, Japan.

Duffy, J.E., C.L. Morrison, and R. Rios. 2000. Multiple origins of eusociality among sponge-dwelling shrimps (*Synalpheus*). Evolution. 54:503–516.

Eberhard, W.G. 1996. Female control. Princeton University Press, Princeton, N.J.

Emlen, S.T. 1982. The evolution of helping. I. An ecological constraints model. American Naturalist 119:40–53.

Faivre, B., A. Gregoire, M. Preault, F. Cezilly, and G. Sorci. 2003. Immune activation rapidly mirrored in a secondary sexual trait. Science 300:103–103.

Foster, K.R., T. Wenseleers, and F.L.W. Ratnieks. 2006. Kin selection is the key to altruism. Trends in Ecology and Evolution 21:57–60.

Gherardi, F., and W.H. Daniels. 2003. Dominance hierarchies and status recognition in the crayfish *Procambarus acutus acutus*. Canadian Journal of Zoology 81:1269–1281.

Gosselin, T., B. Sainte-Marie, and L. Bernatchez. 2005. Geographic variation of multiple paternity in the American lobster, *Homarus americanus*. Molecular Ecology 14:1517–1525.

Griffin, A.S., and S.A. West. 2002. Kin selection: fact and fiction. Trends in Ecology and Evolution 17:15–21.

Hamilton, W.D. 1964. The genetical evolution of social behavior I, II. Journal of Theoretical Biology 7:1–52.

Harrison, M.K., and B.J. Crespi. 1999. A phylogenetic test of ecomorphological adaptation in *Cancer* crabs. Evolution 53:961–965.

Hatchwell, B.J., and J. Komdeur. 2000. Ecological constraints, life history traits and the evolution of cooperative breeding. Animal Behaviour 59:1079–1086.

Heg, D., Z. Bachar, L. Brouwer, and M. Taborsky. 2004. Predation risk is an ecological constraint for helper dispersal in a cooperatively breeding cichlid. Proceedings of the Royal Society of London, Series B 271:2367–2374.

Heppell, S.S., S.A. Heppell, F.C. Coleman, and C.C. Koenig. 2006. Models to compare management options for a protogynous fish. Ecological Applications 16:238–249.

Hoogland, J., and P.W. Sherman. 1976. Advantages and disadvantages of bank swallow (*Riparia riparia*) coloniality. Ecological Monographs 46:33–58.

Horak, P., I. Ots, H. Vellau, C. Spottiswoode, and A.P. Møller. 2001. Carotenoid-based plumage coloration reflects hemoparasite infection and local survival in breeding great tits. Oecologia 126:166–173.

Hughes, W.O.H., and J.J. Boomsma. 2004. Genetic diversity and disease resistance in leaf-cutting ant societies. Evolution 58:1251–1260.

Jensen, P.C., J.M. Orensanz, and D.A. Armstrong. 1996. Structure of the female reproductive tract in the Dungeness crab (*Cancer magister*) and implications for the mating system. Biological Bulletin 190:336–349.

Johnson, V.R. 1977. Individual recognition in the banded shrimp *Stenopus hispidus* (Olivier). Animal Behavior 25:418–428.

Juchault, P., T. Rigaud, and J.P. Mocquard. 1993. Evolution of sex determination and sex ratio variability in wild populations of *Armadillidium vulgare* (Latr.) (Crustacea, Isopoda): a case study in conflict resolution. Acta Oecologica 14:547–562.

Kerfoot, W.C., and L.J. Weider. 2004. Experimental paleoecology (resurrection ecology): chasing van Valen's Red Queen hypothesis. Limnology and Oceanography 49:1300–1316.

Kokko, H. 2005. Treat 'em mean, keep 'em (sometimes) keen: evolution of female preferences for dominant and coercive males. Evolutionary Ecology 19:123–135.

Komdeur, J. 1996. Influence of helping and breeding experience on reproductive performance in the Seychelles warbler: a translocation experiment. Behavioral Ecology 7:326–333.

Komdeur, J., A. Huffstadt, W. Prast, G. Castle, R. Mileto, and J. Wattel. 1995. Transfer experiments of Seychelles warblers to new islands—changes in dispersal and helping-behavior. Animal Behaviour 49:695–708.
Lin, N., and C.D. Michener. 1972. Evolution of sociality in insects. Quarterly Review of Biology 47(2):131–159.
Lively, C.M. 1996. Host-parasite coevolution and sex—do interactions between biological enemies maintain genetic variation and cross-fertilization? Bioscience 46:107–114.
Masta, S.E., and W.P Maddison. 2002. Sexual selection driving diversification in jumping spiders. Proceedings of the National Academy of Sciences, USA 99:4442–4447.
McCurdy, D.G., M.R. Forbes, and J.S. Boates. 2000. Male amphipods increase their mating effort before behavioural manipulation by trematodes. Canadian Journal of Zoology 78:606–612.
Moore, J. 1984. Altered behavioral responses in intermediate hosts: an acanthocephalan parasite strategy. American Naturalist 123:572–577.
Murai, M., T. Koga, and H.S. Yong. 2002. The assessment of female reproductive state during courtship and scramble competition in the fiddler crab, *Uca paradussumieri*. Behavioral Ecology and Sociobiology 52:137–142.
Parker, G.A. 2006. Sexual conflict over mating and fertilization: an overview. Philosophical Transactions of the Royal Society, Series B 361:235–259.
Passaglia, C., F. Dodge, E. Herzog, S. Jackson, and R. Barlow. 1997. Deciphering a neural code for vision. Proceedings of the National Academy of Sciences, USA 94:12649–12654.
Queller, D.C. 1994. Male-female conflict and parent-offspring conflict. American Naturalist 144:S84–S99.
Queller, D.C., and J.E. Strassmann. 1998. Kin selection and social insects: social insects provide the most surprising predictions and satisfying tests of kin selection. BioScience 48:3:165–175.
Rice, W.R., and B. Holland. 1997. The enemies within: intergenomic conflict, interlocus contest evolution (ICE), and the intraspecific Red Queen. Behavioral Ecology and Sociobiology 41:1–10.
Richter, C., M. Wunsch, M. Rasheed, I. Kotter, and M.I. Badran. 2001. Endoscopic exploration of Red Sea coral reefs reveals dense populations of cavity-dwelling sponges. Nature 413:726–730.
Rondeau, A., and B. Sainte-Marie. 2001. Variable mate-guarding time and sperm allocation by male snow crabs (*Chionoecetes opilio*) in response to sexual competition, and their impact on the mating success of females. Biological Bulletin 201:204–217.
Russell, A.F., and B.J. Hatchwell. 2001. Experimental evidence for kin-biased helping in a cooperatively breeding vertebrate. Proceedings of the Royal Society of London, Series B 268:2169–2174.
Saino, N., A.M. Bolzern, and A.P. Møller. 1997. Immunocompetence, ornamentation, and viability of male barn swallows (*Hirundo rustica*). Proceedings of the National Academy of Sciences, USA 94:549–552.
Schmid-Hempel, P. 1998. Parasites in social insects. Princeton University Press, Princeton, N.J.
Selander, R.K. 1964. Speciation in wrens of the genus *Campylorhynchus*. University of California Publications in Zoology 74:1–224.
Spanier, E., J.S. Cobb, and M.-J. James. 1993. Why are there no reports of eusocial marine crustaceans? Oikos 67:573–576.
Strassmann, J.E., and D.C. Queller. 1989. Ecological determinants of social evolution. Pages 81–101 *in*: M.D. Breed and R.E. Page, editors. The genetics of social evolution. Westview, Boulder, Colo.
Strong, D.R., Jr. 1972. Life history variation among populations of an amphipod (*Hyalella azteca*). Ecology 53:1103–1111.

Tarpy, D.R., and T.D. Seeley. 2006. Lower disease infections in honeybee (*Apis mellifera*) colonies headed by polyandrous vs. monandrous queens. Naturwissenschaften 93:195–199.

van Valen, L. 1973. A new evolutionary law. Evolutionary Theory 1:1–30.

Walters, J.R., C.K. Copeyon, and J.H. Carter III. 1992. Test of the ecological basis of cooperative breeding in red-cockaded woodpeckers. Auk 109:90–97.

Wickler, W., and U. Seibt. 1972. Über den Zusammenhang des Paarsitzens mit anderen Verhaltensweisen bei *Hymenocera picta* Dana. Zeitschrift für Tierpsychologie 31:163–170.

Wilson, E.O., and B. Hölldobler. 2005. Eusociality: origin and consequences. Proceedings of the National Academy of Sciences, USA 102:13367–13371.

Wirtz, P., and R. Diesel. 1983. The social structure of *Inachus phalangium*, a spider crab associated with the sea anemone *Anemonia sulcata*. Zeitschrift für Tierpsychologie 62: 209–234.

INDEX